Micro-organisms in Action: Concepts and Applications in Microbial Ecology

Edited by

J.M. LYNCH
BTech, PhD, DSc, CChem, FRSC, CBiol, FIBiol
AFRC Institute of Horticultural Research
Littlehampton

J.E. HOBBIE
BA, MA, PhD
Marine Biological Laboratory
Woods Hole, Massachusetts

BLACKWELL SCIENTIFIC PUBLICATIONS
OXFORD LONDON EDINBURGH
BOSTON PALO ALTO MELBOURNE

Editorial offices:
Osney Mead, Oxford OX2 0EL
(*Orders*: Tel: 0865-240201)
8 John Street, London WC1N 2ES
23 Ainslie Place, Edinburgh EH3 6AJ
Three Cambridge Center, Suite 208
Cambridge, MA 02142, USA
667 Lytton Avenue, Palo Alto
California 94301, USA
107 Barry Street, Carlton
Victoria 3053, Australia

First published 1979 under the title
Microbial Ecology: A Conceptual Approach
Reprinted 1980, 1984
Second edition 1988

Set by Setrite Typesetters,
Hong Kong
Printed in Great Britain
by The Alden Press Ltd,
Oxford

DISTRIBUTORS

USA and Canada
Blackwell Scientific Publications Inc
PO Box 50009, Palo Alto
California 94303
(*Orders*: Tel. (415) 965-4081)

Australia
Blackwell Scientific Publications
(Australia) Pty Ltd
107 Barry Street
Carlton, Victoria 3053
(*Orders*: Tel. (03) 347-0300)

British Library
Cataloguing in Publication Data

Micro-organisms in action: concepts and applications in microbial ecology.
1. Microbial ecology 2. Biotic communities
I. Lynch, J.M. II. Hobbie, John E.
III. Microbial ecology
576′.15 QR100

ISBN 0-632-01652-3
ISBN 0-632-01653-1 (pbk)

Library of Congress
Cataloging-in-Publication Data

Micro-organisms in action: concepts and applications in microbial ecology.
Rev. ed. of: Microbial ecology. 1979.
Bibliography.
1. Microbial ecology. I. Lynch, J.M. (James Michael)
II. Hobbie, John E. III. Microbial ecology.
QR100.M515 1988 576′.15 87-18250

ISBN 0-632-01652-3
ISBN 0-632-01653-1 (pbk)

Contents

List of contributors

DR J.M. ANDERSON
Dept of Biological Sciences, University of Exeter, Prince of Wales Road, Exeter EX4 4PS, UK

DR D.F. BEZDICEK
Dept of Agronomy and Soils, Washington State University, Pullman, Washington 99164-6470, USA

DR M.M. FLETCHER
Center of Marine Biotechnology, University of Maryland, 600 East Lombard Street, Baltimore, Maryland 21202, USA

DR D.M. GIBSON
MAFF Torry Research Station, PO Box 31, 135 Abbey Road, Aberdeen AB9 8DG, UK

DR D. GODWIN-THOMAS
Dept of Applied Biology, University of Wales Institute of Science and Technology, PO Box 13, Cardiff CF1 3XF, UK

PROFESSOR G.W. GOODAY
Dept of Genetics and Microbiology, Marischal College, University of Aberdeen, Aberdeen AB9 1AS, UK

PROFESSOR W.A. HAMILTON
Dept of Genetics and Microbiology, Marischal College, University of Aberdeen, Aberdeen AB9 1AS, UK

DR J.E. HOBBIE
Marine Biological Laboratory, Woods Hole, Massachusetts 02543, USA

DR P.N. HOBSON
Microbial Biochemistry Dept, The Rowett Research Institute, Greenburn Road, Bucksburn, Aberdeen AB2 9SB, UK (*Present address:* 4 North Deeside Road, Aberdeen AB1 7PL, UK)

DR A.C. KENNEDY
Dept of Agronomy and Soils, Washington State University, Pullman, Washington 99164-6420, USA

DR J. LACEY
AFRC Institute of Arable Crops Research, Rothamsted Experimental Station, Harpenden, Hertfordshire AL5 2JQ, UK

PROFESSOR J.M. LYNCH
AFRC Institute of Horticultural Research, Littlehampton, West Sussex BN17 6LP, UK

PROFESSOR C.G. ORPIN
AFRC Institute of Animal Physiology and Genetics Research, Babraham, Cambridge CB2 4AT, UK *and* Dept of Arctic Biology, University of Tromsø, 9001 Tromsø, Norway

DR C.C. PAYNE
AFRC Institute of Horticultural Research, East Malling, Maidstone, Kent ME19 6BJ, UK

DR N.J. POOLE
ICI Plant Protection Division, Jealott's Hill Research Station, Bracknell, Berkshire RG12 6EY, UK

PROFESSOR J.H. SLATER
Dept of Applied Biology, University of Wales Institute of Science and Technology, PO Box 13, Cardiff CF1 3XF, UK

DR J.R. VESTAL
Dept of Biological Sciences, University of Cincinnati, Cincinnati, Ohio 45221-0006, USA

DR A.J. WEIGHTMAN
Dept of Applied Biology, University of Wales Institute of Science and Technology, PO Box 13, Cardiff CF1 3XF, UK

Preface

In the earlier volume [6] to which this is a sequel, the preface indicated the growth in interest of microbial ecology since 1975. Now there have been four International Symposia on Microbial Ecology at Dunedin [5], Warwick [3], East Lansing [4] and Ljubljiana in 1986, with the next scheduled for 1989 in Kyoto, Japan. There have also been several texts published, notably those by Campbell [2] and Atlas and Bartha [1] as well as the series *Advances in Microbial Ecology* and the journals *Microbial Ecology* and *FEMS Microbiology Ecology*.

Constructive critical book reviews are always welcome but these can lead to unsuspected outcomes. When John Hobbie reviewed the earlier volume in *Ecology* in 1981 he indicated the need for a greater ecosystem approach. He became the obvious choice to replace Nigel Poole as co-editor when Nigel found his commitments to his company did not allow him time for the task. We thank the many students, teachers and researchers on microbial ecology everywhere who offered constructive comments. We also thank the authors and publishers listed below for allowing us to reproduce either original or copyright material.

Authors: Anderson & Sørensen (1986) (Fig. 3.2.18); Bazin & Saunders (1973) (Figs. 2.1.6, 2.1.7); Bedford (1981) (Fig. 4.2.6b); Bird & Akhurst (1983) (Fig. 4.2.2); Board (1969) (Fig. 4.3.4); Bunnell *et al.* (1977) (Fig. 3.1.14); Cliff *et al.* (1981) (Fig. 2.2.5); Cooper & MacCallum (1984) (Figs. 2.2.3, 2.2.4); Evans *et al.* (1987) (Fig. 4.1.1). Gibbs & Harrison (1976) (Fig. 2.2.2); Gieskes & Kraay (1984) (Fig. 3.2.6); Gregory (1973) (Figs. 3.5.2, 3.5.3); Griffin (1977) (Fig. 3.1.18); Haddock & Jones (1977) (Figs. 2.5.2, 2.5.6); Hamilton (1985) (Fig. 2.5.9); Harman *et al.* (1980) (Fig. 4.2.12); Jenkinson (1977) (Fig. 3.1.13); Jerlov (1976) (Fig. 3.2.1); Johnson *et al.* (1980) (Fig. 4.2.1); Jørgensen (1980) (Fig. 3.2.13), (1983a) (Fig. 3.2.12), (1983b) (Fig. 3.2.15); Kröger (1977) (Fig. 2.5.5); Lisansky (1984) (Table 4.2.1); Mack *et al.* (1975) (Fig. 4.4.4); Mandelstam *et al.* (1982) (Fig. 2.5.3); Odom & Peck (1981) (Fig. 2.5.8); Oglesby (1977) (Fig. 3.2.7); Parsons *et al.* (1977) (Fig. 3.2.4); Paul & Voroney (1984) (Table 3.1.6); Payne (1986) (Fig. 4.2.6); Payne *et al.* (1981) (Fig. 4.2.7); Ridout *et al.* (1986) (Fig. 4.2.11); Sieburth (1975) (Figs. 2.1.2, 2.1.3, 2.1.4), (1984) (Fig. 3.2.8); Simon (1985) (Fig. 3.4.3); Southward (1982) (Figs. 3.3.9, 3.3.10); Stanier *et al.* (1976) (Figs. 2.5.11, 2.5.12, 2.5.13); Stewart (1984) (Fig. 3.1.17); Thauer & Badziong (1980) (Fig. 2.5.7) Thauer *et al.* (1977) (Fig. 2.5.4); van Veen *et al.* (1984) (Fig. 3.1.16); White (1979) (Figs. 3.1.5, 3.1.15).

Publishers: Academic Press (Figs. 3.2.13, 4.2.6b, 4.2.7, 4.3.4); Am. Phytopathological Soc. (Fig. 4.2.12); Am. Soc. for Microbiol. (Figs. 2.5.2, 2.5.4, 2.5.6, 3.1.17, 3.4.3, Table 3.1.6); Annual Reviews (Fig. 2.5.9); *Appl. Environ. Microbiol.* (Fig. 3.4.3); Chapman & Hall (Figs. 2.2.3, 2.2.4, 3.1.18); CRC Press (Figs. 2.1.6, 2.1.7, 2.5.7); Cambridge University Press (Figs. 2.2.5, 2.5.5); Edward Arnold (Fig. 2.2.2); Elsevier (Figs. 2.5.8, 3.2.1); Forestry Commission (Fig. 3.3.1); *Fortschr. Zool.* (Fig. 4.2.6); Inter-Research (Fig. 3.2.18); *J. Fish. Res. Bd. Canada* (Fig. 3.2.7); *J. Gen. Microbiol.* (Fig. 4.2.11); *J. Invertebr. Path.* (Fig. 4.2.7); *J. Mar. Biol. Ass. U.K.* (Figs. 3.3.9, 3.3.10); S. Karger AG (Fig. 4.2.6a); Leonard Hill (Figs. 3.5.2, 3.5.3); Marine Biological Association (Figs. 3.3.9, 3.3.10); Martinus Nijhoff (Figs. 3.1.16, 4.1.1); *Microbiol. Ecol.* (Fig. 3.4.2); Nat. Research Council of Canada (Fig. 4.2.1); Online Publications (Table 4.2.1); Oxford University Press (Fig. 3.1.14); Pergamon Press (Figs. 2.1.6, 2.1.7, 3.1.14, 3.2.4, 4.2.2); Plenum Press (Figs. 3.2.6, 3.2.8); Prentice-Hall (Figs. 2.5.11, 2.5.12, 2.5.13); Society for General Microbiology (Fig. 4.2.11); Springer-Verlag (Figs. 3.1.19, 4.4.4); University Park Press (Figs. 2.1.2, 2.1.3, 2.1.4); Wiley (Fig. 3.2.12).

In times when there are so many pressures on scientists it can be difficult to secure delivery of manuscripts. Sue Bewsey not only typed the chapters where Lynch and/or Payne were authors, but she also typed many begging letters to authors pleading for their manuscripts; we as editors are extremely grateful to her. We also thank

Robert Campbell, Simon Rallison and Jane Andrew at Blackwell Scientific Publications for their help and support.

Finally, our wives Mary Lynch and Olivann Hobbie are thanked for their tolerance of us during the preparation of the book.

REFERENCES

[1] Atlas R.M. & Bartha R. (1981) *Microbiol Ecology. Fundamentals and Applications.* Addison-Wesley, Reading, Mass.

[2] Campbell R. (1983) *Microbial Ecology*, 2nd edn. Blackwell Scientific Publications, Oxford.

[3] Ellwood D.C., Hedger J.N., Latham M.J., Lynch J.M. & Slater J.H. (eds.) (1980) *Contemporary Microbial Ecology*. Academic Press, London.

[4] Klug M.J. & Reddy C.A. (eds.) (1984) *Current Perspectives in Microbial Ecology*. American Society for Microbiology, Washington DC.

[5] Loutit M. & Miles J.A.R. (eds.) (1978) *Microbial Ecology*. Springer-Verlag, Berlin.

[6] Lynch, J.M. & Poole N.J. (eds.) (1979) *Microbial Ecology. A Conceptual Approach*. Blackwell Scientific Publications, Oxford.

Part 1
Introduction

microbe, n. minute living being, micro-organism (esp. of bacteria causing diseases and fermentation); hence *microb*ial, *microb*ic, adjs. (F. f. Gk. *mikros* small + *bios* life).

ecolog/y n. branch of biology dealing with organisms' relations to one another and to their surroundings [f. G *ökologie* f. Gk *oikos* house, see -LOGY].

From *The Concise Oxford Dictionary of Current English* (1982) 7th edn.

THE DICTIONARY definition of microbial ecology has not changed significantly since that used in an earlier form of this text [1], yet with so much publication and debate on the subject in 9 years it is useful to reflect on the boundaries of microbial ecology.

It is necessary to understand the behaviour of micro-organisms under defined and controlled conditions in the laboratory in order to understand behaviour in natural environments. This book starts with a consideration of the microbial cell followed by some descriptions of interactions between cells of different microbial species. Viruses have frequently been omitted from discussions on microbial ecology and whereas they are non-cellular they are usually bound to cells of other micro-organisms, plants or animals. Recognition of the potential significance of plasmids in nature has increased greatly in the past few years and again it is critical to understand *in vitro* behaviour in order to predict behaviour in nature. Population and community dynamics can certainly be affected by plasmids. Ultimately, it is the metabolism of individual species and communities which determines the primary influence of micro-organisms on the environment and it is often useful to consider this from the energetic standpoint.

The term 'ecosystem' was first used by Tansley [2] to describe 'not only the organism complex but also the whole complex of physical factors, what we call the environment'. The second part of the book considers soil and water as well as animals and plants as environments for micro-organisms. Additionally aerial dispersal between environments and the effect of extreme environmental influences are discussed.

It is sometimes difficult to carry out cost–benefit analyses of scientific endeavour, especially when no obvious commercial benefit accrues but social amenities are improved. For example, the microbial deodorization of animal wastes on farms close to urban conurbations may make the environment more amenable to the residents but it is difficult to place a monetary value on this. Probably more scientific resources have been invested in studies on the fixation of dinitrogen than any other subject area in microbial ecology. Symbiotic processes are of major economic significance in agriculture and natural environments and this is the reason for their prominence in the chapter on dinitrogen fixation; this provides a complementary account to the earlier one [1] which had greater emphasis on free-living systems and physiology/biochemistry. The only other process where inoculation has proved an economic success is the biological control of pests and diseases. These beneficial activities of micro-organisms in the environment should be balanced by harmful microbial activities such as microbial spoilage of feeds and pollution of water. Also the effects of xenobiotic compounds in natural and man-made environments are discussed.

Our conclusion to the book will identifiy some areas of microbial ecology where the new developments in biotechnology might be exploited and studied.

The book is not intended as a comprehensive guide to microbial ecology but hopefully it will equip the reader to investigate chosen fields of study in the discipline and to think about them constructively.

REFERENCES

[1] Lynch J.M. & Poole N.J. (eds.) (1979) *Microbial Ecology. A Conceptual Approach*. Blackwell Scientific Publications, Oxford.

[2] Tansley A.G. (1935) The use and abuse of vegetational concepts and terms. *Ecology* **16**, 284–307.

Part 2
Principles of Microbial Behaviour in Ecosystems

'Yes, I have a pair of eyes,' replied Sam, 'and that's just it. If they wos a pair o' patent double million magnifyin' gas microscopes of hextra power, p'raps I might be able to see through a flight o' stairs and a deal door; but bein' only eyes, you see my wision's limited.'

Charles Dickens 1812–1870
Pickwick Papers

2.1 The potential of the microbial cell and its interactions with other cells

The basic consideration of this chapter will be the individual microbial cell. Much of microbiology considers populations of cells, and often the implicit assumption is that a pure culture, be it batch, chemostat or synchronous, is a collection of uniform cells or at least that any differences are averaged out. This is usually an acceptable generalization as, although there will be considerable phenotypic variations between adjacent cells, the characteristics of such cultures can only be reflections of the capabilities of each individual cell. In natural environments, microbial cells will almost always be in mixed cultures where they have to interact with cells of other species, but again these interactions are limited by the potential of each cell.

2.1.1 THE RANGE OF ORGANISMS

The taxonomic scope of microbial ecology

The microbial ecologist is concerned with a very wide range of organisms. There will likely be many different types of organism in any particular environment, and it is essential to any true understanding of their roles and interactions to be able to identify them as closely as possible. The micro-organisms encompass the three forms of life of the present day: eukaryotes, prokaryotes and viruses. These three groupings have no clear phylogenetic interrelationships, and apart from their chemical make-up the only thing that they have in common is that they are studied by microbiologists. Of the five-Kingdom classification of cellular forms [5, 34], microbiologists rightly claim three Kingdoms: (a) the Monera (Procaryotae, bacteria) [50]; (b) the Protista (protozoa, algae, slime moulds, and flagellated fungi or water moulds); and (c) the Fungi. These are dealt with in this chapter and the viruses and plasmids in the following two chapters.

Prokaryotes

The starting point for any work with prokaryotes must be *Bergey's Manual* [11, 29], which firmly establishes the Kingdom Procaryotae for the bacteria.

With the advent of the use of nucleic acids in the study of the classification of bacteria, in particular the sequencing of 16S ribosomal RNA, new light has been cast onto their phylogenetic relationships. Thus, taxonomy currently is in an exciting state of flux. We can expect many surprises in the next few years, hence the need for simplicity in the meantime. The *Shorter Bergey's Manual* [25] gives three rules of the game:

1 Use all the information available to you.

2 Apply common sense at each step.

3 Use the minimum number of tests to make the identification. It also gives seven practical steps to follow:

1 Make sure you have a pure culture.

2 On the basis of the isolation procedure, establish whether you have a chemolithotrophic autotroph, a photosynthetic organism or a chemoheterotrophic organism (see Chapter 2.5).

3 Examine living cells by phase contrast and Gram-stained cells by light microscopy.

4 Examine gross growth appearances.

5 Test for oxygen requirements.

6 Test the dissimilation of glucose.

7 Complete additional tests appropriate to genera suggested by results from tests listed above.

They add the cautionary note that the most frequent cause of mistaken identity of bacteria is errors in determination of shape, Gram-reaction and motility.

THE EUBACTERIA

These bacteria inhabit virtually every ecological niche in the biosphere. They may be divided conveniently into Gram-negative bacteria, Gram-positive bacteria, and the permanently wall-less mycoplasmas which are mostly pathogenic in animals and plants.

The Gram-negative bacteria consist of the following: green and purple sulphur and non-sulphur bacteria, a diverse group of organisms which are photosynthetic in anoxic environments; pseudomonads, aerobic rods respiring a wide range of organic compounds; aerobic nitrogen-fixing bacteria; facultatively anaerobic rods: dissimilatory sulphate-reducers; spiral and curved aerobic bacteria; anaerobic fermentors; coccobacilli; sheathed bacteria; budding and prosthecate bacteria; chemolithotrophic bacteria; rickettsias; spirochaetes; and gliding bacteria.

The Gram-positive bacteria consist of the following: endospore formers; asporogenous rods and cocci; and actinobacteria.

THE ARCHAEBACTERIA

In recent years an apparently disparate group of bacteria have been shown to have a series of biochemical characteristics that set them completely apart from all other bacteria; they have walls of unusual polysaccharides instead of peptidoglycan, membranes often with unusual glycerol ether isoprenoid lipids instead of phospholipid bilayers, and characteristic ribosomes and ribosomal RNA sequences [48]. Moreover, they characteristically inhabit extreme environments. They include the red halophiles of brine pools and salted fish, the methanogens of anaerobic muds, animal guts and anaerobic digesters, and the thermoacidophiles of hot acid springs and smouldering coal wastes (cf. Chapter 3.4).

THE CYANOBACTERIA

The Cyanobacteria [14] are of great importance as primary producers. They evolve oxygen during photosynthesis and many can fix nitrogen. They occur as major constituents of marine and freshwater blooms, and also thrive in the upper layer of damp nutrient-poor soil and on wet rocks. Thermophilic species grow in clear successions down the progressively cooling effluents of hot springs. Some species produce toxins, and their blooms can cause fish mortalities. Cyanobacteria occur as symbionts; e.g. *Nostoc* spp. in the lichen *Peltigera canina* and in the water fern *Azolla filiculoides*. Their capacity for nitrogen fixation (see Chapter 4.1) plays a major role in the nutrition of these symbioses, and in lichens their photosynthesis is the major source of carbon and energy. Some, called cyanellae, occur as symbionts in some protozoa such as *Cyanophora* spp. These cyanellae still retain a thin peptidoglycan coat, but their DNA complexity is much less than that of free-living cyanobacteria and quite close to that of the algal chloroplasts. In fact, the recently discovered Prochlorophytes are very close to chloroplasts, as they contain chlorophylls *a* and *b* instead of the chlorophyll *a* and phycobiliproteins of cyanobacteria; these were discovered as extracellular symbionts of tropical marine ascidians.

Eukaryotic microbes

Following the five-Kingdom classification introduced above, the eukaryotic micro-organisms are the Fungi and the Protista — the water moulds, slime moulds, algae and protozoa. Also to be included are the lichens which, although symbiotic associations, are classified as discrete species. In contrast to the prokaryotes, the traditional taxonomy of these groups was established using morphological characters by biologists in the 19th century and earlier. For the most part their classification has stood the test of time. However, the phylogenetic relationships both between and within the different groups remain topics of healthy dispute and speculation, chiefly because there is little fossil record except for such organisms as diatoms or coccolithophorids which have silicified or calcified cell components of characteristic morphology.

THE FUNGI

The true fungi have two major cell types, the hypha and the yeast [13]. These two forms have different properties, and they are very important determinants for growth in different environments. Some fungi can only form hyphae, others can only form yeast cells, while others can grow as either depending on conditions. Another characteristic is that fungi have a rigid cell wall. This limits them to being saprophytic on organic substrates, or parasitic on living animals, plants, algae, protozoa or other fungi. The classification of the fungi is based on the pattern of their spore formation. The four phyla are:

1 Zygomycotina: typically ephemeral saprophytic opportunists, spreading quickly through habitats rich in simple carbohydrates or starch, and rapidly sporulating; sporangiospores formed in large numbers, readily dispersed by air current or water droplets; and mycelium coenocytic, i.e. without regular cross-walls. A few species are parasitic on other fungi, animals or plants.

2 Ascomycotina: septate hyphae; saprophytes and parasites. Many cause the soft rots of stored products. The hemiascomycetes (with their ascospores formed in a naked ascus, and not in a structured fruiting body) include the familar yeast, *Saccharomyces cerevisiae*.

3 Deuteromycotina: lacking a known sexual phase, most probably allied to Ascomycotina.

4 Basidiomycotina: mushrooms and toadstools, as ephemeral fruiting bodies, produce large numbers of basidiospores from a parent mycelium made of a continuum of interconnecting hyphae that can be very large and old (e.g. as old as the tree in the case of a mycorrhizal fungus, and sometimes many hundreds of years old in the case of fairy rings); many are important wood rotters, giving rise to brown rots and white rots. Heterobasidiomycetes, with septate basidia, include the smuts and rusts, important plant parasites.

THE PROTISTA

Although forming a coherent Kingdom, these eukaryotic microbes present an enormous variety of form, function and habitat. Their diversity is such that phylogenetic considerations have little relevance to the ecologist. Thus the Euglenophyta contain many phototrophic species, but also many saprotrophic species and phagotrophic species. In this and many other cases, it is best for the ecologist to take a functional view and to call the phototrophs algae and the heterotrophs protozoa.

The algae [44] are of immense importance as the primary producers in the seas and lakes, but are also found in the surface layers of soil and on such substrates as moist tree trunks. The limits of microbiology become blurred when dealing with the algae. For example, a giant kelp many metres long is not usually considered grist for the microbiologist's mill. However, nearly all groups of algae contain species that are solely unicellular, and so are unequivocally microbes. Some algae can live heterotrophically. For example, living diatoms without chlorophyll are found in the darkness of marine sediments, where they survive on dissolved organic matter and may produce cellulases. Some algae occur as phototrophic symbionts of invertebrates. The most important of these in terms of global productivity are the 'zooxanthellae', now known to be dinoflagellates, which are intracellular symbionts in corals. Here they photosynthetically fix carbon dioxide and at the same time control the calcification of the coral skeleton. Some algae which lack a rigid cell wall can also engulf particles of food, and so can practise photosynthetic, absorptive or phagocytic acquisition of nutrients.

The algae have been classified traditionally by pigmentation and life cycle. The principal phyla of interest to microbial ecologists, i.e. those with planktonic or single-celled existence, are summarized below:

1 Rhodophyta, the red algae: principally marine and multicellular, immotile, some unicellular forms occur in sand or encrusting rock pools.

2 Chlorophyta, the green algae: many planktonic species, freshwater and marine, motile by flagella, others multicellular, motile or sessile.

3 Euglenophyta: unicells, motile by a single flagellum and also by 'euglenoid motion', a squirming of the body of the cell; common in nutrient-rich freshwater pools, also found in the sea and in the soil.

4 Bacillariophyta, diatoms: in microplankton of seas and lakes; motile by gliding; with silica frustules which are species-specific, and which readily survive in sediments and fossil deposits to enable a picture of the phytoplankton existing at former points in time to be built up.

5 Dinophyta, dinoflagellates: in microplankton of seas and lakes; motile with two flagella; some cause 'red tides', which can result in fish mortality by the production of toxins, others are luminescent and cause the nocturnal phosphorescence of warmer seas.

6 Chrysophyta; biflagellate cells, chiefly found in fresh water, also the marine silicoflagellates.

7 Haptophyta: biflagellate planktonic algae; many are species of the nanoplankton. The coccolithophorids have a covering of calcified scales; blooms of these species can give the sea a milky appearance and coccoliths from dead cells can build up in the sediments.

There is much information on the occurrence and distribution of many of the algae in the microplankton [46] as diatoms are readily collected in plankton-sampling nets and microscopically identified. However, in many parts of the ocean and in most lakes the most abundant and the most productive algae are the nanoplankton; these forms may be as small as 2 μm in diameter and pass readily through the finest netting. Special techniques such as the use of inverted microscopes, epifluorescence, and electron microscopy, are needed to identify and count them.

The protozoa [37] exhibit a very wide range of form and way of life. In some species the adult cell is very large and has a remarkably complex structure. Many are predators on bacteria, fungi (see Section 2.1.8), algae, yeasts or other protozoa, and many are parasitic or symbiotic in animals (see Chapter 3.3). As mentioned above, many shown close phylogenetic affinities with unicellular algal forms. Some can lead a secondary phototrophic existence by having algal endosymbionts (see Section 2.1.8). The major phyla of interest to the microbial ecologist are summarized below:

1 Sarcodina, the amoeboid protozoa: in a wide variety of habitats; free-living amoebae as predators in the soil; the foraminiferans and radiolarians, with their intricate calcified and silicified skeletons in the marine plankton, accumulating in the bottom ooze to form beds of sediments; human and animal pathogens; and the slime moulds, whose feeding stage is as amoebae (see below).

2 Ciliophora: includes the familiar ciliates of hay infusions, *Tetrahymena* and *Paramecium*, and many other very intricately structured cells; some such as *Vorticella* are important in sewage treatment processes, where bacteria, other microbes and detritus form their food (Chapter 4.4).

Also phylogenetically clearly in the Kingdom Protista, not least by virtue of their having motile stages and '9 + 2' flagella, are the following groups that ecologically can be thought of as fungi:

3 Phycomycota: including the oomycetes, water moulds or pathogens of plants or fish; and the hyphochytridiomycetes, water moulds.

4 Chytridiomycota: water moulds living on detritus such as fallen leaves or parasitic on algae, protozoa or cyanobacteria.

5 Labyrinthomorpha: marine amoeboid cells moving in tubes of their own making; *Labyrinthula macrocystis* has caused considerable coastal ecological changes during this century in Britain by causing wasting disease of the marine eel grass, *Zostera marina*.

6 Slime moulds: phylogenetically they are best thought of as amoebae, but they have been considered as fungi as they produce macroscopic fruiting bodies, either by aggregation of amoebae to form a pseudoplasmodium (e.g. *Dictyostelium*) or by the amoeba forming a multinucleate feeding plasmodium which then forms the fruiting body (e.g. *Physarum*).

2.1.2 THE SIZE OF MICROBES

The small size of a bacterium means that a habitat can be correspondingly small. It also means that a small volume can contain a very large number of organisms or potential propagules. From a practical point of view it also means that only rarely can direct observations be made of the numbers and identities of microbes in natural environments.

The sizes of microbes (Table 2.1.1) show steps greater than an order of magnitude of increasing size between typical viruses, prokaryotes and eukaryotes. There is little variation in the size of the bacteria, most of the cultured forms being in the range 1–3 μm by 0.5–1.5 μm. However, most planktonic bacteria in lakes and oceans are small cocci or rods between 0.2 and 0.6 μm in diameter. Extremes of size include cells of *Mycoplasma* spp. at 0.1 μm in diameter and the purple sulphur bacterium *Thiospirillum jenense* at 40 by 4 μm. There is a much greater range of cell size among the eukaryotic microbes, with a range in the algae represented by the prasinophycean marine alga *Micromonas pusilla* at 1 by 0.7 μm and the giant marine diatom, *Ethmodiscus rex*, at 2000 μm in diameter.

Within any species there can be a marked range of sizes of vegetative cells (see Table 2.1.2) as they tend to increase in size with increasing growth rate. For example, for ten species of bacteria [24] the log phase cells in batch culture were an average of 2.4 times as large as stationary cells. The primary hyphae of *Neurospora crassa* quoted in Table 2.1.2 were growing nearly three times as fast as the secondary branches. Many diatoms also show a range of size, as on cell division each daughter cell forms a new

Table 2.1.1. The sizes of micro-organisms (from various sources)

Organism	Dimensions (μm)	Generation time (h)	Swimming speeds (μm sec^{-1})
Viruses			
Poliovirus	0.03 × 0.03	—	—
Tobacco mosaic	0.3 × 0.02	—	—
T_2-Bacteriophage	0.2 × 0.06	—	—
Prokaryotes			
Marine bacteria	0.2 diam.	?	?
Pseudomonas aeruginosa	1.5 × 0.5	0.5	60
Serratia marcescens	1.7 × 1.0	0.5	30
Bacillus megaterium	7.6 × 2.4	0.3	20
Eukaryotes			
Mucor hiemalis	8 diam.	3	—
Euglena gracilis	50 × 15	7	230
Paramecium caudatum	250 × 50	10	1000
Stentor coeruleus	1000 × 200	34	—

Table 2.1.2. Range of cell size in some species (from various sources)

Species	Size (μm)	Volume (μm^3)	Conditions	Remarks
Salmonella typhimurium	1.43		Doublings h^{-1} 2.73	Size is cell thickness
	0.87		Doublings h^{-1} 0.61	
		1.24	Doublings h^{-1} 2.0	
		0.33	Doublings h^{-1} 0.13	
Nostoc species	5 × 13	240	In oldest leaf	The symbiont of *Azolla*
	4 × 5	40	In youngest leaf	
Neurospora crassa	11.7	(1075)	Primary hyphae	Volume quoted for 10 μm of hypha
strain spco-9	5.7	(255)	Secondary branches	
Stephanodiscus astraea	70.6	62 603		Range of diameters of the diatom in Britain
var *minutula*	18.4	2657		
Saccharomyces cerevisiae		49.1	Doublings h^{-1} 0.33	
		29.0	Doublings h^{-1} 0.14	

silica valve to fit inside its respective parent valve. Thus, the average surface area of the valves of an actively growing population gets progressively smaller, until eventually, after perhaps several years, a new large cell is formed from an auxospore. This decrease in surface area may to some extent be compensated for by the cells becoming thicker, but nevertheless the size and shape of a diatom cell in plankton must be an important determinant for such properties as sinking rate, photosynthetic yield, and growth rate. These properties presumably are not critical for the cells' survival within the limits imposed by the changes in size. Another diatom, *Stephanodiscus astraea*, shows a seasonal change in size, being larger in the winter (34 μm diameter) than in summer (24 μm).

Increasing size has a number of consequences for a microbial cell. Larger cells have the potential for greater adaptability as they have more space for genetic material and the enzymic machinery. There is clearly a lower limit of size for a cell to contain the essential equipment for self-replication, and presumably the smallest free-living bacteria approach this size. Thus, any increase in size for such a modest bacterium would give it the chance of packaging more information, which would permit it to be more adaptable. Increasing size, however, reduces the vital ratio of surface area to the volume as the former is a function of the square of the cell radius and the latter of the cube. This decreasing ratio must limit the upper size of a prokaryotic cell, which is dependent on its cell membrane for all of the processes of uptake of nutrients, transport of ions, control of pH, reduction of electron acceptor, and export of some end products of metabolism and of enzymes and surface components. The eukaryotic cell has solved this problem of decreasing surface to volume ratio by the elaboration of specialized systems of internal membranes. Thus, the surface area available for the terminal respiratory system is located on the mitochondrial membranes, that for synthesis of macromolecules for export in the membranes of the endoplasmic reticulum and Golgi apparatus, and that for photosynthesis in the membranes of the chloroplasts. The eukaryotic cell membrane can thus specialize in the transport of nutrients and ions.

During evolution an increased cell size has necessarily led to an increased generation time (Table 2.1.1), but it also permits a much greater capacity for adaptations and differentiations [9], for storage for food materials to survive periods of temporary starvation, and for motile organisms it allows greater speeds to be attained (see Table 2.1.1). In the fungi, slime moulds and algae we see the first steps towards a multicellular existence, where an aggregation of cells can produce a larger organism to give an increasing exploitation of the advantages of size. In particular, this process allows different cells to specialize in different ways and so give different cell types. For example, the spore cells and stalk cells in *Dictyostelium* collectively allow amoebae to produce a fruiting body several millimetres high.

The size of a microbial cell determines biological properties, such as its desirability to predators, each of which has its own limited capacity or appetite for particular sizes of prey, and physical properties. For example the velocity (dx/dt) at which a cell will fall through its suspending medium is given by Stoke's Law (Section 3.5.3) for an ideal particle:

$$V_t = \frac{2}{9} r^2 g \left(\frac{\rho_p - \rho_a}{\eta} \right)$$

where r is radius, ρ_p and ρ_a are densities of particle and fluid, g is acceleration due to gravity, and η is viscosity of fluid. Thus larger organisms will settle more rapidly than smaller ones. For the smaller organism, below about 4 μm, Brownian movement will become an important factor that will counteract sedimentation. However, if these small cells clump together, or flocculate, they will settle much more rapidly. Larger organisms can counteract the tendency to fall in aqueous environments by a number of means. Some prokaryotic organisms, such as the planktonic cyanobacterium, *Trichodesmium erythraeum*, produce internal gas vacuoles to reduce their density. Some organisms, such as flagellated algae, stay suspended by swimming upwards. Many planktonic algae increase their surface area by producing spines or hair-like outgrowths, which can slow their rate of fall by increasing their viscous drag. Thus the diatom *Thalassiosira fluviatilis* produces about 76 chitinous spines that increase its surface area nearly threefold and nearly halve its rate of sinking in seawater.

As the surface area to volume ratio increases with decreasing size of cell, it follows that surface properties are proportionally much more important for bacteria than for larger organisms. The properties that affect the physical relationships between bacteria and their milieu, and between each other, are surface tension at liquid–solid and liquid–air interfaces, and surface charge. Bacteria differ in the hydrophilic or hydrophobic natures of their surfaces. Bacteria with a relatively high proportion of hydrophobic groups on their surfaces tend to get trapped on water-air interfaces, and so are seen as a surface scum on a growth medium; they can be separated from more hydrophilic cells by trapping them in a foam. Bacteria also differ in surface charge, which is dependent on the pH and ionic strength of the surrounding medium. This can be measured by observing the electrophoretic movements of the cells when placed in an electric field. In general, bacteria are negatively charged at neutral pH, have an isoelectric point at about pH 2–4, and are positively charged at pH values below this. The surface charge of microbes clearly affects their attachment to surfaces such as soil particles and also affects their clumping capability. For example, a major contribution to the flocculation of cells of the brewing yeasts, *Saccharomyces cerevisiae,* may be the formation of calcium ion bridges between adjacent negatively charged cells.

2.1.3 CELL STRUCTURE AND ITS IMPLICATIONS

All living organisms have a common physical structure: they are composed of the microscopic subunits called cells. All cells have a common chemical composition with the invariable occurrence of DNA, RNA and protein, and common chemical activities (their metabolism). However, microbial ecologists have to consider two types of cell, the prokaryotic cell and the eukaryotic cell, which when viewed together represent the greatest and most fundamental phylogenetic discontinuity amongst living organisms. In the following section, these two different types of cells will be functionally compared.

Genetic information and its transcription and translation

The bacterial chromosome is a single closed loop of double stranded DNA. It is tightly coiled into a nuclear body attached to the cell membrane, but has its peripheries as unwound stretches of single-stranded DNA exposed to the cytoplasm; thus for any gene transcription and translation, the synthesis of messenger RNA and protein synthesis can occur simultaneously [26].

Genetic transfer between bacterial cells is always unidirectional from a donor to a recipient; it occurs by conjugation mediated by F factors, transduction mediated by bacteriophage, or transformation mediated by free DNA itself. Usually only part of the donor genome is transferred. It is difficult to assess the extent of these processes in nature, but they must be of importance in the evolution of bacterial species. One process that has proved vitally important to man is the transfer of the R factor plasmids. These extrachromosomal pieces of DNA, which carry genes for antibiotic resistance, are transferred by conjugation from the donor to the recipient; this, in its turn, then becomes a donor. Thus the R factor spreads through the population of cells and can occasionally be transferred to another species of bacterium. The spread of antibiotic resistance from a benign to a pathogenic bacterium is obviously a very dangerous phenomenon and is causing considerable concern in medical and public health microbiology (see Section 2.3.3).

In eukaryotic cells the DNA occurs bound to basic proteins, histones, as chromosomes within the nucleus surrounded by the nuclear membrane. Thus transcription must be followed by the migration of messenger RNA into the cytoplasm, where translation can take place. The processes of mitosis and meiosis are essential to ensure that daughter cells and gametes have the correct complement of chromosomes. The haploid number of chromosomes is characteristic for any particular species, but some strains of cells, particularly those much selected by man (e.g. brewing yeasts), can occur with a certain number of each chromosome (polyploid) or with differing numbers of each chromosome (aneuploid). Recombination of genetic material in eukaryotes takes place chiefly through sexual reproduction, with the fusion of two haploid gametes to give the diploid zygote. A haploid–diploid alternation of generations is common amongst fungi and algae.

Bearing in mind the different physical arrangements of the DNA, the processes of transcription and translation to RNA and protein are biochemically similar in prokaryotes and eukaryotes. Finer differences, such as the subunit make-up of the ribosomes (measured as Svedberg units, S; 30S+50S to 70S in prokaryotes and 40S+60S to 80S in eukaryotes), are reflected in the very important differences in susceptibilities to antibiotics; streptomycin and chloramphenicol inhibit the function of the 70S ribosomes and cycloheximide the 80S. Such specificities dictate the choice of antibiotics as agents for chemotherapy or for use in selective media for isolation.

The cell membrane: barrier to the outside world

The cell membrane is the most important barrier between the cell and the outside environment. By virtue of its selective permeability properties, it is responsible for maintaining the appropriate levels of nutrients, metabolites, salts and pH in the internal environment, and so it is the site of nutrient, ion and proton translocating systems. The membrane is semipermeable, which results in the movement of water in or out of the cell according to the relative osmotic pressures inside and outside. Its properties must control any export or import of macromolecules, for example, the release of hydrolytic enzymes by bacteria. In the prokaryotic cell it is the site of the energy-yielding terminal respiratory system, and in the purple photosynthetic bacteria it is the site of photosynthesis; the surface area available for these two functions can be increased by infoldings, such as are found in *Azotobacter* and *Nitrosomonas* spp. when they are respiring rapidly, and *Rhodospirillum* species when they are photosynthesizing. A further internal elaboration of membrane in bacterial cells, the mesosome, appears to be found only in cells that have been shocked by treatments such as centrifugation, filtration, or fixation for electron microscopy, and may represent the aggregation of a pool of membrane precursors rather than a functioning structural component of the bacterial cell.

Eukaryotic outer cell membranes have the important attribute that they can have direct contact with the membrane systems within the cell. Thus, in an outgoing direction, macromolecules such as surface components or hydrolytic enzymes can be synthesized in the endoplasmic reticulum, modified through the Golgi apparatus, and released outside the cell by fusion of the Golgi-derived vesicles with the cell membrane. Similarly, protozoa that live in fresh water export the excess water taken up by osmosis by means of the contractile vacuole, a membrane system that pumps the water into a vacuole and releases it to the outside at regular intervals by fusion with the cell membrane. In an ingoing direction, many eukaryotic cells that lack a rigid wall can engulf food particles or draw in liquid droplets by phagocytosis or pinocytosis, respectively, by invaginating the cell membrane and so forming an intracellular vacuole. These food particles, such as bacteria, are digested after fusion of the vacuoles with lysosomes, Golgi-derived vesicles containing lytic enzymes. The process of endocytosis, when not followed by digestion, can lead in specialized cases to the phenomenon of intracellular symbiosis. For example, the ciliate *Paramecium bursaria* will take up cells of the alga *Chlorella*, which can become incorporated as photosynthetic 'organelles' (see Section 2.1.8).

Mitochondria and chloroplasts

These organelles of eukaryotes show some autonomy within the cell as they have their own DNA, RNA and protein synthetic machineries; they are capable of synthesizing some, but by no means all, of their specific proteins. Their DNA (as closed circles) and ribosomes (as 70S) appear to be more like those of prokaryotic cells than those in the rest of the eukaryotic cells, and their protein synthesis is inhibited by antibiotics such as streptomycin and chloramphenicol. In some species the mitochondria can become very much reduced in size and complexity in anaerobic conditions. Thus rumen ciliates and fermenting yeast cells characteristically have tiny mitochondria lacking tubular indentations (cristae). The close analogies between mitochondria and aerobic bacteria and between chloroplasts and cyanobacteria have led to the hypothesis that these organelles, and hence the eukaryotic cell, arose by the engulfment of prokaryotes as endosymbionts by a primeval cell that had acquired the property of endocytosis [33]. This is an appealing theory, but is impossible to prove or disprove against the alternative view of the origin of the eukaryotic cell. In this view mitochondria and chloroplasts represent portions of an orginal single progenitor; these portions have been sequestered and thus kept separate during evolution, in this way retaining some of their original characteristics [15].

Surface layers

The form of the microbial cell that we see is a manifestation of its covering, the cell wall. This structure determines many properties of the cell, its shape, its resistance to osmotic stress, its scope of nutrition, and its powers and mechanisms of movement.

Nearly all prokaryotes have a rigid cell wall (the exceptions being such organisms as halophiles, L-forms and mycoplasmas). The wall determines the shape of the cell, as a coccus, rod or filament. Treatment of cells with lysozyme, which causes hydrolysis of the glycosidic bonds in peptidoglycan, may lead to the digestion of the wall and so to the formation of protoplasts or spheroplasts, which are osmotically labile spheres. The peptidoglycan, the characteristic component of prokaryotic walls, is a major component of the Gram-positive wall but a minor component of the Gram-negative wall. Its synthesis is stopped by antibiotics such as penicillin, cycloserine and bacitracin. The components of the wall will be major determinants of the surface properties of the cell, such as charge, binding of ions, immunogenicity, and recognition by bacteriophages. The dynamics of their synthesis and lysis by the septal region of dividing cells will determine if daughter cells form chains or colonies by staying together, or separate by drifting or swimming apart.

Eukaryotic cells have no characteristic wall component. In fungi and most algae the protoplast is completely enclosed in a rigid wall with polysaccharides such as glucans, chitin and cellulose as major components. This wall imposes a fixed form and prevents activities such as phagocytosis and amoeboid movement. Thus the cells are limited to obtaining nutrients as soluble materials passing through wall and membrane. However, the rigidity of the wall makes possible the formation of integral structures such as toadstools and seaweeds, and enables the hyphae of filamentous fungi to penetrate host tissue or other substrates. Many protozoa, the slime moulds and some unicellular algae do not have a rigid covering. For example, *Euglena* spp. have a protein pellicle which is flexible and so allows euglenoid movement via the contraction of microtubules arranged in rows beneath grooves in its surface. Amoebae have just a thin coat, the glycocalyx, which allows amoeboid movement and the phagocytic engulfment of prey. However, cells without a rigid coat require a method of osmoregulation to prevent osmotic damage. Many freshwater organisms can pump out water via contractile vacuoles, while some estuarine algae control the concentration of the intracellular pools of soluble metabolites, such as mannitol in *Platymonas*, in response to changes in the extracellular osmotic pressure. Other coatings of considerable importance to the ecology of micro-organisms are the extracellular slimes and mucilages with which some attach themselves to their substrates.

Motility

When most motile bacteria are viewed in the electron microscope in a shadow-cast or negatively stained preparation, they are seen to have one or more flagella. Each flagellum is a helical strand that self-assembles from subunits of one single protein, flagellin. The composition of the flagellin is characteristic of the particular strain of bacterium, and flagella are antigenic determinants. The flagellum is totally extracellular and is attached to the cell via a hook region; this in turn, is attached to a basal assembly that is inserted through the wall and membrane. The basal assembly is a rotary motor that causes the flagellum to rotate in the water to drive the cell forward while the hook region acts as a universal joint. Another type of prokaryotic motility is spirochaete movement, in which these helical cells screw themselves through the water. The mechanism is homologous to that of flagellated bacteria as the spirochaete has axial filaments, called 'periplasmic flagella', which extend back from either end of the cell but are enclosed in an outer sheath. The rotation of these filaments between cell and sheath results in the propagation of a helical wave through the spirochaete, which thus swims through the water. A third type of motility is gliding, shown by cyanobacteria, myxobacteria and the cytophagas. Its mechanism remains unknown.

The major organelle for motility in algae, protozoa and fungal zoospores is the eukaryotic flagellum and its homologue, the cilium. The eukaryotic flagella and cilia have a universal structure of the '9 + 2' system of microtubules, surrounded by a protrusion of the cell membrane. Other forms of motility are: amoeboid movement, the pushing out of pseudopodia in amoebae and slime moulds; gliding, the secretion of a slime track by some diatoms; and euglenoid movement, the sinuous flexing of the cell body by some protozoa.

2.1.4 CELL REPLICATION

As a microbial cell grows, it takes in nutrients and synthesizes macromolecules. It obviously cannot keep enlarging in size and so it eventually must divide. The process of cell division in bacteria is regulated with precision; under uniform conditions new cells are very similar in both size and content of genetic information. There is an obligate link between cell division and DNA replication so that enucleate cells are not formed. However, it is possible for cells to divide in a time that is shorter than the time for the DNA polymerase to completely replicate the chromosome. This apparent contradiction is explained by the successive initiation of new rounds of DNA replication before each previous one has finished. Hence rapidly growing cells have a higher DNA content per cell. As well as the chromosome replication, there must be a general increase prior to division in cell content of enzymes, membrane and wall [26]. The act of division itself involves the ingrowth of membrane and wall at the midpoint of the cell to separate the two protoplasts. The separation of the two daughter cells is the result of action by autolytic enzymes on the septum that separates them. Gram-negative cells in general separate quite readily, but Gram-positive cells show a greater tendency to remain attached as filaments or clumps. Various treatments, such as high temperature or specific chemicals, can also inhibit separation of daughter cells.

Cells of the yeast, *Saccharomyces cerevisiae*, divide by the formation of a bud. The bud is smaller than the parent cell and grows until it reaches adult size. The release of each bud leaves a 'bud scar' on the parent cell, a ring of chitin-rich material that bordered the channel through which the cytoplasm passed into the bud. New buds are not formed in the region of a bud scar, so we know there is a maximum number of about twenty buds produced from one cell. This is a curious example of physical senescence in a microbe.

Fungal hyphae grow almost exclusively by apical growth. The synthesis of new wall material is predominantly in the tips of growing hyphae, both at the periphery of a colony and also at the sites of the subapical branches. In *Aspergillus nidulans*, as in unicellular organisms, the volume of cytoplasm per genome is held constant under particular growth conditions by the formation of subapical hyphal compartments delimited by septa and by regulated branching [12].

2.1.5 CELL NUTRITION

The availability of suitable nutrients is clearly a major factor in determining whether or not a microbe will grow in a particular environment. As a whole, micro-organisms are nutritionally much more diverse than higher organisms. There has been much interest recently in the microbial degradation of a wide range of natural and artificial chemicals (see Chapter 4.5), both with a view to recycling wastes and with a view to preventing biodeterioration. It is clear that many micro-organisms, particularly the bacteria, have or can acquire the ability to utilize novel or unusual chemicals.

There are four useful ways to classify potential nutrient molecules: as either essential or useful but dispensable; as either building blocks for macromolecules or energy sources, or both; as either macronutrients required in large quantities or micronutrients; and as either macromolecules requiring breakdown before entry to the cell or small molecules readily entering as soluble nutrients.

The major elements of the cell are carbon, hydrogen, oxygen, nitrogen, sulphur, phosphorus and minerals.

The ubiquitous source of carbon is carbon dioxide. This is utilized as sole carbon source only by autotrophs, but all organisms probably incorporate some via anaplerotic pathways (see Section 2.5.7).

Carbohydrates are commonly utilized as sources of carbon. Monosaccharides, such as glucose, are found in low concentrations in nature, probably because they are so rapidly taken up by cells. They are incorporated directly into macromolecules and can act as good energy sources. The utilization of disaccharides, such as lactose and sucrose, depends on the ability of the cell to produce an enzyme to hydrolyse the glycosidic bond. Most of the carbohydrate in nature is in the form of polysaccharides. For example, cellulose and chitin are very abundant in soil but are intractable molecules and are only slowly attacked by hydrolytic enzymes to yield soluble sugars. However, the ability to break down a particular polysaccharide is not widespread. Thus cellulases are common in fungi but more unusual in bacteria; they are produced by myxobacteria and by some species of *Bacillus*, *Cytophaga, Pseudomonas, Clostridium* and the actinomycetes.

Organic acids are readily used directly as carbon sources by most microbes. Fats are digested by lipases to give glycerol and fatty acids, and organisms with this ability are important as food spoilage organisms. Hydrocarbons, from methane through to petroleum, form a discrete group of carbon sources. Hydrocarbon utilizers degrade fuels and block fuel lines, but also break down petroleum wastes and oil spills. Proteins and their constitutent amino acids are utilized as carbon sources by proteolytic organisms such as *Pseudomonas* spp. found as food spoilage organisms and *Staphylococcus* spp. found as human pathogens.

Nitrogen is abundant as gaseous dinitrogen, but only a few prokaryotic organisms, such as *Azotobacter* and *Rhizobium* spp. and some cyanobacteria, are nitrogen fixers (see Chapter 4.1). Nitrate is found in aerobic soils and waters, but not all aerobic organisms have the necessary nitrate reductase. Ammonia is more readily utilized; at higher concentrations it is incorporated via the ubiquitous glutamate dehydrogenase, in lower concentration via the less common glutamate synthase. Amino acids, nucleotides, uric acid and urea can also provide the necessary nitrogen.

Sulphur occurs widely in nature as sulphate; this is reduced via sulphide for incorporation into the sulphur-containing amino acids and cofactors. Some organisms can utilize hydrogen sulphide as source of sulphur. Organic sulphur, as amino acids cysteine and methionine, can also be used.

Phosphates are very common in nature, but are often largely unavailable as they are in the form of insoluble metal phosphates. Probably all organisms can utilize soluble inorganic phosphates, and many produce phosphatases that will hydrolyse organic phosphates in the environment.

Of the minerals, relatively large amounts of potassium, magnesium and iron are required by virtually all cells, while some require calcium (for spore formation in *Bacillus* spp.), sodium (the halophytes) or silicon (diatoms). The requirements and functions of trace metals have proved difficult to determine, as they are commonly present as contaminants in laboratory media. It is doubtful if the presence or absence of the cations commonly required for growth — calcium, zinc, copper, cobalt, manganese, molybdenum — is often an important ecological determinant.

On the other hand, it is likely that the availability of vitamins can be an important ecological factor. Over half the planktonic algae tested require vitamin B_{12} while other algae species as well as bacteria will release it. Thus the cyclings of this vitamin could be controlling factors in the succession of species in plankton blooms.

2.1.6 CELL BEHAVIOUR

Introduction

We call micro-organisms 'simple' and tend to deny them attributes that we freely recognize in 'higher' organisms. However, a closer examination of bacteria and the eukaryotic micro-organisms shows that they respond to a wide range of physical and chemical stimuli with a repertoire of behavioural patterns; these serve to place the cell in a more favourable environment and to co-ordinate concerted growth and differentiation of groups of cells. Indeed, these responses are the reason for motility — without a stimulus to respond to, motility serves no fucntion.

Bacterial chemotaxis

Many bacteria have the ability to move actively towards a more favourable chemical environment. This remarkable property must surely have an effect on the distribution of microbes in a microhabitat, but its true importance is yet to be directly assessed. The phenomenon is readily observed and assayed by counting how many cells swim into a capillary tube containing a chemical when the tube is placed into a suspension of bacteria. Thus *Escherichia coli* responds, positively or negatively, to many specific chemicals, which can be grouped into classes by testing for their competition with each other [2]. It has been found that *E. coli* has at least 20 chemoreceptors, each with particular specificities and sensitivities (see Table 2.1.3).

There is ample evidence that the chemoreceptors detect the presence of the attractant molecules and not their metabolites. For example, some readily metabolized chemicals such as gluconate and glycerol are not attractants, while some non-metabolizable analogues of attractants are attractants themselves. In this way, fucose is sensed by the galactose chemoreceptor but is not a carbon source for *E. coli*. In the case of repellents, most of them are harmful substances but not all harmful substances are repellents. Thus, as with attractants, there are specific chemoreceptors that control the negative chemotaxis.

The process of chemoreception requires the presence of molecules with specific recognition sites for the particular substance being sensed. The obvious candidates are proteins as both enzymes and binding proteins have the requisite specificities. So far the chemoreceptor molecules identified have proved to be proteins that were already known. For galactose, the binding proteins are present in the periplasmic space and, for fructose, they are present in the membrane-bound enzyme II of the phosphotransferase system. In each case the proteins also serve as the primary stages for the specific uptake mechanisms of active transport.

The chemoreceptors do not just sense the presence or absence of an attractant, they must also sense its concentration as a bacterium responds to a gradient of the attractant [28]. The bacterial cells are so small that it is most unlikely that they could measure a concentration difference in a gradient between their anterior and posterior ends. Instead they respond to the change in concentration with time. Thus, in an isotropic solution (no gradient) a bacterium swims randomly for short periods in a straight line, then briefly tumbles, then resumes swimming in a different straight line. If the concentration of the attractant is increased, the pathlength of the runs increases and the frequency of the tumbles decreases. Conversely, a drop in concentration results in an increase in tumbling frequency (see Fig. 2.1.1). Repellents give opposite results. In each case in the new isotropic concentration the changed ratio of runs and tumbles relaxes over a period of time to the original 'normal' ratio.

Thus the bacterium has a 'memory' which enables it to compare its current environment with that a short time ago. If it is swimming in an unfavourable direction,

Table 2.1.3 Chemoreceptors of *Escherichia coli**

Attractants	Threshold concentration (μM)	Other chemicals sensed	Repellents	Threshold concentration (μM)	Other chemicals sensed
N-Acetyl-D-glucosamine	10	—	Fatty acids		
D-Fructose	10	—	e.g. acetate	300	Higher acids
D-Galactose	1	D-Glucose, D-fucose	Alcohols		
D-Glucose	3	D-Mannose	e.g. ethanol	1000	Higher alcohols
D-Ribose	7	—	Hydrophobic amino acids		Isoleu, Val, Try, Phe,
Maltose	3	—	e.g. L-leucine	10	Gln, His
Trehalose	6	—	Indole	1	Skatole
D-Mannitol	7	—	Aromatic compounds		
D-Sorbitol	10	—	e.g. benzoate	100	Salicylate
L-Aspartate	0.06	L-Glutamate	H^+	pH 6.5	—
L-Serine	0.3	L-Cysteine, L-alanine, glycine	OH^-	pH 7.5	—
			Sulphide	3000	2-Propanethiol
Oxygen	—	—	Metal cations		
			e.g. $CoSO_4$	200	$NiSO_4$

* Compiled from Adler [2].

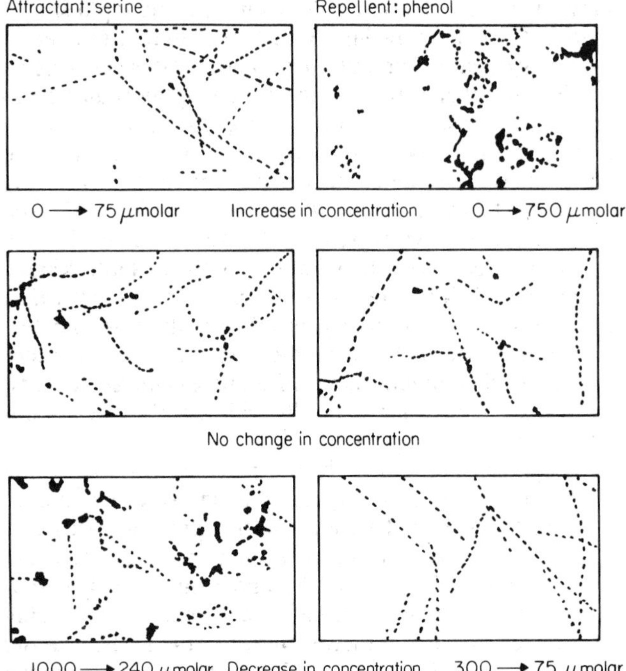

Fig. 2.1.1. Motility tracks of individual cells of *Salmonella typhimurium* (as shown by darkfield microscopy with stroboscopic illumination at 5 pulses s^{-1}), 2–7 s after changing the concentration of an attractant or repellent by adding and rapidly mixing appropriate solutions to suspensions of bacteria. Left-hand side: serine, an attractant; right-hand side: phenol, a repellent. Top: increase in concentration; middle: no change in concentration; bottom: decrease in concentration. Note that increasing attractant or decreasing repellent concentrations results in longer straight runs; decreasing attractant or increasing repellent concentrations results in tumbles and frequent changes in direction. In each case these alterations in swimming behaviour relax back to the normal control patterns as time goes on. Areas represent 165 × 90 μm. After Koshland [27].

tumbling is increased so that it will rapidly try a new direction. It is swimming in a biased random walk. The run results from the flagellar motor rotating in the driving sense; the tumble results from a reversal of the motor so that the flagellum flies awry. The temporal sensing of concentration gradient is transduced to a forward:reverse switch for the motor.

Many bacteria show aerotactic responses. Aerobic bacteria swim towards oxygen, anaerobic bacteria away from it. These responses are demonstrated by observing the pattern produced by bacteria on a microscope slide under a coverslip. Aerobic bacteria congregate at the outer edge and anaerobic bacteria congregate in the centre. In *Salmonella typhimurium* this response is mediated by the electron transport chain so that the cell senses a change in its protonmotive force; lowered oxygen gives lowered respiration and a lowered protonmotive force and results in increased tumbling while increased oxygen gives the opposite. Thus aerotaxis, is a form of 'energy taxis' towards environments that maintain an optimal protonmotive force in the cell [53]. This is less specific than chemotaxis, but interacts with it to prevent cells being trapped in an energy-depleting environment. Without aerotaxis, cells attracted by a readily oxidized substrate such as serine would deplete the oxygen and become trapped by the anaerobiosis. With aerotaxis, they swim to an optimum position for both serine and oxygen concentrations.

Bacteria such as *Pseudomonas* spp. appear to be attracted by the fungi, algae and plants they live on, but in general the specificities of the attractants have not been determined. However, a specific attraction is shown by the host leguminous plants to the symbiotic *Rhizobium* spp. (see Chapter 4.1), which swim to the root hairs and attach themselves to penetrate.

Chemotaxis and chemotropism in eukaryotic microbes

Movement towards nutrients by swimming or crawling (chemotaxis) or by directed growth (chemotropism) is widespread amongst the motile Protists [20]. Thus chemotropism to nutrients such as amino acids is shown by hyphae of the Oomycetes such as *Achlya* and *Saprolegnia* spp. This is in marked contrast to the hyphae of the Fungi Kingdom where there are no reports of tropism to nutrients; instead these organisms rely on random branching to find new food sources [19].

The chemotaxis of several organisms has been investigated in detail. The plasmodia of the slime mould *Physarum polycephalum* will crawl towards nutrients such as the sugars, glucose, galactose, mannose and fructose. The zoospores of the rumen chytrid *Neocallimastix frontalis* are attracted by soluble diffusible materials from the plant tissue, such as barley awns, so that they first invade them and then germinate and grow. Carbohydrates are the likely attractants from the plant tissue. By using capillary tube assays, the zoospores have been shown to have at least four chemoreceptors [39]:

1 The glucose receptor, sensitive to glucose, galactose, xylose, sorbose, fucose and 2-deoxyglucose.

2 The sucrose receptor, sensitive to sucrose, fructose and raffinose.
3 The sorbitol receptor, sensitive to sorbitol and mannitol.
4 The mannose receptor, sensitive to mannose and glucose.

Threshold concentrations for chemotaxis to single carbohydrates were similar to those reported for bacteria, about 5 μM for glucose and fructose and 0.1 μM for sucrose. Barley awns are richest in these three sugars; low concentrations of them act synergistically, actually giving greater chemotaxis than expected from addition of their separate effects.

There is evidence in some cases for the chemical attraction of parasites or symbionts to their hosts. For example, soil-borne zoospores of the important plant pathogens, *Phytophthora* and *Pythium* spp., are attracted by plant roots. This response is elicited by a range of chemicals in root exudates and extracts including amino acids, organic acids and ethanol, but whatever the mechanism this chemotaxis must increase the chances of infection of plants by the pathogenic cells swimming through the interstitial spaces in the soil.

Neighbouring cells of particular species often seem to be aware of each other. This is manifest in a general way in the spacing out of vegetative cells in a primitive form of territorial grazing. Thus the hyphae of a fungal colony grow out away from each other; the amoebae of a slime mould crawl apart across the surface of a substrate. These negative autotropisms and autotaxes are probably not controlled by species-specific chemicals. For example, a growing hypha will be surrounded by a zone of diminished oxygen concentration, and this will act to repel neighbouring hyphae.

In contrast, cells of many species can show very specific movements towards one another. These positive chemotaxes and chemotropisms play an important role in the social lives of the eukaryotic microbes, enabling them to form multicellular structures and to differentiate so as to give different cell types in the one environment. The best understood case is that of the remarkable aggregation of amoebae of the cellular slime mould, *Dictyostelium discoideum* [36]. When well fed, the cells are spread apart but, when they come to the end of their food supply, they form assemblages of many cells. These assemblages aggregate to form slugs, which migrate and eventually form fruiting bodies of stalk cells and spore cells. The aggregation process involves amoebae streaming inwards towards a focus. Each amoeba is responding to outwardly moving pulses of cyclic 3′5′-adenosine monophosphate (cyclic AMP) by crawling towards the source, the centre of the aggregate. The pulse of cyclic AMP is propagated by the cells themselves by the following mechanism: an amoeba responds to cyclic AMP diffusing to it by (a) moving a pseudopodium forward; (b) secreting the enzyme phosphodiesterase to destroy the cyclic AMP; (c) becoming temporarily unresponsive to cyclic AMP; and (d) releasing its own pulse of cyclic AMP, which will then similarly affect the cell behind it. With suitable kinetic parameters defining the rates of these different activities, the final result is an orderly wavelike motion of amoebae to a central point to form the aggregate. Some other species of slime moulds have different specific attractants, and mixed populations of cells will sort themselves out into different aggregates.

Many ascomycete and basidiomycete fungi show the important property of hyphal fusion, whereby neighbouring hyphae of the same species grow towards one another and fuse. This allows mycelia to be formed that are three-dimensional networks, and competition becomes co-operation. The most dramatic example, the fairy ring in soils, can be many metres across. This is the fruiting manifestation of one ring of interconnecting hyphae growing outwards with continual mixing of nuclei and cytoplasm.

Very specific chemical communication is seen in the fungi, algae and protozoa during sexual reproduction [20, 21]. Sex attractants are released from the gametes of some species to diffuse into the surrounding environment and invite the complementary gametes of the same species. In this way the motile female gametes of the alga, *Ectocarpus*, and the fungus, *Allomyces*, release ectocarpen (a C_{11} hydrocarbon) and sirenin (a sesquiterpene), respectively, and each of these chemicals specifically attracts the respective male gametes. Soon the female gametes become surrounded by clusters of male gametes. Another pattern, attraction of sexual cells to hyphae of opposite mating type, is seen in many different fungi. In one manifestation, the male antheridia of the aquatic oomycete, *Achlya*, grow towards the female oogonia because of attraction by the sterol sex hormone (antheridiol) that diffuses from the oogonia. In another species, the zygophores of opposite mating type of a zygomycete, *Mucor*, attract each other to fuse in pairs. In all these examples, specific organic molecules are released to the environment by microbes to signal their presence to other cells with the appropriate chemoreceptors. In the chemical turmoil of soil or stream, a few molecules of the particular compound will help to ensure that cells will come together to form organized structures or that gametes will find each other and fuse together.

Phototaxis and phototropism

Many microbes move towards or away from light of a particular intensity. Unfortunately for the experimenter, these responses are complicated by other responses to light such as when bacteria swim faster in the light than in the dark ('photokinesis') and when the sporangiophore of the fungus *Phycomyces* shows a transient spurt of growth after an increase in light intensity ('photomecism'). Nevertheless it is clear that unidirectional light is a major factor controlling the localized distribution of organisms in many habitats [23]. Thus flagellated algae such as *Euglena* and *Chlamydomonas* will swim towards light, cyanobacteria such as *Anabaena* and diatoms such as *Nitzschia* will glide towards it, the slug of aggregated amoebae of *Dictyostelium* will crawl towards it, and the sporangiophore of the zygomycete *Phycomyces* and the fruiting bodies of the basidiomycete *Coprinus* will grow towards it.

The motile algae typically show a dose-response curve to light intensity. But there is an inversion intensity above which they move away from the light. Their photoreceptor pigments are characteristically not their photosynthetic pigments, but instead are carotenoids, or retinal or flavoproteins.

The purple bacteria, such as *Rhodospirillum* and *Chromatium*, are phototactic, as was shown nearly one hundred years ago by Engelmann. However, they do not immediately swim towards the source of unidirectional light as do many eukaryotic cells. Instead they respond to a change in light intensity with time. If they sense a drop in light intensity, they cease to swim forward and undergo a shock reaction, tumbling to change their direction of swimming. This is clearly analagous to the mechanism of bacterial chemotaxis, and results in bacteria accumulating in illuminated areas by being trapped by the optimum light intensity. As in the case of aerotaxis, the cells are responding to changes in protonmotive force, here those induced by light across their photosynthetic membrane [53]. Their photoreceptor pigments are their photosynthetic pigments, and if a spectrum of light is cast across a suspension of these bacteria they will quickly form sharp bands corresponding to the absorption maxima of their pigments. *Halobacterium* is also phototactic, but the photoreceptor is the bacterial rhodopsin of its purple membrane and the phototaxis is independent of changes in the protonmotive force.

Phototropism is common amongst spore-bearing structures in fungi. Spores stand more chance of being widely dispersed when in the air exposed to wind and rain than when hidden under stones or between soil particles. Thus the sporangiophore of *Phycomyces blakesleeanus* grows towards a unidirectional source of light. The tubular subapical region acts as a lens to focus the light to the farther side, which stimulates its growth relative to that of the near side. The photoreceptor is apparently a membrane-bound flavoprotein. The related fungi, *Pilobolus* spp., are common on dung and show a remarkable mechanism that results in an explosive discharge of a package of spores aimed at the source of light. This is achieved by the directed growth of a sporangiophore with a swollen apex that acts as a lens to focus the light.

Orientated responses to gravity and magnetism

Prokaryotes do not seem to respond to gravity by positive movements but some bacteria and cyanobacteria can counteract gravity by producing gas vacuoles as a flotation mechanism. Among eukaryotes geotropism is widespread, especially in fungi; many have spore-bearing structures which grow upwards, such as the sporangiophore of *Phycomyces*. In mushrooms, the cap is orientated accurately at right angles to the earth's gravitational field so that after release from the basidia the spores can drop unimpeded between the vertical gills to the outside air. In some species of the plant pathogen, *Phytophthora*, the zoospores swim upwards against gravity; this presumably helps them move in the soil to find a new host plant.

The magnetotactic bacteria are a group of microaerophilic aquatic bacteria that contain crystals of magnetite in a membrane-bound organelle, the magnetosome [8]. This responds to the vertical inclination of the earth's magnetic field so that the cells act as compass needles, tilt downwards, and swim into the sediment. An interplay between aerotaxis and magnetotaxis then keeps them in their optimum microaerophilic environment. North-seeking bacteria predominate in the northern hemisphere and south-seeking in the southern.

Orientated responses to contact and pressure

These responses are probably quite widespread, but are difficult subjects for experimentation. Ciliate protozoa such as *Paramecium* will show an avoidance response so that, when they collide with an object, they will back away, turn and then go forwards again. With the use of an electrode inserted into a single cell of *Paramecium*, it has been shown that mechanical stimulation of the anterior end results in a transient increase in membrane permeability to calcium ions, an influx of calcium ions, and thus a reverse in direction of ciliary beating. Stimulation of the posterior end results in an increase in membrane permeability to potassium ions, an efflux of potassium ions, hyperpolarization of the cell, and enhanced ciliary beating to drive the cell forward [30]. With either reaction, the cell attempts to find a clear passage.

Organisms such as myxobacteria and germinating fungal spores sometimes align themselves with striations or irregularities in their substrata. This 'contact guidance' can give rise to patterns of movement or growth.

Orientated responses to heat

Movements through gradients of temperature are probably quite common amongst micro-organisms but have been little studied. When placed in a thermal gradient, cells of the ciliate *Paramecium* spp. will congregate at an optimum temperature. In the same way, the migrating slug of *Dictyostelium discoideum* will turn and crawl towards the warmer side of a Petri dish. This latter observation is remarkable as the slug is apparently sensitive to a difference in temperature between its two sides of only 0.0005°C.

Ecological significance of microbial behaviour

The responses described here must have a profound effect on the distribution of organisms in microhabitats. The position of a motile organism at any particular time will be a consequence of its reactions to stimuli such as the presence or absence of chemicals or light. It is difficult to see how the chemotaxis to a sugar such as galactose would have any effect on the life of *E. coli* in a human intestine, but if its response is to swim a short distance to a richer nutrient source then it perhaps will have a competitive advantage over an organism that does not have this ability.

2.1.7 MICROBIAL DIFFERENTIATION

Differentiation and secondary metabolism

Differentiation is any change in the activities of a microbial cell from those associated with its vegetative growth. It is often a phenomenon which occurs during starvation when cellular materials are redistributed to give new structures. Thus it is commonly observed during the ageing of a batch culture in a fermenter or during the ageing of the centre of an outwardly spreading colony on an agar plate.

A frequent concomitant process is that of the formation of secondary metabolites, which can be thought of as the chemical analogy of morphological differentiation. Secondary metabolites are the major products of industrial fermentation, and include such useful chemicals as antibiotics and vitamins. Nearly half the isolates of *Streptomyces* spp. from soil produce sufficient antibiotic activity to have been worthy of examination for possible commercial production. However, as secondary metabolites become major products of the cell, their formation must compete with the anabolic processes allowing the growth of the cell. Therefore, growth-linked metabolism and secondary metabolism are usually, at least to some extent, mutually exclusive processes.

No generalizations can be made about the roles of secondary metabolites in the lives of their producing organisms. In some cases functions can be suggested: some pigments, such as carotenoids, can be photoprotective; others, such as melanin, can protect an organism from attack by wall-lytic enzymes; antibiotics, such as penicillin or streptomycin, might increase the competitiveness of their producers (see Section 2.1.8). However, there are large numbers of secondary metabolites that have no apparent function; this has led to the idea that it is not the chemical end-product which is primarily important, rather it is the process of its formation which allows some metabolic activity, for example, recycling of coenzymes, to occur when conditions do not allow balanced metabolism and growth. Whatever its rationale, secondary metabolism can be seen as a playground of chemical evolution where nature experiments with all the possibilities of organic chemistry [54].

HORMONES CONTROLLING DIFFERENTIATION

There are many examples of species-specific and sex-specific metabolites controlling sexual differentiation in the eukaryotic microbes, but in only a few cases are the chemical structures of these hormones known [20, 21]. As they often act via the medium between two cells, they are often termed pheromones. Certainly many more will be identified in the future, and hormones will eventually be described which control aspects of asexual differentiation.

However, a few well-studied examples have shown the scope of these very specific and very potent chemicals in co-ordinating the development of neighbouring cells prior to physical contact. When two mycelia of a *Mucor* species of opposite mating type grow towards each other, they communicate by mutual transfer of diffusible specific metabolites; this results in both mating types producing a new metabolite, trisporic acid, which stimulates its own production by increasing the rate of synthesis of its precursors. Trisporic acid diffuses to neighbouring hyphae and causes a switch from the asexual differentiation, with formation of sporangiospores, to sexual differentiation, with formation of sexual hyphae which fuse in mated pairs to produce zygospores. In another species, *Achlya*, when male and female mycelia grow towards each other the male responds to the steroid sex hormone (antheridiol) produced by the female by forming antheridial branches and by releasing a second steroid sex hormone (oogoniol). This hormone, in turn, causes the female to differentiate oogonia and stimulates the release of more antheridiol. In both of these examples the hormones serve as chemical signals so that each cell recognizes that a potential mate is nearby. In both cases there is positive feedback to strengthen the commitment to mating as the cells approach each other. This is analogous to the mating displays of animals, which strengthen the pair bond. The result is an orderly sequence of events leading to successful cell fusion between two compatible cells mixed amongst the varied cells of the soil or a stream.

There is one example of a sex pheromone in bacteria. Recipient strains of *Streptococcus faecalis* secrete an octapeptide that induces conjugation with donor strains at very low concentrations (down to 4×10^{-11} M).

Sporulation

SPORES FOR DISPERSAL AND FOR SURVIVAL

The production of spores is a widespread phenomenon amongst micro-organisms. Spores have two major functions; they act as agents of dispersal and as agents for survival through adverse conditions. In both cases their formation is clearly a major factor determining the distribution of the producer organisms.

By sucking air samples through a spore trap, which impels them on to a microscope slide or an agar plate, it is easily seen that there is a rich 'aerial plankton' of spores (see Chapter 3.5). Spores are also abundant in samples of foam from a river or stream. In both cases most of these will be fungal spores since the fungi are better adapted for the production of dispersal spores than other organisms. The fruiting bodies of mushrooms and puffballs, the conidiophores of *Penicillium* and *Aspergillus* spp. and the sporangia of *Mucor* and *Rhizopus* spp. are all examples of structures producing large numbers of spores to be wafted away by air currents or splashed away by raindrops. Amongst the prokaryotes, members of the actinomycetes are the only organisms that produce such large numbers of spores for dispersal. They form chains of conidia on aerial filaments growing away from the substrate.

Some fungi and algae produce motile zoospores which move away from the parent cells to settle on a new substrate. For example, a newly cleaned ship's bottom will rapidly become covered with a growth of seaweeds after their motile zoospores have attached themselves to it by secreting an adhesive material. Some fungal spores are explosively shot away from the parent mycelium. Thus basidiospores of a mushroom are shot from the basidium into the space between adjacent gills, from where they can fall unimpeded into the air below the cap. Spores of the common leaf-inhabiting yeast *Sporobolomyces* spp. are propelled a considerable distance, so that the lid of a Petri dish will become a mirror image of the culture below.

There is a considerable range in the size and shape of spores. These properties will determine such factors as rate of sinking in air or water and ease of entrapment on substrates. The spores of the aquatic hyphomycetes in streams are characteristically long narrow branched structures, often tetraradiate, which readily become entangled in the organic debris that is the substrate for the vegetative cells. Spores in the air are breathed in by man and animals, and their size is critical to the depth to which they penetrate the respiratory tract. Thus the spores of the allergenic actinomycetes, responsible for widespread farmer's lung disease (see Section 3.5.6), are small, 0.5–1.3 μm in diameter, and readily accumulate in the pulmonary spaces. Many spores have a characteristic ornamentation on their surface. The biological significance of this is usually not clear, but it does serve as a useful taxonomic character.

For survival, spores have a greater resistance to adverse conditions than their parent cells and go through a dormant period after formation [52]. To these ends, they have thick multilayered walls and often require a trigger for germination. The most important examples to man are the endospores of *Bacillus* and *Clostridium* species. These have extreme heat resistance; for example, *Bacillus thermophilus* spores are only killed by heating at 100°C for 4 hours. Their existence means that 20 min at 120°C in a steam autoclave or 4 hours at 160°C in a dry oven are needed as standard sterilization techniques in a microbiological laboratroy. Endospores also have great resistance to desiccation, to ultraviolet light and to chemical and enzymic attack. The spore protoplast contains a large amount of a spore-specific component, calcium dipicolinate, and is surrounded by a thick cortex composed chiefly of a specific peptidoglycan and a hydrophobic proteinaceous spore coat, rich in cysteine and hydrophobic amino acids.

Other resistant spores include the endospores produced by some thermophilic actinomycetes, the zygospores of *Mucor* and *Rhizopus* and the ascospores of *Neurospora*.

CONTROL OF SPORULATION

Characteristically the major factor controlling sporulation is the availability of nutrients. Starvation encourages it; nutrients discourage it. Thus a culture of *Penicillium* growing on an orange forms conidia from the centre of the colony outwards by differentiation of its older mycelium where the nutrients have been depleted. Similarly, when they have eaten the available bacteria, amoebae of *Dictyostelium* aggregate and differentiate to form the stalk cells and spore cells of the fruiting body. A batch culture of a *Bacillus* species starts to form endospores when it has reached stationary phase, i.e. when the medium is depleted of at least one nutrient. The process of endospore formation takes about 8 hours, but during the first 2 or 3 hours the cells can be induced to grow vegetatively by the readdition of nutrients. After this time they are committed to sporulation. This, as befits a process favoured by starvation, is usually to some degree an endotrophic process which involves the redistribution of cellular metabolites to form the spore-specific materials.

Sporulation can be regulated by the production of diffusible morphogens. For example, the formation of akinetes (resting spores) in the cyanobacterium *Cylindrospermum licheniformae* is stimulated by a specific heterocyclic metabolite produced by sporulating cultures [1]. This metabolite is active at low concentrations (<0.3 μM) and diffuses to neighbouring cells to induce their sporulation. But the most complex differentiation shown by bacteria is the formation of fruiting bodies among the myxobacteria. In *Myxococcus xanthus* this involves the aggregation of cells to form stalked fruiting bodies bearing myxospores; the process is remarkably analogous to the sporulation of *Dictyostelium discoideum* discussed above. In *Myxococcus* fruiting body formation only occurs at high cell densities, and this phenomenon is regulated by a diffusible secreted signal molecule which has been identified as adenosine.

SPORE GERMINATION

Many of the spores produced and dispersed in large numbers will germinate readily. In some species germination will occur even without exogenous nutrients, just requiring a moist atmosphere. The resistant resting spores usually require a 'trigger' to break their dormancy [52]. This can be physical, such as the heat treatment of 60°C for 30 min for ascospores of *Neurospora crassa*, or chemical, such as the presence of alanine for endospores of *Bacillus subtilis*. In either case, the breaking of the dormancy of the spore serves as a timing device so that germination occurs when conditions for growth are liable to be favourable; this occurs after a forest fire for *Neurospora* and in the presence of nutrients for *Bacillus*.

In these cases the dormancy is constitutive, i.e. broken by an external agency. In other cases dormancy can be imposed on a spore by the presence of an internal or external factor, so that germination will ensue when it is removed. Examples are the volatile chemicals, produced by mature cultures of many fungi, that prevent the germination of spores in their vicinity. Thus the chances of competition with the parent colony by its own spores are reduced.

Heterocysts of cyanobacteria

Heterocysts are thick-walled cells found at regular intervals in the chains of some blue-green bacteria [1]. When they occur, they are the major site of nitrogen fixation (see Chapter 4.1). In *Anabaena cylindrica* their formation is repressed in the presence of ammonium salts. When such a heterocyst-free filament is moved to a medium free of fixed nitrogen, heterocysts differentiate from newly divided cells at regular intervals. The heterocysts have a very much reduced content of photosynthetic pigments and do not contain the oxygen-evolving photosystem II (see Section 2.5.5). As nitrogenase is very rapidly denatured by oxygen, the clear inference is that the filament is physically separating the oxygen-evolving and the nitrogen-fixing cells and that fixed nitrogen passes from heterocyst to vegetative cell and fixed carbon passes in the other direction.

2.1.8 INTERACTIONS WITH OTHER CELLS

Types of interactions

The previous sections have concentrated on the individual microbial cell. However, as John Donne tells us, 'No man is an *Island*, entire of it self; every *man* is a piece of the *Continent*, a part of the *main*.' Thus microbial cells nearly always occur within complex interacting communities.

The terminology used to label interactions between micro-organisms and between micro-organisms and higher forms of life sometimes causes confusion. Table 2.1.4 is based on the definitions given by Alexander [4]. Associations can also be described in terms of dependency (*obligate* or *facultative*), durability (*persistent* or *transient*), specificity (*specific* or *non-specific*) and location (*ecto*-organisms live outside the cells of the host and may be housed in specially adapted organs of the host while *endo*-organisms live within the host cells in vacuoles in contact with cytoplasm). Symbioses can be beneficial or harmful to either partner.

It is also useful at this stage to define some of the interactions of micro-organisms with the environment (*indigenous* or *autochthonous*, growing in and contributing to the overall metabolic activity within an environment; *exotic* or *allochthonous*, not indigenous, may be imported to an environment) and some aspects of populations (*axenic*, pure culture; *conventional*, natural microflora; *gnotobiotic*, defined microflora; *homeostasis*, capacity to maintain population stability in the face of a variable environment; *invasiveness,* ability to migrate and establish itself in a location possessing resources as yet unexploited; *pioneer*, first species to colonize an environment; *stable population*, population which does not change with time).

In recent years Starr [49] and Lewis [31] have produced schemes to describe all aspects of the interactions between two organisms, but as yet their terminology has not been widely accepted.

In complex ecosystems containing a diversity of organisms the metabolic activity of any individual organism is likely to have an effect on its near neighbour. For example, although the flora of the rumen or gastro-intestinal tract can be regarded as a single metabolically active tissue in symbiosis with its host, interactions of many types are continually taking place between components of the microflora. These interactions are largely self-regulating so that an equilibrium is established between the various beneficial, mutualistic or antagonistic activities. In other ecosystems many of the organisms coexist in close physical contact at an interface with solid material, e.g. on plant fragments, adjacent to gut epithelial cells or within dental plaque. Many others are in a constant state of movement either as a result of their own intrinsic motility or from the forces applied to them.

As an example of a complex ecosystem we can consider the microbial communities developing on surfaces submerged in water (see Figs. 2.1.2, 2.1.3 and 2.1.4). These can range from comparatively simple communities of

Table 2.1.4. Terms used in description of biological interactions

Term	Definition
Characteristic species	Species generally present or abundant in a particular environment
Codominant	Two or more species present in large, but roughly equal, numbers
Commensal	Different species living in close association without much mutual influence
Defence mechanism	Antimicrobial activity either passively or actively employed by a host to resist infection by a pathogen
Host	Animal or plant in which, or on which, another organism lives
Interspecific association	Species residing in close proximity or regularly occurring together
Succession	Replacement of one type of population by another in response to modification of habitat
Symbiosis	An association of two living organisms, or populations, which, in the absence of environmental change, is stable
Mutualism	Both partners gaining benefit
Neutralism	Neither partner benefiting
Antagonism	One or both partners suffering
(a) competition	The situation in which the populations of two species are mutually limiting because of their joint dependence on a common nutrient
(b) amensalism	Repression of one species by toxins produced by another
(c) predation	Ingestion of one organism by another
(d) parasitism	Derivation of nutrients from living tissue or cells
Synergism	Association of organisms having complementary activities resulting in greater formation of products than by component organisms growing alone

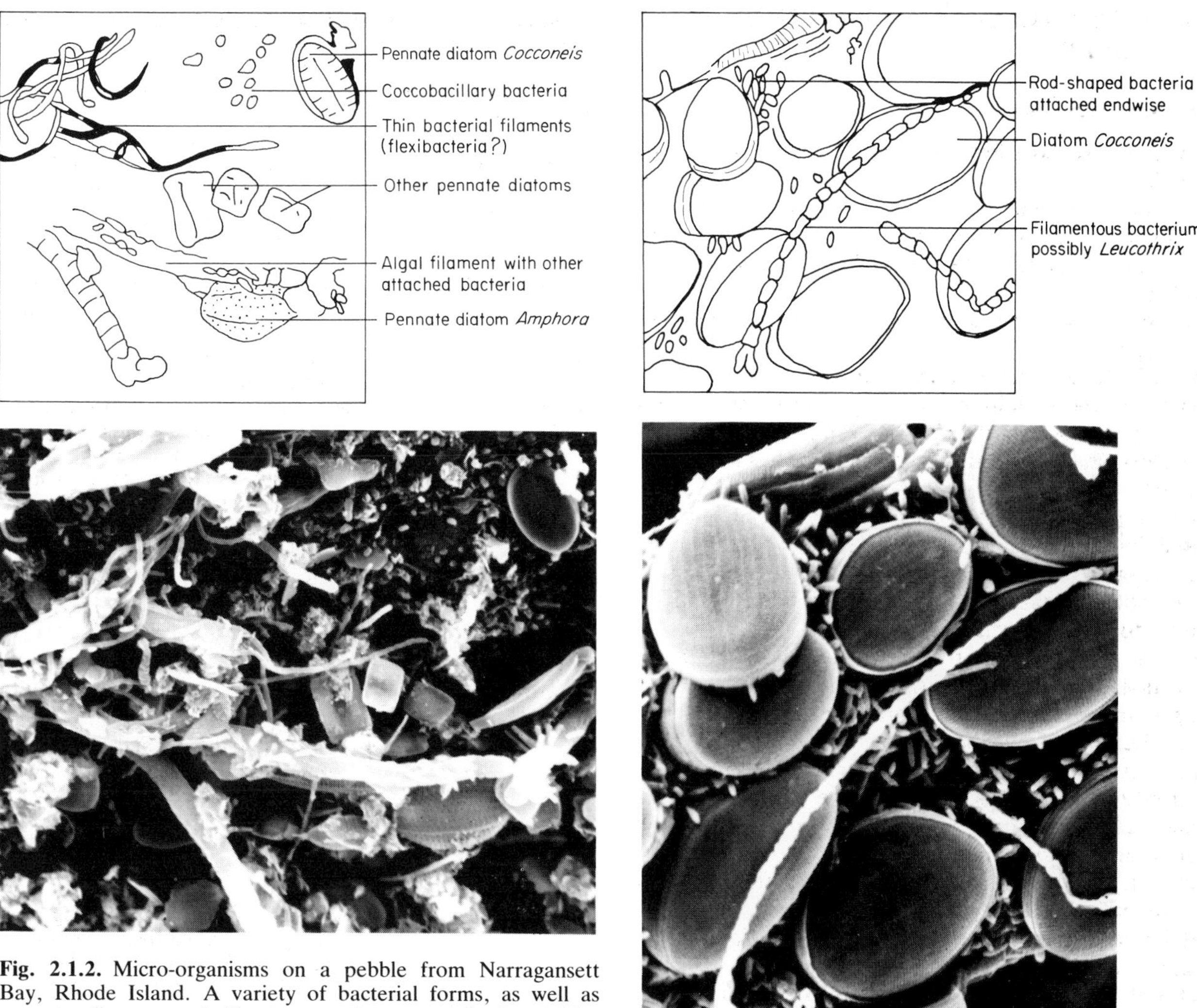

Fig. 2.1.2. Micro-organisms on a pebble from Narragansett Bay, Rhode Island. A variety of bacterial forms, as well as pennate diatoms and filamentous algae, are clearly visible. Magnification × 1700. From Sieburth [45] with permission.

Fig. 2.1.3. The green alga *Cladophora* covered by diatoms (*Cocconeis*) and various filamentous rod-shaped bacteria. Magnification × 2200. From Sieburth [45] with permission.

attached micro-organisms to complex and extensive communities composed of micro-organisms, plants and animals. In lakes, the communities may comprise attached bacteria, unicellular and filamentous algae, protozoans, roundworms, rotifers, annelids, crustaceans and insects [40]. The nature of the community is strongly dependent upon the characteristics of the substratum (particularly when a plant or animal is the underlying surface) as well as on the movement of the surrounding water, its fluctuations in level and its chemistry. When sufficient light and nutrients are available, large concentrations of algal cells (blooms) may occur, and these have a considerable effect upon other micro-organisms present. Since algae can secrete substantial portions of photosynthetically fixed carbon, accumulation of both stimulatory and inhibitory substances during blooms can affect not only the present mixed microbial population but also the subsequent pattern of succession [17].

Many macroalgae have considerable populations of microalgae and bacteria (epiphytes) living on their outer

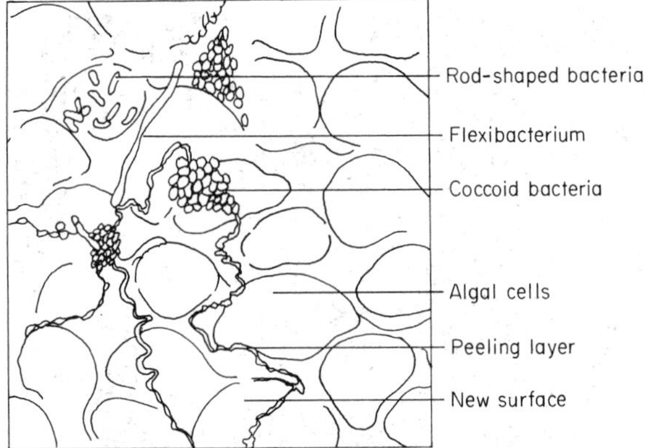

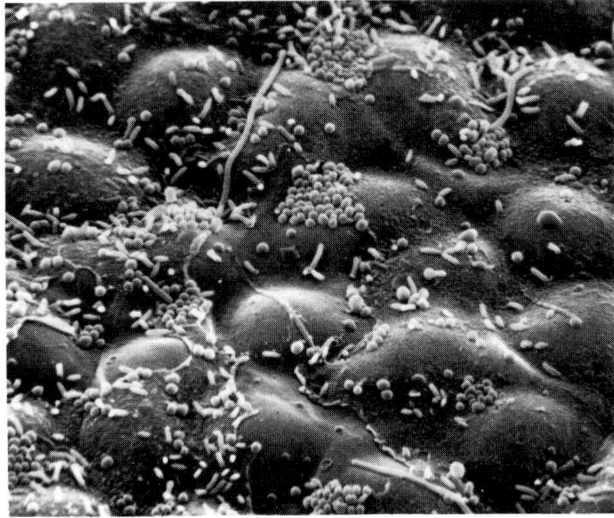

Fig. 2.1.4. The green alga *Ulva lactuca* (sea lettuce) which is covered by a variety of rod-shaped and coccoid bacteria but is free of diatoms. Magnification × 1900. From Sieburth [45] with permission.

surfaces (see Figs. 2.1.3 and 2.1.4), whereas others are relatively free of attached micro-organisms. Some young areas of the algae may remain epiphyte-free because they grow rapidly, while older portions have become heavily populated. Attachment and growth of epiphytes might also be prevented by the production of bacteriocides or bacteriostats [41]; for example, tannin has been shown to be implicated in the reduction of microbial epiphytes on the brown alga *Ascophyllum nodosum* [35]. Another factor affecting epiphytic growth could be the amount of surface mucilage produced by the potential host, since copious production would result in repeated sloughing off of the outer layer and make it difficult for epiphytes to maintain colonization. Brown algae produce relatively large quantities of surface mucilage and they have few permanent epiphytes; red algae have little surface mucilage and they are often heavily colonized [10].

In many cases the relationship between macroalgae and epiphytic bacteria may be beneficial to both. The bacteria may use their host as a support when they assimilate nutrients from the surrounding medium, or they may utilize algal exudates, such as amino acids, sugars and organic acids, or surface mucilages, which are usually sulphuric acid esters of polysaccharides [41]. In turn, the bacteria may provide the algae with necessary vitamins or growth factors.

Thus these apparently simple environments, pebble and algal surfaces in water, can provide us with many types of interactions. Briefly considered below are some more clear-cut examples of specific microbe–microbe interactions. Others, especially competition and predation, are considered in detail in Chapter 2.4. Interactions with plants are considered in Chapters 3.1 and 4.1 and with animals in Chapter 3.3.

Parasitism

An excellent example of parasitism is provided by the interaction between bdellovibrios and their host bacteria. These associations were discovered accidentally during attempts to isolate bacteriophages of *Pseudomonas phaseolicola*. Highly motile, tiny vibroid bacteria were observed which attacked the pseudomonad, causing it to lyse. These and other strains of lytic vibrios were then studied in detail and classified into a new genus, *Bdellovibrio*.

These organisms are active against Gram-negative bacteria, especially pseudomonads and enterobacteria. They are obligate aerobes and are widespread in nature, probably occurring in any aerobic environment that supports growth of suitable host bacteria. Individual cells of *Bdellovibrio* physically collide with individual cells of the bacterium and attach to their cell surfaces. There is no evidence for a specific chemotaxis. Within 5–20 min after attachment, the parasite has lysed a pore through the cell-wall layers of the host so that it comes to lie in the periplasmic space, distorting the host protoplast. This penetration seems to be partly enzymic and partly physical, with the primary impact and a subsequent high-speed rotation (up to 100 rps) of the bdellovibrio aiding the process [51]. After the entry of the bdellovibrio the pore is sealed and the peptidoglycan of the host is enzymically deacetylated. This makes it resistant to lysozyme and provides an exclusion mechanism which prevents an attack by other bdellovibrios [42]. The host cell gradually rounds up to become a 'bdelloplast'; it loses its motility and its metabolic activities, and its cell contents are progressively degraded to provide a supply of nutrients into the periplasmic space. Meanwhile, over 1–3 hours, the *Bdellovibrio* grows into a long spiral cell which finally divides into several daughter cells. These are released to the medium via specific lysis of the modified peptidoglycan. The number of daughter cells depends on the size of the host: about six for *Escherichia coli*, ten for *Pseudomonas fluorescens*, and 25 for *Spirillum serpens*. It is clear from the intricacies of this system that the bdellovibrios are highly evolved parasites with many specific adaptations to their unique habitat — the Gram-negative periplasm.

Fig. 2.1.5. Scanning electron micrograph of spores of *Cochliobolus sativus* with large annular depressions caused by a giant soil amoeba. Magnification × 3500. Photograph courtesy of Dr K.M. Old.

Predation

Predator–prey relationships are considered in detail in Chapter 2.4, but one example of predation will be considered here, that of protozoa and fungal conidia in the soil. When conidia of *Cochliobolus sativus* were incubated in natural soil for several weeks, they became perforated by holes 2–4 μm in diameter and their contents were lysed (see Fig. 2.1.5). Direct microscopic examination proved the culprit to be a giant soil amoeba resembling *Leptomyxa reticulata*. Over a period of 2–4 hours, the amoebal pseudopodia invested a conidium, excised a disc from its wall, lysed its contents, and moved on to another conidium [38]. Later the conidial shell was invaded by bacteria. It seems that many fungal propagules are attacked by these giant soil amoebae, so this type of association may be a means for the control of fungal pathogens.

Amensalism

A major example of intermicrobial amensalism is the production of antibiotics in the soil — or is it? Despite the immense importance of antibiotic formation to man, the ecological significance of this phenomenon remains open to debate. Many soil micro-organisms, especially actinomycetes, *Bacillus* species and fungi, produce specific antibiotics as secondary metabolites when in pure culture (see Section 2.1.7). These antibiotics are potent inhibitors of growth of specific groups of organisms: β-lactams such as penicillins and cephalosporins act against Gram-positive bacteria; streptomycin acts against Gram-positive and Gram-negative bacteria; while cycloheximide acts against eukaryotic cells. It thus seems reasonable to conclude that the attribute of antibiotic production aids the competitiveness of the producing organism in the struggle for existence in the soil [22].

Experiments suggesting this come from a study of the effect of *Streptomyces hygroscopicus* var *geldanus*, which produces the antibiotic geldanamycin, on the plant pathogenic fungus *Rhizoctonia solani* [43]. When the streptomycete was inoculated into sterile soil and two days later *R. solani* and peas were added, the disease (root rot of the pea) was controlled. The saprophytic growth of *R. solani* in the soil was also inhibited by *S. hygroscopicus* and by geldanamycin. Methanolic extracts of the soils that had been incubated for at least two days with *S. hygroscopicus* yielded appreciable amounts of the antibiotic. Thus there was a good correspondence between antibiotic production and disease control. There was no evidence for antagonism between *S. hygroscopicus* and *R. solani* owing to competition for nutrients or to parasitism. For these reasons, the antagonism can be ascribed directly to the streptomycete producing an antibiotic which is active against the fungus.

The debate occurs because experiments carried out in sterile soil are not proof of events in natural soil. Believers in the importance of antagonistic effects of antibiotics in nature state that antibiotic production is very widespread in soil micro-organisms (e.g. over 3000 different antibiotics are produced by *Streptomyces* species); that these organisms will produce antibiotics when inoculated into sterile soil; that antibiotic production in pure culture is associated with stress, for example, shortage of nutrients, a condition which is likely in the soil; and that genes for antibiotic resistance are found in natural populations of organisms that might be exposed to the antibiotic, suggesting that the antibiotic has been a selective pressure in evolution. Non-believers counter with the arguments that antibiotics can rarely be detected in natural soils; that if produced there they would be inactivated by the soil microflora or removed by adsorption to clay surfaces; and that their production in pure culture is usually preceded by rapid growth in a rich nutrient medium, a condition that would be rare in the soil. What is needed to settle the argument is probably technically impossible at the moment — an experimental investigation of antibiotic production and effects in a natural soil microcosm.

Nevertheless, while the academics argue, the idea that highly competitive microbial populations are fruitful sources for new antibiotics has proved to be valuable for the pharmaceutical industry in the last few decades. Recently the spectrum of antibiotic-producing organisms has been widened with the discovery of a range of novel β-lactams produced by Gram-negative bacteria. It has been suggested that these antibiotics play roles in intermicrobial competition among organisms such as the enteric bacteria [32] and in biological control (see Chapter 4.2).

Mutualism

There are many examples of highly evolved mutualistic associations among the microbes. Two types of associations are presented here, although it has been argued that these relationships are not mutualistic but are parasitism of the phototrophs by the heterotrophs. However, as the relationships show many aspects indicative of long periods of coevolution, and as they enable the phototrophs to survive in environments where they otherwise would not live, they will be considered here as good examples of mutualism.

The lichen symbiosis includes about 18,000 species, inhabiting a wide range of environments. The fungus is nearly always an ascomycete, usually one of the species of the lecanorales that occur only in lichens (but make up about one-third of the total number of fungal species known), but very occasionally is a basidiomycete. The phototroph is usually a green alga, mostly of the genus *Trebouxia* (known only from lichens), or a cyanobacterium (often *Nostoc* species). The form of the lichen, much more elaborate than either fungus or alga would produce alone, is a clear case of structural synergism. Although the fungus is the major part of the structure of the lichen, the alga has a distinct influence on the morphology of the thallus; this has been shown by the discovery of chimeral lichens which contain one continuum of fungal mycelium but two different algae in two structurally different parts [47]. The algae are confined to a distinct region of the thallus and are enveloped in fungal hyphae. In some cases fungal haustoria penetrate the algal cells, but this is rare.

Lichens are well known as the first colonizers of harsh environments — tundra, desert, rocks, asbestos roofs — and laboratory studies have shown that they often require stresses such as wetting and drying. It has proved very difficult to resynthesize a lichen in the laboratory from its component parts; this has only been achieved with painstaking techniques [3].

Within the lichen, the phototroph releases at least half of the photosynthetically fixed carbon to the fungus, either as glucose in the case of the cyanobacteria or as a polyol in the case of the green algae. This release is a property of the symbiosis, as freshly isolated algae quickly stop releasing photosynthate. When the phototroph is a cyanobacterium, nitrogen is actively fixed and released from bacterium to fungus. Again the intact structure of the lichen is important in this attribute and the rate of nitrogen fixation quickly falls when the lichen is homogenized.

The lichen thallus very rapidly takes up any minerals, nutrients or other materials that may pass over it. This attribute presumably aids it in scavenging for macro- and micronutrients, but also makes it very susceptible to pollution. Thus the species diversity of lichens is a very sensitive monitor for airborne sulphur dioxide or fluoride pollution, and analyses of lichens give good estimates of pollution due to heavy metals and radioactive fall-out.

Protozoa have also exploited the primary production of algae and cyanobacteria [18]. In these cases the phototrophs are intracellular, sequestered in vacuoles, and the mutualistic relationships have presumably evolved from phagotrophic feeding by the protozoa with the phototroph avoiding digestion.

The ciliate *Paramecium bursaria* is green because it contains several hundred green algae of the genus *Chlorella*. Each algal cell grows and divides in pace with the growth and division of the paramecium. Without other food the green ciliates grow and survive in the light but not in the dark. Thus photosynthesis can provide the ciliate with nutrients. Conversely, when fed with bacteria in the dark, the algal numbers per ciliate are maintained so the host can provide nutrients for the algae. *Paramecium bursaria* is phototactic; the algae are the photoreceptors and the ciliate apparently responds to the oxygen produced during photosynthesis.

Among the amoeboid protozoa there are many examples of algal symbionts in the foraminifera and the radiolaria. Different species of protozoa may have chlorophytes, dinoflagellates, diatoms or rhodophytes as symbionts. In the giant foraminifera the algal photosynthesis is implicated in the calcification of the characteristic test (shell) as it is in the calcification of the symbiotic reef-building corals. In both cases the rate of calcification is much higher in the light due to the alga's utilization of carbon dioxide and the subsequent shifting of the equilibria of the following reactions towards the right:

$$Ca^{2+} + 2HCO_3^- \rightleftharpoons Ca(HCO_3)_2 \quad (2.1.1)$$

$$Ca(HCO_3)_2 \rightleftharpoons CaCO_3 \downarrow + H_2CO_3 \quad (2.1.2)$$

$$H_2CO_3 \rightleftharpoons CO_2 + H_2O \quad (2.1.3)$$

In reaction (2.1.1) the bicarbonate is provided in part by animal respiration; in reaction (2.1.2) the calcium carbonate is deposited as animal exoskeleton; in reaction (2.1.3) the animal's carbonic anhydrase provides the carbon dioxide that is utilized by algal photosynthesis. In both foraminifera and radiolaria, the surface area for their feeding is greatly enlarged by sprays of a myriad protoplasmic projections, the rhizopodial network. A fascinating behavioural response observed in some species is that in the light the algal symbionts are moved out along these rhizopodia, like washing being hung out from a tenement window in the sunshine. Come dusk, they are withdrawn again towards the body of the host cell.

The flagellate *Cyanophora* contains symbiotic cyanobacteria, termed cyanellae. There are two bacteria per cell. When the flagellate divides, one cyanella is moved into each daughter cell, and then divides to restore the norm of two per cell.

Commensalism

An example of commensalism is given by the relationship between the two chemolithotrophic nitrifying bacteria, *Nitrosomonas* and *Nitrobacter*. Together they transform ammonium through to nitrate; one supplies an intermediate product, nitrite, to the other:

$$NH_4^+ \xrightarrow[\textit{Nitrosomonas}]{} NO_2^- \xrightarrow[\textit{Nitrobacter}]{} NO_3^-$$

This relationship does not require proximity between the cells as the nitrite can diffuse between them. The process has been investigated mathematically both in soil [16] and in columns packed with glass beads and inoculated with pure cultures of the two bacteria [7]. The following equations have been derived to describe the interactions:

$$dS_1/dt = D(S_0 - S_1) - \mu m_1 \quad (2.1.4)$$

$$dm_1/dt = \mu m_1 - K_3 m_1^2 \quad (2.1.5)$$

$$dS_2/dt = K_1 m_1 - DS_2 - \frac{\lambda S_2 m_2}{S_2 + L} \quad (2.1.6)$$

$$dm_2/dt = \frac{\lambda S_2 m_2}{S_2 + L} - K_4 m_2^2 \quad (2.1.7)$$

$$dS_3/dt = K_2 m_2 - DS_3 \quad (2.1.8)$$

Where S_1, S_2 and S_3 are the concentrations of ammonium, nitrite and nitrate, respectively, in the column, S_0 is the input concentration of ammonium, m_1 and m_2 are the biomass densities of *Nitrosomonas* and *Nitrobacter*, respectively, μ and λ are the maximum specific growth rates of *Nitrosomonas* and *Nitrobacter*, respectively, L is the saturation constant of *Nitrobacter*, $K_1 - K_4$ are constants, and D is the flow rate per unit void volume of the column.

When this model was tested by analysing the effluent from the glass column, it provided a satisfactory fit for steady state conditions. However, when the steady state was unbalanced by changing the rate of flow through the column, a smooth monotonic transition to a fresh steady state was achieved for an increase in flow (see Fig. 2.1.6) but not for a reduction (see Fig. 2.1.7). The overshoots in the observed concentrations of nitrite and nitrate were not predicted by the model. Although more recent work [7] has produced some explanation for this phenomenon and a better model, the results serve to emphasize both the value of modelling and the difficulty of extrapolat-

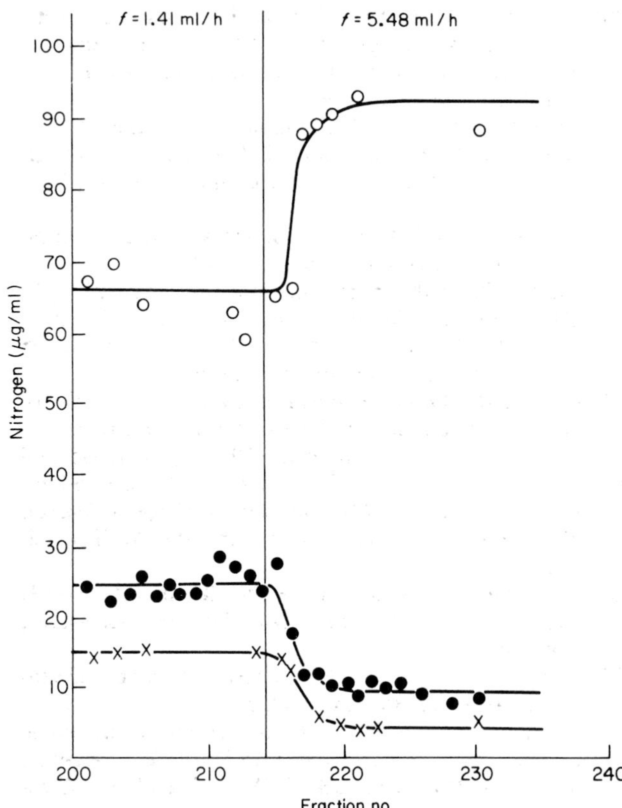

Fig. 2.1.6. Effect of a shift-up in the flow rate (f) on the effluent from a column supplied with ammonium and containing *Nitrosomonas* and *Nitrobacter*. ○——○, ammonium concentration; ●——●, nitrite concentration; ×——×, nitrate concentration. The fractions represent 5 ml volumes taken consecutively. Redrawn with permission from Bazin & Saunders [6].

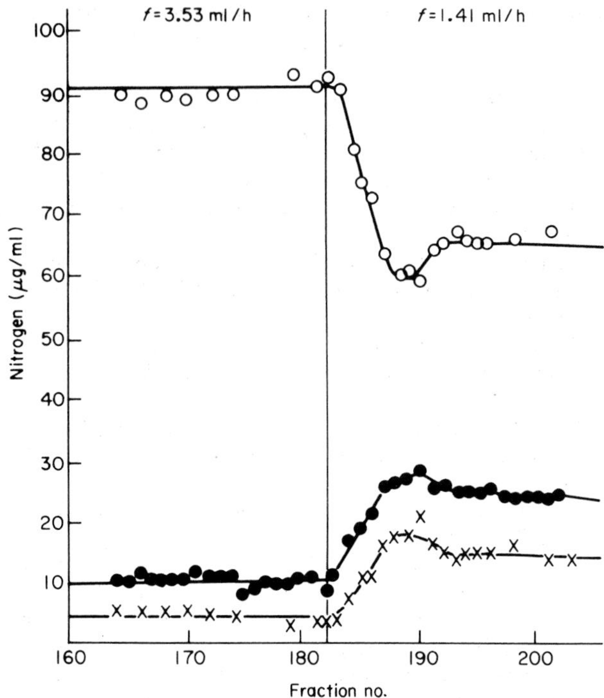

Fig. 2.1.7. Effect of a shift-down in the column described in Fig. 2.1.6. Redrawn with permission from Bazin & Saunders [6].

ing from such models with confidence to the natural ecosystem.

2.1.9 CONCLUSION

This chapter has given an outline of the capabilities of individual cells. However, it will become abundantly clear later in the book that this information on its own does not permit us to predict what the cell's activities actually are in any particular environment, let alone what synergistic activities it might be a part of. The problem of assessing how much of a micro-organism's potential is expressed in an environment in which it is found is central to the advance of microbial ecology as a discipline. There are many physiological attributes that appear to have no ecological significance and there is no reason to suppose that every activity does have a functional role, or that every activity expressed in a particular habitat will be found in every other habitat.

As micro-organisms occur in virtually every habitat in the biosphere, microbial ecologists clearly have much to offer the science of ecology; but, as the potential habitats for micro-organisms are so diverse, microbiologists of all leanings have much to offer the science of microbial ecology.

2.1.10 REFERENCES

[1] Adams D.G. & Carr N.G. (1981) The developmental biology of heterocyst and akinete formation in Cyanobacteria. *CRC Crit. Rev. Microbiol.* **9**, 45–100.

[2] Adler J. (1975) Chemotaxis in bacteria. In *Primitive Sensory and Communication Systems*, p.91 (ed. M.J. Carlile). Academic Press, London.

[3] Ahmadjian V. & Jacobs J.B. (1983) Algal-fungal relationships in lichens; recognition, synthesis and development. In *Algal Symbiosis*, p.147 (ed. L.J. Goff). Cambridge University Press, Cambridge.

[4] Alexander M. (1971) *Microbial Ecology*. John Wiley, London.

[5] Barnes R.S.K. (ed.) (1984) *A Synoptic Classification of Living Organisms*. Blackwell Scientific Publications, Oxford.

[6] Bazin M.J. & Saunders P.T. (1973) Dynamics of nitrification in a continuous flow system. *Soil Biol. Biochem.* **5**, 531–43.

[7] Bazin M.J., Saunders P.T. & Prosser J.I. (1976) Models of microbial interactions in the soil. *CRC Crit. Rev. Microbiol.* **4**, 463–98.

[8] Blakemore R.P. (1982) Magnetotactic bacteria. *Annu. Rev. Microbiol.* **36**, 217–38.

[9] Bonner J.T. (1974) *On Development: The Biology of Form*. Harvard University Press, Cambridge, Mass.

[10] Brock T.D. (1966) *Principles of Microbial Ecology*. Prentice-Hall, New Jersey.

[11] Buchanan R.E. & Gibbons N.E. (eds.) (1974) *Bergey's Manual of Determinative Bacteriology*, 8th edn. Williams & Wilkins, Baltimore.

[12] Bull A.T. & Trinci A.P.J. (1977) The physiology and metabolic control of fungal growth. *Advan. Microbial Physiol.* **15**, 1–84.

[13] Burnett J.H. (1976) *Fundamentals of Mycology*. Edward Arnold, London.

[14] Carr N.G. & Whitton B.A. (eds.) (1982) *The Biology of the Cyanobacteria*, 2nd edn. Blackwell Scientific Publications, Oxford.

[15] Cavalier-Smith T. (1981) The origin and early evolution of the eukaryotic cell. In *Molecular and Cellular Aspects of Microbial Evolution*, p.33 (ed. M.J. Carlile, J.F. Collins & B.E.B. Moseley). Cambridge University Press, London.

[16] Day P.R., Doner M.E. & McLaren A.D. (1978) Relationships among microbial populations and rates of nitrification and denitrification in a Hanford soil. In *Nitrogen in the Environment*, p.305 (ed. D.R. Nielson & J.G. Macdonald). Academic Press, New York.

[17] Ferguson Wood E.J. (1965) *Marine Microbial Ecology*. Chapman & Hall, London.

[18] Goff L.G. (ed.) (1983) *Algal Symbiosis*. Cambridge University Press, Cambridge.

[19] Gooday G.W. (1975) Chemotaxis and chemotropism in fungi and algae. In *Primitive Sensory and Communication Systems*, p.155 (ed. M.J. Carlile). Academic Press, London.

[20] Gooday G.W. (1981) Chemotaxis in the eukaryotic microbes. In *Biology of the Chemotactic Response*, p.155 (ed. J.M. Lackie & P.C. Wilkinson). Cambridge University Press, Cambridge.

[21] Gooday G.W. (1983) Hormones and sexuality. In *Secondary Metabolism and Differentiation in Fungi*, p.239 (ed. J.W. Bennett & A. Ciegler). Dekker, New York.

[22] Gottlieb D. (1976) The production and role of antibiotics in soil. *J. Antibiot.* **29**, 987–1000.

[23] Halldal P. (1980) Light and microbial activities. In *Contemporary Microbial Ecology*, p.1 (ed. D.C. Ellwood, J.N. Hedger, M.J. Latham, J.M. Lynch & J.H. Slater). Academic Press, London.

[24] Herbert D. (1961) The chemical composition of microorganisms as a function of their environment. In *Microbial Reaction to Environment*, p.391 (ed. G.G. Meynell & H. Gooder). Cambridge University Press, London.

[25] Holt J.G. (ed.) (1977) *The Shorter Bergey's Manual of Determinative Bacteriology*, 8th edn. Williams & Wilkins, Baltimore.

[26] Ingraham J.L., Maaløe O. & Neidhardt F.C. (1983) *Growth of the Bacterial Cell*. Sinauer Associates, Sunderland.

[27] Koshland D.E. (1974) Chemotaxis as a model for sensory systems. *FEBS Letters* **40**, 53–9.

[28] Koshland D.E. (1977) Bacterial chemotaxis and some enzymes in energy metabolism. In *Microbial Energetics*, p.317 (ed. B.A. Haddock & W.A. Hamilton). Cambridge University Press, Cambridge.

[29] Krieg N.R. & Holt J.G. (eds.) (1984) *Bergey's Manual of Systematic Bacteriology*, Vol. 1. Williams & Wilkins, Baltimore.

[30] Kung C. & Saimi Y. (1982) The physiological basis of taxes in *Paramecium*. *Ann. Rev. Physiol.* **44**, 519–34.

[31] Lewis D.H. (1974) Micro-organisms and plants: the evolution of parasitism and mutualism. In *Evolution in the Microbial World*, p.367 (ed. M.J. Carlile & J.J. Skehel). Cambridge University Press, Cambridge.

[32] Lorenzo V. de & Aguilar A. (1984). Antibiotics from Gram-negative bacteria: do they play a role in microbial ecology? *Trends Biochem. Sci.* **9**, 266–9

[33] Margulis L. (1975) Symbiotic theory of the origin of eukaryotic cells: criteria for proof. In *Symbiosis*, p.21 (ed. D.H. Jennings & D.L. Lee). Cambridge University Press, Cambridge.

[34] Margulis L. & Schwartz K.V. (1982) *Five Kingdoms: An Illustrated Guide to the Phyla of Life on Earth*. Freeman, San Francisco.

[35] Mitchell R. & Cundell A.M. (1977) *The Role of Microorganisms in Marine Fouling and Boring Processes*. Office of Naval Research Contract N00014-76-C-0042 NR-104-967, Technical Report No.3.

[36] Newell P.C. (1981) Chemotaxis in the cellular slime moulds. In *Biology of the Chemotactic Response*, p.89 (ed. J.M. Lackie & P.C. Wilkinson). Cambridge University Press, London.
[37] Nisbet B. (1984) *Nutrition and Feeding Strategies in Protozoa*. Croom Helm, London.
[38] Old K.M. (1977) Giant soil amoebae cause perforation of conidia of *Cochliobolus sativus*. *Trans. Brit. Mycol. Soc.* **68**, 277–81.
[39] Orpin C.G. & Bountiff L. (1978) Zoospore chemotaxis in the rumen phycomycete *Neocallamastix frontalis*. *J. Gen. Microbiol.* **104**, 113–22.
[40] Reid G.K. & Wood R.D. (1971) *Ecology of Inland Waters and Estuaries*. Van Nostrand, New York.
[41] Rheinheimer G. (1974) *Aquatic Microbiology*. Wiley InterScience, London.
[42] Rittenberg S.C. & Thomashow M.F. (1979) Intraperiplasmic growth — life in a cozy environment. In *Microbiology — 1979*, p.80 (ed. D. Schlessinger). American Society for Microbiology, Washington.
[43] Rothrock C.S. & Gottlieb D. (1984) Roles of antibiotics in antagonism of *Streptomyces hygroscopicus* var *geldanus* to *Rhizoctonia solani* in soil. *Can. J. Microbiol.* **30**, 1440–7.
[44] Round F.E. (1981) *The Ecology of Algae*. Cambridge University Press, Cambridge.
[45] Sieburth J. McN. (1975) *Microbial Seascapes*. University Park Press, Baltimore.
[46] Sieburth J. McN. (1979) *Sea Microbes*. Oxford University Press, New York.
[47] Smith D.C. (1975) Symbiosis and the biology of lichenised fungi: In *Symbiosis*, p.373 (ed. D.H. Jennings & D.L. Lee). Cambridge University Press, Cambridge.
[48] Stackebrandt E. & Woese C.R. (1981) The evolution of prokaryotes. In *Molecular and Cellular Aspects of Microbial Evolution*, p.1 (ed. M.J. Carlile, J.F. Collins & B.E.B. Moseley). Cambridge University Press, London.
[49] Starr M.P. (1975) A generalised scheme for classifying organismic associations. In *Symbiosis*, p.1 (ed D.H. Jennings & D.L. Lee). Cambridge University Press, Cambridge.
[50] Starr M.P., Stolp H., Trüper H.G., Ballows A. & Schlegel H.G. (eds.) (1981) *The Prokaryotes*, Vols. I & II. Springer-Verlag, Berlin.
[51] Stolp H. (1981) The genus *Bdellovibrio*. In *The Prokaryotes*, Vol. I, p.618 (ed. M.P. Starr, H. Stolp, H.G. Trüper, A. Ballows & H.G. Schlegel). Springer-Verlag, Berlin.
[52] Sussman A.S. & Halvorson H.O. (1966) *Spores: Their Dormancy and Germination*. Harper & Row, New York.
[53] Taylor B.L. (1983) Role of proton motive force in sensory transduction in bacteria. *Annu. Rev. Microbiol.* **37**, 551–73.
[54] Zahner H., Anke, H. & Anke T. (1983) Evolution and secondary pathways. In *Secondary Metabolism and Differentiation in Fungi*, p.153 (ed. J.W. Bennett & A. Ciegler). Marcel Dekker, New York.

2.2 Viruses

2.2.1 INTRODUCTION

Viruses are obligate parasites that contain only one type of nucleic acid (RNA or DNA) and rely on the cells of the host to provide part or all of the machinery necessary for their replication. They resemble genetic elements in their capacity to replicate and express their gene functions within cells, but unlike other genetic elements they can exist in a functionally inert extracellular form as a virus particle (virion).

Although some viruses can survive for long periods in an external environment, their ecology is intimately linked with the host and often with a vector that may be involved in the transfer of virus from one host to another. This chapter will concentrate on the genetic diversity of viruses, the consequences of virus infection and the mechanisms of virus transmission and persistence — factors essential to the survival and maintenance of viruses in host populations.

2.2.2 STRUCTURE AND GENETIC DIVERSITY OF VIRUSES

Viruses exhibit an extraordinary diversity in structure, host range and mechanisms of replication. The different groups of viruses isolated from plants, animals and micro-organisms (see Table 2.2.1) have particles with a wide range of structural types based on a rod-shaped or spherical architecture (see Fig. 2.2.1). Certain particularly complex bacteriophages have tail assemblies used in the attachment of the virus to the bacterial cell surface. In all virus particles, the genetic material (genome) is surrounded by a protein coat which, in turn, may be enclosed by an envelope containing other proteins and lipids. There are, however, subviral entities with no protein coat. These consist of a relatively stable, highly base-paired, low molecular weight, circular RNA molecule with no protein coat [31]. These entities are known as 'viroids' and are not formally classified as viruses.

The nature and configuration of virus genomes are equally diverse, and different virus groups contain single or double-stranded RNA or DNA in linear or circular form (see Table 2.2.1). No genetic systems other than viruses and viroids contain only RNA [19]. In almost all cases, DNA viruses may use host enzymes to copy (transcribe) messenger RNA from the viral genome. In some RNA viruses the nucleic acid can, in principle, be used as messenger RNA and translated directly into proteins on host cell ribosomes. The nucleic acid in these examples is referred to as positive (+) sense. In the retrovirus group the virions contain + sense RNA but the genetic material is first copied during infection into a DNA 'provirus' by a virus-associated enzyme (reverse transcriptase). Viruses of other groups contain double-stranded, or negative sense RNA which must be transcribed to produce positive strand molecules. In these and some other viruses, the virus coat is not simply an inert protective structure but contains functional enzymes (e.g. transcriptases) essential in initiating replication.

Replication of viral DNA molecules involves a semi-conservative process in which the genetic information on one strand may be used as the template to correct errors that may arise during copying of the complementary strand. In contrast, the replication of viral RNA is a conservative process and, because of this, RNA copying

Table 2.2.1. The diversity of virus genome structure and host range

Nucleic acid*					
Type	Conformation	Virus groups†	Notable examples	Hosts‡	Vector§
dsDNA	Linear	Pox-, herpes-, irido-, adeno- and tectiviruses; tailed phages, African swine fever virus	Myxoma (myxomatosis), herpes virus, variola (smallpox), varicella-zoster (chickenpox, shingles); phage λ	V, I, B	I
dsDNA	Circular (one molecule)	Baculo-, plasma-, papova- and corticoviruses	Nuclear polyhedrosis, polyoma virus	V, I, B	—
dsDNA	Circular (several molecules)	Polydnavirus	Viruses of parasitic hymenoptera	I	—
dsDNA	Open circular	Caulimoviruses, hepatitis B virus	Hepatitis B (serum hepatitis)	P, V	I
ssDNA	Linear + and − sense	Parvovirus	Dog parvovirus, adeno-associated virus	V, I	—
ssDNA	Circular + sense	Gemini-, micro- and inoviruses	Maize streak virus, øx174 phage	P, B	I
dsRNA	Linear (unsegmented)	Some isometric fungal viruses (totiviruses)	*Saccharomyces* sp. viruses	F	—
dsRNA	Linear (segmented)	Reo-, birna- and cystoviruses; other isometric fungal viruses (including partitiviruses)	Reovirus, rotaviruses, bluetongue, ø6 phage, infectious pancreatic necrosis	V, P, I, B, F	I
ssRNA	Linear (unsegmented) + sense	Toga-, corona, picorna- calici-, luteo-, tombus-, sobemo-, clostero-, carla-, necro-, poty-, potex- and tobamoviruses; Nudaurelia B group, ssRNA phages, maize chlorotic dwarf and maize rayado finovirus groups	Yellow fever, rubella (German measles), poliovirus, foot-and-mouth disease virus, MS2 phage, barley yellow dwarf virus, tobacco mosaic virus, potato viruses X and Y	V, P, I, B	I, F
ssRNA	Linear (dimeric) + sense	Retroviruses	Human immunodeficiency virus (AIDS)	V	—
ssRNA	Linear (segmented) + sense	Tomato spotted wilt virus, nodaviruses, and rice stripe virus group	Nodamura virus	I, P	I
ssRNA	Linear (multicomponent)	Hordei-, ilar-, diantho-, como-, nepo-, tobra-, cucumo- and bromoviruses; pea enation mosaic and alfalfa mosaic virus groups	Cowpea mosaic, tobacco rattle, cucumber mosaic	P	I, N
ssRNA	Linear (unsegmented) − sense	Rhabdo- and paramyxoviruses	Measles, rabies	V, P	I
ssRNA	Linear (segmented) − sense	Orthomyxo-, arena- and bunyaviruses	Influenza, Lassa fever, sandfly fever	V, I	I
ssRNA	Circular	Viroids (these are subviral entities not formally classified as viruses)	Potato spindle tuber viroid	P	—

* Virus nucleic acid type is shown as double-stranded (ds) or single-stranded (ss). Nucleic acid conformation is recorded as linear, circular (= covalently closed circular) or open circular (with discontinuities in both strands). Single-stranded nucleic acid molecules that contain the same base sequence as messenger RNA are referred to as + sense while − sense ss molecules have a base sequence complementary to the + sense strand. Segmented genomes are where two or more pieces of nucleic acid are present in each virion. Multicomponent genomes are where the virus nucleic acid segments are distributed amongst two or more virus particles. A 'dimeric' genome (e.g. retroviruses) contains two copies of the viral nucleic acid in a single particle.

† The English vernacular name is used for the different virus groups. Data taken from Matthews [22] with some updating from decisions taken in 1984 by the International Committee on Taxonomy of Viruses. The grouping of viruses according to nucleic acid type and conformation does not mean that the different virus groups in each category are in any other way related to each other.

‡ The host range recorded indicates that some of the viruses listed in each category have vertebrate (V), invertebrate (I), plant (P), fungal (F) or bacterial (B) hosts.

§ Some of the viruses listed in specific groups have invertebrate (I), fungal (F) or nematode (N) vectors.

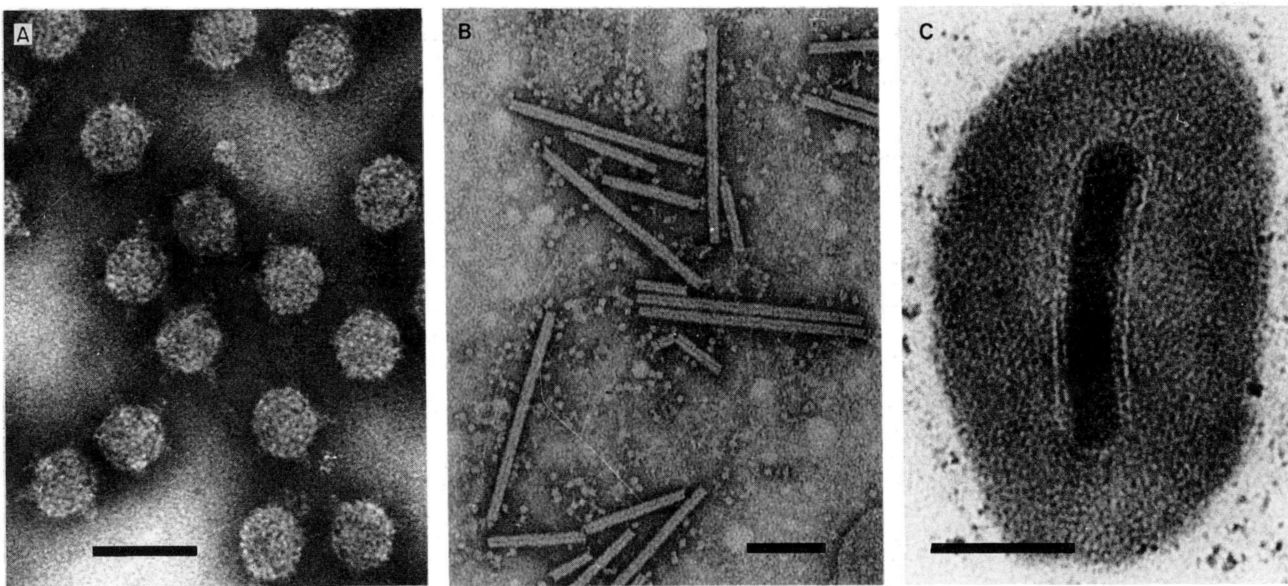

Fig. 2.2.1. Virus particle morphology: some examples of spherical and rod-shaped architecture. Scale bar = 100 nm. (A) A spherical particle with icosahedral symmetry and spikes (cytoplasmic polyhedrosis virus; reovirus group). (B) A rod-shaped virus with no envelope (tobamovirus group) showing both complete and broken virus particles. (C) A rod-shaped virus surrounded by an envelope and occluded within a proteinaceous crystal (granulosis virus; baculovirus group).

lacks the proofreading and editing functions of DNA. As a consequence, the error level on RNA copying is estimated to be 10^5-10^7 times higher than in DNA copying [30]. Viral RNA genes thus have greater survival problems than DNA genes but can, in theory, evolve more rapidly. Despite this, sequence conservation in some RNA genes is remarkable, particularly for gene sequences coding for enzymes involved in viral nucleic acid synthesis in which there may be constraints on variation imposed by the steric requirements of the proteins [23]. Although DNA virus genomes are generally present as one molecule, certain RNA virus genomes may be present in one or more segments of nucleic acid within a single virus particle (segmented genome viruses), or the different segments may each be packaged within separate virus particles (multicomponent viruses).

Reanney [30] thought that the genome segmentation common in RNA viruses aids RNA gene survival because a small unit of genetic information would have a higher chance of being correctly copied than a larger molecule. However, another consequence of segmented virus genomes (where genes are detached from one another) is that mixed infections with related virus strains provide greater opportunities for gene recombination between virus isolates by a process of segment reassortment; this promotes the evolution of new virus types such as some distinct strains of influenza A virus [27].

In multicomponent viruses, individual virus particles do not contain all the required genetic information, so particles are mutually dependent on each other to achieve successful replication. Some viruses ('satellites') are similarly dependent on a co-infecting virus for some replication functions [7] such as occurs between adenovirus and the defective parvovirus, adeno-associated virus. In both these examples, effective co-transmission of different particles is essential to ensure virus survival.

2.2.3 VIRUS INFECTION AND ITS CONSEQUENCES

When an organism encounters a virus, infection may result only if a number of specific interactions between the virus and the target cell occur. In infections of animal and bacterial cells at least, the virus attaches (adsorbs) to receptors on the plasma membrane or cell wall [4, 7].

This attachment process is highly specific and is mediated through the recognition of specific receptors on the host cell by defined sequences of proteins on the virus surface. The presence or absence of appropriate receptors can control the susceptibility of a host to a virus. Poliovirus, for example, binds only to primate cells whereas some togaviruses will bind to and infect both vertebrate and invertebrate cells [4].

With many plant and bacterial viruses only the viral nucleic acid penetrates the target cell after initial adsorption. In most animal viruses, the virion or some subviral particle crosses the cell membrane ultimately by a process of fusion. The partial or total 'uncoating' of nucleic acid occurs within the cell. This uncoating step is another important factor in determining host cell specificity since it will most often occur only in susceptible cells [4].

The consequences to the host of virus infection range from the highly cytocidal (lytic bacteriophage, and baculovirus infections of many insect hosts, for example) to virtual absence of any detectable effect. In most cases relatively few viruses seem to be associated with severe disease. Excessive virulence would, in fact, be evolutionarily self-destructive since hosts may die before virus transmission can occur. Nonetheless, many viruses do cause significant disease epidemics, such as those listed in Table 2.2.1. Amongst virus diseases of vertebrates, the collective economic and health effects attributable to togaviruses alone are massive. This group includes the causal agents of rubella (German measles) and many arthropod-transmitted viruses including yellow fever. Influenza (an orthomyxovirus) is believed to have contributed to the death of 20 million people during the pandemic of 1919. Rabies (a rhabdovirus) is estimated to kill about 15,000 people per year worldwide. A foot-and-mouth disease virus outbreak amongst cattle in the United Kingdom during 1967–68 caused direct and indirect losses of as much as £100 million. In addition, many plant viruses cause serious yield reductions; the luteoviruses, barley yellow dwarf, beet western yellows and potato leaf roll viruses caused annual losses in the USA between 1951 and 1960 of at least $60 million [7].

The physiological status of the host and its capacity to respond to infection are important in minimizing the rate and extent of invasion by viruses. For example, when temperate phages infect bacteria, the cellular level of cyclic AMP will influence whether the virus will become integrated into the chromosomal DNA (lysogenic) or will replicate and lyse the host. The AMP level is controlled by factors including bacterial growth rate [29]. Lysogenic bacteria also tend to be resistant to further virus challenge or 'superinfection'. Interference between different viruses may also modulate the infection process in vertebrate and plant cells.

The best studied response to infection is in vertebrates where the host reacts to virus infection in a multitude of ways. In a recent review [12], Fields identified at least four distinct but interrelated components of the host defences: interferons, macrophages, humoral (serum-mediated) immunity and cellular immunity. These responses place enormous selection pressure on virus survival.

The initial virus infection in vertebrates might be expected to produce acute disease followed by immunity as the virus is eliminated by the developing host defences. However, despite a vigorous host immune response, many viruses are able to persist within the host. Retroviruses in particular escape the immune response by integrating their genetic information as a provirus into the host DNA [36]. This aids persistence and also enables the viruses to recombine with host genes and transfer ('transduce') host DNA, including tumour-inducing 'oncogenes', from one host to another [3].

The immune response, which includes the production of specific antibodies against the protein components (structural antigens) of the virus, places selection pressure on the development of new virus antigenic types. Thus, one method by which influenza A viruses survive in the population is by considerable antigenic variation (known as 'antigenic drift') caused by accumulated mutations [27]. In effect, the virus overcomes host immunity by altering its antigenic structure. These and other means of virus survival are considered further below.

2.2.4 VIRUS TRANSMISSION

The survival of viruses within their ecosystem depends upon both persistence and transmission. Viruses must persist to increase the opportunities for transmission to other susceptible hosts. Likewise, without efficient dispersal to susceptible populations most viruses will not survive in the long term. When viruses are transmitted from host parent to offspring the process is called 'vertical' transmission; all other transmission is called 'horizontal'. Many viruses are transmitted both vertically and horizontally.

Vertical transmission

Effective vertical transmission of a virus enables it to survive in relatively low density host populations [2].

Vertical transmission of certain viruses, such as lysogenic phage and retroviruses [37] in bacteria and vertebrates, respectively, occurs through integration of the viral DNA as provirus within the host DNA. A virus can also be transmitted from adult female to offspring within the egg (as in some mosquito-transmitted viruses [16]) or by virus contamination of the genital tract which can expose the vertebrate foetus or invertebrate egg to infection. This external contamination occurs in the transfer of herpes viruses in vertebrates as well as in the transmission of certain insect baculovirus infections [7, 26].

Up to 130 plant viruses are transmitted vertically in seeds; the phenomenon is particularly widespread amongst virus groups for which no vector is known. In lower plants, resting spores are probably crucial in disseminating and maintaining viruses in the fungal population [7].

Horizontal transmission — by contact

Horizontal transmission of virus disease amongst animals occurs through the gastro-intestinal, urinogenital and respiratory tracts, by vector inoculation and by entrance through wounds. Virus transmission in the food is the commonest method of spread of viruses among leaf-feeding insects [26]. In plants, wounding is an almost essential prerequisite of infection [7]. Contact transmission in the absence of a vector requires that viruses be shed into the environment by infected individuals. Skin lesions produced by herpes and pox viruses provide the clearest example of this in vertebrates but virus is also transmitted in other body fluids. For example, promiscuous homo- and heterosexual activities increase the risk of transmission of human immunodeficiency virus (the causal agent of acquired immune deficiency syndrome (AIDS), a retrovirus), which is also spread through virus contamination of blood, leading to higher disease incidence amongst drug addicts, haemophiliacs and other individuals requiring regular blood transfusions [5, 20, 21]. Serum hepatitis (hepatitis B) can spread through contaminated blood as well as by mouth [39]. Influenza, measles (paramyxovirus) and picornaviruses (including foot-and-mouth disease virus and common cold viruses) are most frequently spread in aerosols via the respiratory tract.

The spread of viruses in water provides the natural means of transmission for virus infections in aquatic hosts, but faecal contamination of water supplies by viruses infecting the gastro-intestinal tract could also be a particularly important source of virus contamination. Adenoviruses, rotaviruses and enteric picornaviruses (including poliovirus and hepatitis A) are among the most stable viruses in the aquatic environment [7, 13].

In those viruses spread by contact, the crowding together of susceptible hosts improves the prospects of virus dissemination. Plant monoculture in agricultural systems enhances virus spread [18] and unusually large assemblies of animals promote the spread of infection, as probably occurred in the influenza A pandemic at the end of the First World War. The proportion of phages which attach to and infect their bacterial hosts is also directly related to the host concentration [29]. Where host density is not high, viruses which are spread exclusively by contact require a greater capacity to survive outside the host than viruses spread by other methods. It is generally true that viruses transmitted by contact are capable of surviving for long periods in the external environment. Extreme examples include the survival of tobacco mosaic virus in dried plant material for 50 years [7] and the persistence of the baculovirus of the Douglas fir–tussock moth in forest soils for a similar period [33].

Horizontal transmission — by vectors

Arthropods are important vectors of many virus diseases of plants and vertebrates. Of some 320 or more vectored plant viruses (about 75% of the total known plant pathogenic viruses), 93% are transmitted by arthropods [17]. Although more than half of these have aphid vectors, whiteflies, mealybugs and mites are also important vectors. In many situations, the plant viruses are transmitted in a manner in which the vector quickly loses its capacity to transmit. For example, with many aphid-transmitted viruses (such as clostero- and potyviruses), the insect acts as a flying syringe and the virus is held as a mechanical contaminant for a short period after the insect feeds on an infected plant [17]. In this form of transmission, referred to as 'non-circulative', the vector is able to transmit immediately after feeding (see Fig. 2.2.2). The vector may lose the capacity to transmit the disease within minutes, in which case the virus is termed 'non-persistent' (e.g. beet mosaic virus, potato virus Y). 'Semi-persistent' viruses (e.g. sugar beet yellows virus) may be retained for longer periods though the capacity for the vector to transmit these viruses lasts only for hours or at most 1–2 days (see Fig. 2.2.2).

In contrast, a 'circulative' or 'persistent' association occurs when the virus can invade the haemolymph of the vector ('non-propagative', as in geminivirus transmission by planthoppers and malva yellows virus by *Myzus persicae*) or can directly infect the cells of the vector ('prop-

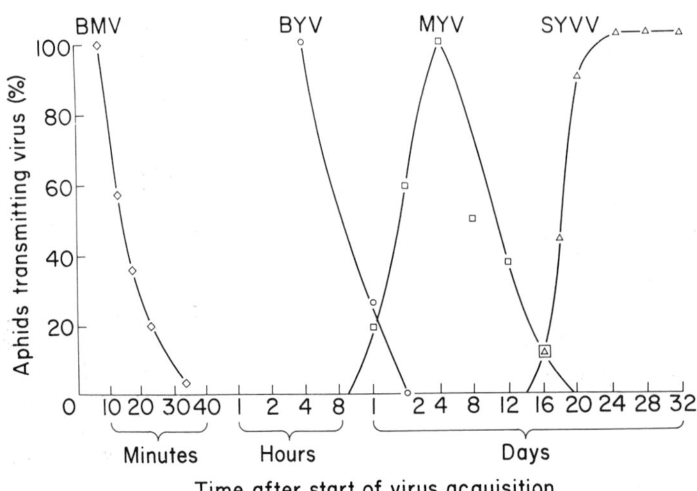

Fig. 2.2.2. Transmission of four aphid-borne viruses, illustrating the different retention periods for different viruses. Non-persistent transmission: beet mosaic virus (BMV). Semi-persistent transmission: beet yellows virus (BYV). Non-propagative transmission: malva yellows virus (MYV). Propagative transmission: sowthistle yellow vein virus (SYVV). Reproduced with permission from Gibbs & Harrison [14]; compiled from information in Watson [35], Costa *et al.* [8] and Duffus [9].

agative', as in the transmission of sowthistle yellow vein virus and other plant rhabdoviruses and reoviruses). Circulative transmission only occurs after an incubation period of several days (see Fig. 2.2.2) but the vector may transmit virus for life [6, 17].

Regardless of the mechanism of transmission, the associations between plant viruses and their vectors are generally very specific. Thus, potyviruses induce the production in infected plants of a proteinaceous 'transmission factor' which seems essential for the retention and release of virus particles from the vector aphids. The properties of the coat protein of the virus particle may also play an important role in transmission specificity [18].

Vector specificity is also important in the transmission of certain vertebrate virus diseases (particularly those caused by togaviruses and certain reoviruses) by vectors such as mosquitoes and ticks. The relationship between virus and vector is usually propagative and the virus replicates within the vector without apparent harmful effects. One of the best-studied examples is the transmission of yellow fever virus (togavirus) by mosquitoes (see Fig. 2.2.3). This is a complex cycle between a wild animal reservoir for the virus in the forests of South America and Africa and a vector which is, for the most part, arboreal. Incursions by man into this environment during deforestation programmes allows transmission by mosquito bite to humans and subsequent spread of the virus to cities where other mosquitoes can act as vectors [7]. Although yellow fever virus is propagatively associated with its vector, nonpropagative association of vertebrate viruses with their vectors can occur. The best-known example is the spread of myxoma virus (a poxvirus causing myxomatosis) by a wide range of biting insects including fleas and mosquitoes.

Nematodes and fungi are essential vectors for certain soil-borne plant-pathogenic viruses. Trichodorid nematodes are known to retain the capacity to transmit tobraviruses (e.g. tobacco rattle virus) for periods up to 10 months, whereas transmission of nepoviruses by longidorid nematodes occurs over periods of days or weeks only [7]. It is generally believed that the viruses are carried as extracellular contaminants and do not replicate in the nematode. Amongst fungi, two classes of obligate root parasites are well-documented vectors of virus disease; these are Chytridiales (e.g. *Olpidium* sp. — a vector of tobacco necrosis virus) and Plasmodiophorales (e.g. *Polymyxa betae* — a vector of beet necrotic yellow vein virus) [7].

The ecosystem for vector-transmitted viruses becomes that of the vector and differences in vector mobility influence the rate of virus spread (see Fig. 2.2.4). As a result, viruses transmitted by soil-inhabiting fungi or

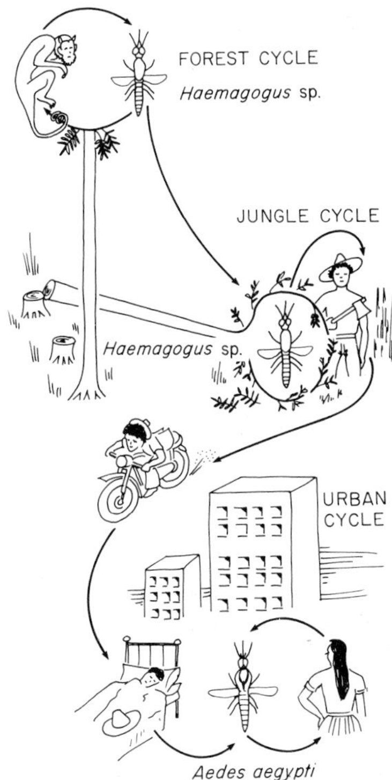

Fig. 2.2.3. Diagrammatic representation of the reservoirs of yellow fever virus in South America. The arboreal mosquito *Haemagogus spegazzinii* is the main vector in forests while the virus is disseminated by other vectors (e.g. *Aedes aegypti*) in cities. Redrawn with permission from Cooper & MacCallum [7].

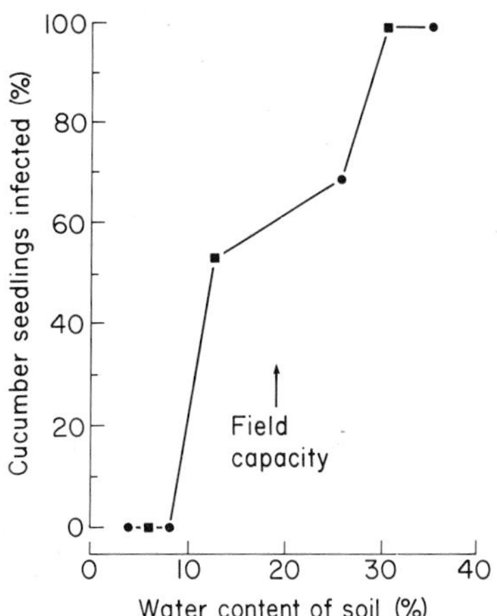

Fig. 2.2.4. Effect of soil moisture content on transmission of tobacco rattle virus to cucumber roots by *Trichodorus* nematodes. Successful transmission depends on the movement of nematodes in films of water and is diminished in dry soils. Soil water and 'field capacity' are discussed in Section 3.1.2. Reproduced with permission from Cooper & MacCallum [7].

nematodes are much slower to spread to new sites than those transmitted by arthropods. Finally, birds may serve as important passive carriers of virus diseases of both vertebrates and invertebrates. Insectivorous birds are known to spread large amounts of infectious virus in insect virus disease epizootics [11, 26].

2.2.5 VIRUS PERSISTENCE

The persistence of viruses through environmentally unfavourable periods can be achieved within the host, the vector or the external environment.

Persistence in the host

Persistent virus infections in vertebrate hosts can take several forms. Persistence by incorporation of the viral genome into that of the host has already been mentioned. Some viruses can cause persistent disease within a single individual without producing infectious virus or transmitting virus to their progeny. This occurs in the case of sub-acute sclerosing panencephalitis, a rare and fatal progressive disease of the central nervous system associated with persistent measles virus infection [24, 32]. Such persistence is not a part of the cycle of infection as it represents an evolutionary dead-end unless the virus can be transmitted in some other way, such as contact transmission in normal measles virus infection. Other viruses including members of the herpes virus group such as varicella-zoster and herpes simplex persist in a latent form. The viruses are reactivated under certain conditions as when immune responses to the virus are weakened in older individuals, and infectious virus is then shed from the host [38]. Extended host persistence combined with prolonged shedding of virus also occurs in some infections. For example up to 10% of adults infected with hepatitis B can produce and shed virus for several months or longer [39].

The mechanism by which such extended persistence can occur in the presence of a host immune response is not always clear. Integration of provirus is known to occur with retroviruses [3], some herpesviruses [10] and papovaviruses [15] at least. Some viruses overcome host immunity by antigenic change and/or by the production of defective virus particles that compete with normal virus for cellular functions and create persistent infections within cells [32]. Others, such as human immunodeficiency virus, may evade the immune defences by invading the tissues which generate the immune response [24].

The degree of specificity for the host is another adaptive feature of viruses. Amongst plant viruses, some (such as potato virus X) have only a narrow host range [18] and are thus particularly fitted for survival in cultivated plant populations where contact spread between closely-spaced individuals is possible. In contrast, other plant viruses effectively increase their chances of persistence by infecting a wide range of host species. This happens in particular when the natural hosts are short-lived and the viruses need to be more opportunistic to survive. Thus, alfalfa mosaic virus infects at least 150 species in 23 plant families under natural conditions [7]. A number of viruses of vertebrates and invertebrates also infect a range of host species. The capacity of influenza A virus to replicate in a range of vertebrates, particularly birds, must provide an extensive reservoir of virus even if transmission between different species is infrequent [27].

Although most bacteriophage infections can be lytic under optimal laboratory conditions, persistent infections may also occur frequently in nature. Phage persistence in host populations can be achieved by lysogeny and by other potential modifications of the virus–host interaction. These include high-frequency mutation of phage-sensitive bacteria to resistant forms; the presence of few virus receptor sites per bacterium allows a low probability of infection per cell and previously infected cells can produce sufficient virus receptor-destroying enzyme to render much of the population resistant [29].

Persistence in the vector population

Outside the primary host, viruses can persist within their vector when they are transmitted propagatively. As adult ticks live for up to 10 years, certain tick-borne diseases can survive for several years in the tick population. Sugar beet leaf curl virus rapidly loses infectivity within plant sap and its propagation and persistence in its heteropteran (plant bug) vector is essential to its survival [7]. Viruses may also be transmitted vertically by the vector. Certain bunyaviruses and other mosquito-borne viruses are known to overwinter in the eggs of the vector [16] while soil-borne wheat mosaic virus survives for several years in resting spores of its plasmodiophoromycete vector, *Polymyxa graminis* [18].

Persistence in the external environment

The need to survive in the external environment is particularly important for mechanically transmitted viruses (where a vector is not involved). Virus inactivation is caused by a number of factors including adverse temperatures, sunlight (particularly the ultraviolet component), exposure to air–water interfaces, extremes of pH and enzymic degradation.

Virus infections of Lepidoptera and other insects often need to survive within a discontinuous population as, in many cases, only the larval stage is susceptible to infection. Several distinct groups of insect-pathogenic viruses, baculoviruses, entomopoxviruses and cytoplasmic polyhedrosis viruses, have evolved a mechanism for environmental survival by the production of large numbers (10^9–10^{11} per larva) of stable proteinaceous inclusion bodies each of which encloses one or more virus particles (see Fig. 2.2.1C). The combination of protection and the large number of infectious propagules ensures that the viruses can survive for prolonged periods in soil and other protected environments [11, 26].

Survival of a wide range of viruses is enhanced by their adsorption to biologically inert surfaces in the environment. Adsorbents of viruses in soil and water include sand, clays, bacterial cells, naturally occurring colloids, silts and sediments [13]. Adsorptive behaviour of different viruses is probably governed by the combined surface charge and hydrophobic properties of the adsorbent and the virion. Once adsorbed, virion inactivation rates are reduced significantly. Thus, survival of phage T7 at 4°C was extended from 5 to 31 weeks by binding to clays [13]. In a similar way, aquatic sediments prolong survival of enteroviruses and rotaviruses; poliovirus is inactivated more than 40 times faster in seawater than in marine sediments. The protective effect of adsorption arises by protection from proteolytic enzymes and other inactivating compounds, increased stability of the virus coat (capsid), prevention of virus aggregate formation and blocking of ultraviolet radiation [13].

Virus population biology

To survive within the host population, all viruses need a constant supply of susceptible hosts except when they are effectively transmitted vertically or have good survival outside the host. Depending on the extent of persistence and the rate of transmission, there is a critical host population density below which infection cannot be maintained. Thus, with its capacity to persist within the host and reactivate, a virus such as varicella-zoster can survive in a host population of 10^3 compared with a population threshold of about 3×10^5 for measles virus (see Fig. 2.2.5) [36].

Recently, Anderson and May [1, 2] have extended the concept of theoretical models used by ecologists to characterize interactions between predators and their prey to assess the relationship between viruses and their hosts. If no virus is present, the equilibrium level of the population (N^*) is:

$$N^* = A/b$$

where A is the rate at which individuals are introduced to the populations and b is the rate of natural mortality. If viruses are present, then the change in population level is:

$$\mathrm{d}N/\mathrm{d}t = A - bN - \alpha Y$$

where α is the mortality rate due to disease and Y is the number of infected individuals. If the population level falls too much then the host and the disease die out. The threshold level of host for this mutual survival is:

$$A/b > (\alpha + b + v)/B$$

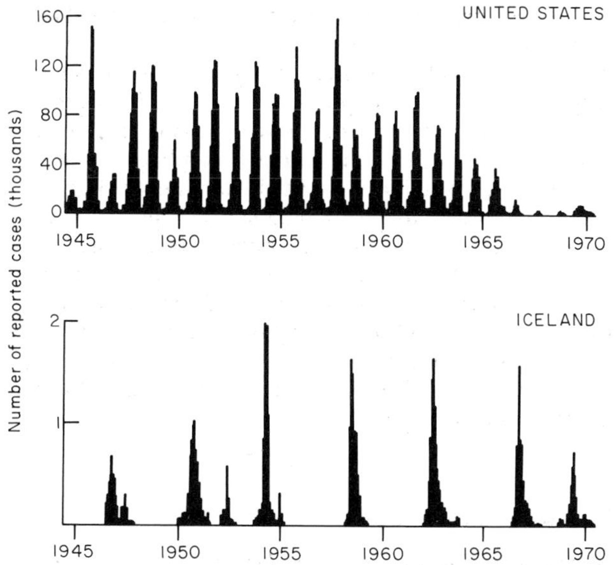

Fig. 2.2.5. Reported cases of measles per month, 1945–70, for the USA and Iceland. The survival of the virus in the large USA population is characterized by annual outbreaks diminishing after 1964 because of vaccination. In Iceland, the population is too low to maintain the disease in an endemic state and outbreaks are sporadic. Redrawn with permission from Cliff *et al.* [6].

where v is a recovery rate and B is a transmission coefficient [1].

Such a model, which is supported by data on virus epidemiology, has considerable predictive value in the analysis of virus–host interactions.

2.2.6 VIRUS ECOLOGY AND DISEASE CONTROL

Except in those circumstances where viruses are used for the biological control of pests (see Chapter 4.2), investigators are usually concerned with the development of control methods for limiting the spread of virus infection within the host population. Although considerable advances have been made in the chemical treatment of certain virus infections in the host, the intimate association between virus and cell increases the difficulty of eliminating virus replication without causing irreparable cell damage. For such reasons, the control of virus disease at present is largely preventive rather than curative. However, virus disease can be controlled by interrupting the virus–host cycle.

A key component of virus survival is the requirement for a constant supply of susceptible hosts. Reducing the size of the susceptible host population will reduce the infection rate. In certain vertebrates, this may be achieved by increasing the immunity of individual hosts through vaccination (see Fig. 2.2.5) with inactivated viruses or viruses of low pathogenicity [34]. Although in plants viruses are not eliminated by an immune-type response, crop varieties can be bred which are resistant to virus infection, and infection with related viruses of low virulence can provide a measure of 'cross-protection' by interference with the replication of more virulent strains. It is also often possible to reduce the reservoirs of alternative hosts for both plant and animal viruses.

The transmission cycle in viruses spread by mechanical means can be broken by interrupting or reducing contact between infected and susceptible hosts. Foot-and-mouth disease outbreaks in the United Kingdom are controlled by restricting movement of farm animals and slaughtering infected individuals [34]. Less drastic methods are necessary for the control of human disease. Patient isolation is appropriate if the virus is spread readily by contact as was the case with smallpox, the only virus disease of humans to be eradicated. This achievement was greatly helped by the absence of alternative hosts and vectors for the virus [34]. Current control methods for AIDS are aimed at preventing the spread of infection by creating public awareness of the risks of transmission by promiscuous sexual activity and at the screeening of blood products for the presence of viral antigen and antibodies to the virus.

In water-borne infections, adsorption of viruses to solids plays a major role in removing them during water treatment [13]. Thus, waste water treatments practised in the developed world, involving coagulation, flocculation and sedimentation, remove approximately 99% of contaminating viruses [25]. Where drinking water supplies are obtained from water sources likely to be contaminated with viruses, further treatment is necessary. Chlorine, which was largely introduced for the control of coliform bacteria, is not always an adequate virus disinfectant. Iodine, bromine and ozone seem more effective but are not yet used on a wide scale [7].

Vector transmission introduces an additional element into virus control. The vector itself may be amenable to direct control by pesticides. However, if the vector population is highly mobile and very effective in virus transmission, the level of population reduction required to significantly reduce disease incidence may be difficult to achieve or may require unacceptably high inputs of pesticide. In certain plant agricultural ecosystems, the spread of infection can be restricted by planting crops in areas where vector activity is reduced, such as seed potatoes in Scotland, and by changing the time of planting to avoid the times of vector immigration (as in the early autumn immigration of aphids carrying barley yellow dwarf virus into early-planted cereal crops) [28].

Man has had considerable influence on the emergence and spread of virus diseases. Ecological disturbance can result in the transfer of viruses to new hosts and new vectors. The spread of yellow fever to man is a classic example (see Fig. 2.2.3) but the same applies in plant virus ecosystems. More than 30% of the 62 known viruses of the Solanaceae are aphid-transmitted. In Europe, most are transmitted by vectors that did not encounter potatoes until some 400 years ago when potatoes were introduced from the Americas [17]. To avoid such problems in future, and the frequently adverse disruption of an equilibrium between a virus and its natural host reservoir, it is important to continue to improve our understanding of virus ecology.

2.2.7 REFERENCES

[1] Anderson R.M. & May R.M. (1979) Population biology of infectious diseases: Part I. *Nature (Lond.)* **280**, 361–7.

[2] Anderson R.M. & May R.M. (1981) The population dynamics of microparasites and their invertebrate hosts. *Phil. Trans. Roy. Soc. Ser.* B **291**, 451–524.

[3] Bishop J.M. (1984) Exploring carcinogenesis with retroviruses. In *The Microbe 1984. I: Viruses*, p.121 (ed. B.W.J. Mahy & J.R. Pattison). Symposia of the Society for General Microbiology, Vol. 36, Part I. Cambridge University Press, Cambridge.

[4] Bukrinskaya A.G. (1982) Penetration of viral genetic material into host cell. *Adv. Virus Res.* **27**, 141–204.

[5] Clark M. (1985) AIDS: two British blood tests launched. *Nature (Lond.)* **316**, 474.

[6] Cliff A.D., Haggett P.H., Ord J.K. & Versey G.R. (1981) *Spatial Diffusion*. Cambridge University Press, Cambridge.

[7] Cooper J.I. & MacCallum F.O. (1984) *Viruses and the Environment*. Chapman & Hall, London and New York.

[8] Costa A.S., Duffus J.E. & Bardin R. (1959) Malva yellows, an aphid transmitted virus disease. *J. Amer. Soc. Sug. Beet Tech.* **10**, 371–93.

[9] Duffus J.E. (1963) Possible multiplication in the aphid vector of sowthistle yellow vein virus, a virus with an extremely long insect latent period. *Virology* **21**, 194–202.

[10] Epstein M.A. (1982) Persistence of Epstein–Barr virus infections. In *Virus Persistence*, p.169 (ed. B.W.J. Mahy, A.C. Minson & G.K. Darby). Symposia of the Society for General Microbiology, Vol. 33. Cambridge University Press, Cambridge.

[11] Evans H.F. & Harrap K.A. (1982) Persistence of insect viruses. In *Virus Persistence*, p.57 (ed. B.W.J. Mahy, A.C. Minson & G.K. Darby). Symposia of the Society for General Microbiology, Vol. 33. Cambridge University Press, Cambridge.

[12] Fields B.N. (1984) Mechanisms of virus–host interactions. In *The Microbe 1984. I: Viruses*, p.197 (ed. B.W.J. Mahy & J.R. Pattison). Symposia of the Society for General Microbiology, Vol. 36, Part I. Cambridge University Press, Cambridge.

[13] Gerba C.P. (1984) Applied and theoretical aspects of virus adsorption to surfaces. *Adv. Appl. Microbiol.* **30**, 133–68.

[14] Gibbs, A.J. & Harrison B.D. (1976) *Plant Virology, the Principles*. Edward Arnold, London.

[15] Griffin B.E. (1982) How do Papova Viruses persist in their hosts? In *Virus Persistence*, p.227 (ed. B.W.J. Mahy, A.C. Minson & G.K. Darby). Symposia of the Society for General Microbiology, Vol. 33. Cambridge University Press, Cambridge.

[16] Grimstad P.R. (1983) Mosquitoes and the incidence of encephalitis. *Adv. Virus Res.* **28**, 357–438.

[17] Harris K.F. (1983) Sternorrhynchous vectors of plant viruses: virus–vector interactions and transmission mechanisms. *Adv. Virus Res.* **28**, 113–40.

[18] Harrison B.D. (1981) Plant virus ecology: ingredients, interactions and environmental influences. *Ann. Appl. Biol.* **99**, 195–209.

[19] McGeoch D.J. (1984) The nature of animal virus genetic material. In *The Microbe 1984. I: Viruses*, p.75 (ed. B.W.J. Mahy & J.R. Pattison). Symposia of the Society for General Microbiology, Vol. 36, Part I. Cambridge University Press, Cambridge.

[20] Maddox J. (1984) AIDS casts a longer shadow. *Nature (Lond.)* **312**, 97.

[21] Maddox J. (1986) Further anxieties about AIDS. *Nature (Lond.)* **319**, 9.

[22] Matthews R.E.F. (1982) Classification and nomenclature of viruses. *Intervirology* **17**, 1–199.

[23] Matthews, R.E.F. (1985) Viral taxonomy for the nonvirologist. *Ann. Rev. Microbiol.* **39**, 451–74.

[24] Mims C.A. (1982) Role of persistence in viral pathogenesis. In *Virus Persistence*, p.1 (ed. B.W.J. Mahy, A.C. Minson & G.K. Darby). Symposia of the Society for General Microbiology, Vol. 33. Cambridge University Press, Cambridge.

[25] Olson B.H. & Nagy L.A. (1984) Microbiology of potable water. *Adv. Appl. Microbiol.* **30**, 73–132.

[26] Payne C.C. (1982) Insect viruses as control agents. *Parasitology* **84**, 35–77.

[27] Pereira M.S. (1982) Persistence of influenza in a population. In *Virus Persistence*, p.15 (ed. B.W.J. Mahy, A.C. Minson & G.K. Darby). Symposia of the Society for General Microbiology, Vol. 33. Cambridge University Press, Cambridge.

[28] Plumb R.T. (1983) Barley yellow dwarf virus — a global problem. In *Plant Virus Epidemiology: The Spread and Control of Insect-Borne Viruses*, p.185 (ed. R.T. Plumb & J.M. Thresh). Blackwell Scientific Publications, Oxford.

[29] Primrose S.B., Seeley N.D. & Logan K.B. (1982) Methods for the study of virus ecology. In *Experimental Microbial Ecology*, p.66 (ed. R.G. Burns & J.H. Slater). Blackwell Scientific Publications, Oxford.

[30] Reanney D. (1984) The molecular evolution of viruses. In *The Microbe 1984. I: Viruses*, p.175 (ed. B.W.J. Mahy & J.R. Pattison). Symposia of the Society for General Microbiology, Vol. 36, Part I. Cambridge University Press, Cambridge.

[31] Sänger H.L. (1984) Minimal infectious agents; the viroids. In *The Microbe 1984. I: Viruses*, p.281 (ed. B.W.J. Mahy & J.R. Pattison). Symposia of the Society for General Microbiology, Vol. 36, Part I. Cambridge University Press, Cambridge.

[32] Ter Meulen V. & Carter M.J. (1982) Morbillivirus persistent infections in animals and man. In *Virus Persistence*, p.97 (ed. B.W.J. Mahy, A.C. Minson & G.K. Darby). Symposia of the Society for General Microbiology, Vol. 33. Cambridge University Press, Cambridge.

[33] Thompson C.G., Scott D.W. & Wickman B.E. (1981) Long-term persistence of the nuclear-polyhedrosis virus of the Douglas fir tussock moth, *Orgyia pseudotsugata* (Lepidoptera: Lymantriidae) in forest soil. *Envir. Ent.* 10, 254–5.

[34] Tyrrell D.A.J. (1984) The eradication of virus infections. In *The Microbe 1984. I: Viruses*, p.269, (ed. B.W.J. Mahy & J.R. Pattison). Symposia of the Society for General Microbiology, Vol. 36, Part I. Cambridge University Press, Cambridge.

[35] Watson M.A. (1946) The transmission of beet mosaic and beet yellows viruses by aphids; a comparative study of a non-persistent and a persistent virus having host plants and vectors in common. *Proc. Roy. Soc. Ser. B* **133**, 200–19.

[36] Weiss R.A. (1982) The persistence of retroviruses. In *Virus Persistence*, p.268 (ed. B.W.J. Mahy, A.C. Minson & G.K. Darby). Symposia of the Society for General Microbiology, Vol. 33. Cambridge University Press, Cambridge.

[37] Weiss R.A. (1984) Viruses and human cancer. In *The Microbe 1984. I: Viruses*, p.211 (ed. B.W.J. Mahy & J.R. Pattison). Symposia of the Society for General Microbiology, Vol. 36, Part I. Cambridge University Press, Cambridge.

[38] Wildy P., Field H.J. & Nash A.A. (1982) Classical herpes latency revisited. In *Virus Persistence*, p.133 (ed. B.W.J. Mahy, A.C. Minson & G.K. Darby). Symposia of the Society for General Microbiology, Vol. 33. Cambridge University Press, Cambridge.

[39] Zuckerman A. (1982) Persistence of Hepatitis B virus in the population. In *Virus Persistence*, p.40 (ed. B.W.J. Mahy, A.C. Minson & G.K. Darby). Symposia of the Society for General Microbiology, Vol. 33. Cambridge University Press, Cambridge.

2.3 Plasmids

2.3.1 INTRODUCTION

The resolution of microbial population and community structure and composition falls short of the realization that a given microbial species is not necessarily a stable entity. It is true, as a general statement, that microbial 'genes and species boundaries are strenuously maintained: a *Flavobacterium* sp. isolated from one environment is much like that isolated somewhere else, even with respect to those properties that are not used as criteria for identification' [18]. That is, the basic genetic structure of a micro-organism is stable. It is, however, also true that microbial characteristics can vary at significant frequencies; microbial ecologists need to be aware of the genetic basis of variations that may have profound effects on overall population and community ecology.

It is now quite clear that many mechanisms contribute to genetic variation and instability in microbial populations [15]. One mechanism operates on a cell's chromosome by point mutations or translocations. Another group of mechanisms involve the movement of a gene or a whole block of genes within the same microbial population or between different species. These mechanisms, based on conjugation, transformation and transduction, result in genetic recombination and the emergence of a novel organism when there is an appropriate selection pressure [6, 14, 15, 19]. Since *plasmid-mediated gene transfer* via conjugation is quantitatively the most significant process involved in generating novel genotypes, it is important to understand the basic structure and properties of plasmids.

2.3.2 PLASMIDS — SECONDARY GENETIC ELEMENTS

The fundamental genetic information required to produce a cell is located in the chromosome. Prokaryotic cells have a single circular chromosome. In *Escherichia coli* the DNA molecule contains about 4×10^6 bases coding for about 900–1000 different genes. In eukaryotic micro-organisms more than one chromosome may be present and they are linear.

Most bacterial species carry *extra-chromosomal elements*, additional DNA molecules separate from the chromosome. These molecules, smaller than the chromosome, are sometimes referred to as *minichromosomes*, *plasmids* or *episomes* (see Fig. 2.3.1). They are circular DNA molecules which replicate inside the cell separately from the host chromosome. Their replication is linked to cell growth and division which ensures that succeeding generations of cells receive at least one copy of the replicated plasmid DNA. Since plasmids are molecules of DNA they can carry genes that code for any cellular function, although rarely for properties indispensable to the cell (see Section 2.3.3). Plasmids are replicons (i.e. capable of controlling their own replicaton) which exist separately from the chromosome (i.e. they are said to be extra-chromosmal).

Bacterial plasmids range in size from 2000 base pairs (about 0.05% of the total chromosome size, able to code for about three genes) to more than 800 000 base pairs (about 20% of the cell's chromosome, able to code for 200 genes) [1]. The larger plasmids have been less well studied because of difficulties in the methods used to isolate intact plasmids from cells.

A given bacterial species may contain more than one plasmid (see Table 2.3.1). Plasmids smaller than 10 kb are generally present in multiple copies within each cell and are termed *relaxed plasmids*. For example, plasmid RSF1030, a 5.6 MDa plasmid which encodes ampicillin resistance (β-lactamase), has between 20 and 40 copies of itself in each cell. Plasmid ColEl, a 4.2 MDa plasmid encoding colicin E1 production in *E. coli* strains, normally

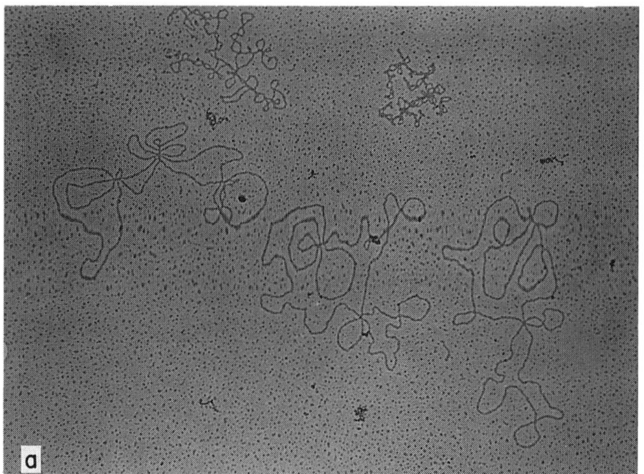

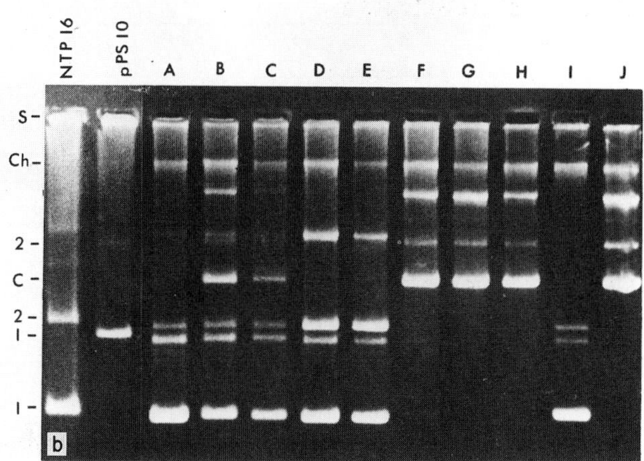

Fig. 2.3.1. (a) Electron micrograph of purified DNA of the plasmid NTP16 (molecular weight 5.6 × 10^6). Magnification x 15 000. The circular nature of the DNA molecule is clearly apparent. Three of the molecules appear as relaxed circles, while two appear as tightly-coiled super-coils. (b) Agarose gel electrophoresis of plasmid DNA. The tracks on the left show different forms (monomer, 1; dimer, 2) of the plasmids NTP16 and pPS10. Tracks A–J show plasmids isolated from cells containing both types, while others contain co-integrates (C) in which the two plasmid molecules have joined together. Tracks A–J also contain some contaminating DNA from the bacterial chromosome (Ch). The positions of the slots (S) in the agarose, from which the samples were electrophoresed, are also shown. Photographs courtesy of Peter Strike.

has 10–15 copies per cell. Larger plasmids are present with one to three copies per cell and are termed *stringent plasmids*.

Plasmids were originally recognized because of their capacity to transfer phenotypic characters from one cell to another cell. Initially plasmid transfer was observed between cells of the same species but it is now clear that in many instances plasmids can transfer between different species (see Section 2.3.4). Plasmids that are self-transmissible in this way are known as *conjugative* plasmids and transfer during conjugation between two cells. Conjugative plasmids need to carry a set of genes known as the *transfer genes*, or *tra* genes, which promote the formation of conjugation pili and facilitate movement of plasmid DNA from one cell to another. Generally plasmids which carry the *tra* genes are larger than 20 kb and are capable of self-transmission through bacterial populations.

Smaller plasmids are normally unable to carry out conjugation and are therefore known as *non-conjugative* or *non-transmissible plasmids*. They do not encode the *tra* genes. Non-conjugative plasmids can transfer from one cell to another by means of the conjugative system set up between two cells which also contain a conjugative plasmid: that is, non-conjugative plasmids can 'piggy-back' between cells under these conditions by a mechanism which is generally referred to as *mobilization*.

Plasmids can also be categorized on the basis of their ability to coexist within the same cell. If two different plasmids are able to exist simultaneously within one cell they are said to be *compatible*; if they cannot coexist they are said to be *incompatible*. Thus plasmids can be categorized into *incompatibility groups* (*Inc* groups) such that compatible plasmids belong to different incompatibility groups and incompatible plasmids belong to the same incompatibility group. For *E. coli* over 25 incompatibility groups have been recognized, and for *Staphylococcus aureus* seven plasmid incompatibility groups are known [13].

Most plasmids can only exist and replicate in a limited number of closely related species and are known as *narrow host range plasmids*. In a few instances, especially plasmids of incompatibility group P (*IncP* plasmids), they are able to exist in a wide range of host bacterial species.

Some plasmids have the ability to integrate into the chromosome, thereby losing their autonomous replication ability, and are termed *episomes*. In this state the episome behaves in exactly the same way as any other set of chromosomal genes. Plasmids that can integrate also have the ability to reverse the process and revert to an

Table 2.3.1 The types and number of plasmids in some bacteria

Organism	Plasmids		
	Number	Size (MDa)	Functions
Streptococcus lactis	5	1.0, 2.0, 5.5, 33, 60*	*Lactose fermentation, cellular aggregation, conjugation
Rhizobium leguminosarum strain TOM	4	180, 220, 220*, 500	*Nitrogen fixation
Rhizobium leguminosarum strain 55310a	3	400, 550, 600*	*Nitrogen fixation
Rhizobium leguminosarum strain 300	5	>200, >200, 210*, 180, 100+	*Nitrogen fixation +Conjugation
Bacillus thuringenesis strain berliver 1715	17	3.9, 5.4, 7.5, 9.5, 15, 21, 25, 28, 39, 42, 51, 60, 77, 100, 120, 180	
Bacillus thuringenesis strain kurstaki MD	8	1.5, 5.2, 5.6, 9.3, 10, 30*, 47*, 54	*Endotoxin
Streptococcus cremoris strain Wg2	5	1.5, 2.9, 6.1, 11.5, 16*	*Protease

autonomous, self-replicating entity within the cell. The best-known example of such an episome is the F plasmid (*fertility* or *sex factors*) which has a size of 62 MDa. It carries the *tra* genes and is therefore capable of conjugative transfer. In a mixed population of F^+ cells (cells containing the F plasmid and also known as *donor cells*) and F^- cells (cells which lack the F plasmid and also known as *recipient cells*) transfer of the F plasmid occurs via the F-pilus linking conjugating pairs of F^+ and F^- cells. The entire population rapidly becomes F^+ as a result of replication and transfer of the F plasmid. In such a mating some chromosomal genes will also transfer at very low frequencies (10^{-7}) as a result of the bacterial chromosome utilizing the F-pilus transfer mechanism. If the F plasmid integrates into the chromosome it produces a donor cell referred to as a *high frequency of recombination* (Hfr) strain because of the highly increased frequency of chromosomal gene transfer. In Hfr strains the mechanism for F plasmid transfer is retained and in this case some of the chromosomal genes are carried along with the F plasmid. Some genes are transferred at frequencies of 10^{-1} into F^- cells. Indeed if conjugation in *E. coli* populations is allowed to proceed for 100 minutes or so, then the complete chromosome may be transferred. Further details of this system of gene transfer may be found elsewhere [2, 3, 8].

Occasionally, when an integrated F plasmid excises to return to the autonomous replication state, part of the chromosome may be incorporated and a plasmid is produced that is known as an *F prime factor*. Since F primes carry a piece of chromosomal DNA, this can be transferred between F^+ and F^- cells at the same frequency as the transfer of a single F factor. This transfer represents an additional method by which a small segment of chromosomal DNA can be moved through bacterial populations. Many other plasmids, such as IncP plasmids, have the capacity to mobilize chromosomal genes and form plasmids termed *R primes* (R = resistance, since many of these plasmids were first identified by the drug resistance genes they carried) [9].

2.3.3 FUNCTIONS ENCODED BY PLASMIDS

All plasmids carry genes for their own replication and stable inheritance in their bacterial host cell [11, 12]. In addition, larger plasmids carry the genes involved in the *tra* functions. In many instances plasmids have been observed in bacterial cells with no apparent phenotypic functions encoded on them. In such cases the plasmids are termed *cryptic*. Although it may be that no functional genes are encoded on a cryptic plasmid, it is more likely that insufficient experimental work has been carried out to deduce what genes are present.

Many phenotypic functions are known to be encoded by genes located on plasmids (see Table 2.3.2). The first indication of the importance of plasmids came after antibiotics began to be used in large quantities. It was shown that the rapid appearance of micro-organisms resistant to certain antibiotics was associated with plasmids carrying genes that encoded the resistance mechanisms. These plasmids were referred to as *R plasmids* (*resistance plasmids*). The ecological significance of carrying genes on transferable plasmids is clear: under conditions of continuous or excessive use of a particular antibiotic, only the cells that contain the R plasmid survive. Moreover,

Table 2.3.2 Phenotypic characteristics encoded on plasmids

Resistance functions	
Antibiotics	Ampicillin, chloramphenicol kanamycin, tetracyline, streptomycin, spectinomycin, sulphanilamide, etc.
Heavy metals	Mercury, nickel, cobalt, tellurium, boron, chromium, arsenic, cadmium
Ultraviolet light	
Cryoresistance	
Bacteriocins	Colicins, etc.
Bacteriophage	X, T3, T7
Toxins	Enterotoxins, plant toxins
Catabolic functions	
Sugar catabolism	Lactose, sucrose, raffinose
Aromatic compound catabolism	Salicylate, napthalene, xylene, toluene, nicotine, chlorobenzoate, phenylacetate
Xenobiotic compound catabolism	Pesticides — parathion, dalapon — nylon oligomers, PCB, 2,4D
Catabolic enzymes	Dehalogenases, haemolysins, pectinases, proteases, oxygenases, hydrogenases
Antibiotic detoxification	
Opine utilization	Octopine, nopaline
RNA catabolism	

Table 2.3.2 (Continued)

Functions associated with interactions between bacteria and eukaryotes	
Animals	α-Haemolysins, serum resistance, ST and LT enterotoxins, colonizing-factor antigen (CFA), exfoliature toxin, adhesion K88 antigen
Insects	Delta toxin
Plants	Gall induction, tumour formation, nitrogen fixation, auxin production, opine synthesis, host–bacterium nodulation, halo blight, phaseotoxin, hairy root disease
DNA-associated functions	
Sex pili	
Restriction endonucleases	*Eco*RI
DNA modification enzymes	Methylase
DNA polymerases	
Transfer genes	
DNA mobilization genes	
Surface exclusion properties	
Phage exclusion	B39

the selective conditions (i.e. presence of antibiotic) ensure the rapid spread, through cell-to-cell contact, of the R plasmid through the whole population. In many instances it has been shown that the same R plasmid can appear in different bacterial species because of an ability to cross genus and species boundaries. Resistance plasmids are now known for many of the drugs and antibiotics used in medical and veterinary treatment (see Table 2.3.2).

Many enzymes and whole pathways for the breakdown of unusual carbon compounds, such as aromatic compounds, are coded for by genes located on plasmids. These plasmids are frequently referred to as *degradative* or *catabolic plasmids*. Novel mechanisms for the degradation of different aliphatic, aromatic or polyaromatic hydrocarbons, alkaloids and chlorinated compounds are normally associated with degradative plasmids. *Pseudomonas* species in particular are a rich source of degradative plasmids. In some cases the catabolic genes on the plasmid appear to be readily transferable between the chromosome and plasmid DNA. For example, copies of the genes involved in the toluene and xylene pathway — the TOL plasmid — can be inserted into the chromosome. In some cases the catabolic plasmid may not carry all the genes for the complete mineralization of certain compounds. For example, the CAM plasmid, which encodes for camphor degradation, also requires chromosome-located genes to completely degrade one of the products, isobutyric acid. In some areas complex interactions exist between genes of the chromosome and of the plasmid and frequent interchange of genes or groups of genes can occur.

Other functions known to be associated with plasmids are shown in Table 2.3.2. There seems to be no limit to the types of genetic information which can be plasmid-encoded, especially when it is realized that certain plasmids (e.g. plasmid R68–45 in *Pseudomonas* species) have a function (*chromosome-mobilizing ability*) which enables any small piece of chromosomal DNA to be integrated on to a plasmid and then transferred with the plasmid.

Many genes are carried by special structures termed *transposons*. These are genes bordered at each end by special sequences of DNA known as *insertion sequences* or *IS elements*. The presence of IS elements enables transposons to move readily between chromosome and plasmid DNA and between different plasmids [4, 17].

2.3.4 PLASMID ECOLOGY

Plasmids play a central role in the reorganization of bacterial genetic material that leads to the rapid dispersal of genes between different individuals of the same microbial population, as well as between different species. The same conjugative plasmids may be isolated from many different genes and species taken from different habitats. For example, plasmid RP1, an IncP group plasmid, has been found in 17 different genera [15]. Researchers have developed complex networks showing the probable pathways of plasmid and gene transfer in natural populations (see Fig. 2.3.2).

The potential for gene transfer in processes mediated by plasmid transfer has been calculated, making various assumptions about the number of individuals in a population and the frequencies of plasmid transfer and chromosomal gene transfer (see Table 2.3.3). The transfer of genes by plasmids ensures the rapid dissemination of

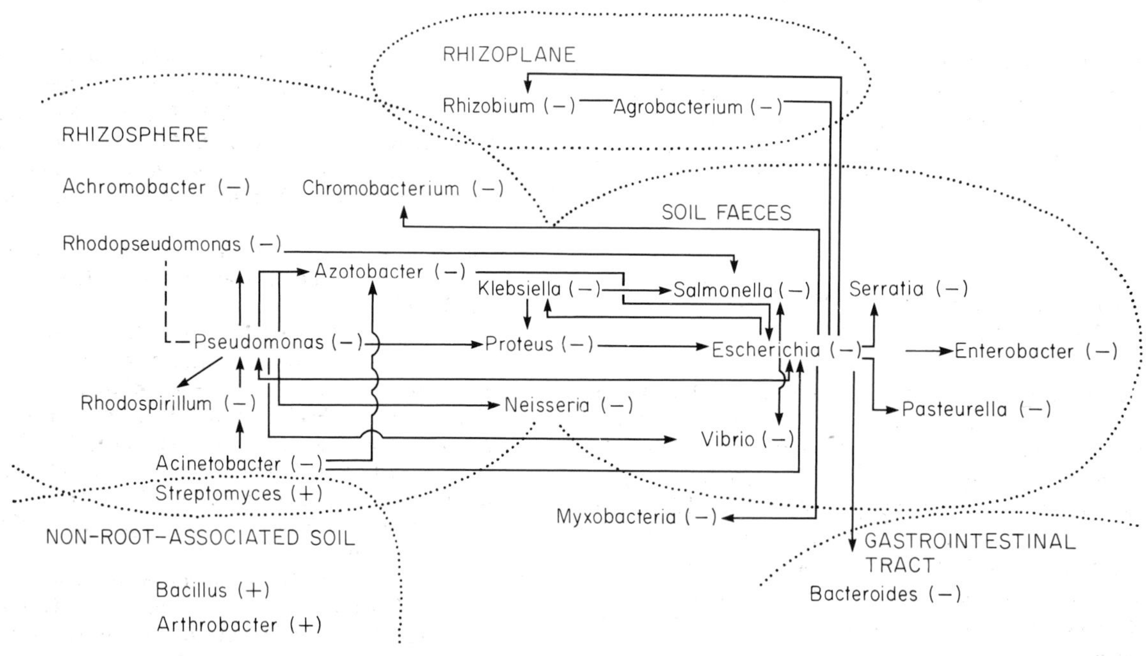

Fig. 2.3.2. Known pathways of intergeneric genetic exchange among bacteria in several ecosystems. The rhizoplane is the surface of plant roots, and the rhizosphere is the zone surrounding the plant. Enteric organisms have been grouped under the heading 'soil faeces' since the huge population of bacteria present both in faeces and in rhizosphere soil have many opportunities to exchange genes when faecal matter is reintroduced into the soil environment. Most of the DNA transfers shown have been carried out under laboratory conditions. There is no certainty that the same transfers occur in nature. The transfers listed were all mediated by conjugation and transduction. No attempt has been made to include transfers made possible by transformation. The (–) and (+) refer to the reaction of the taxa shown to the Gram stain. After Reanney *et al.* [16].

Table 2.3.3. Potential for gene transfer mediated by plasmids in rhizobia

Determination	Expected frequencies of plasmid transfer		
	10^{-2}	10^{-6}	10^{-10}
Total no. of matings	10^{13}	10^{9}	10^{5}
No. of plasmid genes transferred	10^{15}	10^{11}	10^{7}
No. of matings in which chromosomal DNA is transferred			
Assuming frequency is 10^{-4} per plasmid transferred	10^{9}	10^{5}	10
Assuming frequency is 10^{-6} per plasmid transferred	10^{7}	10^{3}	10^{-1}
Assuming frequency is 10^{-8} per plasmid transferred	10^{5}	10	10^{-3}
No. of chromosomal genes transferred			
Assuming frequency is 10^{-4} per plasmid transferred	3×10^{11}	3×10^{7}	3×10^{4}
Assuming frequency is 10^{-6} per plasmid transferred	3×10^{9}	3×10^{5}	3×10^{2}
Assuming frequency is 10^{-9} per plasmid transferred	3×10^{7}	3×10^{3}	3×10^{-1}

Potential for gene transfer by plasmids in matings between rhizobia contained in 1 km^2 of a pure crop of legume. Estimates are based on the following assumptions: (a) that there are 10^6 legumes per km^2 and 10^9 rhizobia per legume (10^{15} rhizobia per km^2); (b) that the rhizobium chromosome carries about 3000 genes; (c) that the average sized plasmid is 1/30 the size of the chromosome; and (d) that the average sized fragment of chromosomal DNA transferred is 1/10 of the total chromosome. No assumption is made about the frequency at which bacteria can mate with each other in the soil. In the laboratory, pair formation and plasmid transfer seldom take as long as 30 min. After Beringer & Hirsch [1].

necessary information that, under appropriate selective conditions, may be vital to the survival of that population [10]. Plasmid transfer and the measurement of genes have been described as a horizontal evolution event [5] and are most clearly seen in the evolution of novel catabolic pathways [15].

2.3.5 PLASMIDS IN EUKARYOTIC ORGANISMS

Much less is known about extra-chromosomal elements in eukaryotic micro-organisms [7]. Yeasts contain an autonomously replicating circle of DNA termed the *2 μ circle*. It is present in the nucleoplasm and there may be up to 50 copies per cell. Its function is not known although it may be involved in drug resistance. There are some indications that fungi have plasmids and that both fungi and protozoa have virus-like particles containing extra-chromosomal DNA and replicating independently of the chromosomes.

2.3.6 REFERENCES

[1] Beringer J.E. & Hirsch P.R. (1984) Genetic adaptation to the environment. In *Current Perspectives in Microbial Ecology*, p.63 (ed. M.J. Klugg & C.A. Reddy) American Society for Microbiology, Washington, DC.

[2] Broda P. (1979) *Plasmids*. W.H. Freeman, San Francisco.

[3] Bukhari A.I., Shapiro J.A. & Adhya S.L. (1977) *DNA Insertion Elements, Plasmids and Episomes*. Cold Spring Harbour Laboratory, Cold Spring Harbour, New York.

[4] Campbell A. (1981) Evolutionary significance of accessory DNA elements in bacteria. *Ann. Rev. Microbiol.* **35**, 55–83.

[5] Clarke P.H. & Slater J.H. (1986) Evolution of enzyme structure and function. In *The Bacteria*, Vol. X (ed. J.R. Sokatch). Academic Press, Orlando.

[6] Cohen S.N. (1976) Transposable genetic elements and plasmid evolution. *Nature (Lond.)* **263**, 731–8.

[7] Esser K. Kück U., Lang-Hinrichs C., Lemke P., Osiewacz H.D., Stahl U. & Tudzynski P. (1986) *Plasmids of Eurkaryotes. Fundamentals and Applications*. Springer-Verlag, Berlin.

[8] Hayes W. (1970) *The Genetics of Bacteria and Their Viruses*. Blackwell Scientific Publications, Oxford.

[9] Holloway B.W. (1979) Plasmids that mobilize bacterial chromosome. *Plasmid* **2**, 1–19.

[10] Levy S.B. (1985) Ecology of plasmids and unique DNA sequences. In *Engineered Organisms in the Environment: Scientific Issues*, p.180 (ed. H.O. Halvorson, D. Parmer & M. Rogul). American Society for Microbiology, Washington, DC.

[11] Meacock P.A. & Cohen S.N. (1980) Partitioning of

bacterial plasmids during cell division: a cis-acting locus that accomplishes stable plasmid inheritance. *Cell* **20**, 529–42.

[12] Nordstrom K., Molin S. & Light J. (1984) Control of replication of bacterial plasmids: genetics, molecular biology and physiology of the plasmid R1 system. *Plasmid* **12**, 71–90.

[13] Old R.W. & Primrose S.B. (1981) *Principles of Gene Manipulation: An Introduction to Genetic Engineering*. Blackwell Scientific Publications, Oxford.

[14] Reanney D.C. (1976) Extrachromosomal elements as possible agents of adaption and development. *Bact. Revs.* **40**, 552–90.

[15] Reanney D.C., Gowland P.C. & Slater J.H. (1983) Genetic interactions among microbial communities. In *Microbes in their Natural Environment*, p.379 (ed. J.H. Slater, R. Whittenbury & J.W.T. Wimpenny). Cambridge University Press, Cambridge.

[16] Reanney D.C., Kelly W.J. & Roberts W.P. (1982) Genetic interactions among microbial communities. In *Microbial Interactions and Communities*, Vol. 1, p.287 (ed. A.T. Bull & J.H. Slater). Academic Press, London.

[17] Shapiro J.A. (1985) Intercellular communication and genetic change in bacterial populations. In *Engineered Organisms in the Environment: Scientific Issues,* p.63 (ed. H.O.H. Halvorson, D. Parmer & M. Rogul). American Society for Microbiology, Washington, DC.

[18] Slater J.H. (1984) Genetic interactions in microbial communities. In *Current Perspectives in Microbial Ecology*, p.87 (ed. M.J. Klugg & C.A. Reddy). American Society for Microbiology, Washington, DC.

[19] Slater J.H. & Sommerville H.J. (1979) Microbial aspects of waste treatment with particular attention to the degradation of organic compounds. In *Microbial Technology: Current Status and Future Prospects*, p.221 (ed. A.T. Bull, C.R. Ratledge & D.C. Ellwood). Cambridge University Press, Cambridge.

2.4 Microbial population and community dynamics

2.4.1 INTRODUCTION

Analysis of the growth of micro-organisms is an important aspect of microbiology, including microbial ecology. Monod [25, 26], who first enunciated the fundamentals of microbial growth dynamics, imputed its importance to the fact that it 'is the basic method of microbiology'. But as Meadow and Pirt [23] have observed, 'There is probably no other technique whose principles are so often ignored with the result that experiments are seriously limited if not meaningless.' Frequently, little attention is paid to the nature of a microbial population's growth beyond perhaps the single requirement that it may be cultured, despite the abundant evidence that a micro-organism's composition and structure and its physiological capabilities may be radically influenced by the method of growth and the status of the growth environment. Pure culture studies have shown that the basic macromolecular composition of an organism depends markedly on the rate at which it grows and the nature of the growth conditions, particularly the effect of different substrate limitations or excesses [3, 17]. Many micro-organisms exhibit a range of phenotypes in response to different environmental conditions; in some cases, these result in the expression of a number of completely different morphological forms (see Chapter 2.1). This has led to the erroneous identification of a number of different 'species'. It is also now well established that the growth environment influences the expression of such cellular components as enzymes and affects their levels and activities [9]. Furthermore, it has recently been shown that more than one pathway capable of fulfilling the same metabolic function may exist in the same organism, and which mechanism operates depends on the growth environment of the organism. For example, *Klebsiella pneumoniae* uses the relatively low affinity mechanism, based on the enzyme glutamate dehydrogenase, for assimilating ammonium ions from environments where the ions are in excess. Under conditions of low ammonium ion concentration the organism switches to a different, high affinity mechanism where assimilation is carried out by the operation of the enzymes glutamine synthetase and glutamate amidotransferase [2]. The extension of such principles to studies of the ecology of micro-organisms has yet to be pioneered.

Microbial ecologists also have a need for suitable mathematical models to assess and predict the development of a population of microbes. These models must incorporate the effect of the physicochemical parameters on the dynamics of the population and use defined constants which have a biological relevance to the system being modelled.

In the past much emphasis has been placed on understanding the growth of populations containing single species of micro-organisms. It is, however, obvious that in nature populations of micro-organisms do not often grow in complete isolation from each other [34, 35]. Instead, populations interact in various ways, for example competitively [16, 41]. Furthermore, it is now well established that stable associations of micro-organisms, termed microbial communities, also occur in nature [33, 34] and, although at the present time little quantitative experimentation has been undertaken, the techniques now exist whereby the microbial ecologist may begin to assess the role and importance of these associations and use them as model systems.

This chapter is concerned with the fundamentals of population and community growth kinetics in both closed and open growth systems. Closed (or batch) growth systems represent the traditional method of growing micro-organisms where there is initially an excess of all the required growth nutrients. As a result, for a period of time, growth proceeds at the maximum rate. Furthermore, once the cultures have been established there are no major additions to or removals from the culture. In open growth systems additions to and removals from the culture take place either continuously or discontinuously. This simulates the conditions occurring in natural environments. In one type of open growth system with continuous culture conditions, populations may even be grown at sub-maximal rates under substrate-limited conditions [27, 28].

2.4.2 MICROBIAL GROWTH IN CLOSED ENVIRONMENTS

Growth in an unlimited environment — basic equations

The simplest system to analyse is the growth of a single micro-organism in a completely favourable environment which provides all the required growth resources in unlimited quantities and maintains constant physicochemical conditions. The microbe utilizes the nutrients and, in a series of orderly, complex and interrelated biosynthetic sequences, produces new cell material or *biomass*. Consequently, the organism increases in size and, after a period of time during which the biomass doubles, cell division occurs and a population is established containing two individuals. Each new organism repeats the whole process and, after a second generation has been completed, the population doubles in size again to contain four individuals. Thus for unicellular organisms growth is a process which results in an increase in both the number of individuals and biomass. In contrast, growth of multicellular micro-organisms increases first biomass and the number of cells within an individual and only subsequently does the size of the population increase. The many methods which have been used to estimate population size or biomass concentration are reviewed elsewhere [24, 31].

The fundamental features of microbial growth dynamics may be described in two different, but related, ways depending on the starting viewpoints.

In the first approach growth is as described in Table 2.4.1 (and in Fig. 2.4.1), where x is the number of individual organisms inoculated into the closed environment.

The rate of increase of both biomass and the number of individuals accelerates with time, and the rate of acceleration depends on the composition and physical nature of the growth environment as well as the intrinsic capacity of the micro-organisms to synthesize new biomass at a given rate. Clearly, in an environment which contains many preformed precursors of protein, nucleic acid and other macromolecules, the growth of the population is more rapid than in an environment where these metabolites have to be synthesized. However, under constant environmental conditions the time taken to complete each generation and double the size of the population is constant. This characteristic period is known as the *culture doubling time*, t_d (unit: time — conventionally hours) although sometimes it is referred to as the *mean generation time*.

The size of a population after a period of growth, x_t, depends on the initial population size, x_0, and the period of time for growth, t, and may be mathematically formulated by:

$$x_t = x_0 \cdot 2^{t/td} \tag{2.4.1}$$

Equation (2.4.1) is the mathematical representation of Table 2.4.1 since, when $t = t_d$, $x_t = 2x_0$ and the growth time is that interval needed to complete one generation.

Table 2.4.1. Cyclical growth processes

	Number of complete cell cycles or number of generations							
	0	1	2	3	4	5	6	. . . n
Number of individual cells in the population at the beginning of each generation	x	$2x$	$4x$	$8x$	$16x$	$32x$	$64x$	. . . ax
	2^0x	2^1x	2^2x	2^3x	2^4x	2^5x	2^6x	. . . 2^nx
Population size when:								
$x = 1$ organism	1	2	4	8	16	32	64	. . . a
$x = 1 \times 10^6$ organisms	1×10^6	2×10^6	4×10^6	8×10^6	1.6×10^7	3.2×10^7	6.4×10^7	. . . $a \times 10^6$
Rate of cell number increase per generation	—	x	$2x$	$4x$	$8x$	$16x$	$32x$	. . . $(a-b)x$
Multiplication factor per generation	2	2	2	2	2	2	2	. . . 2

Moreover, equation (2.4.1) describes an exponential function and characterizes the fundamental behaviour of a microbial population in an ideal growth environment.

It is usually more convenient to derive a linear form of the exponential growth equation by taking the natural logarithms of both sides of equation (2.4.1):

$$\ln x_t = \ln x_0 + \ln 2 \,.\, t/t_d \qquad (2.4.2)$$

and this may be represented graphically by plotting ln$x$ against time (see Fig. 2.4.2). The intercept on the abscissa shows the initial population size, whilst the slope of the line is $0.693/t_d$ and may therefore be used to determine the culture doubling time.

The growth of a microbial population may be analysed from an alternative viewpoint. Consider an initial population size, x_0, existing at time, $t = 0$. After a small time interval, δt, a small fraction of the organisms present within x_0 complete their individual cell cycles and divide, thereby increasing the population size by δx. It is important to remember that in a normal microbial population there is a distribution of organisms at various stages of the cell cycle. That is, some fraction of the population is small young individuals, the recent products of their parents' cell division, whilst another fraction represents organisms which have completed their doubling in biomass and are on the point of dividing (see Fig. 2.4.3). In certain circumstances a population may contain individuals all at the same stage in the cell cycle and so divide in phase with each other (see Fig. 2.4.4). These cultures are known as *synchronous populations* showing stepwise increases in population size as depicted in Fig. 2.4.1. In naturally occurring populations, synchrony is rare since natural variations in the length of the cell cycles serve to randomize the times at which cell divisions occur. Clearly the magnitude of δx (assuming that δt remains constant) depends on the size of x_0 and hence the rate of change of the population size (that is, the rate of growth) is directly proportional to the initial population size. Thus:

$$\frac{dx}{dt} \propto x_0$$

and

$$\frac{dx}{dt} = \mu x_0 \qquad (2.4.3)$$

where μ is a proportionality constant termed the *specific growth rate* and is a measure of the amount of new

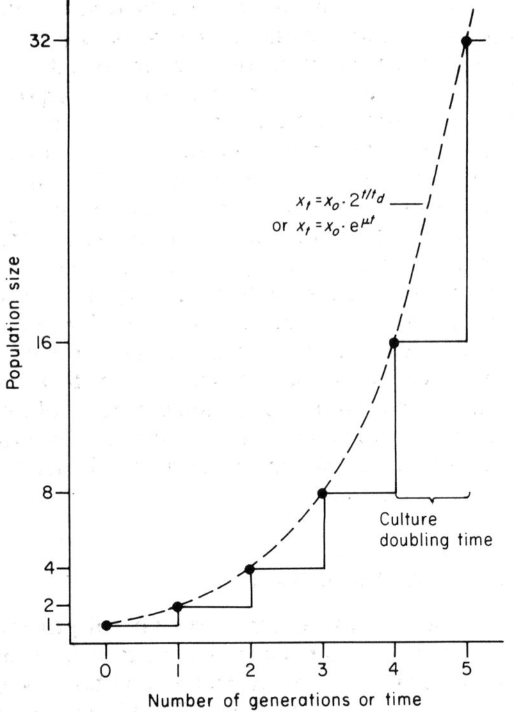

Fig. 2.4.1. Exponential growth in a closed, unlimited environment. The smooth curve is described by the basic growth equations (2.4.1) and (2.4.4) and is the form of growth obtained if the population size is large. The stepwise line is obtained in a synchronized population and seen if the population size is measured in terms of cell number.

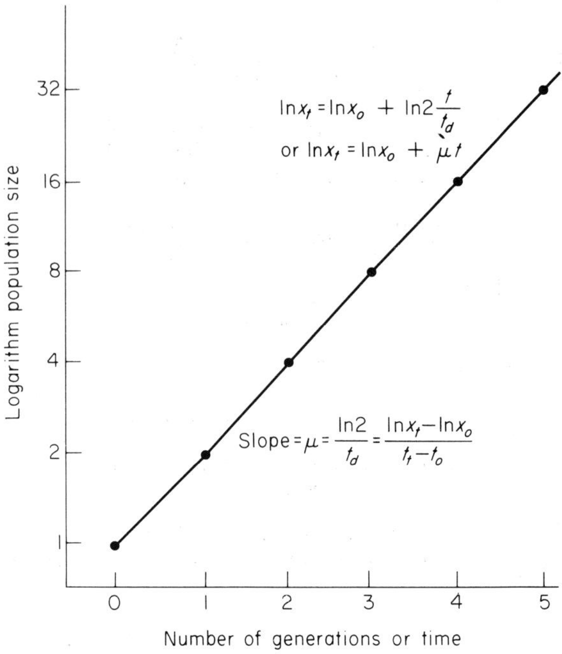

Fig. 2.4.2. Exponential growth in a closed, unlimited environment. This is the linear representation of exponential growth derived from equations (2.4.2) and (2.4.5).

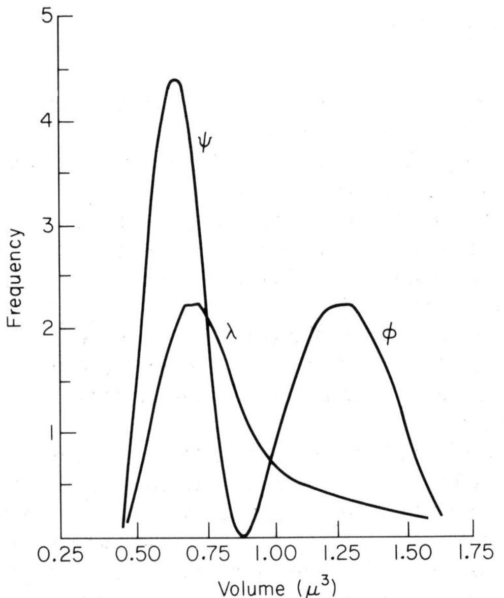

Fig. 2.4.3. Frequency function of the distribution of volume of newly formed cells (ψ), extant cells (λ) and dividing cells (ϕ) of *Escherichia coli* strain B/—/ growing in glucose minimal medium at 30°C. After Marr *et al.* [22].

biomass (or individual organisms) produced by unit amount of existing biomass (or individual organisms) in unit time. In a strict sense μ has units of g new biomass (g existing biomass)$^{-1}$ h^{-1} but this is usually simplified to units of reciprocal time, in this case h^{-1}. In common with the doubling time, the value of the specific growth rate depends on the type of micro-organism and the exact physicochemical conditions under which it is growing. However, in an optimum environment with all the growth requirements in excess, the rate of growth of the population is at a maximum and so μ becomes the *maximum growth rate*, μ_{max}.

Equation (2.4.3) has the solution:

$$x_t = x_0 \, . \, e^{\mu t} \tag{2.4.4}$$

which is analogous to equation (2.4.1) and describes the same exponential curve (see Fig. 2.4.1). Similarly equation (2.4.4) has a linear form represented by:

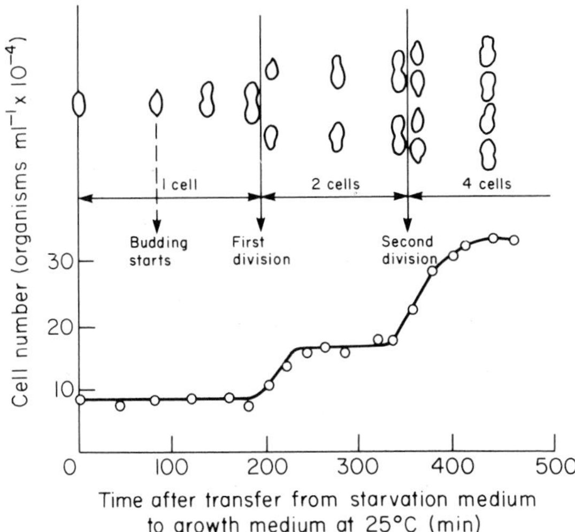

Fig. 2.4.4. Growth of a synchronized yeast culture (*Saccharomyces cerevisiae*) on inoculation into synthetic medium. After Williams & Scopes [44].

$$\ln x_t = \ln x_0 + \mu t \qquad (2.4.5)$$

By comparing equations (2.4.2) and (2.4.5) or by considering the slopes in Fig. 2.4.2, it can be seen that:

$$\mu = \frac{\ln 2}{t_d} = \frac{0.693}{t_d} \qquad (2.4.6)$$

and so the specific growth rate is inversely proportional to the culture doubling time.

Equations (2.4.1) and (2.4.4) describe exponential growth and are known as the *basic growth equations*. The maximum specific growth rate and the culture doubling time are two of the basic growth parameters. These equations have been developed here in the context of a microbial population growing by binary fission. It is important to note, however, that the basic growth equations have a fundamental importance and can be applied to any organism, including filamentous fungi, growing in an unlimited environment.

These two methods of deriving the basic growth equations (2.4.1) and (2.4.4) have been formulated by considering microbial populations as numbers of individual cells and the rates of population increase as increases in the number of cells in the population. The size of the population can be considered in many ways, and parameters such as biomass per unit volume, ATP content, lipopolysaccharide content, nucleic acid content, etc. can be used to measure changes in population size with time [6, 18]. Whilst suitable cellular (and hence population) parameters such as these may be used it is important to remember that many cellular components, such as DNA, ATP or protein per cell, alter in response to changes in environmental conditions, and may not be entirely reliable parameters for estimating real changes in population size. In most practical situations these 'variable' parameters change in response to changes in population specific growth rate and are only useful indicators of population change under balanced growth conditions. Balanced growth occurs when all cellular components increase evenly and in proportion with each other during growth. Since such conditions rarely apply in nature, because population specific growth rates are not normally constant for long enough periods of time, it is generally preferable to measure population changes in terms of an overall parameter such as cell number or biomass concentration.

Growth in a limited environment

It is self-evident that in reality exponential growth can only continue for a limited period of time in a closed environment, an axiom which Stanier and his colleagues illustrated by a simple calculation [37]. If a single microorganism with a doubling time of 0.33 h was grown exponentially for 48 h, the final population would contain $2^{48/0.33}$ or 2.23×10^{43} individuals. If it is assumed that each individual has a dry weight of 10^{-12} g, then the total weight of the final population would be 2.2×10^{28} kg — a mass which is approximately 400 times that of the earth.

Growth in a closed environment is a self-limiting process for many possible reasons but principally because of either the exhaustion of one of the essential nutrients required for growth or the accumulation of metabolic waste products which inhibit growth. As a result the simple basic growth equations do not provide an adequate model of the overall pattern of growth in a closed environment.

In natural or laboratory closed cultures, microbial growth usually follows a characteristic sequence of events which together are known as the *closed* (or *batch*) *culture growth cycle* (see Fig. 2.4.5). In laboratory cultures, for example, in simple shake flasks, the various phases of growth are easily observed since many of the potential variables, such as temperature or the rate of aeration, are kept relatively constant. Nevertheless, the main features of the growth cycle apply also to natural closed cultures. One of the assumptions made in formulating the basic growth equations was that growth began immediately and

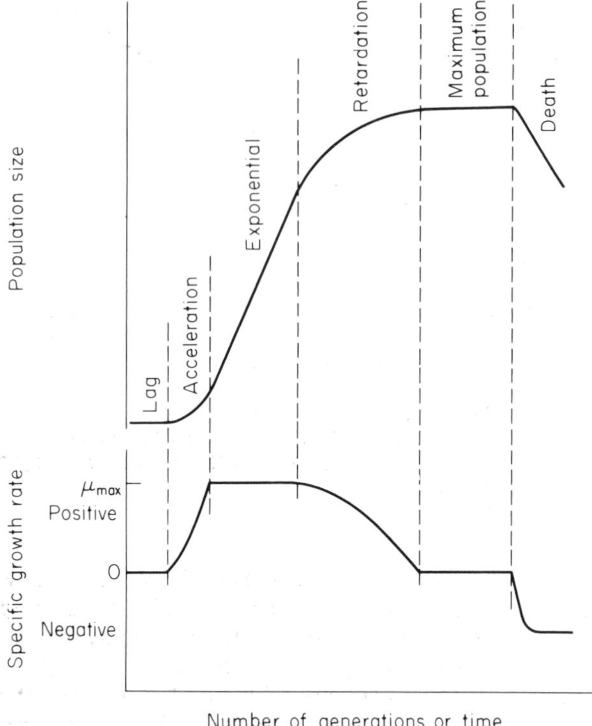

Fig. 2.4.5. The phases of growth occurring during a closed culture growth cycle in a limited environment, showing the changes in the organism's specific growth rate. After Bull [3].

proceeded at the maximum specific growth rate predicted for those growth conditions. Usually, however, the initial phase of the closed culture growth cycle shows no detectable increase in the population size and so the specific growth rate is zero.

Many reasons may be advanced to explain the *lag phase*, such as the need for the population to adapt to the fresh growth conditions. This process might involve the synthesis of a specific enzyme required to catabolize the provided carbon and energy source or the requirement to elaborate a complete biosynthetic pathway for the synthesis of a metabolite.

The *acceleration phase* marks the detectable beginnings of growth and a gradual increase in the population's specific growth rate. The acceleration phase ends once the growth rate reaches the maximum characteristic of the *exponential phase*. The exponential phase is the only part of the growth cycle which may be described by the basic growth equations and frequently growth in this phase is referred to as balanced growth, since all the individual processes involved in biomass production increase at the same rate. As previously indicated, unattenuated exponential growth cannot continue indefinitely, and indeed this ideal growth stage may be comparatively short.

The *retardation phase* is established once the culture's specific growth rate begins to decline and continues until growth ceases. In contrast to the exponential phase, the retardation phase may be quite long and, indeed, the major part of the growth cycle; for example, in the case of the growth of the filamentous fungus, *Aspergillus nidulans*, shown in Fig. 2.4.12, the retardation phase comprises nearly three-quarters of the three growth stages (that is, the acceleration, exponential and retardation phases) whereas the exponential phase is less than a quarter. The progressive decline of the specific growth rate in this stage is due to the gradually increasing effects of the depleting growth resources or the accumulation of toxic waste products.

In the *maximum population phase* (sometimes known as the *stationary phase*), although active growth has ceased the population remains metabolically active, retaining the potential for continued growth should new growth conditions be established. During this phase, reserve materials may be mobilized in order to maintain viability for as long as possible, but clearly the capacity to survive under these conditions is limited and eventually cell death and lysis result. During the *death phase* the viable count decreases, although the exact kinetics of this phase are complex depending greatly on the type of organism, its previous culture history and the nature of the physical environment during this time.

The logistic growth model

One of the more useful models, which has been known for many years and which is frequently applied to the growth of a population in a limited, closed environment, is the *logistic equation* [30, 42]. This description of growth may be used from the beginning of the exponential phase through to the end of the maximum population phase and is basically a simple modification of the basic growth equation (2.4.3). For most micro-organisms growing in closed culture, this part of the growth cycle is an S-shaped curve and may be empirically described by:

$$\frac{dx}{dt} = \mu_{max} \cdot x \left[1 - \frac{x}{x_f}\right] \tag{2.4.7}$$

where μ_{max} and x are the same terms as previously defined and x_f represents the size of the population achieved in the maximum population phase. The significant feature of the logistic equation is the term $[1 - (x/x_f)]$ which described the reduction of the specific growth rate as the exponential growth phase gives way to the retardation phase. That is, it models the decrease in the rate of increase of the population as the population density increases. When the population density, x, is small at the beginning of exponential growth, then x/x_f is also small, tending towards zero, and hence equation (2.4.7) simplifies to the basic growth equation for exponential growth (2.4.3). However, with increasing growth time, x becomes larger and a significant fraction of x_f. Thus x/x_f tends towards unity and $[1 - (x/x_f)]$ becomes significantly less than one, effectively reducing the value of the maximum specific growth rate. At the end of the retardation phase $x = x_f$ and so $dx/dt = 0$, which characterizes the maximum population phase.

The maximum population size, x_f, reflects the potential of the growth environment to support that population and is said to represent the *carrying capacity* of the environment. The logistic equation often accurately describes the observed pattern of growth in closed cultures (see Figs. 2.4.11 and 2.4.12) but it needs to be emphasized that it is an empirical model. There are many other parameters besides the maximum population size which are likely to be implicated in the reduction of the population's specific growth rate and it is unlikely that these are directly accounted for by the carrying capacity.

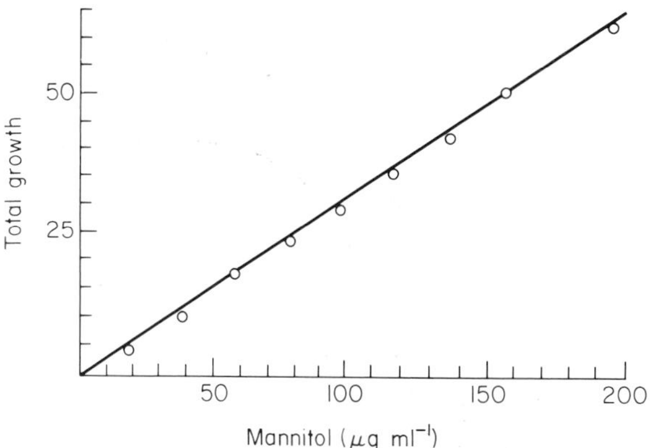

Fig. 2.4.6. Total growth of *Escherichia coli* in synthetic medium with the organic source (mannitol) as the limiting factor. Ordinate: arbitrary units, where one unit is equivalent to 0.8 μg dry weight ml $^{-1}$. After Monad [26].

Growth limitation by exhaustion of the limiting substrate — observed growth yield

If growth ceases as the result of the depletion of one particular nutrient, the *growth-limiting substrate*, then the final population size, x_f, is directly proportional to the initial concentration of the limiting nutrient, s_R:

$$x_f \propto S_R$$

and

$$x_f = YS_R \qquad (2.4.8)$$

where Y is the *observed growth yield* and is defined as that quantity of biomass produced in unit time as the result of utilization of unit amount of the limiting nutrient in the same time. If x_f is measured as grams of biomass per litre, then the observed growth yield has units of grams of biomass produced per gram of growth-limiting substrate utilized. An example of the relationship between x_f and amount of growth-limiting substrate used is shown in Fig. 2.4.6.

The observed growth yield is a term which depends on two fundamental metabolic processes. The yield term is a variable which depends on the balance between: (a) how much of the available growth-limiting substrate is used for new biomass produced; and (b) how much of the substrate is consumed by the growing cells for energy to drive biosynthetic reactions. Clearly the greater the efficiency of energy generation or the less energy required for biomass production, then the less substrate has to be dissimilated for this purpose and the more substrate is available for biomass production. This can be clearly demonstrated for a facultative anaerobe, *Streptococcus faecalis*, growing aerobically (good energy production from unit amount of substrate consumed) and anaerobically (poor energy production from unit amount of substrate consumed). *S. faecalis* growing aerobically on glucose has a Y value of 0.32 g biomass (g glucose used)$^{-1}$, whereas anaerobically Y is 0.12 g biomass (g glucose used)$^{-1}$. Under anaerobic conditions a greater fraction of the glucose has to be used to produce the same amount of ATP for biosynthesis, thus affecting the observed growth yield.

The value of the observed growth yield also depends on the nature of the substrate limiting growth, since different substrates are fed into the cell's central metabolic pathways at different points, which in turn depends on the amount of energy which may be generated for each molecule of substrate consumed. In general terms the lower the redox level of the substrate the greater the energy content. For example, for *Klebsiella aerogenes* grown aerobically on glucose, $Y = 0.39$ g biomass (g glucose used)$^{-1}$; for aerobic growth on succinate, Y = 0.23 g biomass (g succinate used)$^{-1}$.

Yield terms have widely been regarded as a 'fundamental' constant of microbial growth. Many different ways have been developed to express the observed growth yield in efforts to find a standard term which enables adequate comparisons to be made between different organisms grown on different substrates under different, and often variable, conditions. There is no single satisfactory solution, and each method of expressing Y values has its faults.

Y_{sub} is the *molar growth yield* with units of grams of biomass produced per mol of substrate consumed. The subscript indicates the particular substrate for which the yield has been determined. Molar growth yield is a widely used form because it relates the efficiency of substrate conversion to biomass production, and is similar to the common way of expressing substrate concentrations. It does not standardize the term for substrate carbon content and substrate energy content, both of which are crucial if comparisons between different substrates are to be made.

Y_{oxygen} is a molar growth yield term with units of g biomass produced per mol of oxygen consumed. This method of expression is based on the reasonable expectation that the amount of oxygen consumed is directly related to the amount of energy generated, and hence the amount of substrate used directly for energy generating purposes. However, linkage between oxygen uptake and energy consumption is not fixed or constant and can vary significantly.

Molar growth yields may also be based on the gram carbon content of substrate and biomass but this method is not widely used, largely because the carbon content of biomass is rarely determined directly. Y_{carbon} has units of grams of cell carbon per gram of substrate carbon consumed, and is sometimes expressed as a percentage known as the *carbon conversion efficiency*. The limitation of this method is that the amount of energy generated from a particular substrate is not dependent on the content but on the redox potential of the substrate.

The molar growth yield for ATP, Y_{ATP}, has been advocated to overcome the problem of exactly measuring the amount of substrate used for energy generation. Y_{ATP} has units of grams of biomass produced per mol of ATP generated and in principle ought to give reliable estimates of observed growth yields since it is based on the amounts of energy actually available for biosynthetic purposes. For many anaerobic organisms, generating ATP via substrate level phosphorylation, it is possible to calculate accurately the amount of ATP produced from the known pathways of the fermentation. In general terms, Y_{ATP} values of 10.5 g biomass produced per mol ATP are commonly observed. For aerobic organisms, the amount of ATP produced cannot be accurately determined or estimated, and so this term suffers from the same limitations as Y_{oxygen}.

As this brief discussion has indicated, determination of observed growth yields is difficult and comparisons are hard to make with quantitative certainty.

Influence of growth-limiting substrate on the specific growth rate

In many closed environments, the decline in a population's specific growth rate is related to the depletion of one particular nutrient and this relationship can be used to develop a second simple closed culture growth model (see above). Monod was the first to recognize clearly that, if there was a decrease in the concentration of a growth substrate, then there was a reduction in the population's specific growth rate (see Fig. 2.4.7) [25, 26]. That is, the growth rate is restricted by the concentration of the growth-limiting substrate. Moreover, Monod established that the form of the relationship was very similar to the effect of substrate concentration upon the rate of an enzyme-catalysed reaction. Indeed the relationship can be adequately described by an equation analogous to the Michaelis–Menten enzyme kinetics equation, namely:

$$\mu = \mu_{max} \cdot \frac{s}{(K_s + s)} \tag{2.4.9}$$

where s is the concentration of the growth-limiting substrate, μ is the specific growth rate and μ_{max} is the maximum specific growth rate obtained in the absence of any substrate limitation. The term K_s is a constant known as the *saturation constant* and is defined as that growth-limiting substrate concentration which allows the organism to grow at half the maximum specific growth rate (see Fig. 2.4.7). The saturation constant is the third basic growth parameter and has the same units as substrate concentration. It is a measure of the affinity the organism has for the growth-limiting substrate; that is, the lower the value of K_s, the greater the organism's affinity for the substrate and the greater the capacity to grow rapidly in an environment with low growth-limiting substrate concentrations.

In most closed cultures, particularly laboratory batch cultures, the saturation constant has little significance in the kinetics of growth since, for the majority of the growing period, all the growth substrates are in excess and the dominant growth parameter is the maximum specific growth rate. This is not the case, however, for organisms with poor affinities for the limiting substrate where comparatively high substrate concentrations,

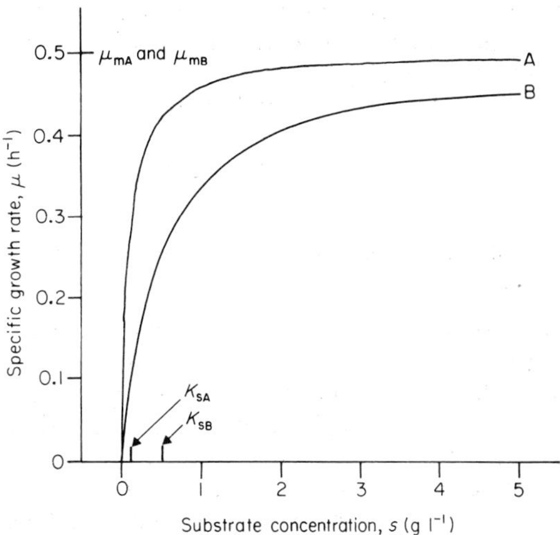

Fig. 2.4.7. The relationship between the specific growth rate and the growth-limiting substrate concentration. The Monod relationship, equation (2.4.9), is shown for two organisms: A, $\mu_{max} = 0.5\ h^{-1}$, $K_s = 0.1$ g substrate l^{-1} and B, $\mu_{max} = 0.5\ h^{-1}$, $K_s = 0.5$ g substrate l^{-1}.

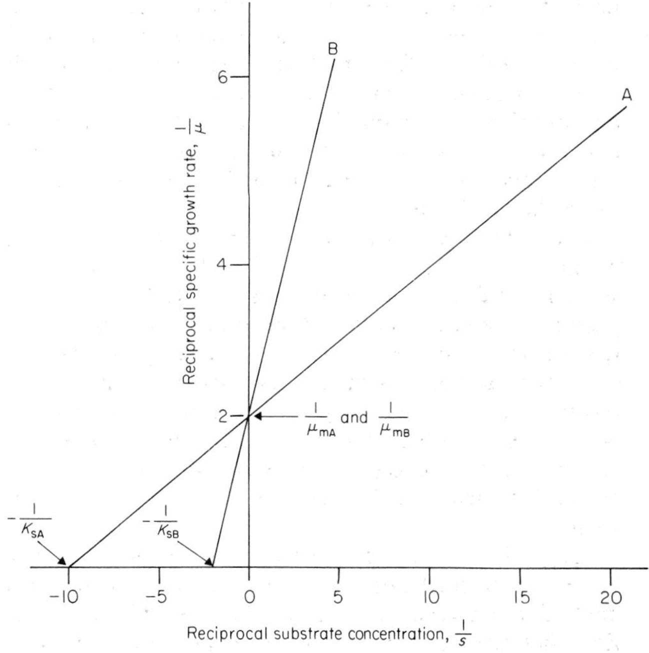

Fig. 2.4.8. The linear form of the Monod equation (2.4.10) for the same organisms as in Fig. 2.4.7.

occurring for significant periods of time towards the end of a closed culture growth cycle, have a measurable effect on the culture's growth rate. Moreover, in many natural environments (and in some forms of open culture systems) low nutrient concentrations are normal and so the saturation constant becomes an important growth parameter (see Section 2.4.3).

Equation (2.4.9) describes a hyperbolic function with μ approaching μ_{max} asymptotically as s increases and may be rearranged to give a linear function:

$$\frac{1}{\mu} = \frac{1}{s} \cdot \frac{K_s}{\mu_{max}} + \frac{1}{\mu_{max}} \qquad (2.4.10)$$

A plot of $1/\mu$ against $1/s$ yields a straight line (see Fig. 2.4.8) intercepting the abscissa at $1/\mu_{max}$ and the ordinate at $-1/K_s$.

In theory the value of the saturation constant can be determined from closed culture kinetics. The specific growth rate is measured in a number of separate cultures growing under identical conditions but with various concentrations of one growth-limiting substrate. The practical difficulty is that, for many substrates, the K_s values are very small and so extremely low initial substrate concentrations are required in order to restrict the organism's growth rate. Obviously under such conditions a slight increase in biomass concentration (needed to obtain a growth curve) results in the utilization of significant quantities of the limiting substrate which immediately decreases the specific growth rate still further. It is therefore difficult to achieve a sufficiently prolonged period of growth during which the specific growth rate is constant. Normally, saturation constants are estimated from growth in open systems (see Section 2.4.3), but there too, analytical difficulties exist in determining low substrate concentrations.

Whilst most growth conditions in nature normally depend on microbial responses to low growth-limiting substrate concentrations, it must be recognized that in certain situations high substrate conditions exist and these can have an adverse effect on population growth rates (see Fig. 2.4.9). Many modifications of the Monod equations have been suggested [12] to include an appropriate *inhibition constant*, K_i, such as that used in the Haldane equation:

$$\mu = \frac{\mu_{max} \cdot s}{(K_s + s)(1 - s/K_i)} \qquad (2.4.11)$$

Rate of growth-limiting substrate utilization

It follows from equation (2.4.8) that in a small time inter-

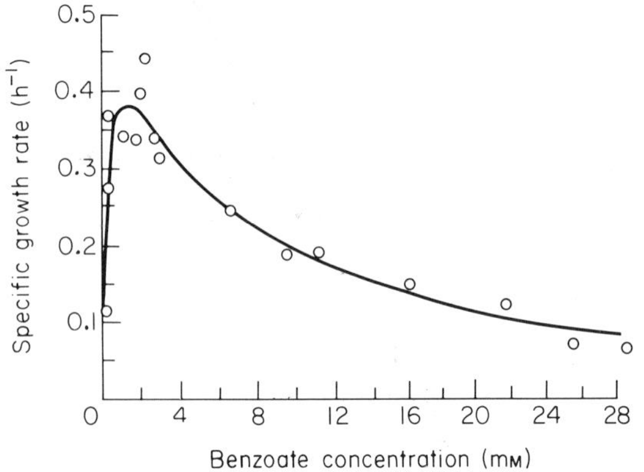

Fig. 2.4.9. Inhibition of the specific growth rate of *Klebsiella aerogenes* growing on sodium benzoate (Edwards [13]).

val, δt, the population increases by a small amount, δx, utilizing a small amount of the growth-limiting substrate, δs. Thus:

$$-\frac{dx}{ds} = Y \tag{2.4.12}$$

and, by substituting for dx by equation (2.4.3) and rearranging, the rate of substrate utilization is given by:

$$-\frac{ds}{dt} = \frac{\mu}{Y} \cdot x \tag{2.4.13}$$

The negative sign indicates that the growth-limiting substrate concentration decreases with time of population growth.

Under conditions where the population's specific growth rate remains constant, μ/Y is a constant term given the symbol q and is known as the *specific metabolic rate* or *metabolic quotient*. This constant defines the rate of uptake of the growth-limiting substrate by a unit amount of biomass per unit time (unit: g substrate (g biomass)$^{-1}$ h^{-1}). It must be realized, however, that rarely in nature are conditions sufficiently constant for μ to remain fixed and that Y itself is not necessarily constant (see next section).

Thus if μ is considered to be variable (but Y constant) then:

$$-\frac{ds}{dt} = \frac{s}{(K_s + s)} \cdot \frac{x}{Y} \tag{2.4.14}$$

when substituting equation (2.4.10) for μ in equation (2.4.13).

Variations in the observed growth yield — maintenance energy

The above description of the observed growth yield has assumed that Y is constant for a given organism under constant conditions and for a given substrate. However, this assumption is not true since it is known that Y varies particularly with changing specific growth rate of the organisms. The main observation is that the value of Y is substantially lower when determined with organisms growing at significantly sub-maximal growth rates, compared with the same organism growing on the same substrate but at higher growth rates. This variation in Y is due to the fact that Y is a complex term accounting for two basic cellular processes: (a) it allows for a fraction of the dissimilated substrate used to produce energy which is directly used for subsequent biosynthetic reactions; and (b) it contains a fraction of substrate used for energy production for maintaining certain key metabolic processes which do not directly contribute to biomass production [31]. These key processes include energy required for maintaining solute gradients, motility, maintenance of internal pH, turnover of macromolecules, etc.; collectively the energy requirement for these purposes is known as *maintenance energy*. The *maintenance energy coefficient*, m, is defined as the amount of substrate for maintenance energy used per unit amount of biomass in unit time. Thus the overall rate of growth-limiting substrate utilization may be expressed as:

Overall rate of substrate utilization	=	Rate of substrate use for biomass production	+	Rate of substrate use for maintenance

$$\frac{ds}{dt} = \left[\frac{ds}{dt}\right]_{\text{growth}} + \left[\frac{ds}{dt}\right]_{\text{maintenance}}$$

From equation (2.4.13) we have:

$$\frac{\mu}{Y} \cdot x = qx = q_{\text{growth}} \cdot x + q_{\text{maintenance}} \cdot x \tag{2.4.15}$$

where $q_{\text{growth}} = \mu/Y_G$ and Y_G is the *true* or *maximum growth yield*, and $q_{\text{maintenance}} = m$, the *maintenance coefficient*. Thus:

$$\frac{\mu x}{Y} = \frac{\mu x}{Y_G} + mx \tag{2.4.16}$$

and rearranging:

$$\frac{1}{Y} = \frac{1}{Y_G} + \frac{m}{\mu} \tag{2.4.17}$$

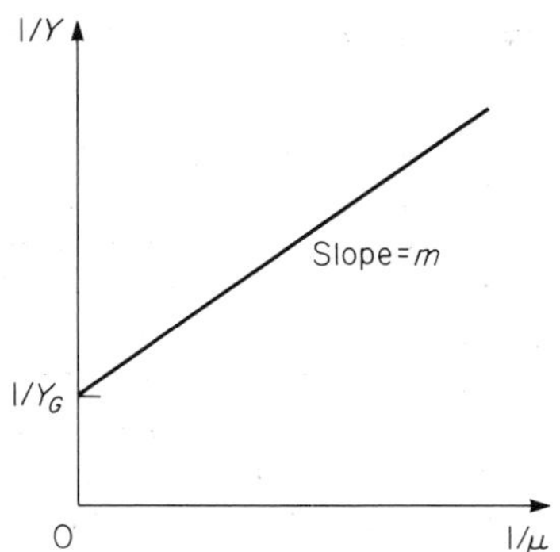

Fig. 2.4.10. Determination of the true growth yield (Y_G) and maintenance coefficient (m). After Pirt [31].

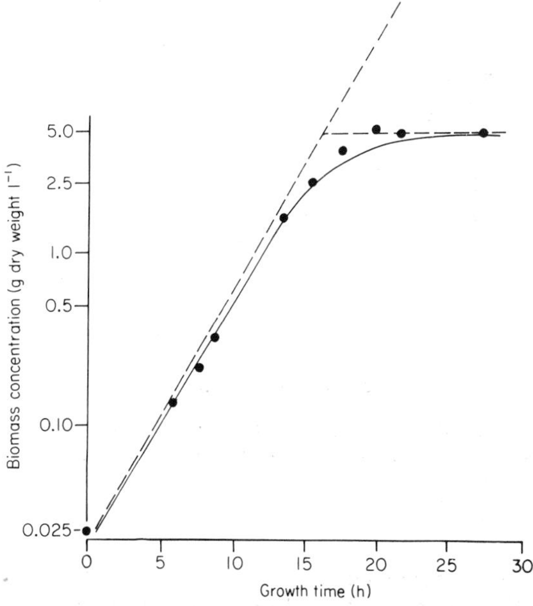

Fig. 2.4.11. A model of the growth of *Geotrichum candidum* in a closed culture with glucose as the limiting substrate (S_R = 10 g glucose l^{-1}). The observed biomass concentrations (g dry weight l^{-1}) are shown by ●. The logistic model (——) was fitted using μ_{max} = 0.353 h^{-1} and x_f = 4.9 g dry weight l^{-1}. The saturation model (---) was fitted using the following constants: μ_{max} = 0.353 h^{-1}, K_s = 0.024 g glucose l^{-1} and Y = 0.40 g dry weight (g glucose)$^{-1}$. The continuation line (---) indicates continued exponential growth in an unlimited environment. (Data from Trinci [39] and modelled according to Bull *et al.* [4].)

Thus the deviations in the observed growth yield can be quantified by examining their changes in response to changes in the specific growth rate of the population (see Fig. 2.4.10). Clearly Y_G values are always smaller than values of Y. In general terms m values are in the order of 10–100 mg carbon substrate used (g biomass)$^{-1}$ h^{-1}. This analysis assumes that m is constant, but again it seems likely that the value of the maintenance coefficient does vary with growth conditions and growth rate; to date no quantitative models have been proposed to account for these suspected variations.

The saturation growth model

This is the second model which describes the growth of a population in a closed, limited environment and it may be applied to the same stages of growth as the logistic model. The model depends on the prediction of biomass and growth-limiting substrate concentration by simultaneously solving the following two equations by a simple iterative computation. The growth-limiting substrate concentration is given by equation (2.4.14) and the biomass concentration is given by a composite equation derived from equations (2.4.3) and (2.4.9):

$$\frac{dx}{dt} = \mu_{max} \cdot \frac{s}{(K_s + s)} \cdot x \qquad (2.4.18)$$

Growth curves may be calculated from a knowledge of the initial biomass, the growth-limiting substrate concentrations, the maximum specific growth rate for the particular conditions in use, the organism's substrate affinity and the yield. In some cases, for example the growth of the fungus *Geotrichum candidum* on glucose (see Fig. 2.4.11), the model provides a reasonable fit to the observed data. In this case the fit is almost as good as that of the logistic model. Frequently, however, the saturation growth model over-stimates the rate of biomass production and, as a result, the time taken by the culture to reach the maximum population phase. This is amply illustrated for *Aspergillus nidulans*, another filamentous fungus, growing on glucose, where obviously the saturation model provides a much poorer fit to the observed data than the logistic model (see Fig. 2.4.12).

One of the more obvious problems in modelling microbial growth is that a number of simplifying assumptions have to be made in developing a mathematical description of growth which is relatively simple to solve and furthermore which contains constants that have biological meaning. In the case of the saturation model it is assumed that the restriction of exponential growth depends only on the concentration of the growth-limiting substrate; many other parameters which may affect the

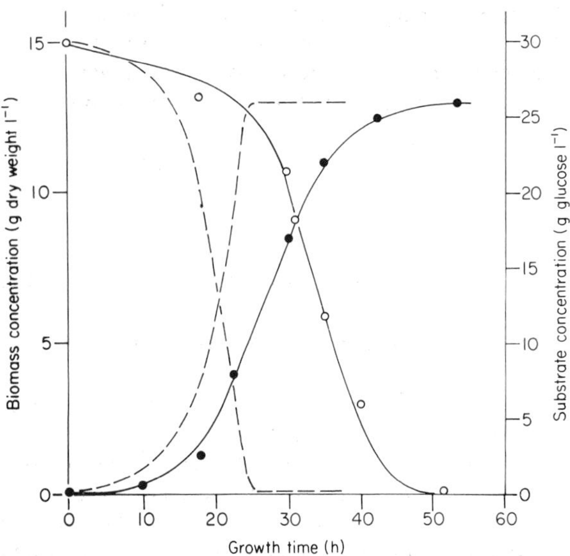

Fig. 2.4.12. A model of the growth of *Aspergillus nidulans* in a closed culture with glucose as the limiting substrate (S_R = 30 g glucose l^{-1}). The observed biomass concentrations (g dry weight l^{-1}) are shown by • and the observed glucose concentrations (g glucose l^{-1}) by ○. The logistic model was fitted using μ_{max} = 0.2 h^{-1} and x_f = 13.0 g dry weight l^{-1}. The saturation model was fitted using the following constants: μ_{max} = 0.2 h^{-1}, K_s = 0.072 g glucose l^{-1} and Y = 0.43 g dry weight (g glucose)$^{-1}$. (Data from Carter & Bull [8] and modelled according to Bull *et al.* [4].)

specific growth rate and which may vary in a complex fashion during the growth cycle are ignored. Obviously some simplifications must be made in order to develop a practicable and experimentally useful model, and indeed with the present state of knowledge many potentially significant growth parameters probably remain unidentified and cannot therefore be quantitatively included in more complex models. For example *Aspergillus nidulans* may deviate from the predicted saturation model as a result of the accumulation of toxic waste products or oxygen limitation. These have a major effect on the rate of growth, which is more accurately reflected in the logistic growth model than in the saturation model. In this context it is interesting to note the better fit of the *Geotrichum candidum* growth curve to the saturation model, possibly because the initial glucose concentration is some three times lower than for *Aspergillus nidulans*. This lower population will produce fewer wastes and use less oxygen.

Deviations from growth in a limited environment

The Monod equation (equation 2.4.9) assumes that the growth rate declines gradually as the growth-limiting substrate concentration approaches zero. At zero this description of growth assumes that μ = 0. However, it is commonly observed in natural environments that this is not the case: the 'growth-limiting' substrate reaches zero (or perhaps more accurately reaches a value beyond the capability of the analytical system) well before growth ceases, and in some instances before exponential growth ceases. Most interpretations suggest that growth can proceed, often at maximal rates, because of the intracellular accumulation of significant concentrations of the growth-limiting substrate. It is important to recognize, therefore, that the external (environmental) concentrations of the growth-limiting substrate may not be the value which dictates population growth rates. It is, of course, very much harder to measure the intracellular concentrations of the growth-limiting substrate.

2.4.3 MICROBIAL GROWTH IN OPEN ENVIRONMENTS

In open growth systems there is a continuous input of growth substrates and removal of waste products, cells and unused substrates. These systems are known as continuous-flow cultures [36]. In continuous-flow cultures the exponential growth phase is prolonged indefinitely so that steady state conditions are established. Unlike batch cultures, where to a large extent the behaviour of cells at a particular point in time is influenced by the culture's previous history, steady state conditions remove the influence of transient conditions. There are substantial additional advantages to open growth: specific growth rate may be directly controlled by external factors, continued substrate-limited growth may be established, sub-maximal growth rates can be imposed, and biomass concentration may be set independently of the growth rate. These advantages are discussed in more detail by Pirt [31] and Bull [3]. The most widely used continuous culture system is the *chemostat* which is characterized by growth control exercised through a growth-limiting substrate.

Chemostat growth kinetics

A chemostat is a culture vessel containing a fixed volume, V, of growing culture (unit: volume). For both aerobic and anaerobic systems, agitation maintains a homogeneous culture and, in the case of aerobic cultures, ensures adequate oxygen transfer. A defined growth medium containing one of the constituent substrates, S_R, at a concentration which is known to be growth-limiting, is pumped into the growth vessel at a constant rate F (unit: volume time^{-1}). By such mechanisms as a weir overflow or another pumping system, the culture, the unused substrates and the products are removed at exactly the same flow rate F. It is assumed that in an adequately mixed culture the incoming fresh medium is instantaneously mixed with the resident culture. Within the vessel the culture grows to a certain biomass concentration, x (this also gives the biomass concentration in the effluent culture), and reduces the initial concentration of the growth-limiting substrate, S_R, to a value of s.

The dilution rate

It has already been shown that an organism's specific growth rate depends on the concentration of the limiting nutrient. Clearly in a chemostat culture the concentration of the limiting substrate depends both on the rate at which fresh substrate is supplied and on the dilution factor as the fresh substrate is dispersed throughout the culture vessel. That is, the growth-limiting substrate concentration depends on the flow rate, F, and the culture volume, V, and in particular on the ratio of these two parameters which is known as the *dilution rate*, D:

$$D = \frac{F}{V} \qquad (2.4.19)$$

The dilution rate has units of reciprocal time (normally h^{-1}) and is a measure of the number (or fraction) of the culture volume changes achieved in unit time. Thus a dilution rate of 1.0 h^{-1} means that there is one complete volume change per hour. Moreover, it is important to understand that the dilution rate has the same units as the specific growth rate and under appropriate conditions equals the organism's specific growth rate. The reciprocal of the dilution rate is the *mean residence time* and this parameter gives the average time an organism is likely to remain within the culture vessel.

The dilution rate and biomass concentration

Within the culture vessel growth is occurring, tending to increase the biomass concentration, but at the same time biomass is lost from the culture vessel through the effluent. Thus there is a biomass balance in the culture vessel which may be expressed as:

The rate of change of biomass concentration in the culture vessel	=	The rate of biomass production (growth)	−	The rate of biomass removal (washout)
$\frac{dx}{dt}$	=	μx	−	Dx

or:

$$\frac{dx}{dt} = x\,(\mu - D) \qquad (2.4.20)$$

Subsituting for μ by equation (2.4.9), we have:

$$\frac{dx}{dt} = x\left[\frac{\mu_{max} \cdot s}{(K_s + s)} - D\right] \qquad (2.4.21)$$

With reference to equation (2.4.20), there are three different situations which need to be considered.

1 If $\mu > D$, then dx/dt is positive and the biomass concentration in the culture vessel increases since the rate of biomass production exceeds the rate of biomass washout.

2 If $\mu < D$, then dx/dt is negative and the biomass concentration decreases since the rate of culture washout is greater than biomass synthesis.

3 If $\mu = D$, then $dx/dt = 0$ and the biomass concentration remains constant and the culture is said to be in a steady state. This third situation is the preferred, stable mode of operation and in time all chemostat continuous cultures reach a steady state, provided that the dilution rate is less than μ_{max}.

The dilution rate and growth-limiting substrate concentration

A similar growth-limiting substrate balance equation can be formulated as follows:

The rate of change of growth-limiting substrate concentration in the culture vessel	=	The rate of input of fresh substrate	−	The rate of substrate removal (washout)	−	The rate of substrate utilization by the organism (growth use)
$\frac{ds}{dt}$	=	DS_R	−	Ds	−	$\frac{\mu x}{Y}$

or:

$$\frac{ds}{dt} = D(S_R - s) - \frac{\mu x}{Y} \qquad (2.4.22)$$

where the final term is derived from equation (2.4.13).

Again, there are three possible situations which need to be considered.

1 If $\mu > D$, then ds/dt is negative and the growth-limiting substrate concentration decreases. The biomass concentration is increasing in this condition and therefore utilizing more of the available substrate.

2 If $\mu < D$, then ds/dt is positive and the growth-limiting substrate concentration increases.

3 Finally if $\mu = D$, then $ds/dt = 0$ and the growth-limiting substrate concentration reaches a constant, steady state value at the same time as the biomass concentration.

The biomass and growth-limiting substrate concentrations in the steady state

As already noted, one of the most important features of chemostat continuous-flow culture is that steady state growth conditions are always achieved and that it is a self-balancing system, maintaining unique values for the biomass and growth-limiting substrate concentrations so long as the dilution rate is kept constant. In the steady state culture, $\mu = D$ and dx/dt and ds/dt are given by equations (2.4.21) and (2.4.22). Thus for equation (2.4.21):

$$0 = \tilde{x}\left[\frac{\mu_{max}\cdot\tilde{s}}{(K_s + \tilde{s})} - D\right]$$

where $\tilde{s}$ and $\tilde{x}$ indicate the steady state growth-limiting substrate and biomass concentrations, respectively. Hence:

$$D = \frac{\mu_{max}\cdot\tilde{s}}{(K_s + \tilde{s})}$$

and

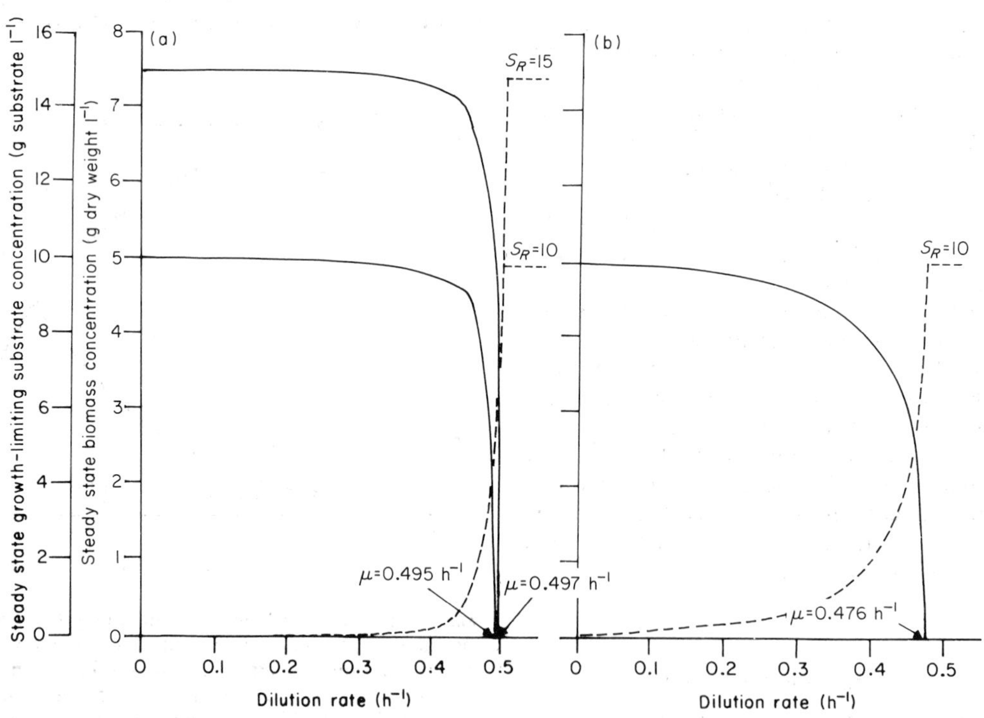

Fig. 2.4.13. The theoretical steady state biomass and growth-limiting substrate concentrations for two organisms, *A* and *B*, in a chemostat continuous-flow culture. (a) Organism *A* ($\mu_{max} = 0.5\ h^{-1}$, $K_s = 0.1$ g substrate l^{-1}, $Y = 0.5$ g biomass (g substrate)$^{-1}$) is growing in two different cultures with various concentrations of the growth-limiting substrate. When $S_R = 10$ g substrate l^{-1}, the lower line results and $D_{crit} = 0.495\ h^{-1}$. When $S_R = 15$ g substrate l^{-1}, the upper line results and $D_{crit} = 0.497\ h^{-1}$. (b) Organism *B* ($\mu_{max} = 0.5\ h^{-1}$, $K_s = 0.5$ g substrate l^{-1}, $Y = 0.5$ g biomass (g substrate)$^{-1}$), differing from *A* only in the affinity for the limiting substrate, is growing with an $S_R = 10$ g substrate l^{-1} and $D_{crit} = 0.476\ h^{-1}$.

$$\tilde{s} = \frac{DK_s}{(\mu_{max} - D)} \tag{2.4.23}$$

Also, for equation (2.4.22):

$$0 = D(S_R - \tilde{s}) - \frac{\mu\tilde{x}}{Y}$$

and

$$\tilde{x} = Y(S_R - \tilde{s}) \tag{2.4.24}$$

Equations (2.4.23) and (2.4.24) enable us to predict the steady state concentrations at any dilution rate provided that the initial growth-limiting substrate concentration and the three basic growth parameters, namely, μ_{max}, K_s and Y, are known. Equation (2.4.23) shows that the growth-limiting substrate concentration in the steady state, for a particular organism for which μ_{max} and K_s are obviously constant for particular conditions, depends only on the dilution rate chosen and is even independent of the initial growth-limiting substrate concentration, S_R (see Fig. 2.4.13). Also, it is the value of the residual growth-limiting substrate concentration, s, which in fact determines the organism's growth rate, and so in selecting a particular dilution rate the experimenter ensures that the population's growth rate is equal to the dilution rate. Equation (2.4.24) indicates that the steady state biomass concentration depends not only on the value of s (and its associated dependent parameters) but also on the yield and the initial growth-limiting substrate concentration (see Fig. 2.4.13). Normally steady state cultures are analysed by determining biomass and growth-limiting substrate concentrations and equations (2.4.23) and (2.4.24) are used to calculate the basic growth parameters, particularly the saturation constant and the yield. The μ_{max} has to be separately determined either from growth in closed culture or from washout kinetics. It must be noted that frequently it is difficult to determine the residual substrate concentrations since these may be extremely low, particularly if the organism has a high affinity for the substrate in question.

Establishing and maintaining the steady state

Equations (2.4.23) and (2.4.24) describe the situation which exists once the steady state has been established. However, a central feature of chemostat systems is that, from whatever initial conditions of biomass and substrate concentrations, the system adjusts itself to the unique steady state concentrations provided that the dilution rate is set at a constant value. Consider a culture vessel running at a dilution rate D_1 and which is newly inoculated with a small volume of cell suspension to give an initial biomass concentration x_1 and substrate concentrate s_1, a value which is slightly less than the initial substrate concentration S_R (see Fig. 2.4.14). The substrate concentration is much greater than the expected steady state concentration $\tilde{s}_1$ for D_1 and hence is greater than the concentration for which the organism's specific growth rate equals D. Thus with $s_1 > \tilde{s}_1$, then $\mu > D_1$ and, as we have previously seen, dx/dt is positive and ds/dt is negative. With time the substrate concentration must decline as growth proceeds and s_1 approaches $\tilde{s}_1$ with the result that the specific growth rate declines until $\mu = D_1$ at $\tilde{s}_1$.

Now when it comes to establishing a different steady state culture consider the situation where D_1 is instantaneously increased to D_2 and so immediately afterwards $s_2 (= \tilde{s}_1) < \tilde{s}_2$ and $x_2 (= \tilde{x}_1) > \tilde{x}_2$. Hence it must be that $\mu < D_2$ and so dx/dt is negative and ds/dt is positive. The biomass concentration declines concomitantly with an increase in the growth-limiting substrate concentration until

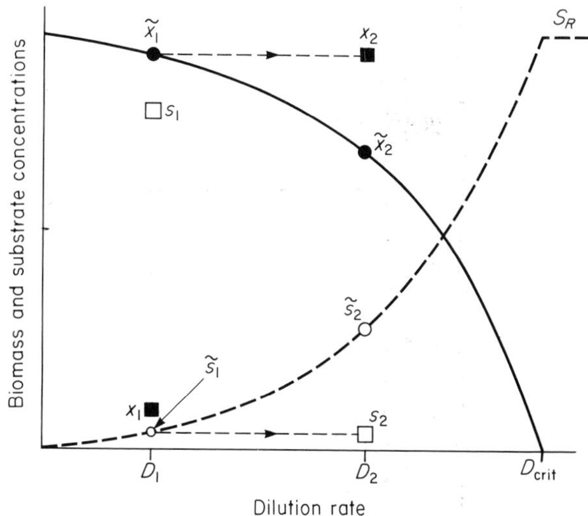

Fig. 2.4.14. Establishing and maintaining the steady state chemostat culture. (a) From an initial inoculation of the culture vessel at a dilution rate of D_1: ■ = initial biomass concentration, x_1; □ = initial growth-limiting substrate concentration, s_1; ● = steady state biomass concentration, $\tilde{x}_1$; and ○ = steady state growth-limiting substrate concentration, $\tilde{s}_1$. (b) After a change in the dilution rate from D_1 to D_2: ■ = initial biomass concentration, x_2; □ = initial growth-limiting substrate concentration, s_2; ● = steady state biomass concentration x_2 at D_2 and ○ = steady state growth-limiting substrate concentration s_2 at D_2.

$\mu = D_2$, $dx/dt = 0$ and $ds/dt = 0$ and the culture is again in a steady state. Furthermore, small alterations in the biomass and growth-limiting substrate concentrations set up opposing changes in the specific growth rate of the culture such that the original, desired steady state is restored.

The critical dilution rate and the calculation of the maximum specific growth rate from culture washout kinetics

From equation (2.4.20) it can be deduced that there is an upper limit to the dilution rate above which steady state cultures cannot be established. This is because the organism's specific growth rate has a maximum value, μ_{max}, which is genetically determined and therefore cannot be exceeded. Thus if $D > \mu_{max}$ then dx/dt must be negative, a steady state culture cannot be obtained and the culture is said to *washout*. There is a unique dilution rate, known as the *critical dilution rate*, D_{crit}, below which steady state cultures are theoretically possible and above which culture washout occurs. In fact D_{crit} is reached when $s = S_R$ and so from equation (2.4.23) substituting for $\bar{s}$ and D and rearranging:

$$D_{crit} = \frac{\mu_{max} \cdot S_R}{(K_s + S_R)} \tag{2.4.25}$$

If $K_s \ll S_R$, equation (2.4.25) simplifies to:

$$D_{crit} = \mu_{max}$$

However, D_{crit} is not a constant, unlike μ_{max}, but depends on the type of organism and the initial concentration of the growth-limiting substrate. In Fig. 2.4.13 organisms A and B are identical except that $K_{sA} = 0.1$ g substrate l^{-1} whilst $K_{sB} = 0.5$ g substrate l^{-1}. Under identical growth systems D_{crit} for $A = 0.495\ h^{-1}$ when $S_R = 10$ g l^{-1} but is significantly higher than for B, $0.476\ h^{-1}$, when $S_R = 10$ g l^{-1}. For organism A, D_{crit} is 0.99 μ_{max} at $S_R = 10$ g l^{-1} but when $S_R = 15$ g l^{-1} then $D_{crit} = 0.994\ \mu_{max}$.

During washout from a culture vessel when $D > \mu_{max}$, the organisms grow at μ_{max} provided that $K_s \ll S_R$. Thus, the kinetics of washout may be used to calculate μ_{max} values since:

$$\frac{dx}{dt} = x\,(\mu_{max} - D)$$

where D has known, fixed values greater than μ_{max}. This has the solution:

$$x_t = x_0 e^{(\mu_{max} - D)t}$$

where x_0 is the concentration of the organism in a steady state at time $t = 0$ when the dilution rate is increased stepwise to a value $D > \mu_{max}$. Thus:

$$\ln x_{t0} = \ln x_0 + (\mu_{max} - D)t$$

and

$$\mu_{max} = \frac{\ln x_t - \ln x_0}{t} + D \tag{2.4.26}$$

This method can only be used as long as the organism is capable of growth at μ_{max} immediately after increasing the dilution rate.

This method of determining μ_{max} assumes that it is immediately reached by the population when the dilution rate is stepped up to $D > \mu_{max}$. This is not the case and in shift-up experiments of this sort a period of time equal to the C and D time for the cell cycle must elapse before the new, higher growth rate is established [21]. This can be clearly seen in washout experiments with the cyanobacterium *Anacystis nidulans* [19] (see Fig. 2.4.15).

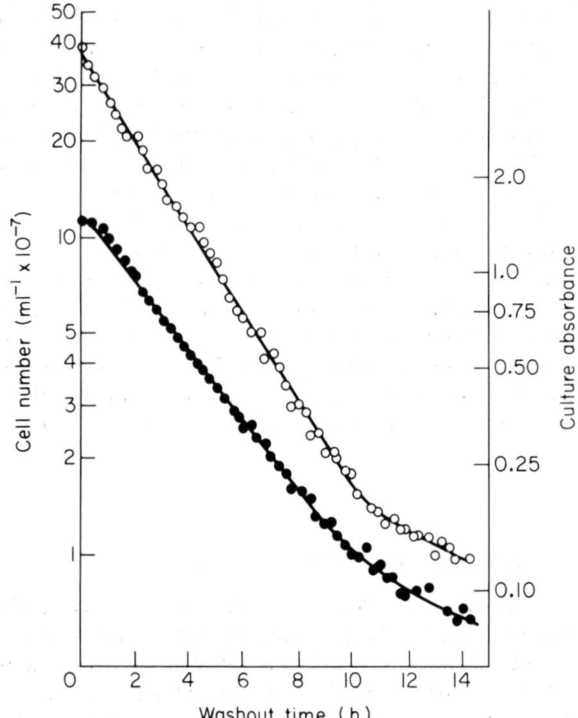

Fig. 2.4.15. Washout curves for *Anacystis nidulans* in terms of cell number (○) and absorbance (●) for a light-limited culture grown at an initial dilution rate of 0.03 h^{-1}. At $t = 0$ the dilution rate was instantaneously increased to 0.37 h^{-1}. After Karagouni & Slater [19].

2.4.4 THE KINETIC PRINCIPLES OF COMPETITION BETWEEN MICROBIAL POPULATIONS

There are many types of interaction between two or more microbial populations [5, 7, 35] but, from the point of view of the industrial exploitation of micro-organisms, competition creates the most problems. These range from the problem of contaminants in pure culture fermentations [32] to the behaviour of industrial fermentations with unstable recombinant DNA culture [29]. Moreover, competition is the most important interaction between populations in nature, and understanding the kinetic principle is important in primary studies involving the enrichment and isolation of particular micro-organisms.

Competition between two or more micro-organisms occurs when the component populations are restricted, in terms of either their growth rates or their final population sizes, as the result of a common dependence on an external factor. Competition can, within the terms of this widely accepted definition, occur either in closed culture, where growth is ultimately limited by the availability of a particular growth resource, or in open culture, where growth is continuously limited. Bazin [1] has erroneously attacked the view that competition can occur in batch culture, misunderstanding the substantial significance of competitive events which influence final population sizes. Ecologically, this type of competition is important and is entirely consistent with the above definition.

Many mathematical models have been proposed (e.g. [1, 14, 29]) and the models given below are presented simply to illustrate the basic principles. The following analyses consider two populations, x_A and x_B, growing in an environment with a single growth-limiting substrate, and assume that apart from the competition interaction there are no other interactions between the two populations.

Free competition in closed environments

In addition to the foregoing assumptions, consider the growth of two different populations in an environment in which all the nutrients are initially in excess but where one of the nutrients is ultimately exhausted and is not immediately replenished. This may be the case, for example, in a soil environment in which there could be a localized increase in the concentration of all the major nutrients normally present at low, growth-limiting concentrations. The growth of the two populations in the unlimited environment is given by:

$$\frac{dx_A}{dt} = \mu_{mA} \cdot x_A \text{ and } \frac{dx_B}{dt} = \mu_{mB} \cdot x_B$$

where μ_{mA} and μ_{mB} are the maximum specific growth rates for A and B respectively. Hence:

$$\ln x_{At} - \ln x_{A0} = \mu_{mA} \cdot t \qquad (2.4.27)$$

and

$$\ln x_{Bt} - \ln x_{B0} = \mu_{mB} \cdot t \qquad (2.4.28)$$

Rearranging these two equations:

$$\mu_{mA} - \mu_{mB} = \frac{\ln R_t - \ln R_0}{t} \qquad (2.4.29)$$

where R_t is the ratio of population A to population B after a period of growth, t, and R_0 is the ratio of the two populations at the beginning of growth of the two populations. Thus the greater the difference between the maximum specific growth rates, the greater the rate of change of the ratio between the two populations (see Fig. 2.4.16).

The combined final population size is dictated by the concentration of the growth-limiting nutrient and so the faster growing organisms are able to utilize a greater proportion of the available limiting nutrient than the slow growing population. For a single growth cycle this competitive advantage may not be particularly significant since, after all, the slower growing population has increased in size. However, in the event of an influx of fresh nutrients, the next starting ratio of the two populations is different from that at the beginning of the first growth cycle and in favour of the fast growing population. Thus, the greater the difference between the two populations' maximum specific growth rates, the higher the chance the faster growing organism has of becoming the dominant population. Under these conditions the capacity for a higher maximum specific growth rate is a competitive advantage. This concept is relevant to the concept of the competitive saprophytic ability of an organism.

Free competition in open environments

Open growth systems usually simulate natural environments more precisely than do closed environments so the kinetics of competition under substrate-limited growth conditions may be particularly relevant to the process as it occurs under natural conditions. As shown previously,

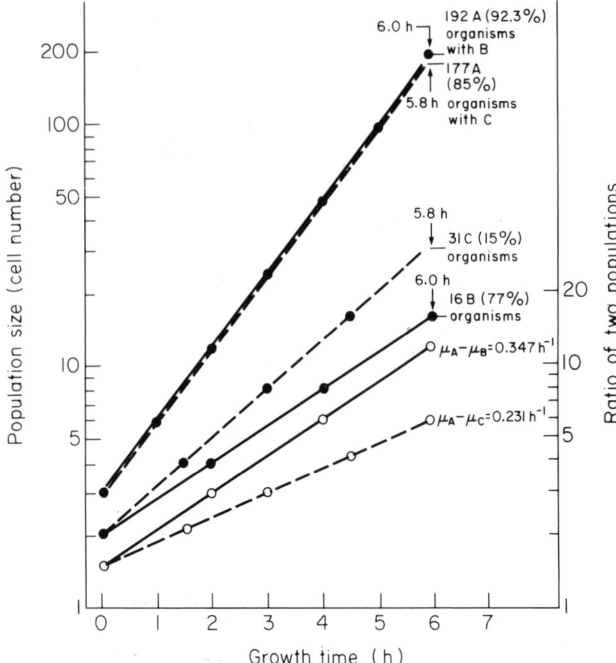

Fig. 2.4.16. Competition between two pairs of micro-organisms, *A* and *B* or *A* and *C*, in a closed culture system. For both mixed cultures the starting population sizes were 3:2, *A* to *B* or *A* to *C*. (●——●) growth curves for *A* and *B* together where $\mu^a_{max} = 0.693\ h^{-1}$ and $\mu^b_{max} = 0.347\ h^{-1}$, (○——○) logarithm of the ratio of population *A* to population *B*; (-----) growth curves for *A* and *C* together where $\mu^a_{max} = 0.659\ h^{-1}$ and $\mu^c_{max} = 0.462\ h^{-1}$; (○----○) logarithm of the ratio of population *A* to population *C*. After Slater & Bull [35].

substrate limitation restricts the growth rate of an organism so that the population able to sustain the highest growth rate has the competitive advantage. Furthermore, in open culture systems it is inevitable that those populations which are the least competitive are eliminated from the growth environment. In this case the second growth parameter, the saturation constant, usually becomes the most important factor determining the outcome of competitive growth.

Consider an organism *A* growing in a chemostat culture at a fixed dilution rate *D* and in a steady state at which the growth-limiting substrate concentration is *s*. So from equation (2.4.23) we have:

$$D = \frac{\mu_{mA} \cdot \tilde{s}}{(K_{sA} + \tilde{s})} \qquad (2.4.30)$$

Now, in the event of a second organism, *B*, being introduced into the growth vessel, initially as a minor population compared with population *A*, then during the first stages of growth of the two-membered culture, the growth of population *B* is given by:

$$\frac{dx_B}{dt} = (\mu_B - D)\, x_B$$

and so

$$\frac{dx_B}{dt} = \left[\frac{\mu_{mB} \cdot \tilde{s}}{(K_{sB} + \tilde{s})} - D\right] x_B \qquad (2.4.31)$$

There are three basic cases to consider in assessing whether or not the growth of population *B* is more or less competitive than that of the established population *A*.

1 For the new population *B* to succeed in ousting population *A*, dx_B/dt has to be positive, an eventuality which is achieved if $\mu B > D$. From equations (2.4.30) and (2.4.31):

$$\frac{\mu_{mB} \cdot \tilde{s}}{(K_{sB} + \tilde{s})} > \frac{\mu_{mA} \cdot \tilde{s}}{(K_{sA} + \tilde{s})}$$

This pertains if either $\mu_{mB} > \mu_{mA}$ (see Fig. 2.4.17a) or $K_{sB} < K_{sA}$ (see Fig. 2.4.17b). However, it must be noted that it is the combined effect of these which is important, determining whether or not organism *B* is more competitive than organism *A*. It is possible to envisage a situation in which $\mu_{mB} > \mu_{mA}$ but $K_{sB} > K_{sA}$ (see Fig. 2.4.17c). For this pair of organisms, at any growth-limiting substrate concentration organism *B* is more competitive, sustaining a higher growth rate than organism *A* at all substrate concentrations. Initially, the growth rate of organism *B* is determined by the steady state conditions established by organism *A*; that is, at a dilution rate of *D*, the growth-limiting substrate concentration is $\tilde{s}$. Gradually, as the proportion of the two populations begins to change in favour of population *B*, *s* begins to decrease and tend towards $\tilde{s}_1$ (see Figs. 2.4.17a and b) which is the growth-limiting substrate concentration which supports a growth rate of $\mu_B = D$. At this substrate concentration ds_A/dt must be negative and so population *A* is unable to grow at the imposed dilution rate and must continue to wash-out of the culture vessel.

2 Population *B* does not replace population *A* if $\mu_B < D$ and so dx_B/dt is negative, a situation which results if:

$$\frac{\mu_{mB} \cdot \tilde{s}}{(K_{sB} + \tilde{s})} < \frac{\mu_{mA} \cdot \tilde{s}}{(K_{sA} + \tilde{s})}$$

and happens if either $\mu_{mB} < \mu_{mA}$ (see Fig. 2.4.17d) or $K_{sB} > K_{sA}$ (see Fig. 2.4.17e). The reverse reasoning to

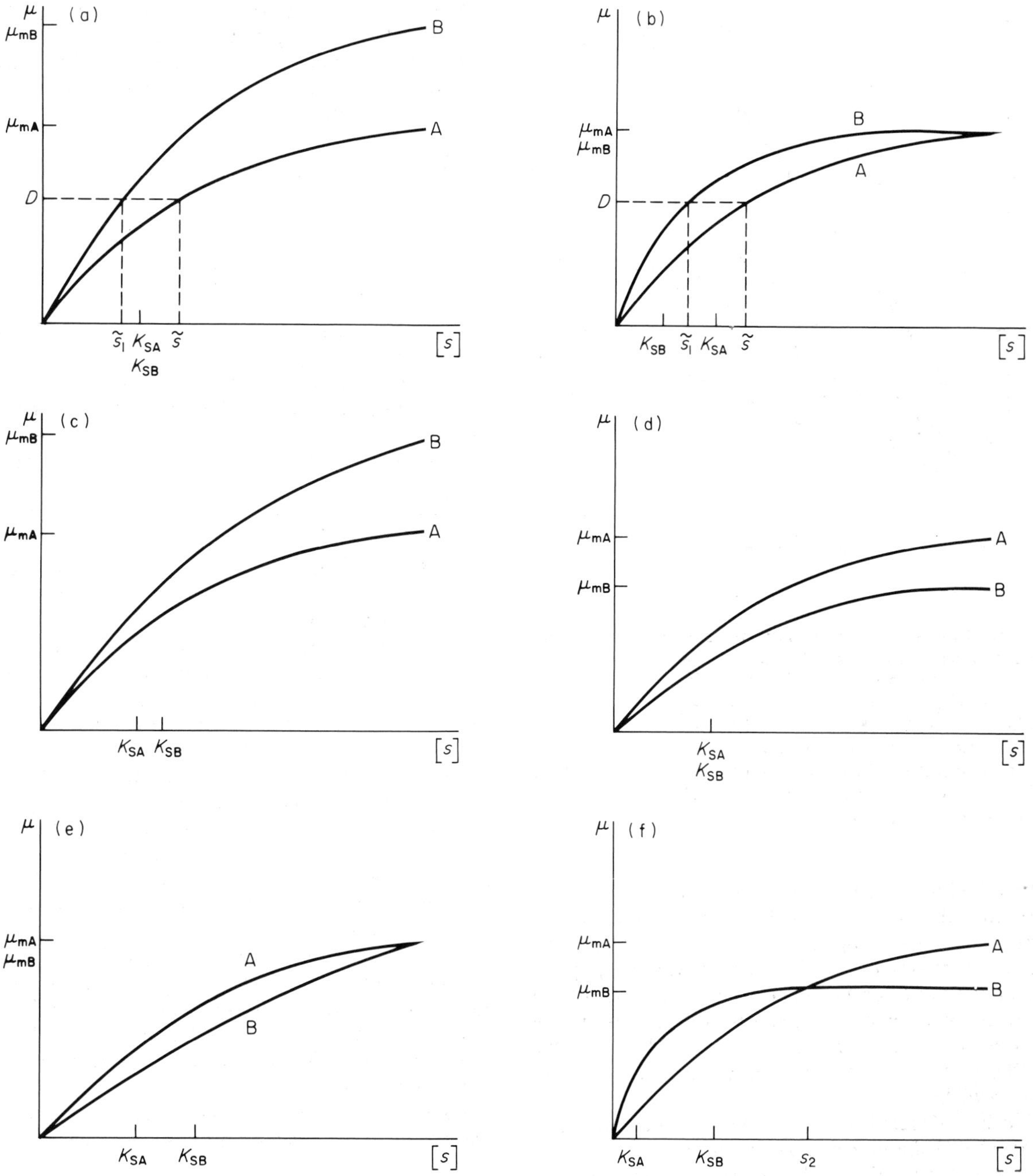

Fig. 2.4.17. The various possible Monod relationships between two organisms, *A* and *B*, used to predict the outcome of free competition between them under substrate limited growth. After Slater & Bull [35].

the first case applies, accounting for the non-competitive growth of population B.

3 It is theoretically possible that a stable mixture of two different populations may occur growing on a single growth-limiting substrate because

$$\frac{\mu_{mB} \cdot \tilde{s}}{(K_{sB} + \tilde{s})} = \frac{\mu_{mA} \cdot \tilde{s}}{(K_{sA} + \tilde{s})}$$

This is unlikely because it would mean that $\mu_{mA} = \mu_{mB}$ and $K_s\hat{A} = K_s\hat{B}$. What is more likely, however, is that at one particular substrate concentration the growth rates of the two organisms may be identical even though $\mu_{mA} \neq \mu_{mB}$ and $K_{sA} \neq K_{sB}$ (see Fig. 2.4.17f). For such a special case to occur, the two organisms must exhibit crossover Monod kinetics. Thus, at substrate concentration below s_2, organism B has the growth rate advantage basically because it has the greater affinity for the limiting substrate. Alternatively, at substrate concentrations above s_2, organism A is able to grow more rapidly than organism B. Even so, a stable mixed culture is not likely to be established even in the most carefully controlled growth systems because of difficulties in maintaining a constant dilution rate.

2.4.5 THE KINETIC PRINCIPLES OF PREY–PREDATOR RELATIONSHIPS

Predation in closed environments

Lotka [20] and Volterra [43] separately formulated the earliest models describing prey–predator dynamics and these have become known as the Lotka–Volterra equations. Originally they were derived assuming that the prey population, on its own, increased in a manner which could be described by the logistic equation (see Section 2.4.2). However, it is convenient to consider first the simpler case of prey–predator dynamics in an unlimited environment where the only restriction on the development of the prey population is the rate of predation. Prey–predator interactions are important relationships since they represent a simple food chain and so the initial stages in developing a trophic structure. The following description does not fully explore all the mathematical concepts which have been established; these may be found elsewhere [10].

Prey–predator dynamics in an unlimited environment

It is necessary to consider two basic equations which describe the development of the two populations.

The rate of change of the prey population	=	The rate of prey biomass production (growth)	−	The rate of prey biomass removal (predation)
$\frac{dH}{dt}$	=	μH	−	fP

$$f = \beta H \qquad (2.4.32)$$

where β is a complex constant incorporating a growth yield term. In other words the rate at which the prey is consumed is directly proportional to the product of the prey and predator concentrations. Hence:

$$\frac{dH}{dt} = \mu H - \beta HP \qquad (2.4.33)$$

The rate of change of predator population	=	The rate of predator biomass loss (death)	+	The rate of predator biomass production (predation)
$\frac{dP}{dt}$	=	$-dP$	+	$f'P$

where d is the *specific death rate* of the predator and f' is the specific growth rate of the predator. The f and f' are closely related since the growth of the predator must depend on the rate at which it can consume its prey and it can again be assumed that the predator's growth rate depends on the concentration of the prey. Thus:

$$f' = \beta' H \qquad (2.4.34)$$

where β' is another complex constant which includes the growth yield term and so:

$$\frac{dP}{dt} = -dP + \beta' HP \qquad (2.4.35)$$

Equations (2.4.33) and (2.4.35) describe completely the related change of the two populations and may be solved simultaneously by a simple computation. The numerical solution shows that both populations oscillate in an undamped fashion with constant amplitudes and that the predator population lags behind the change in the prey population size (see Fig. 2.4.18). Indeed a more extensive mathematical analysis shows that the two populations are exactly one-quarter out of phase with each other.

Now from equations (2.4.33) and (2.4.35) we have:

$$\frac{dH}{dt} = 0 \text{ when } P = \frac{\mu}{\beta}$$

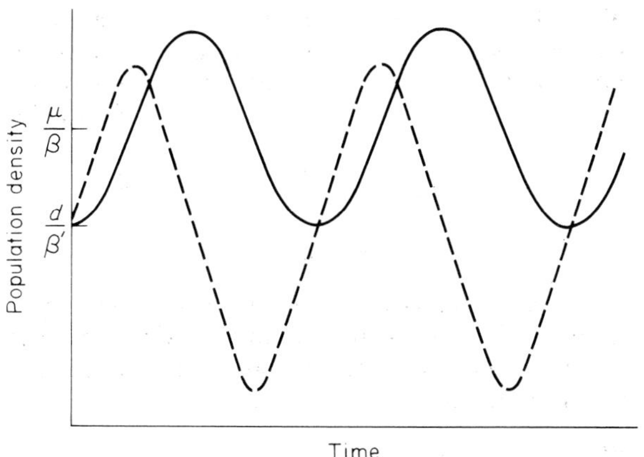

Fig. 2.4.18. Prey–predator dynamics in a closed, unlimited environment. Prey concentration (---) and predator concentration (——).

and

$$\frac{dP}{dt} = 0 \text{ when } H = \frac{d}{\beta'}$$

It is often convenient to represent the changes in the two population sizes by eliminating the time element and constructing a phase space (or phase plane) diagram. At any point in time the state of the mixed culture can be described by coordinates of H and P (see Fig. 2.4.19). The values for H and P are obtained by dividing equation (2.4.33) by equation (2.4.35) giving:

$$\frac{dH}{dP} = \frac{H\,(\mu - \beta P)}{P\,(-d + \beta' H)}$$

which has the solution:

$$\beta' H - d\log H = \mu\log P - \beta P + c \qquad (2.4.36)$$

where c is a constant of integration and has a value which depends on the starting concentrations of the prey and the predator. Thus it is an important feature of the Lotka–Volterra set of equations that they describe a family of closed circles which are dependent upon the initial conditions. That is, a culture initiated close to its theoretical steady state has small amplitude oscillations whereas the further away P is from μ/β and/or H is from d/β', the greater is the amplitude of the oscillations.

Prey–predator dynamics in a limited environment

Under these conditions the original Lotka–Volterra

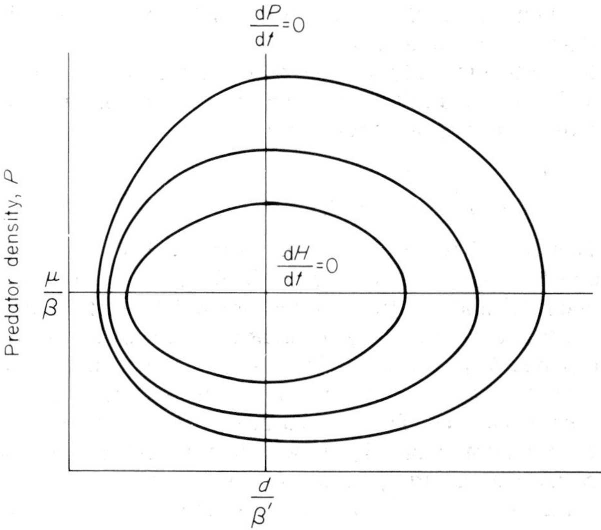

Fig. 2.4.19. Phase space diagram for prey–predator dynamics in a closed, unlimited environment using the Lotka–Volterra equations (2.4.33) and (2.4.35). The family of closed circles indicates that the amplitude of the population oscillations depends on the starting concentrations.

equations incorporated the logistic self-limitation function (equation 2.4.7), restricting the rate of increase of the prey populations, so that equation (2.4.33) may be written as:

$$\frac{dH}{dt} = \mu H - \mu\frac{H^2}{H_f} - \beta HP \qquad (2.4.37)$$

where H_f is the maximum prey population size determined by the carrying capacity of the growth environment. The predator population is described by the same function as before (equation 2.4.35) since this population continues to be influenced, according to this model, only by the concentration of the prey.

Therefore, as in the previous section:

$$\frac{dP}{dt} = 0 \text{ when } H = \frac{d}{\beta'}$$

but now $dH/dt = 0$ as the result of the combined influence of the rate of predation (and hence the size of the predator population) and its own self-limitation. So:

$$\frac{dH}{dt} = 0 \text{ when } P = \left[\frac{\mu}{\beta} - \frac{\mu H}{\beta H_f}\right]$$

Two extreme cases must now be considered, both of which occur when $\mathrm{d}H/\mathrm{d}t = 0$ (see Fig. 2.4.20). Firstly, if the prey population is small, then there is little self-limitation and the term $\mu H/\beta H_f$ becomes negligible. Thus:

$$\frac{\mathrm{d}H}{\mathrm{d}t} = 0 \text{ when } P = \frac{\mu}{\beta}$$

and the size of the predator population is the only factor determining the rate of change of the prey population. On the other hand, if P is small, predation is not noticeable but $\mathrm{d}H/\mathrm{d}t$ can still be zero because the prey's growth rate is dominated by its own limitation. Thus:

$$\frac{\mathrm{d}H}{\mathrm{d}t} = 0 \text{ when } H = H_f$$

There are two major conclusions which can be drawn from this model. First, if $H_f < d/\beta'$ then even the maximum prey population size is unable to support the predator population. Second, these equations mean that the mixed culture tends towards a stable steady state at which $\mathrm{d}H/\mathrm{d}t - \mathrm{d}P/\mathrm{d}t = 0$. As this is achieved, both populations oscillate with gradually decreasing amplitudes (damped oscillations). This is illustrated in Fig. 2.4.20 as a spiral path converging on the steady state point.

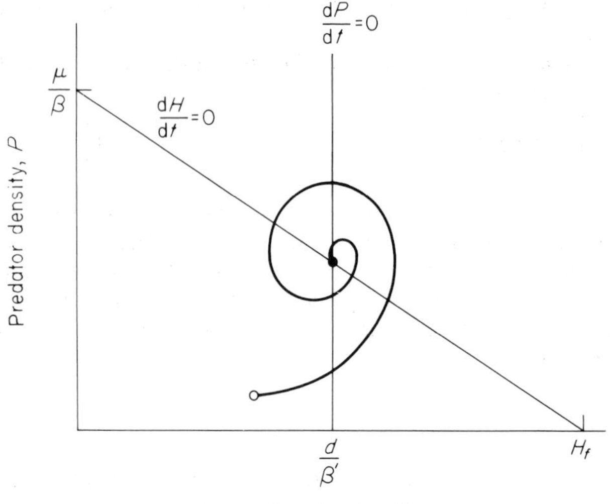

Fig. 2.4.20. Phase space diagram for pre-predator dynamics in a closed, limited environment using the Lotka–Volterra equations (2.4.35) and (2.4.37). In this case for $\mathrm{d}H/\mathrm{d}t = 0$ depends on both the predator population and the carrying capacity of the growth environment for the prey population. ○ = starting ratio of the prey and predator populations. ● = final, stable population sizes achieved from whatever starting conditions. The spiral line indicates the path taken, through a number of damped oscillations, to obtain the steady state populations.

Predation in open environments

Prey–predator relationships have been most successfully analysed in continuous-flow culture systems [10, 11, 40] and it is possible to develop more elaborate, and accurate, models describing the pattern of growth of the two populations by including the Monod relationship (equation 2.4.9). In the first place, this term can be included to describe the growth rate limitation substrate for the growth of the prey population. Furthermore, the prey is itself a limited substrate for the growth of the predator population, whose dynamics may therefore be modified in a similar way. Finally, it must be remembered that open growth systems effectively have a non-specific death rate term due to culture washout which affects both populations and has to be included in the model. Thus:

The rate of change of the predator population	=	The rate of predator biomass production (growth)	−	The rate of loss from culture vessel (washout)

$$\frac{\mathrm{d}P}{\mathrm{d}t} = \left[\frac{\mu_{\max P} \cdot H}{(K_{sP} + H)}\right] P - DP \qquad (2.4.38)$$

where $\mu_{\max P}$ is the maximum specific growth rate of the predator in the presence of saturating prey and K_{sP} is the predator's saturation constant for the prey. Equation (2.4.38) differs from the equivalent equation (2.4.35) for closed culture growth in this respect. In continuous culture there is always a continuous nutrient supply but this may not be rapid enough to maintain $\mathrm{d}P/\mathrm{d}t$ as a positive term. Now, however, the decline in the predator population is due to washout rather than death.

The rate of change of the prey population	=	The rate of prey biomass production (growth)	−	The rate of culture loss (washout)	−	The rate of prey biomass utilization by the predator (predation)

$$\frac{\mathrm{d}H}{\mathrm{d}t} = \left[\frac{\mu_{\max H} \cdot s}{(K_{sH} + s)}\right] H - DH - \beta HP \qquad (2.4.39)$$

where $\mu_{\max H}$ is the maximum specific growth rate of the prey, K_{sH} is the saturation constant of the prey and s is the growth-limiting substrate concentration.

Now equation (2.4.39) is analogous to equation (2.4.22) and so:

1 The rate of supply of the growth-limiting nutrient, DS_R, is equivalent to:

$$\left[\frac{\mu_{\max H} \cdot s}{(K_{sH} + s)}\right] H$$

2 The rate of unused growth-limiting substrate removal by culture washout, Ds, is equivalent to DH.

3 More importantly, the rate of growth-limiting substrate utilization, βHP, is equivalent to $\mu x/Y$. Thus (and it can be shown in other ways) βHP may be replaced in equation (2.4.39) by:

$$\left[\frac{\mu_{\max} \cdot H}{(K_{sP} + H)}\right] \frac{P}{Y}$$

where Y is the growth yield term for the predator with the prey as the substrate (units: g predator biomass (g prey biomass)$^{-1}$, see equation 2.4.11). Hence:

$$\frac{dH}{dt} = \left[\frac{\mu_{\max H} \cdot s}{(K_{sH} + s)}\right] H - DH - \left[\frac{\mu_{\max P} \cdot H}{(K_{sP} + H)}\right] \frac{P}{Y} \qquad (2.4.40)$$

Equations (2.4.38) and (2.4.40) again describe oscillating population sizes, one-quarter out of phase with the predator lagging behind the prey population but, in contrast to the Lotka–Volterra equations, this model does not depend on the starting population sizes. That is, for a given dilution rate, a stable, oscillating culture is established with constant and characteristic amplitudes which are independent of the previous history of the culture.

2.4.6 CONCLUSIONS

This chapter has sought to outline the basic principles of microbial production and community growth. These concepts form the basis of one of the more important developing areas of microbial ecology, namely, the formulation of reliable mathematical models which describe the behaviour of microbial populations in nature. It is now becoming realistic to formulate models, sometimes in considerable and complex detail, in terms of meaningful biological parameters and to assess their validity both in the laboratory and in the natural environment.

Indeed it is perhaps inevitable that microbial ecologists will progress from a predominantly qualitative, descriptive era into a primarily quantitative phase. This development, however, needs to be tempered with caution since a purely mathematical analysis of a particular system achieves little in furthering our understanding of why a particular population behaves in a particular way in a given environment (see Section 3.1.5). It is of paramount importance to be able to assign biological relevance to the various terms included in the mathematical analysis.

2.4.7 REFERENCES

[1] Bazin M.J. (1981) Mixed culture kinetics. In *Mixed Culture Fermentations*, p.25 (ed. M. E. Bushell & J.H. Slater). Academic Press, London.

[2] Brown C.M., MacDonald-Brown D.S. & Meers J.L. (1974) Physiological aspects of microbial inorganic nitrogen metabolism. *Advan. Microb. Physiol.* **11**, 1–52.

[3] Bull A.T. (1974) Microbial growth. In *Companion to Biochemistry*, p.415 (ed. A.T. Bull, J.R. Lagnado, J.O. Thomas & K.F. Tipton). Longmans, London.

[4] Bull A.T., Bushell M.E., Mason T.G. & Slater J.H. (1975) Growth of filamentous fungi batch culture: a comparison of the Monod and logistic models. *Proc. Soc. Gen. Microbiol.* **3**, 62–3.

[5] Bull A.T. & Slater J.H. (1982) *Microbial Interactions and Communities*, Vol. 1. Academic Press, London.

[6] Burns R.G. & Slater J.H. (eds.) (1982) *Experimental Microbial Ecology*. Blackwell Scientific Publications, Oxford.

[7] Bushell M.E. & Slater J.H. (eds) (1981) *Mixed Culture Fermentations*. Academic Press, London.

[8] Carter B.L.A. & Bull A.T. (1969) Studies of fungal growth and intermediary carbon metabolism under steady and non-steady state conditions. *Biotechnol. Bioeng.* **11**, 785–804.

[9] Clarke P.H. & Lilly M.D. (1969) The regulation of enzyme synthesis during growth. In *Microbial Growth*, p.113 (ed. P.M. Meadow & S.J. Pirt). Cambridge University Press, London.

[10] Curds C.R. & Bazin M.J. (1976) Protozoan predation in batch continuous culture. In *Advances in Aquatic Microbiology*, p.115 (ed. M.R. Droop). Academic Press, London.

[11] Dent V.E., Bazin M.J. & Saunders P.T. (1976) Behaviour of *Dictyostelium discoideum*, amoeba and *Escherichia coli* grown together in chemostat culture. *Arch. Microbiol.* **109**, 187–94.

[12] Dijkhuizen L. & Harder W. (1975) Substrate inhibition in *Pseudomonas oxalaticus*, OX1: a formate using extended cultures. *Antonie van Leeuwenhoek* **41**, 135–46.

[13] Edwards V.H. (1970) The influence of high substrate concentrations on microbial kinetics. *Biotechnol. Bioeng.* **12**, 696–712.

[14] Fredrickson A.G. (1977) Behaviour of mixed cultures of microorganisms. *Annu. Rev. Microbiol.* **31**, 63–87.
[15] Gause G.F. (1934) *The Struggle for Existence*. Williams & Wilkins, Baltimore.
[16] Harder W., Kuenen J.G. & Matin A. (1977) A review: Microbial selection to continuous culture. *J. Appl.* Bact. **43**, 1–24.
[17] Herbert D. (1961) The chemical composition of microorganisms as a function of their environment. In *Microbial Reaction to the Environment*, p.391 (ed. G.G. Meynell & H. Gooder). Cambridge University Press, London.
[18] Jones J.G. (1979) *A Guide to Methods for Estimating Microbial Numbers and Biomass in Fresh Water*. Freshwater Biological Association Scientific Publications No. 39, Ambleside, Cumbria.
[19] Karagouni A.D. & Slater J.H. (1978) Growth of the blue-green alga *Anacystis nidulans* during washout from light and carbon dixoide-limited chemostats. *FEMS Microbiol. Lett.* **4**, 295–9.
[20] Lotka A.J. (1925) *Elements of Physical Biology*. Williams & Wilkins, Baltimore.
[21] Mandelstam J., McQuillen K. & Davies I. (1982) *Biochemistry of Bacterial Growth*. Blackwell Scientific Publications, Oxford.
[22] Marr A.G., Painter P.R. & Nilson A.M. (1969) Growth and division of individual bacteria. In *Microbial Growth* (ed. S.J. Pirt & P.M. Meadow). Cambridge University Press, Cambridge.
[23] Meadow P.M. & Pirt S.J. (eds.) (1969) *Microbial Growth*. Cambridge University Press, London.
[24] Meynell G.G. & Meynell E. (1970) *Theory and Practice in Experimental Bacteriology*. Cambridge University Press, London.
[25] Monod J. (1942) *Recherches sur la croissance des cultures bactériennes*. Hermann, Paris.
[26] Monod J. (1949) The growth of bacterial cultures. *A. Rev. Microbiol.* **3**, 371–94.
[27] Monod J. (1950) La technique de culture continue: théorie et applications. *Ann. Inst. Pasteur* **79**, 390–410.
[28] Novick A. & Szilard L. (1950) Experiments with the chemostat on spontaneous mutants of bacteria. *Proc. Nat. Acad. Sci.* (*USA*) **36**, 708–18.
[29] Ollis D.F. (1982) Industrial fermentations with (unstable) recombinant cultures. *Philos. Trans. Roy. Soc. London, Ser. B.* **297**, 617–29.
[30] Pearl R. & Reed L.J. (1920) On the rate of growth of the population of the United States since 1790 and its mathematical representation. *Proc Nat. Acad. Sci. (USA)* **6**, 275–88.
[31] Pirt S.J. (1975) *Principles of Microbe and Cell Cultivation*. Blackwell Scientific Publications, Oxford.
[32] Powell E.O. (1958) Criteria for growth of contaminants and mutants in continuous culture. *J. Gen. Microbiol.* **18**, 249–68.
[33] Senior E., Bull A.T. & Slater J.H. (1976) Enzyme evolution in a microbial community growing on the herbicide Dalapon. *Nature* **263**, 476–9.
[34] Slater J.H. (1978) The role of microbial communities in nature. In *The Oil Industry and Microbial Ecosystems*, p.137 (ed. H.J. Sommerville & K.F. Chater). Heyden, London.
[35] Slater J.H. & Bull A.T. (1978) Interactions between microbial populations. In *Companion to Microbiology*, p.181 (ed. A.T. Bull & P.M. Meadow). Longmans, London.
[36] Slater J.H. & Hardman J.D. (1982) Microbial ecology in the laboratory: experimental systems. In *Experimental Microbial Ecology*, p.225 (ed. R.G. Burns & J.H. Slater). Blackwell Scientific Publications, Oxford.
[37] Stanier R.Y., Doudoroff M. & Adelberg E.A. (1966) *General Microbiology*, 2nd edn. Macmillan, London.
[38] Stouthamer A.H. (1977) Energetic aspects of the growth of microorganisms. In *Microbial Energetics,* p.285 (ed. B.A. Haddock and W.A. Hamilton). Cambridge University Press, Cambridge.
[39] Trinci A.P.J. (1971) Influence of the width of the peripheral growth zone on the radial growth rate of fungal colonies on solid medium. *J. Gen. Microbiol.* **67**, 325–44.
[40] Tsuchiya H.M., Drake J.F., Jost J.L. & Fredrickson A.G. (1972) Prey–predator interactions of *Dictyostelium discoideum* and *Escherichia coli* in continuous culture. *J. Bact.* **110**, 1147–53.
[41] Veldkamp H. & Jannasch H.W. (1972) Mixed culture studies with the chemostat. *J. Appl. Chem. Biotechnol.* **22**, 105–23.
[42] Verhulst P.F. (1839) Notice sur la que la population suit dans son accroissement. *Corr. Math. et Phys. Pubr. para A* **10**, 113–21.
[43] Volterra V. (1926) Variazione e fluttuazini del numero d'individui in specie animali conviventi. *Mem. Accad. Nazionale Lincei* **2**, 31–113.
[44] Williamson D.H. & Scopes A.W. (1961) Synchronization of division in cultures of *Saccharomyces cerevisiae* by control of the environment. In *Microbial Reaction to the Environment* (ed. G.G. Meynell & M. Gooder). Cambridge University Press, Cambridge.

2.5 Microbial energetics and metabolism

The importance of micro-organisms in the natural environment lies in their ubiquity, their diversity and above all their activity. In microbial ecology what is important is not so much the organisms themselves but the chemical and physical changes in the environment resulting from their nutritional needs and biochemical activities. Any study in microbial ecology, whether it be of nutrient cycles, polluted estuaries, diseased patients or rotting fruit, will inevitably lead to a consideration of the biochemical capabilities of the micro-organism involved.

It is against the background of such a view of microbial ecology that this chapter will examine the biochemistry of micro-organisms as a basis for comprehending the intricacies of their ecology. The integration and balance of metabolism will be considered with particular reference to energetics and with an emphasis on the nature of the cell's interaction with its environment.

2.5.1 CLASSIFICATION ACCORDING TO NUTRITION

Organisms can be classified on the basis of their sources or energy, carbon and reducing equivalents. *Phototrophs* obtain their energy directly from the sun by photophosphorylation, while the *chemotrophs* oxidize organic or inorganic materials by substrate and oxidative phosphorylation. *Autotrophs* synthesize their cell carbon from simple compounds such as CO_2, whereas *heterotrophs* require fixed organic sources of carbon. *Lithotrophs* produce the reducing equivalents required for cell synthesis from inorganic materials such as H_2S and ferrous iron, but *organotrophs* can only obtain their reducing power from the oxidation of organic nutrients.

Such a classification is particularly useful here as it bypasses the concept of species and genus and concentrates rather on the chemical activities of the organisms. Like most taxonomic schemes, however, there are limits to its usefulness. These limits are most often overstepped when too rigid a definition of a term is proposed. For example, the terms autotrophy and lithotrophy have often been a source of confusion rather than of clarification because the distinction between inorganic sources of carbon and of reducing equivalents is seldom made; the situation is aggravated by attempting to simplify the description of an organism's characteristics by compound words such as 'chemoautotroph'. In defining autotrophy itself, authorities differ in whether they accept CO_2 as the 'sole' or merely as the 'main' source of carbon. More rigorous definitions have sought to identify possession of the enzymes ribulose-1,5-bisphosphate carboxlyase and ribulose-5-phosphate kinase of the Calvin cycle as being the distinguishing characteristic of autotrophy. On the other hand, the more pragmatic approach of Whittenbury and Kelly [26] has led to the suggestion that autotrophy should be broadened in meaning to include all organisms capable of obtaining the bulk of their cell carbon from CO_2 or other one carbon (C_1) compounds such as formate or methane. A further possibility would be to set up three groupings of organisms dependent upon their carbon nutrition: heterotroph, autotroph and methylotroph (see Section 2.5.4). These suggestions derive from a consideration of the metabolic problems, primarily energetic, faced by such organisms, and the mechanisms whereby they are overcome (see Section 2.5.7).

Dynamic flow of cell metabolism

The real value of a nutritional classification lies in the focus it brings to the three main components of the dynamic flow of cell metabolism: energy, carbon and reducing equivalents. It is the appreciation of these three that makes the detailed mechanisms of catabolism and anabolism, their integration, and the overall control of cell function and growth most readily understandable. A diagrammatic representation of the bare bones of cell metabolism is given in Fig. 2.5.1.

Heterotrophic micro-organisms, such as protozoa, fungi and the majority of bacteria, are capable of utilizing a vast array of organic compounds as a source of carbon.

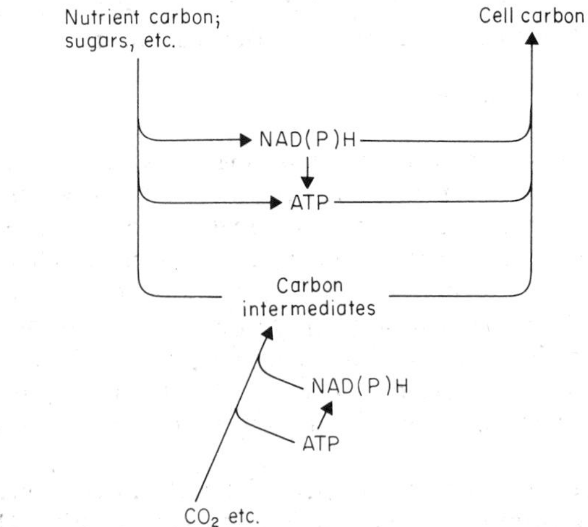

Fig. 2.5.1. Cell metabolism.

These vary from simple sugars, alcohols, acids and amino acids to hydrocarbons and complex carbohydrate, protein and lipid polymers. The processes of catabolism convert these potential nutrients to a relatively few metabolic intermediates which are common to all living tissues. These intermediates are compounds such as pyruvate, phosphoenol pyruvate, acetyl coenzyme A, and α-oxoglutarate which are also the building blocks for the synthesis of the cell polymers. Like the intermediates, the anabolic pathways of cell synthesis are essentially the same in all organisms, whether they have phototrophic, autotrophic or other modes of nutrition.

Since these metabolic intermediates are in general more oxidized than either the nutrients or the cell polymers, the processes of catabolism are predominantly oxidative whereas those of anabolism are reductive. This results in the production of reducing equivalents in catabolism and their utilization in anabolism. These reducing equivalents are generally in the form of reduced pyrimidine nucleotides, either nicotinamide adenine dinucleotide (NAD) or nicotinamide adenine dinucleotide phosphate (NADP), and are shown on Fig. 2.5.1 as NAD(P)H. The oxidative reactions of catabolism are also associated with the release of free energy (exergonic reactions), some of which is captured by the cell through the process of substrate level phosphorylation and stored in a biologically useful form as adenosine triphosphate (ATP). In a complementary fashion, the reductive anabolic reactions and the energy requirement of polymerization utilize NAD(P)H and ATP in cell synthesis.

Those organisms which possess an electron transport chain, or terminal respiratory system as it is sometimes called, are able to convert reducing equivalents, in the form of reduced pyrimidine nucleotides or flavoproteins, to energy, in the form of ATP. With heterotrophic organisms in the presence of oxygen or of an alternative electron acceptor such as nitrate, this capability is most usually expressed as the extra production of ATP by oxidative or respiratory chain-linked phosphorylation.

For the autotrophs, the nutrient carbon as CO_2 is more oxidized than the metabolic intermediates. The reactions leading to the synthesis of these intermediates are therefore anabolic in character and both reductive and energy-requiring (endergonic). In the eukaryotic algae and the cyanobacteria, both ATP and NAD(P)H are produced by non-cyclic photophosphorylation. Bacterial phototrophs, however, have a restricted capability in this regard and can usually produce only ATP by a cyclic photophosphorylation mechanism. (These photophosphorylation systems are closely analogous in both components and mechanism to the respiratory chain-linked phosphorylation in that both have the same basic mechanism of electron transport.) In bacterial photosynthesis, reducing equivalents are obtained from inorganic reductants, such as reduced forms of sulphur, by ATP-driven reversed electron transport. This mechanism also operates in the chemolithotrophic bacteria but they can obtain their ATP from the same reductant through its oxidation linked by the respiratory chain to the reduction of oxygen or an alternative terminal electron acceptor.

The details of the individual reaction mechanisms can now be examined in order to obtain a complete, and hopefully clear, picture of cell function in biochemical terms and its influence on the relationships between the microbial cell and its environment. Such important functions as nitrogen fixation and the assimilation of nitrogen and sulphur into cell components which are omitted from Fig. 2.5.1 will also be dealt with in the discussion of cellular energetics.

2.5.2 MICROBIAL ENERGETICS

For the microbial ecologist, the study of cell physiology is very much the study of bioenergetics. This is true because both the chemical composition of microbial cells and the metabolic sequences involved in their syntheses are essentially the same as in all living tissues, although the range of organic and inorganic nutrients used by the micro-organisms greatly exceeds that found in the plant and animal Kingdoms. On the other hand, the basis for the diversity of chemical activities found among the micro-organisms, and for the extensive array of extreme environments in which they are to be found, lies in the unusual and ingenious methods they have of obtaining the energy to drive that metabolism. In this respect one can mention, for example, the phenomenon of anaerobic growth, in which various genera have developed particular mechanisms of energy generation: namely, photosynthesis, fermentation, and anaerobic respiration with electron acceptors alternative to oxygen.

The central importance of bioenergetic mechanisms and the fact noted earlier that electron transport systems are intimately linked with the interconversion of reducing equivalents and ATP lead to two decisions concerning this discussion of microbial biochemistry: (a) reducing equivalents and energy can be considered together as two sides of the same coin; and (b) the systems of electron and hydrogen transport associated with both respiration and photosynthesis must be considered first.

Oxidation/reduction reactions and potentials

As has already been noted, the reactions of the catabolic pathways are generally both oxidative and exergonic. In biological reactions, oxidation most usually takes the form of the removal of hydrogen or electrons, these being passed on to an acceptor, which is itself therefore reduced. In this way we can refer to the compound being oxidized as a hydrogen and/or electron donor, and the reaction sequence can be represented as:

$$AH_2 \searrow \swarrow B \qquad A \swarrow \searrow BH_2$$

where AH_2 and B are respectively the hydrogen donor and the hydrogen acceptor.

Such a representation stresses two important features of oxidation–reduction (or redox) reactions: each oxidation is accompanied by a reduction; and the two reactions are coupled through the transfer of reducing equivalents in the form of hydrogen or electrons.

Each redox couple such as AH_2/A has a finite tendency to either donate reducing equivalents and be oxidized ($AH_2 \rightarrow A$) or accept them and be reduced ($A \rightarrow AH_2$). When two couples are combined in a complete redox reaction, the net flow of the reaction is determined by the relative tendency of each couple to donate or accept reducing equivalents. This tendency, or potential, can be measured and quantified by comparison with a standard redox couple. The standard redox couple is that present at the hydrogen electrode where hydrogen gas is in contact with hydrogen ions (protons) in solution in the presence of platinum as a catalyst. The reaction is

$$H_2 \rightleftharpoons 2H^+ + 2e^-$$

and the tendency to donate reducing equivalents, in this case as electrons, is measured as the voltage or potential of the electrical current generated when the electrode is coupled in series with another redox couple electrode. At 25°C, 1 atmosphere of hydrogen and pH 0, the redox potential of the hydrogen electrode is zero. At pH 7 however, the potential of the redox couple $H_2/2H^+ + 2e^-$ is −420 mV. The symbol E_0' is used for the redox potential under the latter conditions which are referred to as standard.

A couple of low redox potential will always donate reducing equivalents to a couple of higher potential. The couple $O^{2-}/\frac{1}{2}O_2 + 2e^-$ has an E_0' of 820 mV so that in combination with the hydrogen electrode the complete redox reaction is given by:

$$H_2 + \tfrac{1}{2}O_2 = H_2O$$

with the hydrogen donating electrons and being itself oxidized, while the oxygen accepts the electron pair and is therefore reduced.

During the oxidation of a substrate, therefore, reducing equivalents are transferred in the direction of increasing redox potential. This transfer is accompanied by the release of free energy. The magnitude of the standard free energy change is given by the relationship:

$$\Delta G^{0\prime} = -nF\Delta E_0'$$

where $\Delta G^{0\prime}$ is the standard free energy change, n is the number of electrons transferred, F is the Faraday constant (96.649 kJ V^{-1} mol^{-1}) and $\Delta E_0'$ is expressed in V. The oxidation of hydrogen is therefore associated with a redox difference of 1.24 V which results in a free energy change of −240 kJ mol^{-1}.

In the case of the reaction of molecular hydrogen with oxygen this release of free energy may be violently explosive, but living tissues have evolved mechanisms to conserve this energy and transduce it into biologically useful forms. These mechanisms are of course hydrogen and electron transport, and electron transport-linked phosphorylation. An important property of these biological systems is their reversibility. With the appropriate input of energy, the electron transport chains can transfer reducing equivalents in the reverse direction from high to lower redox potential with the resultant production of NAD(P)H.

The terminal respiratory system

The electron transport chains associated with the oxidation of organic and inorganic substrates by microbial cells (and also found of course in the mitochondria of higher cells) are composed of a series of components, each of which can exist in the oxidized or reduced form. Each component therefore represents a redox couple characterized by a particular redox potential. Of these components, the pyridine nucleotides, flavoproteins and quinones are hydrogen carriers, while the ferredoxins, non-haem iron–sulphur proteins and cytochromes are electron carriers. The chemistry of these and their organization in the respiratory chain is dealt with in most biochemistry texts, but particularly lucidly in the monograph and articles by Jones [10, 11], Haddock & Jones [5], and Ingledew & Poole [9].

The standard potentials of a number of redox couples are shown in Table 2.5.1. In the heterotrophic catabolic sequences that are found in most tissues of both prokaryotic and eukaryotic organisms, the majority of the substrate dehydrogenation reactions have E_0' values in the region of −300 mV, and are capable therefore of donating their hydrogen and electron pair to NAD^+ or $NADP^+$. Certain reactions, however, such as the dehydrogenation of succinate to fumarate, have an E_0' around +20 mV and therefore donate their reducing equivalents to the flavoproteins. These are proteins with either flavin mononucleotide (FMN) or flavin adenine dinucleotide (FAD) as prosthetic group; these cofactors are redox carriers capable of accepting and donating hydrogens, and as a family the flavoproteins span a considerable range of redox potentials.

Once the reducing equivalents from the substrate dehydrogenation reaction have been transferred to an acceptor which is a component of the respiratory chain, they may then be transferred from carrier to carrier within the chain. The transfer is in the direction of increasing redox potential until the final reaction with oxygen catalysed by an oxidase. Oxygen is therefore acting as a terminal electron acceptor. In this manner a single multicomponent respiratory system can channel and control the ultimate reaction with oxygen of hydrogens removed from a myriad of substrates by the action of various specific dehydrogenases. In the eukaryotic protozoa, fungi and algae this activity is localized in the mitochondria, whereas in the more primitive prokaryotic bacteria and cyanobacteria the terminal respiratory system is obligatorily associated with the cytoplasmic cell membrane.

In all cell types the sequential organization and functioning of the components of the respiratory chain is very much as is indicated by their thermodynamically deter-

Table 2.5.1. The standard oxidation–reduction potentials of a number of redox couples of interest in biological systems

Redox couple	E_0' (mV)
$H_2/2H^+ + 2e^-$	−420
Ferredoxin red./oxid.	−410
$NADPH/NADP^+$	−324
$NADH/NAD^+$	−320
Flavoproteins red./oxid.	−300 to 0
Cyt. *b* red./oxid.	+30
Ubiquinone red./oxid.	+100
Cyt. *c* red./oxid.	+254
Cyt. a_3 red./oxid.	+385
$O^{2-}/\tfrac{1}{2}O_2 + 2e^-$	+820

mined ordering in Table 2.5.1. In detail, however, there is considerable variety in the identity of individual components and in the presence or absence of sections of the chain in particular organisms; this can even occur in the same organism under different growth conditions.

The actual organization of the components of the respiratory chain within the mitochondrial or the cell membrane has been the subject of considerable research. As was noted earlier, certain components of the respiratory chains are hydrogen carriers while others are electron carriers. Also it has been experimentally demonstrated that respiration is associated with the expulsion of protons from the active cell or organelle. These facts have led to the proposal that the hydrogen and electron carriers might function alternately in the operation of the chain and that they might be arranged structurally within the membrane so that the hydrogen carriers would transport hydrogen atoms outward across the membrane while the electron carriers would transport electrons back inwards. Such an organization of a hydrogen and an electron carrier is often referred to as a 'loop'. The result of this organization would be the ejection of protons into the surrounding medium. Such a scheme is illustrated in Fig. 2.5.2, with reference to *Escherichia coli*. Each proton-translocating loop in the electron transport chain is equated with a site of phosphorylation; site I is associated with the oxidation of NAD(P)H-linked substrates, and site II with the flavoprotein-linked substrates such as succinate. The possession of a site III (absent from *E. coli*) is associated with the presence of cytochrome *c*. In certain strains of *E. coli* it has been found that growth under conditions of sulphate limitation can lead to loss of the non-haem iron–sulphur protein, with a consequent loss of proton translocation and energy coupling with site I.

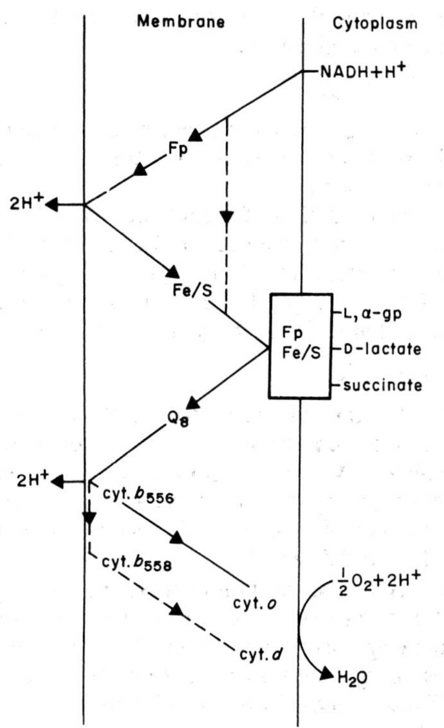

Fig. 2.5.2. Proposed functional organization of the redox carriers responsible for aerobic electron transport in *E. coli*. The scheme includes the various routes for aerobic electron transport in *E. coli*, with the dashed lines indicating alternative pathways for reducing equivalents. Abbreviations: L,α-gp, L-α-glycerophosphate; F_p, flavoprotein; Fe/S, non-haem iron–sulphur protein; Q_8, ubiquinone. Redrawn with permission from Haddock & Jones [5].

The mechanism of energy transduction

The significance of such a looped organization of alternate hydrogen and electron carriers is that it relates directly to the mechanism of energy transduction. The consensus of opinion is that the mechanism of oxidative phosphorylation can best be understood in terms of Mitchell's chemiosmotic hypothesis. Interestingly the hypothesis refers not only to phosphorylation but to all membrane-linked energy transductions, and the theory was originally developed from studies of nutrient and ion transport in bacterial systems. Particularly clear and concise accounts of chemiosmosis have been given by Harold [8] and by Garland [4].

The essential feature of the chemiosmotic hypothesis is that the free energy released during transfer of reducing equivalents 'down' the respiratory chain in the direction of increasing redox potential is conserved in the form of a transmembrane gradient of protons. This gradient is called the protonmotive force, or pmf, and it has both a chemical component (due to protons, the inside of the cell or organelle is alkaline with respect to the outside) and an electrical component (the inside is negative due to the transport of H^+ outwards and e^- inwards).

The electron transfer associated with photosynthesis in the chloroplasts of the eukaryotic algae and the cell membrane of the prokaryotic cyanobacteria and phototrophic bacteria generates an exactly analogous pmf. As with the respiratory pmf in eukaryotic mitochondria and in bacteria, the polarity is such that the inside of the cell is alkaline and negative for the cyanobacteria and the phototrophic bacteria; in chloroplasts, however, the structural organization of the electron transport components is reversed so that the flux of protons is directed inwards.

A pmf is also developed across the cytoplasmic membrane of fungal and algal cells and of bacteria that lack a respiratory chain or are being grown in the absence of a terminal electron acceptor. In these cases the ejection of protons is driven by an H^+-translocating ATPase (analogous to the Na^+/K^+-ATPase of higher cell cytoplasmic membranes). That is to say, all functional membranes have the capacity to develop a pmf, either through respiratory chain- or photosynthesis-linked electron transport or through the action of an H^+-ATPase.

Furthermore, each of these systems is reversible, as are the H^+-ATPases which are also located in the mitochondrial and chloroplast membranes. (The polarity of the ATPase is reversed in the chloroplast in parallel with that of the electron transport systems). Through the primary generation of a pmf, therefore, respiration can drive ATP synthesis in both eukaryotes and prokaryotes. The hydrolysis of ATP can also drive electron transport in reverse with the generation of NAD(P)H from an electron donor of higher redox potential. In addition, photosynthesizing systems can harness the radiant energy of the sun to the production of both ATP and NAD(P)H. Electron transfer reactions and the hydrolysis of ATP can each independently, again through the generation of a pmf, drive the transport of ions and other nutrients as well as the motility of the cell itself where this occurs.

It is pertinent to point out, however, that several concepts of 'classical' chemiosmosis are currently the subject of controversy, including: the existence of proton pumps rather than hydrogen and electron carrier loops; the stoichiometry of protons translocated per e^- transported or ATP hydrolysed; the status of the proposed equilibrium between protons within the membrane and those in the bulk aqueous phase.

None the less it is through the appreciation of electron transport chains and the protonmotive force that the important conclusion is reached that in biological systems reducing equivalents and energy must be considered as interchangeable.

Summary

In this section the factors determining the flow of energy and reducing equivalents associated with the catabolic pathways in aerobic heterotrophic organisms, be they animal, plant or micro-organism, have been identified and can now be summarized.

The breakdown of organic nutrients is generally oxidative and exergonic. This process begins when various specific dehydrogenases remove hydrogens from their substrates and donate them to one of a number of possible acceptors. Most often the acceptor is one of the pyrimidine nucleotides, NAD^+ or $NADP^+$, but the flavoproteins can also play the role of primary hydrogen acceptor. The reduced forms of these primary acceptors are the 'currency' in which reducing equivalents are transferred within the reactions of metabolism. The regeneration of $NAD(P)^+$ requires that NAD(P)H be reoxidized by one of two general mechanisms. In one mechanism the coenzyme donates its reducing equivalents through the reduction of a second organic substrate; examples of this are found in fermentation and in the biosynthetic sequences of anabolism. Alternatively, the reducing equivalents can be donated to the next carrier in the terminal respiratory chain with consequent transduction of the redox energy into ATP.

The redox couples $NADH/NAD^+$ and $NADPH/NADP^+$ are formally equivalent, having almost identical E_0' values, and most organisms possess a transhydrogenase which catalyses the transfer of reducing equivalents between the two. As a general rule of thumb, however, NADH is usually associated with catabolic reactions and the operation of the respiratory chain, while NADPH is the form in which reducing equivalents are donated in reductive anabolic reactions. The pyrimidine nucleotides are *coenzymes* which *couple* oxidation–reduction reactions by the mechanisms of *group-transfer*; the group in this case is reducing equivalents in the form of two hydrogen atoms.

In respiratory chain-linked phosphorylation, redox energy is converted, via the protonmotive force, into the free energy of hydrolysis of ATP, which is thus the principal energy 'currency' of the cell. The potential energy of ATP is a consequence of the whole molecular structure, and is not a feature exclusively of the so-called 'high-energy bond'. Phosphoenol pyruvate (PEP) and acetyl coenzyme A are examples of other high energy compounds important in biological systems. Although the energy of these compounds is measured thermodynamically in terms of their free energies of hydrolysis, they are in fact only seldom hydrolysed to, for example, ADP and Pi. Most commonly the donation of energy in a reaction occurs through the transfer of a phosphoryl group or perhaps of an acyl group. In parallel with the pyrimidine nucleotides, therefore, it is instructive to consider

ATP and other high energy compounds as *coenzymes* which *couple* hydro–dehydration reactions by the mechanism of *group transfer*; the groups in this case are phosphoryl, acyl, etc.

It is the central thesis of this and other [6] presentations that the principles and concepts elaborated here in the analysis of aerobic heterotrophic metabolism in a single organism are equally relevant to our understanding of the nutritional and energetic interactions between organisms that are the essential characteristic of microbial ecosystems.

2.5.3 ANAEROBIOSIS

The ability to grow non-phototrophically in the absence of oxygen is confined to certain bacterial, fungal and protozoal species. The essence of the problem facing such cells is that they must have some means, other than the oxygen-linked terminal respiratory system, of reoxidizing the NAD(P)H produced by catabolic reactions in excess of that required for anabolism. There are two general mechanisms, namely, fermentation and anaerobic respiration. These mechanisms have been the subject of detailed and comprehensive reviews by Thauer *et al.* [24] and by Thauer & Morris [25].

While certain species are classified as facultative anaerobes in that they are able to adapt their metabolism to grow either in the presence of oxygen or in its absence, other species are obligate anaerobes and can only grow in an oxygen-free (anoxic) environment. Such organisms have the additional problem that they are inhibited or even killed by oxygen [16].

Fermentation

Although the term fermentation is often used to describe various microbial processes, mainly industrial, in the present context it refers to the process in which reducing equivalents are directly transferred through the coupling of the oxidation of one substrate to the reduction of a second. The second substrate is in fact generally a product of the catabolic pathway leading from the oxidized substrate, and so such a fermentation pathway is internally balanced, with neither a net production of nor a net requirement for reducing equivalents. Fermentation mechanisms are found in yeasts and in a number of bacterial groups and genera including lactic acid and propionic acid bacteria, Enterobacteriaceae, *Staphylococcus, Bacillus* and *Clostridium*. All the *Clostridium* spp. are obligate anaerobes; they can only grow fermentatively and are sensitive to even very small quantities of oxygen. The propionic acid bacteria are also strict anaerobes, although they do exhibit some degree of oxygen tolerance and there is evidence that they may be able to operate a limited respiratory chain-linked phosphorylation. The other organisms listed above are all facultative anaerobes.

Although the metabolic capabilities of individual organisms are limited by the stricture imposed by the fermentative mechanism for the reoxidation of NAD(P)H, only alkanes and certain aromatic compounds are resistant to anaerobic microbial attack. The favoured substrates of fermentation processes, however, both in the laboratory and in nature, are sugars.

There are four pathways known for the catabolism of sugars. These are the Embden–Meyerhof or glycolytic pathway, the hexose monophosphate or pentose phosphate cycle, the Entner–Doudoroff pathway, and the phosphoketolase pathway (see Mandelstam *et al.* [15] and Stanier *et al.* [22]). The pentose phosphate cycle is only found in aerobic metabolism (although sections of the cycle are always required for the synthesis of ribose for the nucleic acids), but the other three degradative routes all operate in various of the fermentative organisms. These pathways catalyse the breakdown of glucose, and ultimately also the breakdown of other monosaccharides or oligo- and polysaccharides that are first converted to glucose itself or to one of the intermediates of the catabolic pathways. Each of the catabolic sequences has in common a terminal series of reactions leading from glyceraldehyde 3-phosphate to pyruvate, which is thus the principal end-product of all three pathways. At earlier steps in the case of the Entner–Doudoroff and phosphoketolase pathways, and at the glyceraldehyde 3-phosphate to pyruvate stage in all three pathways, reducing equivalents are produced in the form of reduced pyrimidine nucleotides, both NADH and NADPH. The fermentation balance therefore becomes dependent upon converting pyruvate to more reduced final products.

Fermentation products (see Fig. 2.5.3) are produced in amounts equivalent to the substrate consumed and often determine the reaction of that organism with its environment. Particular fermentation products are often used in the identification of organisms, e.g. the coliform bacteria (see Section 4.4.3). Many of the products, such as ethanol, are also of considerable economic importance.

In Fig. 2.5.3, 2H represents two hydrogen atoms being donated in a reductive step. In most cases the donor will be one of the reduced pyrimidine nucleotides, previously

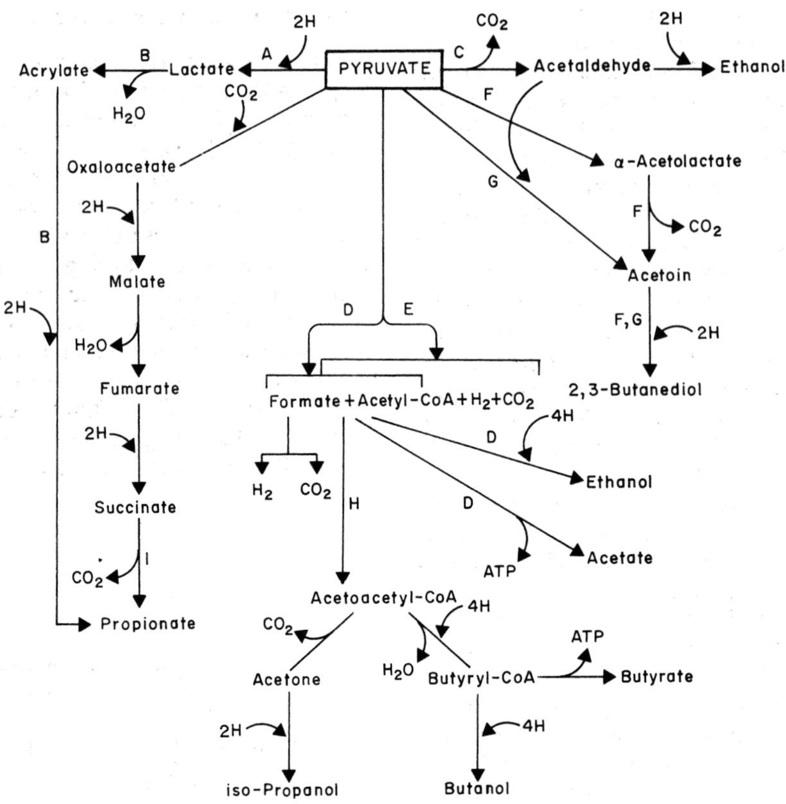

Fig. 2.5.3. Bacterial fermentation products of pyruvate. Pyruvate formed by the catabolism of glucose is further metabolized by pathways which are characteristic of particular organisms and which serve as a biochemical aid to identification. **A**, *Streptococcus* spp., *Lactobacillus* spp.; **B**, *Clostridium propionicum*; **C**, yeast, *Acetobacter* spp., *Zymomonas* spp., *Sarcina ventriculi, Erwinia amylovora*; **D**, Enterobacteriaceae; **E**, *Clostridium* spp.; **F**, *Klebsiella* spp.; **G**, yeast; **H**, *Clostridium* spp. (butyric, butylic organisms); **I**, propionic acid bacteria. Redrawn with permission from Mandelstam *et al.* [15].

represented as NAD(P)H. While most products are formed from pyruvate in one reductive step with the addition of a single pair of hydrogens, certain products and pathways require a greater degree of reduction. Examples of such a situation are the production from acetyl coenzyme A of ethanol by the Enterobacteriaceae and of butyrate and butanol by *Clostridium* spp., and the production from pyruvate of succinate by the Enterobacteriaceae and of propionate by *Clostridium* spp. and the propionic acid bacteria.

Members of the Enterobacteriaceae as well as *Clostridium* spp. are capable of forming acetyl coenzyme A by a reaction traditionally, but incorrectly, referred to as the phosphoroclastic cleavage. With the Enterobacteriaceae, the other product is formate, which, under acid conditions, may be further converted to hydrogen and CO_2. *Clostridium* spp., however, form hydrogen and CO_2 directly without formate as intermediate.

It should also be noted that the CO_2, incorporated at the step pyruvate → oxaloacetate and removed at the step succinate → propionate en route to the synthesis of propionate by the propionic acid bacteria, is not free gaseous CO_2. Rather it is held and transferred by an enzyme–biotin complex between the two reactions, which are thus coupled by the mechanism of group transfer. In the synthesis of succinate by the Enterobacteriaceae, on the other hand, the CO_2 is not bound to an enzyme–biotin complex, and the most likely route of carboxylation is not pyruvate → oxaloacetate but rather phosphoenol-pyruvate → oxaloacetate. Another enzyme catalysing a pyruvate/malate reaction most probably functions in the direction of decarboxylation in cells using intermediates of the tricarboxylic acid cycle as a source of carbon.

Organisms growing fermentatively according to the schemes being outlined here can only synthesize ATP from ADP by the mechanism of substrate level phosphorylation. One example of this involves two of the intermediates of glycolysis which are high energy compounds in the same sense as ATP. These compounds are 1, 3-diphosphoglycerate and phosphoenol-pyruvate and their conversion to, respectively, 3-phosphoglycerate and pyruvate is coupled by phosphoryl group transfer to the synthesis of ATP from ADP.

Other compounds, however, such as acetyl coenzyme A, are also high energy compounds and, in the conversion of acetyl coenzyme A to acetate and of butyryl coenzyme A to butyrate (see Fig. 2.5.3), this energy is transduced with the synthesis of ATP by substrate phosphorylation through the intermediate production of acetyl and butyryl phosphate.

These are the only energy-generating steps possible in the fermentation reactions from pyruvate. However, certain species of the Enterobacteriaceae, and possibly also the propionic acid bacteria, are able to carry out a limited respiratory chain-linked phosphorylation coupled to the reduction of fumarate to succinate (see p.86).

Organic acids, amino acids and other nitrogen-containing compounds can also serve as substrates for fermentation pathways; the products formed from such pathways are similar to those formed from sugars. The full extent of microbial fermentative capability is discussed by Thauer *et al.* [24].

The common features of all organisms growing fermentatively are: (a) the internal redox balance of each fermentation pathway; (b) the comparatively low yield of energy, and hence of growth, for the amount of nutrient consumed; and (c) the formation and excretion into the surrounding medium of reduced fermentation products, often in amounts equimolar with the substrates being fermented.

Two mechanisms, discussed below, are responsible for variation in the pattern of fermentation pathways and products. At the root of both mechanisms is the need for the cell to balance its production and its requirement for energy and reducing equivalents.

Branched fermentation pathways

Where a fermentation pathway is branched, as for example in the production of acetate and butyrate from the saccharolytic fermentation of *Clostridium pasteurianum* (see Fig. 2.5.4), the two branches may differ in their net balance of ATP and NAD(P)H. If acetate were the sole fermentation product the net yield would be two ATPs per two pyruvates, which gives a total of four ATPs per glucose fermented, allowing for the two ATPs per glucose from glycolysis. At the same time, the two pairs of reducing equivalents produced at the glyceraldehyde 3-phosphate dehydrogenase and the so-called phosphoroclastic cleavage would be converted to four molecules of hydrogen via ferredoxin. On the other hand, if fermentation was exclusively geared to the production of butyrate, the ATP yield would be reduced to three per glucose.

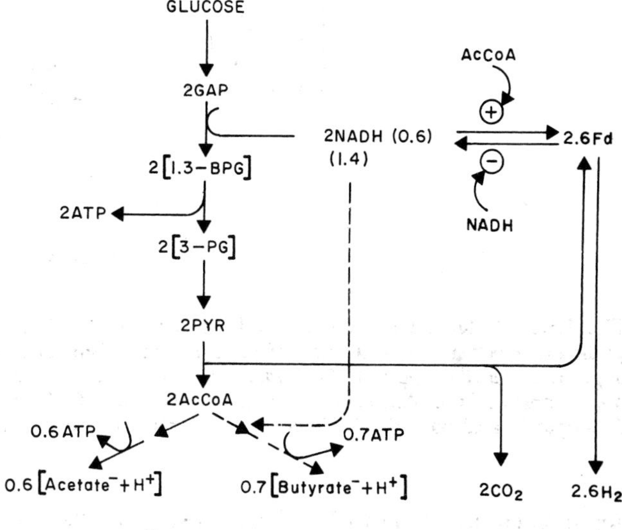

Fig. 2.5.4. Variable thermodynamic efficiency in a branched glycolytic fermentation (*Clostridium pasteurianum*). The ATP gain is not stoichiometrically coupled to glucose catabolism; it may be shifted between 3 ATP/glucose, when glucose is fermented solely to butyrate (NADH: ferredoxin oxidoreductase inactive), and 4 ATP/glucose, when glucose is catabolized solely to acetate (NADH: ferredoxin oxidoreductase fully active). The observed ATP gain is 3.3 ATP/glucose. Abbreviations: GAP, glyceraldehydephosphate; 1,3-BPG, 1,3-biphosphoglycerate; 3-PG, 3-phosphoglycerate; PYR, pyruvate; AcCoA, acetyl coenzyme A; Fd, ferredoxin. Redrawn with permission from Thauer *et al.* [24].

Two molecules only of hydrogen would be evolved since the NADH produced in glycolysis would be reoxidized in the reduction of acetoacetyl coenzyme A to butyryl coenzyme A. In fact the organism maintains a balance between these two routes, largely through control of NADH : ferredoxin oxidoreductase.

Fermentation in mixed cultures

A particularly significant variation in fermentation pathways, of considerable importance in ecological terms, arises from the interactions between micro-organisms in mixed cultures. For example, *Ruminococcus albus* ferments glucose to acetate, ethanol, CO_2 and hydrogen in a scheme analogous to that shown in Fig. 2.5.4. In a mixed culture with *Vibrio succinogenes*, however, ethanol is no longer found as a fermentation product, while the amount of acetate produced increases with a consequent increase in the ATP yield from fermentation. As with *Cl. pasteurianum*, the mechanism of this effect is that all the reducing equivalents produced in the *R. albus* fermentation are converted to molecular hydrogen. This reaction, however, is only possible when the partial pressure of hydrogen is maintained at a very low level, which is achieved in this mixed culture by *V. succinogenes* using the hydrogen as a source of energy through an electron transport-linked phosphorylation associated with the reduction of fumarate to succinate (see p.86).

This *interspecies hydrogen transfer* is therefore of benefit to both organisms in that the hydrogen-producing strain achieves a higher energy yield from its fermentation while the hydrogen-utilizing strain is supplied with a source of energy. In many instances, in fact, the growth of the first organism is only possible as a result of a negative free energy change from its fermentation which is dependent upon the removal of hydrogen. Interspecies hydrogen transfer is a key feature of the nutritional interdependence of many anaerobic organisms and is thus of major significance to the carbon and energy fluxes in anaerobic microbial ecosystems. These characteristics are particularly important with regard to the sulphate-reducing bacteria and methanogens; as an example, one can cite the anaerobic degradation of cellulose by the mixed culture of *Clostridium thermocellum* and *Methanobacterium thermoautotrophicum*.

Anaerobic mixed cultures, or microbial ecosystems which utilize interspecies hydrogen transfer, therefore demonstrate a number of important features:

1 A wider range of nutrients may be fermented; propionate and benzoate, for example, are often only degraded anaerobically by mixed cultures with sulphidogenic or methanogenic bacteria.

2 There is a greater extent of substrate utilization than is possible with pure cultures of the individual organisms involved.

3 The fermentation products differ both quantitatively and qualitatively when the organisms are grown together in mixed culture.

4 More growth is obtained from each organism than is possible in monoculture.

5 Energetically unfavourable reactions, e.g. the fermentation of ethanol to acetate by the so-called S organism in *Methanobacterium omelianski*, are 'pulled through' by the removal of hydrogen in the metabolism and growth of the second organism.

Acetogenic bacteria

A particularly important group of fermenting bacteria are the acetogens, which can be divided into hydrogen-producing acetogens and homoacetogens. The hydrogen-producing acetogens ferment alcohols and acids to acetate plus hydrogen. A key feature of such fermentations is that they are endergonic and only capable of supporting growth in mixed culture with a second hydrogen-utilizing species. Examples of this are the butyrate-fermenting *Syntrophomonas wolfei* and the propionate-fermenting *Syntrophobacter wolinii*, both of which can only grow in association with a hydrogen-oxidizing anaerobe such as a sulphate-reducing or methanogenic bacterium. These are therefore examples of obligate interspecies hydrogen transfer, for which the term syntrophy is used.

The homoacetogens, e.g. *Acetobacterium woodii* and *Clostridum thermoaceticum*, ferment compounds such as the sugars fructose and glucose to acetate by a mechanism involving the direct production of two molecules of acetate from carbon atoms 1 & 2 and 5 & 6, with a third molecule of acetate arising from carbon atoms 3 & 4 by a C_1-C_1 condensation mechanism involving folic acid and vitamin B_{12}. The details of this pathway have recently become clear [3, 27]. The enzyme carbon monoxide dehydrogenase is central to the mechanism, which is referred to as the acetyl-CoA pathway. These organisms also contain electron transport cofactors in the form of ferredoxin, quinones and cytochromes and there is evidence of electron transport-linked phosphorylation. This is particularly evident during the growth of hydrogenase-positive strains such as *A. woodii* and *Cl. aceticum* on CO_2 as sole carbon source, with its reduction by hydrogen again to acetate as metabolic end product.

Anaerobic respiration

The components and organization of the terminal respiratory system from NAD(P)H to the terminal oxidase have already been discussed (e.g. Fig. 2.5.2). The system operates to transfer reducing equivalents from a substrate, or donor, to oxygen as the terminal electron acceptor. The ordering of the sequence of transfers and the mechanism of energy transduction are dependent on the redox potentials of the ultimate donor and acceptor, and of the individual carriers in the chain. As is seen, for example, with succinate dehydrogenase, it is possible for donors and acceptors to interact with the respiratory chain at any point compatible with their redox potentials. A large number of bacteria demonstrate this flexibility; some use various inorganic compounds as donors of both reducing equivalents and energy (chemolithotrophy) while others can use alternative terminal electron acceptors in the absence of oxygen (anaerobic respiration).

The ability to couple biological oxidations to terminal electron acceptors other than oxygen is confined to the bacteria. The principal acceptors are nitrate, nitrite, fumarate, sulphate and CO_2. The redox potentials for these acceptors, as well as for various donors, are given in Table 2.5.2. A complete oxidation sequence, with the release of useful free energy as ATP, depends upon (a) the bacterium having the necessary substrate dehydrogenase, and (b) the potential of the donor redox couple being more negative than that of the acceptor redox couple. The number of ATPs produced per electron pair will ultimately depend on the magnitude of the free energy available (given by the relationship $\Delta G^{0\prime} = -2F\Delta E_0{}'$) and on the number of redox loops present (see Fig. 2.5.2). The requirement for the synthesis of one molecule of ATP is one redox loop and $\Delta G^{0\prime}$ of approximately 44 kJ. Thauer and Morris [24] have discussed the relationship between thermodynamics and mechanisms in energy metabolism and considered the possibility of fractional yields of ATP.

Table 2.5.2. The standard oxidation–reduction potentials of various acceptor and donor redox couples

Redox couple	$E_0{}'$ (mV)
Acceptor	
$\frac{1}{2}O_2/H_2O$	+820
NO_3^-/NO_2^-	+433
NO_2^-/NO	+350
Fumarate/succinate	+33
SO_4^{2-}/SO_3^{2-}	−60
CO_2/CH_4	−244
Donor	
$H_2/2H^+$	−420
$H.COOH/HCO_3^-$	−416
$NADH/NAD^+$	−320
Lactate/pyruvate	−197
Malate/oxaloacetate	−172
Succinate/fumarate	+33

Nitrate and nitrite as terminal electron acceptors

The ability to reduce nitrate as a component of the assimilation of inorganic nitrogen into the protein and other nitrogen-containing compounds of the cell is widespread among the micro-organisms and is also present in plants. The ability to use nitrate as an alternative electron acceptor in anaerobic respiration is, however, confined to the bacteria. Nitrate is a genuine alternative to oxygen in that the organisms concerned are generally facultative anaerobes and the mechanism for nitrate reduction is only synthesized in the absence of oxygen, which is therefore the preferred terminal electron acceptor. *Nitrate respiration* is the term used for the reduction of nitrate to nitrite. This reaction is widespread among bacterial genera including *Pseudomonas, Bacillus, Clostridium* and various members of the Enterobacteriaceae. Denitrification is the term for the reduction through nitrite, nitric oxide and nitrous oxide to nitrogen. This process is confined to a limited number of species. Among these are *Paracoccus denitrificans* and *Thiobacillus denitrificans*. Both of these organisms can use nitrate as terminal electron acceptor and CO_2 as a carbon source; *P. denitrificans* can obtain its energy from the oxidation of hydrogen and *T. denitrificans* from the oxidation of reduced sulphur compounds. These two, along with acetogenic, methanogenic and certain sulphur-reducing bacteria, are the only known non-photosynthetic anaerobic autotrophs; they can obtain all of their requirements for carbon, energy and reducing equivalents from inorganic sources.

In both nitrate respiration and denitrification the first step is the reduction of nitrate to nitrite. The enzyme nitrate reductase has been extensively studied, particularly in *E. coli* and *P. denitrificans* (see Haddock & Jones [5] and Ingledew & Poole [9]). The enzyme is membrane-bound and couples with the respiratory chain through a specific cytochrome *b*. It seems likely that the respiratory chain is functionally organized into two loops such that oxidation of NAD-linked substrates can generate two molecules of ATP. Under nitrate respiring conditions, however, *E. coli* makes considerably less use of the tricarboxylic acid (TCA) cycle for energy generation.

Instead, the principal electron donors with carbohydrate-grown cells are NADH derived from glycolysis and formate derived from the so-called phosphoroclastic cleavage of pyruvate. Succinate, lactate, glycerophosphate and hydrogen, however, can also serve as electron donors.

In denitrification, specific *c*-type cytochromes appear to be involved in all steps of reduction of nitrite to nitrogen. These are carried out by the enzymes nitrite reductase, nitric oxide reductase and nitrous oxide reductase. It is possible that all reactions are coupled to ATP synthesis.

The presence of a soluble NADH-dependent nitrite reductase in such facultative fermenting bacteria as *Klebsiella, Citrobacter* and *Achromobacter* results in ammonia being the end product of nitrate and nitrite reduction. Although nitrite reduction is not linked directly to ATP synthesis, it does fulfil the role of electron sink and so allows the formation of more oxidized fermentation products; for example, acetate may be formed with the concomitant synthesis of ATP by substrate level phosphorylation.

Apart from the value of these reactions to the organism concerned, nitrate respiration, denitrification and ammonia production are critical to the functioning of the nitrogen cycle and the preservation of the nitrogen balance of the biosphere (see Chapters 3.1, 3.2 and 4.1).

Fumarate as terminal electron acceptor

Reference has already been made to the ability of the propionic acid bacteria and *V. succinogenes* to derive useful energy from electron transport-linked oxidations where the terminal step is the reduction of fumarate to succinate. Fumarate reduction is found in a broad range of bacteria, including various species of the Enterobacteriaceae and of the genera *Bacteroides, Haemophilus, Pasteurella, Vibrio, Desulfovibrio, Bacillus, Staphylococcus, Clostridium* (see Kroger [13]). The activity is also found in a few protozoa and helminths.

The reduction of fumarate to succinate ($E_0{}' = +33$ mV) is associated with the acceptance of electrons from either menaquinone ($E_0{}' = -74$ mV) or occasionally desmethylmenaquinone ($E_0{}' = +25$ mV). Where in other organisms the same reaction operates in the opposite direction to oxidize succinate to fumarate as part of the TCA cycle, its reducing equivalents are generally donated to ubiquinone ($E_0{}' = +113$ mV). In both cases, therefore, the flow of reducing equivalents is in the direction of increasing redox potential and is associated with the net release of free energy. The fumarate reductase itself is an iron–sulphur flavoprotein with FAD as prosthetic group; a cytochrome *b* is involved in the transfer of electrons from the menaquinone.

In many anaerobic bacteria which lack a complete TCA cycle, the fumarate reductase serves an anabolic role in supplying succinate for tetrapyrrole synthesis. In anaerobic energy metabolism, however, fumarate reduction may also serve as an electron sink and in this way modify fermentation patterns (as with ammonia production, above). In addition, the reduction may give rise to ATP directly through an electron transport-linked phosphorylation.

As noted earlier, fermentation reactions leading from pyruvate are of two broad types. Reactions such as the formation of lactate or the reduction of acetyl coenzyme A to ethanol take up reducing equivalents and in the process allow the recycling of the pyrimidine nucleotides. On the other hand, the formation of acetate from acetyl coenzyme A, and the parallel production of butyrate from butyryl coenzyme A by *Clostridium* spp., are non-reducing energy-generating reactions associated with the synthesis of ATP. In the case of *Cl. pasteurianum* (see Fig. 2.5.4), and with the mixed culture of *R. albus* and *V. succinogenes*, mechanisms for the reoxidation of NADH other than the formation of a more reduced fermentation product can give rise to a fermentation pathway with a higher overall energy yield. Similarly, in the glucose fermentation of *Streptococcus faecalis*, fumarate serves as a high-potential sink for reducing equivalents. In the presence of fumarate the normal homolactate fermentation is altered to one giving acetate, with a consequent increase in ATP yield.

In organisms possessing cytochromes and other respiratory chain hydrogen and electron carriers, this role of fumarate reduction as an electron sink is combined with electron transport-linked phosphorylation. *E. coli*, for example, can derive energy for its anaerobic growth from the oxidation of hydrogen, lactate, formate, glycerol or glucose, with fumarate as terminal electron acceptor. The closely related organism *Proteus rettgeri*, however, has the remarkable ability of using fumarate as the sole source of energy, of reducing equivalents and of carbon. In the initial reaction sequence, seven molecules of fumarate are converted to six of succinate plus four of CO_2, with an apparent yield of six molecules of ATP. The metabolic pathways involved are shown in Fig. 2.5.5.

The yield of one ATP per molecule of fumarate reduced to succinate is computed from the growth yield of cells and is compatible with the difference in redox

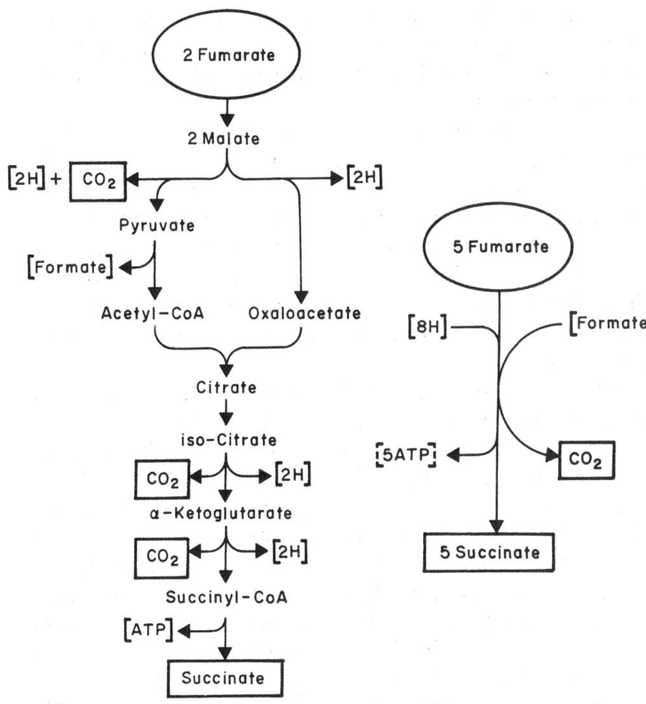

Fig. 2.5.5. Metabolic pathways for the fermentation of fumarate as the only substrate for the anaerobic growth of *P. rettgeri*. Redrawn with permission from Kroger [13].

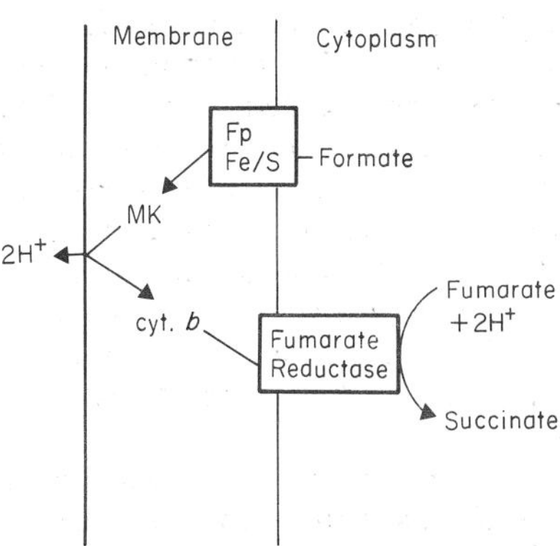

Fig. 2.5.6. Proposed functional organization of the redox carriers responsible for anaerobic electron transport with fumarate as terminal acceptor in *E. coli*. Abbreviations: MK, menaquinone; Fp, flavoprotein; Fe/S non-haem iron–sulphur protein. Redrawn with permission from Haddock & Jones [5].

potentials of the fumarate/succinate couple and electron donors such as hydrogen, formate or NADH. Evidence has also been obtained from *V. succinogenes*, and to a lesser extent from *E. coli*, that fumarate reduction is accompanied by uncoupler-sensitive proton extrusion. It seems likely that the functional organization of formate–fumarate oxidoreductase in *V. succinogenes* and *E. coli* results in the extrusion of one pair of protons for each electron pair passed from the dehydrogenase across the membrane via menaquinone and cytochrome *b* to the reductase at the inner face (see Fig. 2.5.6). It now seems more likely, however, that the menaquinone acts as the cofactor for formate dehydrogenase which has its active site at the outer face of the membrane. Such an organization would be an 'arm' rather than a loop but would still give a proton gradient from the oxidation of formate.

Sulphate-reducing bacteria

The sulphate-reducing bacteria are a diverse collection of organisms which form a coherent group in physiological and ecological terms by virtue of their obligate anaerobiosis and their ability to use sulphate as an alternative electron acceptor. As many strains can also use thiosulphate or other oxidized forms of inorganic sulphur, and since several organisms have now been characterized as requiring elemental sulphur as electron acceptor, it is clear that the common property is the formation of sulphide and thus the term sulphidogenic would more accurately describe the group.

In recent years our knowledge of the nutrition and metabolic capability of the sulphate-reducing bacteria, and hence our understanding of their ecological impact, has increased dramatically [19, 20, 21]. At least nine genera are how recognized; they embrace a wide range of morphological characteristics and cover a broad nutritional spectrum from autotrophy to growth on long-chain fatty acids and aromatic compounds such as benzoate. The sulphate-reducing bacteria can be divided into two groups: (a) those capable only of partial oxidation of a limited range of nutrients such as lactate or malate, with acetate as metabolic end product; and (b) those able to use a much more extensive range of sources of carbon and energy. This latter group can be subdivided again into those capable only of partial oxidation to acetate and those with the capacity for complete catabolism to CO_2. In common with many fermenting species and in particular the acetogens, acetate is therefore a common metabolic product of the sulphate-reducing bacteria. Hydrogen is also formed by these organisms, both in the absence of sulphate as terminal electron acceptor during fermentation of, for example, pyruvate, and in the presence of sulphate with substrates such as lactate or pyruvate.

It is now clear, however, that both acetate and hydrogen can also be oxidized by a number of species and it has been suggested that in sulphate-rich anaerobic ecosystems these are likely to be the principal sources of carbon and energy for the sulphate-reducing population; even potential nutrients such as lactate may be first converted to acetate by fermenting organisms [7, 21].

The sulphate-reducing bacteria contain a normal complement of hydrogen and electron carriers including ferredoxin and flavodoxin, menaquinone, and *b*- and *c*-type cytochromes. The oxidation of hydrogen has been intensively studied with regard to the periplasmic hydrogenase and its electron acceptor, cytochrome c_3, and the development of a transmembrane protonmotive force (see Fig. 2.5.7) [18, 21]. The intriguing idea has also been put forward that energy coupling from substrates such as lactate might arise by a mechanism of *hydrogen cycling* (see Fig. 2.5.8). It is suggested that the carbon catabolism is located in the cytoplasm and is fermentative; reducing equivalents are converted to hydrogen by an intracellular hydrogenase. This hydrogen can then diffuse across the membrane and be oxidized by the periplasmic hydrogenase and so give rise to electron transport-linked phosphorylation. Such a mechanism within a single cell is formally equivalent to interspecies hydrogen transfer between two cells of different species.

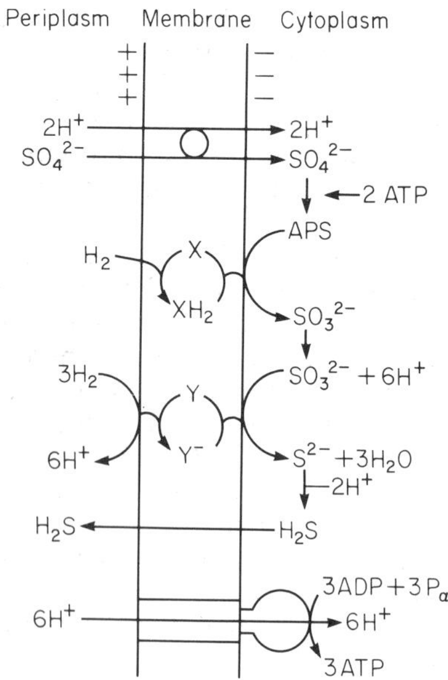

Fig. 2.5.7. A chemiosmotic model for the coupling of dissimulatory sulphate reduction with hydrogen to ATP synthesis. Abbreviations: X, a hydrogen carrier; Y, an electron carrier. Redrawn with permission from Thauer & Badziong [23].

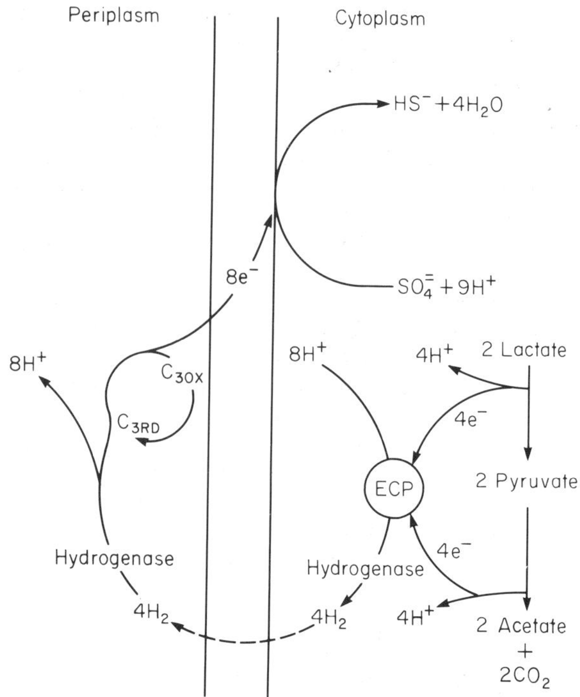

Fig. 2.5.8. A schematic representation of enzyme localization, hydrogen cycling and vectorial electron transfer by *Desulfovibrio* with lactate and sulfate as substrates. Abbreviations: ECP, electron carrier proteins; c_3, cytochrome c_3 (M_r = 13 000). Redrawn with permission from Odom & Peck [17].

The oxidation of acetate by sulphate-reducing bacteria also highlights unusual features of these organisms [21, 25]. In some genera catabolism is by a TCA cycle which shows two unique features: acetate activation is by an acetate:succinyl CoA transferase (with consequent loss of the substrate level phosphorylation step normally associated with succinyl CoA to succinate); and the succinate to fumarate dehydrogenation (E_0' = +33 mV) is coupled to the reduction of menaquinone (E_0' = −74 mV) and ultimately to that of sulphate (E_0' = −60 mV) and sulphite (E_0' = −170 mV). It is possible that ATP-driven reversed electron transport may be an essential component of this reaction. More usually, however, acetate oxidation in sulphate-reducing bacteria is by the acetyl-CoA pathway.

A unique feature of the use of sulphate as terminal

electron acceptor is the requirement to form an activated derivative before reduction can take place. The activated derivative is adenosine phosphosulphate (APS), which is formed from ATP and sulphate with the release of pyrophosphate. The mid-point potential of −60 mV given above and in Table 2.5.2. for the reduction of sulphate in fact refers to the reduction of APS with the production of sulphite and AMP; the true figure for the sulphate/sulphite couple is −516 mV, which would render sulphate quite unable to accept electrons in an exergonic biological process.

These requirements for reversed electron transport and activation of sulphate further limit the relatively low energy yields available to the sulphate-reducing bacteria from their catabolic activities. None the less, these organisms are of considerable ecological and economic importance in sulphate-rich anaerobic environments where they have been implicated in the formation of sulphur and oil deposits, the souring of hydrocarbon fuels, and the anaerobic corrosion of iron and steel [7, 20].

Methanogenic bacteria

The methanogenic bacteria share a number of general characteristics with the sulphate-reducing bacteria: they are obligate anaerobes; they constitute a group of organisms closely related in terms of their physiology and ecology; their nutrition is markedly restricted, with the oxidation of acetate and hydrogen being of key importance; they have a major impact on anaerobic ecosystems; and our knowledge of their cell physiology, while still rudimentary, is growing rapidly.

In two major respects, however, the methanogens are quite different from the sulphate-reducing bacteria or any others we have so far considered. The methanogens are Archaebacteria with, for example: unusual and varied cell wall structures lacking peptidoglycan; non-saponifiable ether-linked polyisoprenoid membrane lipids; a complex intracellular membrane system; and a mitochondria-like atractyloside-sensitive adenine nucleotide translocase. Secondly, with some minor exceptions such as FAD in *Methanobacterium bryantii* and a cytochrome *b* in *Methanosarcina barkeri*, the methanogens appear to lack the components of a standard electron transport chain, and CO_2 therefore does not function as a terminal electron acceptor in a mechanism analogous to nitrate and sulphate in other bacterial anaerobic respiratory systems. In common with the acetogens, the methanogens do have high concentrations of the C_1 transformation cofactors, folic acid (pteridine) and vitamin B_{12} (corrinoid). They also possess a number of other more unusual cofactors such as F_{420}, a flavin electron carrier found also in other bacteria including *Streptomyces griseus*, with an E_0' of around −350 mV and functionally equivalent to ferredoxin. There are also cofactors unique to the methanogens; these are coenzyme M, which is 2-mercaptoethane sulphonic acid ($HS.CH_2.CH_2.SO_3H$), and F_{430}, a nickel tetrapyrrole which is involved in methyl-CoM reductase.

All methanogens can grow on hydrogen and CO_2, where CO_2 is both carbon source and electron acceptor for the energy-generating oxidation of hydrogen with the formation of methane. Growth on formate is probably preceded by conversion to hydrogen plus CO_2. The most nutritionally versatile genus is *Methanosarcina* which grows also on acetate, methanol, and mono-, di- and trimethylamines. In common with the sulphate reducers, CO_2 must first be 'activated' before it can fulfil its electron acceptor role in energy generation. Reduction of CO_2 to the level of formaldehyde is an endergonic process, while the transfer of 'HCHO' to CH_4 is associated with a sufficient decrease in free energy for the theoretical production of up to two molecules of ATP. The absence of any free intermediates between CO_2 and methane excludes the possibility of substrate-level phosphorylation, and, although study of the ATPase suggests a normal chemiosmotic mechanism for ATP synthesis, it remains unclear at the present time how the pmf is generated during catabolism in these organisms.

The routes of methane formation in energy generation and of CO_2 fixation in the assimilation of cell carbon are closely interwoven and best considered together [28, 29]. It seems likely that a bound form of carbon monoxide, [CO], is of crucial importance [24]. This may be formed from CO_2 by the action of carbon monoxide dehydrogenase, or from acetate via the intermediate production of acetyl-CoA which gives rise directly to [CO] and methyl-X. While X has not yet been identified unequivocally, the methyl derivatives of CoM are known to be intimately involved in these carbon transformations. Methyl-X can also be formed from CO_2 in a reaction pathway not yet fully characterized but likely to involve folic acid derivatives. Methyl-X is then reduced to methane in an energy-generating reaction involving F_{430} as cofactor and F_{420} as carrier of electrons from the oxidation of [CO] to carbon dioxide by carbon monoxide dehydrogenase.

In the pathway of CO_2 assimilation, it is likely that [CO] reacts with methyl-X to give rise to acetyl-X and thence to acetyl-CoA. In this case X is likely to be vitamin B_{12}, with methyl-CoM and methyl vitamin B_{12} in equilibrium and both represented as methyl-X in the present formalism. Pyruvate is formed from acetyl-CoA by an F_{420}-dependent reductive carboxylation. In the case of *Methanosarcina barkeri*, which possesses an isocitrate dehydrogenase, α-oxoglutarate for biosynthesis is formed by the oxidative arm of the TCA cycle. *Methanobacterium thermoautotrophicum*, however, lacks isocitrate dehydrogenase and α-oxoglutarate has to be formed via malate and fumarate in a reductive arm of the TCA cycle, with F_{420} again being involved in the reductive carboxylation of succinate to α-oxoglutarate. In no methanogen is there a fully operational TCA cycle.

The methanogenic bacteria are to be found in a wide range of anoxic environments, including the rumen, anaerobic digesters and freshwater sediments. It appears that in the rumen hydrogen is the main energy source used by the methanogens, while in digesters and sediments it is likely that the principal nutrient is acetate. In marine sediments and other sulphate-rich environments, however, hydrogen and acetate are generally metabolized by sulphate-reducing bacteria and there is only limited evidence of methanogenesis. Since the effect of the metabolic activity of either group of strict anaerobes is to maintain hydrogen and acetate at low levels within the ecosystem in question, the explanation for the sulphate reducers out-competing the methanogens rests with the higher affinity for these two substrates demonstrated by the sulphate reducers.

Microbial consortia and biofilms

From this consideration of the cell physiology of facultative and obligate anaerobic organisms it has become evident that they are comparatively restricted in terms of their nutritional and metabolic spectra of activities. A direct consequence of this fact is that anaerobic species are generally dependent upon the activities of other organisms for both their supply of nutrients and the establishment of the physicochemical conditions necessary for their growth. Anaerobic microbial ecosystems represent communities of organisms, or consortia, in which the capabilities of individual species overlap and are integrated in such a way that the consortium as a whole has considerably greater potential than do any of the single species on their own. Such consortia may have macroscopic dimensions and may exist, for example, in the form of biofilms on solid surfaces or at interfaces. By virtue of the oxygen-scavenging activity of aerobic macro- or micro-organisms associated with consortia, anaerobic conditions may develop within a microbial community which itself exists in an aerobic environment.

The biodegradative properties and carbon and energy fluxes through microbial consortia have been considered to involve a functional hierarchy of organisms [6, 29]. The *hydrolytic* species are aerobic or facultative and may include bacteria, yeast and fungi, and on occasion higher forms as, for example, in marine fouling. They are responsible for the initial breakdown to various monomers and metabolic intermediates of the primary nutrients, which may be virtually any substance including biopolymers or hydrocarbons. The *fermentative* organisms are predominantly bacterial and they produce a range of low molecular weight fermentation products. One particular group of fermentative bacteria, the *acetogens*, convert the metabolic intermediates and fermentation products to acetate and hydrogen, the principal nutrients for the terminal oxidative reactions carried out by either *methanogenic* or *sulphate-reducing bacteria*, depending upon the availability of sulphate.

The essential nature of the nutrient chain dependence and interspecies hydrogen transfer displayed by these consortia is that each individual species is only capable of partial or incomplete oxidation of its carbon and energy sources with the formation of reduced metabolic products. These are organic products or hydrogen in the case of organisms using fermentation, and inorganic products with those employing anaerobic respiration. These reduced products are then capable of further oxidation and so serve as sources of carbon, energy and reducing equiva-

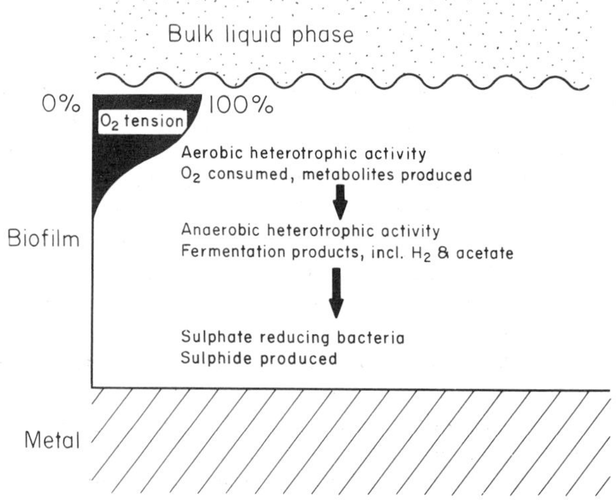

Fig. 2.5.9. Conditions of aerobiosis and nutrient succession within a microbial biofilm. Redrawn with permission from Hamilton [7].

lents for a second organism. However, in the case of the ammonia, hydrogen sulphide and methane produced by *Klebsiella* and related genera during nitrate respiration, by sulphate-reducing bacteria and by methanogenic bacteria, respectively, it is necessary for the reduced gas to diffuse to an aerobic zone before reoxidation can occur (see below). It is for this reason that maximal carbon turnover in anaerobic ecosystems is found close to the aerobic/anaerobic interface. An exception to this rule is that hydrogen sulphide can act as a source of reducing equivalents (but not of energy) for phototrophic bacteria within the anaerobic environment.

In addition to their increased metabolic throughput and carbon flux, microbial consortia can be of major importance in particular practical situations. The role of a microbial biofilm in generating both the physicochemical and the nutrient conditions necessary for the growth of sulphate-reducing bacteria at a metal surface is depicted in Fig. 2.5.9. It is by this means that these organisms can cause anaerobic corrosion [7].

2.5.4 CHEMOLITHOTROPHY

Chemolithotrophy is the term used to describe the mode of nutrition in which energy is obtained from the oxidation of inorganic electron donors. Most organisms displaying this activity are also autotrophic in that their source of carbon is CO_2 or a similar C_1 compound. The routes of carbon assimilation (see Section 2.5.7) will be considered after dealing with photosynthesis and nitrogen fixation.

The relatively high redox potentials of the inorganic electron donors (e.g. Fe^{2+}/Fe^{3+} is +772 mV) means that they couple with the respiratory chain in the region of the cytochromes rather than at the level of the pyrimidine nucleotides or flavoproteins. This results in the need for ATP-driven reversed electron transport for the production of NAD(P)H, and it also means that the capacity for energy generation is considerably reduced. Generally oxidative phosphorylation is confined to site III, or occasionally sites II and III. The chemolithotrophic organisms are strict aerobes, with the exception of the nitrate reduction noted earlier with *Paracoccus denitrificans* and *Thiobacillus denitrificans*, for example, and the sulphate and CO_2 reduction in sulphate-reducing and methanogenic bacteria growing on hydrogen. Chemolithotrophy is confined to the bacteria.

In anaerobic respiration the electron transport chain operates with the 'normal' electron donors of the substrate dehydrogenases associated with carbon metabolism; the electron acceptors, however, are alternatives to the 'normal' oxygen. Chemolithotrophy, on the other hand, represents a mechanistic mirror image where the electron acceptor is oxygen but the electron donors are a series of inorganic oxidants, unique to this mode of metabolism. However, these mechanisms are not mutually exclusive, and a number of bacteria can demonstrate chemolithotrophy and anaerobic respiration at the same time. In fact, as noted above, all organisms capable of anaerobic respiration with nitrate, fumarate, sulphate or CO_2 as the terminal electron acceptor can grow on hydrogen as a source of energy and reducing equivalents.

The subject of chemolithotrophy has been dealt with extensively in the review by Aleem [1].

Hydrogen-oxidizing bacteria

Hydrogen is the most extensively studied electron donor in aerobic chemolithotrophic metabolism. The so-called hydrogen-oxidizing bacteria are taxonomically diverse but form a distinct group on the basis of their physiology; they include species belonging to the genera *Pseudomonas*, *Alcaligenes* and *Hydrogenomonas*. None of the hydrogen-oxidizing bacteria are obligate chemolithotrophs; they are capable of growing on hydrogen or heterotrophically, depending on the environmental conditions.

There are at least two hydrogenases. The particulate hydrogenase is a membrane-bound enzyme and a component of the respiratory chain. In *Hydrogenomonas eutropha* and *Paracoccus denitrificans* the oxidation of hydrogen with oxygen as electron acceptor is associated with phosphorylation at sites I, II and III. In *Pseudomonas saccharophilia*, however, only site II is operational (site III is functional during heterotrophic growth). The soluble hydrogenase, on the other hand, is responsible for the reduction of NAD^+ to NADH. This enzyme is lacking in *Hydrogenomonas facilis* and *Ps. saccharophilia*, and these organisms must generate their reducing equivalents by ATP-driven reversed elecron transport from the ubiquinone/cytochrome *b* region where the particulate hydrogenase couples with the respiratory chain in these bacteria.

The two hydrogenases are constitutive but their activity may be modulated. In the absence of CO_2 the rate of oxidation of hydrogen by lithotrophically growing cells is considerably reduced. Under these conditions the intracellular content of ATP is found to be raised by as much as 30% and the $NADH/NAD^+$ ratio goes from 0.5 to

1.4; it is considered that the levels of the adenine and pyrimidine nucleotides control the hydrogenase activity.

In the absence of growth, excess CO_2 is converted to poly-β-hydroxybutyrate as an endogenous reserve; by this means the cell makes use of any excess ATP or NADH produced from the hydrogen oxidation.

This balance between catabolism (hydrogen oxidation) and anabolism (growth or poly-β-hydroxybutyrate synthesis) and the role of the adenine and pyrimidine nucleotides neatly encapsulate the essence of the problems faced by all living systems and the mechanisms they have developed to solve these problems.

Lithotrophically growing cells lack the enzymes of the Entner–Doudoroff pathway, but when hydrogen is absent the enzymes can be induced by addition of, for example, fructose. That is to say, the enzymes required for heterotrophic growth are repressed by hydrogen and induced by an appropriate organic nutrient. Certain enzymic pathways are not, however, affected by growth on hydrogen, and the cells are therefore capable of 'mixotrophic' growth on hydrogen and such compounds as lactate, pyruvate, succinate, urate or tryptophan.

The functioning of the Entner–Doudoroff pathway (and of the pentose phosphate cycle if present) is also controlled at the level of enzyme activity. Glucose 6-phosphate dehydrogenase, for example, is an allosteric enzyme with ATP acting as an inhibitor; NADPH also acts as a competitive inhibitor at the $NADP^+$ binding site.

Nitrifying bacteria

The nitrifying bacteria play a key role in the operation of the nitrogen cycle, as they are responsible for the conversion of ammonia to nitrite, and of nitrite to nitrate (see Section 2.1.8). The nitrifying bacteria are obligate lithotrophs.

Ammonia is oxidized to nitrite by organisms belonging to the genera *Nitrosomonas*, *Nitrosospira, Nitrosococcus* and *Nitrosolobus*. They contain cytochromes *b* and *c*, with cytochromes *a* and *o* as terminal oxidases. In addition they have a cytochrome P_{460} which is probably involved in the addition of molecular oxygen to ammonia by a mixed function oxidase. The product of this activitiy is hydroxylamine (NH_2OH), which is then oxidized to nitrite. The hydroxylamine dehydrogenase appears to be a flavoprotein in close association with a cytochrome *b*. Electrons may be donated through cytochromes *a* and *o* to oxygen, with the production of ATP, or to the pyrimidine nucleotides by ATP-driven reversed electron transport. Although from consideration of the redox potentials one might expect sites II and III to be involved in phosphorylation, in fact only site III is operational.

The oxidation of nitrite to nitrate is used as the source of energy and reducing equivalents by the *Nitrobacter, Nitrospira* and *Nitrococcus* spp. In this case the oxidation occurs by the addition of water followed by the removal of hydrogen. The dehydrogenase donates electrons to cytochrome a_1, which in turn donates to cytochrome *c*, cytochrome a_3 and finally oxygen. The phosphorylation yield from this oxidation is compatible with the functioning of site III only. The redox potentials of cytochromes a_1 and *c* are such, however, that the transfer of electrons from a_1 to *c* is endergonic and requires the input of some energy.

This is illustrated by the oxidation of NADH (with phosphorylation at sites I, II and III), which is stimulated by uncouplers (but, of course, with the loss of phosphorylation), whereas the oxidation of nitrite is inhibited by uncouplers.

Some of the ATP generated in the oxidation of nitrite can be used to drive reversed electron transport from cytochrome *c* through ubiquinone and flavoprotein to pyrimidine nucleotides.

Sulphur-oxidizing bacteria

Those bacteria capable of gaining their energy and reducing equivalents from the oxidation of sulphur, sulphide, and other reduced sulphur compounds form an important link in the sulphur cycle. The cycle also includes the sulphate reducers and the phototrophic sulphur bacteria which use reduced sulphur compounds as their source of reducing equivalents. The sulphur-oxidizing bacteria as a group contain examples of both obligate and facultative lithotrophy. In addition, while the majority are strictly aerobic and autotrophic, *Thiobacillus denitrificans* can use nitrate as terminal electron acceptor and *T. intermedius* can assimilate organic carbon.

The *Thiobacillus* spp. are the most widespread sulphur-oxidizing organisms, but the filamentous gliding bacteria of the *Beggiatoa–Thiothrix* group also oxidize sulphide with the accumulation of large amounts of intracellular sulphur. When the sulphide concentration in the environment is low, this stored sulphur can be further oxidized to sulphate. *Sulfolobus* spp. are a group of sulphur-oxidizing thermophilic organisms.

A characteristic of all sulphur-oxidizing bacteria is their resistance to the acidic conditions caused by the sulphuric acid they themselves produce. Certain *Thiobacillus* spp.

can grow at pH values as low as one. As a consequence, we find these organisms are responsible for the biodeterioration of sulphur-containing materials like vulcanized rubber, stone and concrete, as well as the leaching of metals from old mine workings and pyrites dumps. Ultimately all acid-tolerant ecosystems are dependent on the CO_2-fixing ability of autotrophic *Thiobacillus* spp. for the supply of carbon.

The details of the routes of oxidation of sulphide, sulphur, thiosulphate and sulphite to sulphate are not yet clearly established. It is likely that coupling with the respiratory chain is at the level of the cytochromes, in particular cytochrome *c*. In the thiosulphate ($S_2O_3^{2-}$) to tetrathionate ($S_4O_6^{2-}$) oxidation there is evidence of ATP synthesis at site III, but, with the oxidation of sulphide (S^{2-}) to sulphur (S_o), and sulphite (SO_3^{2-}) to adenosine phosphosulphate (APS), there is a possibility that both site II and site III are involved. Clearly, in all these reactions NAD(P)H can also be generated by ATP-driven reversed electron transport.

The unique requirement for sulphate when used as a terminal electron acceptor to be first activated by ATP with the formation of APS, which can then be reduced to sulphite, has already been discussed. Exactly the same reactions occur in the reverse direction in the energy-generating oxidation of sulphite to sulphate by the *Thiobacilli*. The reaction sequence is:

$$AMP + SO_3^{2-} \rightarrow APS$$

$$APS + Pi \rightarrow ADP + SO_4^{2-}$$

$$2ADP \rightarrow AMP + ATP$$

The terminal reaction is catalysed by adenylate kinase. These reactions constitute the only substrate level phosphorylation encountered in the chemolithotrophic mode of metabolism.

More detailed information on these so-called colourless sulphur bacteria can be had from several chapters in the Royal Society symposium book [21].

Iron-oxidizing bacteria

In addition to oxidizing sulphide, *Thiobacillus ferrooxidans* can also oxidize ferrous (Fe^{2+}) to ferric (Fe^{3+}) iron. This ability is also found in *Ferrobacillus ferrooxidans* and the stalked bacteria *Gallionella* spp. The filamentous *Sphaerotilus* and *Leptothrix* spp. can also oxidize Fe^{2+} but, as these organisms are heterotrophs, the true significance of this inorganic oxidation is not clear.

Again, the iron oxidation appears to be coupled with the respiratory chain at the level of the cytochromes, and so is able to give rise to both ATP synthesis and ATP-driven reversed electron transport. The potential of the Fe^{2+}/Fe^{3+} redox couple is extremely high at +772 mV, however, and it is not immediately clear how the iron oxidizing bacteria can generate sufficient energy for the synthesis of ATP. This problem may be eased by the mass action effect resulting from the removal of the reaction product due to the extreme insolubility of ferric hydroxide, particularly in alkaline environments. This property is responsible for the characteristic ferric hydroxide-impregnated sheaths and stalks associated with iron-oxidizing bacteria. It is also the cause of the so-called bog iron deposits and of the 'rusty' appearance of many natural streams.

Methylotrophs

In all the chemolithotrophic organisms discussed above, the mode of carbon assimilation is, with very few exceptions, autotrophic with the cell carbon coming from CO_2. As with the phototrophic autotrophs (see Section 2.5.5), the route of CO_2 assimilation is by the Calvin pentose bisphosphate cycle. This is not true, however, of those bacteria capable of obtaining their energy, reducing equivalents and cell carbon from the oxidation of methane and related C_1 compounds.

The obligate methylotrophs can grow only on methane, methanol or dimethylamine. The first isolated methylotroph was *Methylomonas methanica* but in recent years a number of other organisms have been identified. They include *Pseudomonas methanica* and various *Methylococcus* and *Methanomonas* spp.

The facultative methylotrophs cannot grow on methane but can utilize methanol, methylamine, formaldehyde and formate. They include bacteria belonging to the genera *Bacillus, Pseudomonas* and *Vibrio*. The budding bacterium *Hyphomicrobium* is not only a facultative methylotroph, able to grow on methanol, for example, but is also a denitrifier able to use nitrate rather than oxygen as the terminal electron acceptor.

In the oxidation of methane to methanol, molecular oxygen is added, most probably through a mixed function oxidase. Since a greater yield of cells is obtained from growth on methane than on methanol, useful biological energy must be generated at this stage. The mechanism of this energy generation is not known, however, at the present time. The methanol, formaldehyde and formic dehydrogenases couple directly with the respiratory chain. The immediate electron acceptor for the methanol dehydrogenase has not been identified, but the other two are most probably NAD-linked enzymes. Unlike the CO_2-reducing methanogenic bacteria, the methylotrophs contain a normal complement of cytochromes. They are thus capable of satisfying their requirements for energy and reducing equivalents by the standard methods of electron transport and ATP-driven reversed electron transport.

Whereas the other autotrophic organisms incorporate their cell carbon from CO_2, the routes of carbon assimilation in the methylotrophs start from formaldehyde (see Section 2.5.7).

The methylotrophs have been the subject of two recent monographs [2, 14].

2.5.5 PHOTOSYNTHESIS

Eukaryotic algae and cyanobacteria

Photosynthesis, the process whereby organisms are able to grow using the energy of the sun's radiation to power the fixation of atmospheric CO_2, is most commonly associated with the green plants. In the terminology used in this book, such organisms are phototrophic (energy from the sun's radiation), lithotrophic (reducing equivalents from an inorganic source, water) and autotrophic (cell carbon from CO_2). Such a photosynthetic mechanism is found among the micro-organisms in the eukaryotic algae and in the prokaryotic cyanobacteria. The photosynthetic eubacteria differ significantly from these, however, in that inorganic materials other than water are the source of reducing equivalents. Certain species can also use organic compounds as the source of reducing equivalents, or even of carbon during heterotrophic growth.

The key to the transduction of radiant light energy into biologically useful energy in the form of ATP lies in the chlorophyll pigments which are present in all photosynthetic organisms. The chlorophylls are complex porphyrin structures containing bound magnesium (the structurally related cytochromes and haemoglobin contain iron). Individual types of chlorophyll are characterized by different aliphatic side-chains attached to the porphyrin nucleus. These in turn cause the various chlorophylls to absorb light at different wavelengths. The action spectrum of particular organisms, i.e. the wavelengths of light absorption, is often extended by the presence of other light-absorbing pigments such as carotenoids or the phycocyanins found in the cyanobacteria. These secondary pigments do not take part in the primary photochemical event but serve to transmit the energy they have absorbed to the chlorophyll at the photochemical reaction centre. The green colour characteristic of plants and most algae is due to the chlorophyll while the other colours found, for example, in the red and brown algae, the cyanobacteria and the purple photosynthetic bacteria are the consequence of various light-absorbing secondary pigments.

The primary photochemical event is the absorption of a photon of light by the chlorophyll molecule. The consequent excitation of the molecule causes the ejection of an electron with a high redox energy. The reaction centre also has a manganese-containing complex that catalyses the photolysis of water to oxygen, protons and electrons. These elecrons are then taken up by the chlorophyll and so return that molecule to its ground state ready for another photochemical excitation cycle.

The photochemical reaction centre in the algae is referred to as photosystem II; it contains chlorophyll *a*, absorbs light with a wavelength around 680 nm, and has a redox potential of +910 mV.

The ejected electron reduces a primary acceptor, often referred to as Z. The identity of Z is uncertain, but it has a redox potential around −100 mV and is possibly a non-haem iron–sulphur protein. In a manner exactly analogous to the terminal respiratory system, the electron originally ejected from the excited chlorophyll molecule now passes along a series of redox carriers in the direction of increasing potential. These electron carriers have been identified as plastoquinone, a cytochrome *b*, cytochrome *f* (analogous to a cytochrome *c*) and often a copper-containing plastocyanin. Between cytochromes *b* and *f* there is a phosphorylation site for the synthesis of one molecule of ATP from ADP for each electron pair.

From the plastocyanin the electrons are donated to a second reaction centre, somewhat confusingly called photosystem I. This system also contains chlorophyll *a*, but it absorbs light at 700 nm and has a redox potential of +400 mV. Absorption of light is again associated with excitation of the chlorophyll and the ejection of an electron, which in this case reduces an acceptor, X, which is most probably again an iron–sulphur protein with a potential of −530 mV. The electrons then pass to ferredoxin and finally through a flavoprotein NADP-reductase to NADPH.

This so-called Z-scheme, therefore, uses radiant light energy and water to supply the cell's needs for ATP and

NAD(P)H. Since oxygen is produced from the photolysis of water, the eukaryotic algae and cyanobacteria (and the higher plants) are aerobic organisms.

Bacterial photosynthesis

The eubacterial phototrophs are anaerobic organisms which do not carry out the photolysis of water with the production of oxygen. Instead their mechanism of photophosphorylation is cyclic, producing energy only in the form of ATP. The formation of reducing equivalents in these organisms is by ATP-driven reversed electron transport with various inorganic or organic compounds as electron donors.

The photochemical reaction centre contains a characteristic bacterial chlorophyll which absorbs light at a higher wavelength, 870 nm, than does algal chlorophyll. This photosystem is analogous to the algal photosystem I and has a potential of around +490 mV. Again the primary acceptor of the electron is not clearly identified, but it appears to be either an iron–protein or an iron–quinone complex with a potential around −160 mV. Ferredoxins do not seem to be involved in bacterial photosynthesis, and the electron passes from the primary acceptor through ubiquinone, cytochrome *b* and cytochrome *c* and finally returns to the chlorophyll itself. ATP is generated at the cytochrome *b* to cytochrome *c* stage. By this cyclic mechanism, therefore, the bacterial phototrophs generate ATP from radiant light energy.

The various genera are characterized by their mode of NAD(P)H production, which is almost entirely by ATP-driven reversed electron transport. In green sulphur bacteria such as *Chlorobium* spp. sulphide is the electron donor, often with the consequent deposition of extracellular sulphur. The purple sulphur bacteria such as *Chromatium* spp. can oxidize sulphide and sulphur to sulphate. Therefore, these organisms are lithotrophs, as of course are the algae.

The non-sulphur purple bacteria, such as *Rhodopseudomonas* and *Rhodospirillum* spp., cannot be classified so simply, however. Under anaerobic conditions they can grow phototrophically with either hydrogen or sulphide as the electron donor and CO_2 as the carbon source. More often, however, they will use a simple organic compound such as lactate as a source of both reducing equivalents and cell carbon. Furthermore, these bacteria have the ability to grow chemoheterotrophically under aerobic conditions in the dark. The route of electron transport from substrate dehydrogenase to oxygen appears to be through the same quinone and cytochromes as operate in cyclic photophosphorylation; only the terminal oxidase, of either cytochrome *o* or cytochrome a/a_3 type, must be synthesized.

One possible exception to the rule that NAD(P)H is generated in bacterial photosythesis by reversed electron transport is found in *Chlorobium limicola*. The primary acceptor, X, in the photosystem of this bacterium is considerably more reduced than in other related organisms. With a redox potential of −450 mV it is feasible that it could donate its electrons directly to the pyrimidine nucleotides. There is evidence that such a process does operate in *Ch. limicola* with sulphide, thiosulphate, malate or succinate being possible electron donors. The electron transfer would be from the donor, through the cytochrome *c*, the photochemical reaction centre and the acceptor X, to NAD(P)H.

A detailed account of these photosynthetic mechanisms has been given by Jones [12].

Halobacterium halobium

There is, however, one photosynthetic bacterium that apparently breaks all the rules. The archaebacterium *Halobacterium halobium* is an extreme halophile, unable to grow at salt concentrations less than 3 M and found in environments such as drying salt-pans and salted fish. *H. halobium* is an aerobic chemoheterotroph, but under conditions of decreased oxygen light stimulates the synthesis of a purple membrane and the development of a phototrophic mode of metabolism. Unlike all other systems so far studied, however, the energy transduction during this phototrophic growth is not mediated by a looped series of hydrogen and electron carriers.

The purple membrane of *H. halobium* contains a single protein, bacteriorhodopsin. Recent research has shown that the molecule of bacteriorhodopsin spans the membrane, and that it contains seven α-helices arranged at right angles to the plane of the membrane and forming a pore through the centre of the molecule. Instead of chlorophyll, this pore contains a molecule of the visual pigment retinaldehyde. The absorption of light by bacteriorhodopsin causes a cyclic bleaching and regeneration of the chromophore or 'purple complex', and associated with this photochemical cycle is the ejection of protons from the cell. Therefore, by a mechanism fundamentlly different from the electron transport systems found in all respiratory and other photosynthetic organisms, *H. halobium* nevertheless conserves radiant light energy in the form of a transmembrane gradient of chemical and electrical potential, i.e. the protonmotive force. That the pmf produced by *H. halobium* is equivalent in every sense to that in other bacteria, or even in mammalian mitochondria, has been demonstrated by reconstituting bacteriorhodopsin into a membrane vesicle along with ATPase from rat liver mitochondria. Illumination of the vesicle system resulted in the synthesis of ATP.

2.5.6 NITROGEN FIXATION

The economic consequences of nitrogen fixation by the prokaryotic cyanobacteria and bacteria are considered in Chapter 4.1 of this book, but the central importance of energy and reducing equivalents to the process dictate that its biochemistry should be examined in this section also.

The reduction catalysed by the enzyme nitrogenase has a high energy requirement of as many as 15 ATP per molecule of nitrogen reduced. The reduction of dinitrogen to ammonia requires the addition of six hydrogens at high energy level, or low redox potential. The reducing equivalents are donated as protons from solution and electrons from ferredoxin or flavodoxin or both.

Ferredoxin is an iron–sulphur protein characterized by a very low redox potential of −410 mV. In organisms such as *Clostridium pasteurianum* and the rumen bacterium *Peptostreptococcus elsdenii*, the absence of iron stimulates the synthesis of an alternative flavodoxin which has flavin mononucleotide as cofactor. The aerobe *Azotobacter vinelandii*, on the other hand, uses both its species-specific ferredoxin and azotoflavin for nitrogen fixation.

There exist a number of ways in which various organisms generate reduced ferredoxin from their metabolism. In the aerobic *Azotabacter* spp., the TCA cycle enzyme isocitrate dehydrogenase constitutes up to 1% of the soluble protein in the cell. This reaction is used to produce a plentiful supply of NADPH. In *Chromatium* spp. and *Rhodotorula rubrum* it seems that pyruvate dehydrogenase might fulfil the same role. Normally, though, NADPH would be produced in these phototrophic organisms by ATP-driven reversed elecron transport. The reduction of ferredoxin by NADPH would, on its own, be energetically unfavourable, but, in combination with the active production of NADPH and the ATP-driven irreversible donation of electrons from the reduced ferredoxin to nitrogen, the flow of electrons is maintained from NADPH through ferredoxin to nitrogen.

Reduced ferredoxin can also be produced without the involvement of the pyrimidine nucleotides. In the cyanobacteria, ferredoxin is reduced directly in photosystem I of the non-cyclic photophosphorylation. Also in cyanobacteria, in *Chromatium* spp. and in the phosphoroclastic cleavage found in *Clostridium* spp., ferredoxin is reduced by the pyruvate–ferredoxin oxidoreductase. Bacteria belonging to the genera *Clostridium, Azotobacter, Chromatium* and *Chlorobium* can also reduce ferredoxin directly from hydrogen.

During nitrogen fixation there is also often some evolution of molecular hydrogen. This is catalysed by the nitrogenase, and under conditions of metabolism leading to a build-up of energy and reducing equivalents (e.g. excess catabolic activity or lack of CO_2 for anabolism in autotrophs) the levels can be reduced through an energetically wasteful evolution of hydrogen gas.

The nitrogenase enzyme is particularly sensitive to oxygen and the various nitrogen-fixing organisms have evolved different means of coping with this problem. It is a problem which is particularly acute for aerobic *Azotobacter* spp. In these organisms the nitrogenase is protected by its inclusion in a complex system of intracellular membranes. Also, under conditions of excess oxygen, cytochrome *d* serves as the terminal oxidase in *Azotobacter vinelandii*. The effect of this is the loss of phosphorylation at site III, which requires the involvement of cytochromes c_4 and c_5. Consequently, in order to satisfy the same energy requirements for growth, the cells need to oxidize proportionately more substrate with a corresponding greater utilization of oxygen. By this means, therefore, *Azotobacter* spp. can decrease the oxygen content of their immediate environment and so protect their nitrogenase.

From this summary of nitrogen fixation we can see that an appreciation of the mechanisms of the energy-linked transfer of reducing equivalents lies at the heart of our understanding of the functioning, control and protection of a key activity associated with the life process in individual cells and with the balance of complex ecosystems within the biosphere.

2.5.7 PATHWAYS OF CARBON METABOLISM

In this section the mechanisms and pathways of carbon metabolism in heterotrophs and autotrophs are considered. Throughout this discussion of microbial metabolism, nutritional terms have been used to described particular reactions, rather than any attempts being made to define a particular organism or group of organisms. This approach is especially significant when dealing with carbon metabolism.

Heterotrophs

The pathways of heterotrophic metabolism leading from the organic nutrients, through the carbon intermediates, to the molecules and polymers of cell carbon (see Fig. 2.5.1) are to be found in all standard texts of microbial physiology and biochemistry. The carbon intermediates of metabolism are such compounds as acetyl-CoA, pyruvate, phosphoenol pyruvate (PEP), oxaloacetate (OAA) and α-oxoglutarate. The possession of a functional TCA cycle, with the condensation of OAA and acetyl-CoA to give citrate, and of the related catabolic pathways leading from such nutrients as the sugars and amino acids to the cycle means that heterotrophic organisms apparently face no particular problem in producing from their catabolism the carbon intermediates required for their anabolic processes.

These pathways, however, particularly the TCA cycle, fulfil a dual function in supplying the cell with energy and reducing equivalents (a catabolic function) and with the carbon intermediates for growth (effectively an anabolic function). The detailed mechanisms involved in satisfying these two functions are not necessarily identical, as is discussed below and summarized in a schematic form in Fig. 2.5.10.

The energy-yielding catabolism of sugars and polysaccharides by glycolysis and related pathways leads to pyruvate. Similarly the catabolism of the amino acids alanine, threonine, glycine, serine and cysteine gives rise to pyruvate. In the next reaction common to all these catabolic sequences, one of the carbons of pyruvate is lost due to decarboxylation to acetyl-CoA. Acetyl-CoA also arises directly from the β-oxidation of fatty acids and from the breakdown of the amino acids leucine, tryptophan, phenylalanine, tyrosine and lysine. Acetyl-CoA then condenses with OAA to give citrate, and the TCA cycle functions to remove the two carbons of the acetate in two decarboxylation steps and to recycle a molecule of OAA. That is, the catabolism of carbohydrate, lipid and a major part of the protein in any organism's diet supplies the cell with energy and reducing equivalents but reduces the carbon skeletons to CO_2.

Two questions arise. What are the catabolic sequences for intermediates of the TCA cycle, and for amino acids whose degradation leads directly to one of these intermediates? And, secondly, how does the cell organize its

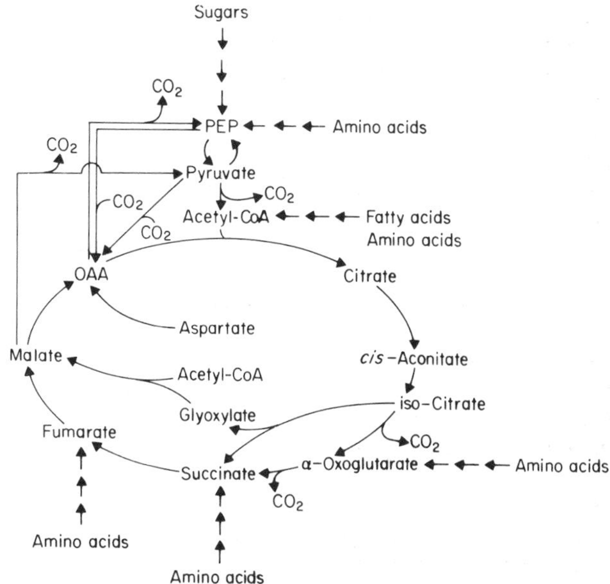

Fig. 2.5.10. Schematic representation of mechanisms whereby central metabolic pathways fulfil catabolic and anabolic roles.

pathways of catabolism so that they supply energy, reducing equivalents *and* the carbon intermediates of growth? The answer to both these questions lies in the cyclic nature of the TCA cycle and in the requirement for both OAA and acetyl-CoA in the condensation to citrate.

The catabolic breakdown of arginine, histidine, proline and glutamate gives rise to α-oxoglutarate; isoleucine, methionine and valine to succinate; phenylalanine and tyrosine to fumarate; and aspartate to OAA. The TCA cycle intermediates themselves can also often serve as carbon nutrients. In order for the cycle to operate in these conditions there must be a mechanism for the production of acetyl-CoA. This is achieved either by the NADP-requiring decarboxylation of malate to pyruvate catalysed by the malic enzyme, or by the ATP-requiring decarboxylation of OAA to PEP by PEP carboxykinase. In addition to supplying acetyl-CoA for the TCA cycle, these two reactions also generate PEP, pyruvate and acetyl-CoA for utilization in the anabolic sequences leading to carbohydrate and nucleic acid sugar derivatives, particular amino acids, and lipids.

ANAPLEROTIC PATHWAYS

Phosphoenol pyruvate, pyruvate and acetyl-CoA can also

readily be formed from, for example, sugar breakdown. But how does the TCA cycle function if α-oxoglutarate is being used in its anabolic role as the first intermediate en route to the synthesis of amino acids, since under these conditions OAA cannot be recycled for condensation with acetyl-CoA? The answer is that OAA is formed directly by a carboxylation or CO_2-fixation mechanism. In yeasts and a few bacteria there is evidence that the reaction is pyruvate + CO_2 → OAA, catalysed by pyruvate carboxylase, but in most cases the reaction is PEP + CO_2 → OAA + Pi and is catalysed by PEP carboxylase.

It is a feature of many anaerobic and autotrophic organisms that they lack the enzyme α-oxoglutarate dehydrogenase and consequently a functional TCA cycle. In these organisms the synthesis of OAA and succinate for anabolic routes leading to various amino acids and the porphyrins is also achieved by the carboxylation of PEP to OAA, followed by the reactions of a reductive arm of the TCA cycle.

The last step in glycolysis, PEP → pyruvate, is essentially irreversible; consequently organisms growing on a carbon source, such as alanine, which gives rise to pyruvate directly require the synthesis of a second enzyme, PEP synthase. This enzyme catalyses the reaction, pyruvate + ATP → PEP + AMP + Pi.

These reactions and sequences which are required for the catabolic pathways to satisfy also their anabolic role have been given the description *anaplerotic*.

Where cells are growing on acetate, or on fatty acids or amino acids which are degraded to acetate, the anaplerotic pathway which operates is known as the glyoxylate cycle. Instead of the two decarboxylation steps leading from isocitrate to succinate in the TCA cycle, growth on acetate induces the synthesis of the enzyme isocitratase, which cleaves the isocitrate to succinate and glyoxylate (CHO.COOH). The second enzyme of the cycle, malate synthetase, catalyses the condensation of this glyoxylate with acetyl-CoA to give malate. The malate can then be converted to OAA, which will condense with a second molecule of acetyl-CoA to give citrate and so allow the continued functioning of either the TCA or the glyoxylate cycle, or both. Through the combined use of these two cycles, cells growing on acetate can supply all their requirements for energy, reducing equivalents and carbon.

Heterotrophic micro-organisms, therefore, have the metabolic capability to grow on sources of carbon that are C_2 (acetate), C_3 (pyruvate), C_4 (succinate), C_5 and C_6 (sugars) compounds, and on more complex molecules and polymers giving rise to such intermediates. The ability to grow on C_1 units is, however, the characteristic feature of the autotrophs.

Autotrophs

Autotrophy is generally defined as the ability to use CO_2 as the main source of cell carbon. Often, however, the term is also loosely used to cover an inorganic source of reducing equivalents (properly described as lithotrophy). Moreover, there are differences of opinion regarding the precise nature of autotrophy and whether the term should include methylotrophic organisms. Rather than directly discuss these largely semantic problems, this section considers all the recognized mechanisms for assimilation of C_1 units in microbial metabolism.

CALVIN CYCLE

The Calvin or pentose bisphosphate cycle (Fig. 2.5.11) is ubiquitously distributed in all plants and micro-organisms capable of fixing CO_2, i.e. algae, cyanobacteria and photosynthetic and chemolithotrophic bacteria. Even in methylotrophs and *Chromatium* spp., where there are other major routes of C_1 assimilation, the Calvin cycle is found still to operate.

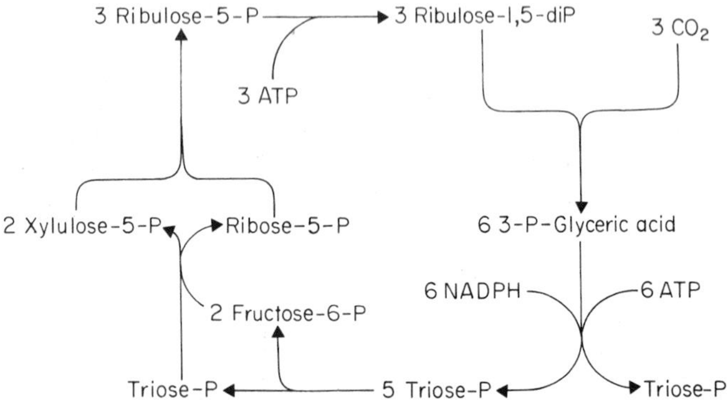

Fig. 2.5.11. A simplified diagram of the Calvin cycle. Redrawn with permission from Stanier *et al.* [22].

REDUCTIVE CARBOXYLIC ACID CYCLE

A second route of carbon assimilation is the so-called reductive carboxylic acid cycle. This involves the reversed cycling of the TCA cycle, along with the reactions acetyl-CoA → pyruvate → PEP → OAA. The key carboxylation steps, succinyl-CoA → α-oxoglutarate and acetyl-CoA → pyruvate, are catalysed by α-oxoglutarate and pyruvate synthetases. Reducing equivalents are donated from ferredoxin. This mechanism of CO_2 fixation is found in the photosynthetic bacteria, although doubt remains as to what proportion of the cell carbon derives from the reductive carboxylic acid cycle and what from the Calvin cycle, which is also operational in the purple sulphur bacteria. In the green sulphur bacteria, however, the only route of CO_2 fixation is by the reductive carboxylic acid cycle.

ACETYL-CoA PATHWAY

The third mechanism of CO_2 fixation is the C_1-C_1 condensation described earlier for acetogenic and methanogenic bacteria. It seems likely that this may also be the route of carbon assimilation in autotrophic sulphate- and sulphur-reducing bacteria.

PENTOSE PHOSPHATE AND SERINE PATHWAYS

The principal routes of C_1 assimilation in the methylotrophs start from formaldehyde (CH_2O) rather than from CO_2. Among the obligate methylotrophs, some bacteria possess the pentose phosphate or allulose pathway (see Fig. 2.5.12). These organisms lack α-oxoglutarate dehydrogenase and consequently do not have a functional TCA cycle.

Other obligate and all facultative methylotrophs, on the other hand, incorporate formaldehyde into their cell carbon by means of a combination of a serine pathway and the glyoxylate cycle (see Fig. 2.5.13). The formaldehyde is first converted to an 'active' C_1 unit, possibly methylene tetrahydrofolic acid. Such organisms also possess a fully functional TCA cycle.

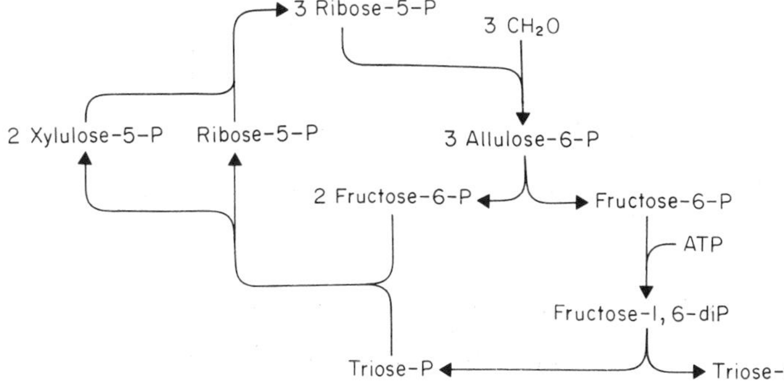

Fig. 2.5.12. A simplified diagram of the pentose phosphate cycle for assimilation of formaldehyde in some obligate methylotrophs. Redrawn with permission from Stanier *et al.* [22].

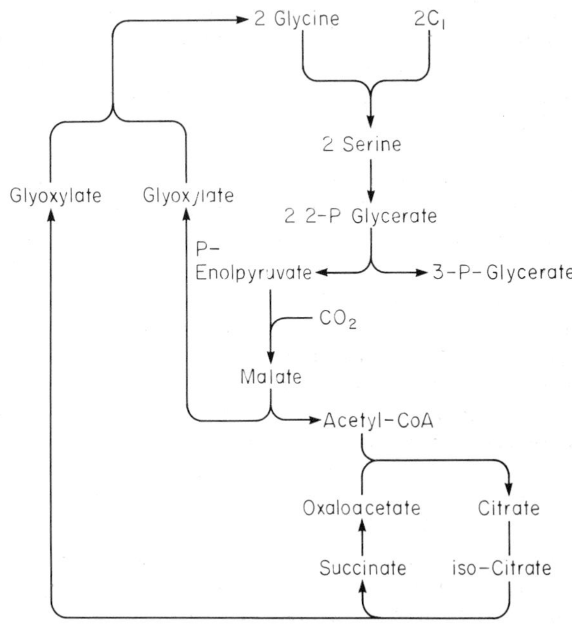

Fig. 2.5.13. The serine pathway coupled with the glyoxylate cycle, used for the assimilation of formaldehyde by some methylotrophs. Redrawn with permission from Stanier *et al.* [22].

2.5.8 CONCLUSION

In this chapter on microbial biochemistry and cell physiology, the chemical activities associated with micro-organisms have been described from the point of view of the importance of these activities to the individual organism. The corner-stone of the structured view of metabolism that has been presented is the identity of the separable but closely interrelated nutritional requirements for energy, reducing equivalents and carbon. This nutritional approach to the intracellular reactions of metabolism has led automatically to consideration of the particular environments in which individual micro-organisms might exist and the likely effects their metabolic activities would have on these environments. Since this is the very essence of microbial ecology, the overview of cellular biochemistry presented here must be central to any conceptual approach to that subject.

2.5.9 REFERENCES

[1] Aleem M.I.H. (1977) Coupling of energy with electron transfer reactions in chemolithotrophic bacteria. In *Microbial Energetics*, p.351 (ed. B.A. Haddock & W.A. Hamilton). Cambridge University Press, Cambridge.

[2] Anthony C. (1982) *The Biochemistry of Methylotrophs*. Academic Press, London.

[3] Fuchs G. (1986) CO_2 fixation in acetogenic bacteria; variations on a theme. *FEMS Microbiol. Revs*. **39**, 181–213.

[4] Garland P.B. (1977) Energy transduction and transmission in microbial systems. In *Microbial Energetics*, p.1 (ed. B.A. Haddock & W.A. Hamilton). Cambridge University Press, Cambridge.

[5] Haddock B.A. & Jones C.W. (1977) Bacterial respiration. *Bacteriol. Revs*. **41**, 47–99.

[6] Hamilton W.A. (1985) Energy sources for microbial growth: an overview. In *Aspects of Microbial Metabolism and Ecology*, p.35 (ed. G.A. Codd). Academic Press, London.

[7] Hamilton W.A. (1985) Sulphate-reducing bacteria and anaerobic corrosion. *Annu. Rev. Microbiol*. **39**, 195–217.

[8] Harold F.M. (1972) Conservation and transformation of energy by bacterial membranes. *Bacteriol. Revs*. **36**, 172–230.

[9] Ingledew W.J. & Poole R.K. (1984) The respiratory chains of *Escherichia coli*. *Microbiol. Revs*. **48**, 222–71.

[10] Jones C.W. (1977) Aerobic respiratory systems in bacteria. In *Microbial Energetics*, p.23 (ed. B.A. Haddock & W.A. Hamilton). Cambridge University Press, Cambridge.

[11] Jones C.W. (1982) *Bacterial Respiration and Photosynthesis*. Van Nostrand Reinhold (UK), Wokingham.

[12] Jones O.T.G. (1977) Electron transport and ATP synthesis in the photosynthetic bacteria. In *Microbial Energetics*, p.151 (ed. B.A. Haddock & W.A. Hamilton). Cambridge University Press, Cambridge.

[13] Kroger A. (1977) Phosphorylative electron transport with fumarate and nitrate as terminal hydrogen acceptors. In *Microbial Energetics*, p.61 (ed. B.A. Haddock & W.A. Hamilton). Cambridge University Press, Cambridge.

[14] Large P.J. (1983) *Methylotrophy and Methanogenesis*. Van Nostrand Reinhold (UK), Wokingham.

[15] Mandelstam J., McQuillen K. & Dawes I. (eds.) (1982) *Biochemistry of Bacterial Growth*, 3rd edn. Blackwell Scientific Publications, Oxford.

[16] Morris J.G. (1985) Changes in oxygen tension and microbial metabolism of organic carbon. In *Aspects of Microbial Metablism and Ecology*, p.59 (ed. G.A. Codd). Academic Press, London.

[17] Odom J.M. & Peck H.D. Jr (1981) Hydrogen cycling as a general mechanism for energy coupling in the sulfate-reducing bacteria, *Desulfovibrio* sp. *FEMS Microbiol. Lett*. **12**, 47–50.

[18] Odom J.M. & Peck H.D. Jr (1984) Hydrogenases, electron-transfer proteins, and energy coupling in the sulphate-reducing bacteria *Desulfovibrio*. *Annu. Rev. Microbiol*. **38**, 551–92.

[19] Pfennig N., Widdel F. & Trüper H.G. (1981) The dissimilatory sulfate-reducing bacteria. In *The Prokaryotes*, p.926 (ed. M.P. Starr, H. Stolp, H.G. Trüper, A Balows & H.G. Schlegel). Springer-Verlag, Berlin.

[20] Postgate J.R. (1984) *The Sulphate-Reducing Bacteria*, 2nd edn. Cambridge Unviersity Press, Cambridge.

[21] Postgate J.R. & Kelly D.P. (eds.) (1982) *Sulphur Bacteria*. The Royal Society, London.

[22] Stanier R.Y., Adelberg E.A. & Ingraham J.L. (1976) *The Microbial World,* 4th edn. Prentice-Hall, Englewood Cliffs, New Jersey.

[23] Thauer R.K. & Badziong W. (1980) Respiration with sulfate as electron acceptor. In *Diversity of Bacterial Respiratory Systems*, Vol. 2, p.65 (ed. C.J. Knowles). CRC Press, West Palm Beach.

[24] Thauer R.K., Jungermann K. & Decker K. (1977) Energy conservation in chemotrophic anaerobic bacteria. *Bacteriol. Revs*. **41**, 100–80.

[25] Thauer R.K & Morris J.G. (1984) Metabolism of chemotrophic anaerobes: old views and new aspects. In *The Microbe 1984*, p.123 (ed. D.P. Kelly & N.G. Carr). Cambridge University Press, Cambridge.

[26] Whittenbury R. & Kelly D.P. (1977) Autotrophy: a conceptual phoenix. In *Microbial Energetics*, p.121 (ed. B.A. Haddock & W.A. Hamilton). Cambridge University Press, Cambridge.

[27] Wood H.G., Ragsdale S.W. & Pezacka E. (1986) The acetyl-CoA pathway: a newly discovered pathway of autotrophic growth. *Trends Biochem. Sci*. **11**, 14–18.

[28] Zeikus J.G. (1983) Metabolism of one-carbon compounds by chemotrophic anaerobes. *Advan. Microb. Physiol*. **24**, 215–99.

[29] Zeikus J.G. (1983) Metabolic communication between biodegradative populations in nature. In *Microbes in Their Natural Environments*, p.423 (ed. J.H. Slater, R. Whittenbury & J.W.T. Wimpenny). Cambridge University Press, Cambridge.

Part 3
Micro-organisms in their Natural Environments

Naturam expellas furca, tamen usque recurret.

(You may drive out nature with a pitchfork, yet she'll be constantly running back.)

Horace 65–8 BC
Epistles

3.1 The terrestrial environment

3.1.1 COMPOSITION OF THE SOIL

Soil governs the productivity of plants in agriculture, forestry and natural ecosystems. Water, sunlight and nutrients control its productivity. It is composed of various proportions of inorganic and organic components: the results of interactions between many complex processes such as the weathering of rocks, the decomposition of plant materials and the redistribution of materials by water movement. Detailed accounts of these processes and of the chemical and pedological properties of soils are given in textbooks on soil science [28, 78]. Other background volumes concerning plant microbiology [20], soil biotechnology [57] and biological interactions in soils [27, 90] have recently been published.

Soil classification

Soils can be divided into a number of major types which are associated with climatic and vegetation zones. For example, one type occurs on the cold and humid upland areas of the British Isles where organic matter tends to accumulate as peat. Major soil differences are usually associated with parent material. Thus soils derived from chalk tend to be shallow, light and freely drained, whereas those derived from clays tend to be heavy and liable to waterlogging.

Soil classification systems make use of all these differences and can reflect either the agricultural potential of the soil, as in land use classifications, or its geological and pedological properties. Texture is an important classification feature (see Fig. 3.1.1) which describes the size distribution of three sizes of soil: clay, silt and sand. The consistency properties of soil are concerned with the power required for its cultivation. In terms of adhesion and resistance to deformation or rupture these are scaled as 1, harsh; 2, friable; 3, plastic; 4, sticky; 5, viscous.

Soil structure and tilth

Some of the soil particles form aggregates, clods or crumbs. The way in which these are arranged spatially

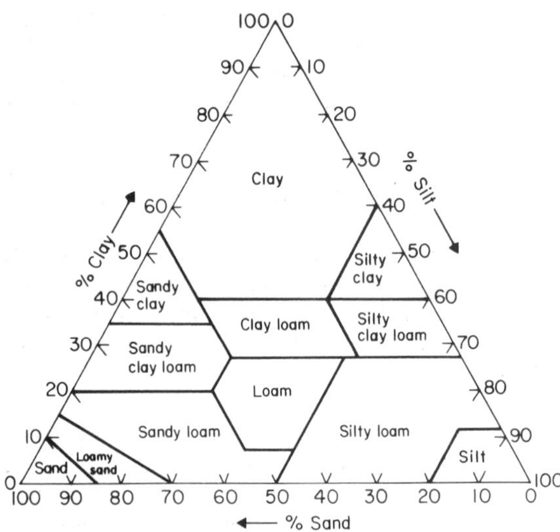

Fig. 3.1.1. The textural classes of soil.

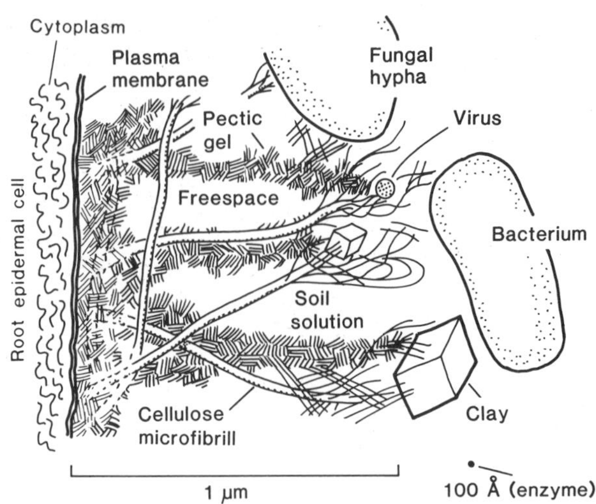

Fig. 3.1.2. Schematic of the root–soil boundary region. After Jenny [49].

gives soil its structure. The tilth of a soil is related closely to this and is dependent on the size distribution of the particles which have become associated and on their mellowness or rawness. A soil is considered to have a good structure when it holds sufficient water to prevent moisture deficits around plant roots during dry periods but on the other hand allows sufficient drainage to prevent waterlogging during wet periods. An extension of this is that there should be sufficient gas-filled pores to allow gas exchange with the atmosphere and reduce the chances of localized anaerobic pockets developing. Structure is also important for seeding; the aggregates in the seed bed must be sufficiently large to prevent wind erosion but must not be so large as to provide a mechanical barrier to the germinating seed.

Micro-organisms can contribute to the structure of soil by producing gums and cements but these can also damage structure by blocking pores. Some of the best soil structures are formed in grassland pastures where there is a large and continuous input of root materials. However, it is unclear if plant tissues and their metabolites or microbiological breakdown products of these provide the major source of soil stabilizers [58]. Fig. 3.1.2 depicts the root region with associated micro-organisms and attached clay particles; this schematic is based on SEM (scanning electron microscope) and TEM (transmission electron microscope) pictures.

Plant residues are other sources of substrates for micro-organisms to produce binding agents (see Fig. 3.1.3).

Micro-organisms appear to be involved in the production of stable aggregrates in two ways. Firstly the microbial cells themselves hold soil particles together by adhesion or mechanical binding and secondly they form extracellular polysaccharides or other organic compounds which produce a similar effect. In general the cells induce the formation of aggregrates whereas the products act as stabilizers. Polysaccharides are major agents in soil aggregation but other polymers are also involved [21]. Waksman and Martin [94] first demonstrated that bacteria produced slimy substances which aggregated sand–clay mixtures. Many bacteria can improve soil structure; this is not related necessarily to the number of micro-organisms actually present but more to their polysaccharide-forming ability.

If polysaccharides are the major agents of aggregation, periodate oxidation followed by treatment with borate should destroy them and hence the structure also. This treatment has been applied to woodland, forest and agricultural soils, but it only destroyed the structure of the latter. The reasons for this anomaly have still to be explained satisfactorily.

Tillage provides more oxygen for the soil microflora to synthesize polysaccharide gums but it also induces the oxidative decomposition of the organic matter which controls structure. Furthermore, it can make both the micro-

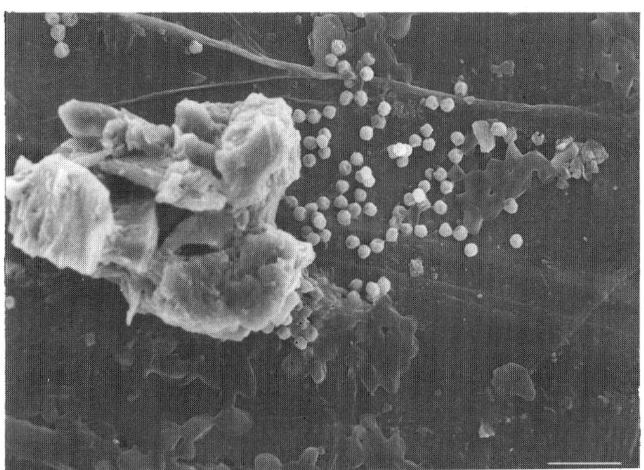

Fig. 3.1.3. Slimy materials produced by bacteria aggregating soil particles to straw in soil. Magnification ×775.

organisms responsible for structure formation and the structure itself more susceptible to damage by frost. It is possible, therefore, that soil structure may be improved in the long term when seeds are drilled directly into uncultivated soil.

The soil profile

Laboratory experiments with plants and micro-organisms on solid media usually involve sieved soil or sand packed into pots. This is unrepresentative of conditions in the field where natural soil-forming processes give rise to more or less distinct layers or horizons, which collectively form the soil profile. These horizons can be differentiated by colour, texture, physical structure, porosity, the amount and kind of organic matter, root growth and microbial and other organisms present. The mineral composition affects the colour of the horizons; for example, grey and ochrous colours are due to the reduced and oxidized forms of iron when there is seasonal waterlogging. Cultivation can cause the horizons to overlap.

3.1.2 PHYSICOCHEMICAL CONDITIONS

Surfaces

Clay particles are colloids and potentially can interact with micro-organisms. Kaolinite, montmorillonite and illite are the principal clay minerals found in soil and have been widely used for experimental studies of particle micro-organism interactions. The sand and silt particles of soil are relatively large, have surfaces which are fairly inert and carry little moisture or organic matter. Accordingly there is little microbial activity on them. A scanning electron micrograph showing micro-organisms on the surface of a quartz grain is shown in Fig. 3.1.4. It is likely that they are attached to the surface because the quartz is coated in places by charged clay colloids or organic matter. This patchy coating may explain the uneven distribution of cells on the grain. It is important to realize that micro-organisms sorb to clay particles but adhere to organic matter.

Clay colloids are coated with metal hydroxides and sesquioxides and the surfaces carry polarized but net electronegative charges. Micro-organisms are similar in some respects to clay colloids since the macromolecules lying on the surface of the cell wall or the capsule possess

Fig. 3.1.4. Colonization of a quartz grain by bacteria in soil as demonstrated by scanning electron microscopy (magnification ×2700). Note the adhesion and/or sorption of a sparse population of rod-shaped bacteria and organic particles which have probably originated from the more densely colonized root in the right foreground. Photograph courtesy of Dr R. Campbell, University of Bristol and Dr A.D. Rovira, CSIRO, Adelaide, Australia.

charged groups, which are again polarized and have a net electronegative charge. The net surface charge is known as the zeta potential and the polarization of charges causes micro-organisms to be sorbed to the clay colloids. It is also possible, however, that negative areas of charge on the clay and micro-organisms will be linked by the intervention of a metal ion which carries a positive charge of not less than two, hence forming a 'bridge'.

A clay particle interacts with water because it has exchangeable ions on or near its surface. When the surrounding water is relatively free from electrolytes, some of the exchangeable ions are replaced by H^+ ions and so leave the clay surface and enter the water layers close to the clay. As the resultant layer of exchanged ions and the water surrounding it both have net positive charges which are distributed throughout a volume rather than over a surface, they form what is known as a diffuse double layer [11].

Water

Micro-organisms are generally considered to be aquatic but lack of water often limits microbial activity in terrestrial systems. The important physical concepts governing soil water are covered in a text on agricultural physics [76] and references specifically for the microbiologist are also available [38, 69].

Gas and water are contained in the voids between soil particles. The gravimetric water content of a soil, ω, is expressed as g water per g soil dried for 24 hours at 105°C whereas the volumetric water content, θ, is expressed as cm^3 water per cm^3 soil. The former is the easiest to determine but the latter is usually a more useful measure because it allows for the shrinkage and expansion of soils which accompany changes in water content. Another useful measure is the bulk density, ρ_b, which is expressed as g solids per cm^3 total volume of solids, liquids and gases. The various expressions are related thus:

$$\theta = \omega\rho_b$$

Microbiologists frequently use the empirical gravimetric determination and express the water content as a percentage. On the other hand physicists more often use the volumetric or bulk density determinations. The volumetric content can be determined non-destructively with a neutron moisture meter. In this technique a probe is inserted into the soil through an access tube and fast neutrons are emitted. The backscatter of thermal neutrons is counted and the count rate is proportional to the volumetric determinations.

None of the above techniques give an idea of the water which is actually available to plants and micro-organisms. This is the 'moisture potential' or 'water potential' and is composed of two major components, the matric and osmotic potentials. These terms are similar to water activity (see Section 4.3.3), which is seldom used to describe water availability in soil.

The matric potential is concerned with the retention of water in the voids between soil particles. These potentials are governed by the surface tension forces and are proportional to the radii of menisci of the water; in order to absorb water the micro-organism or plant root has to exert a suction to overcome these forces. This can be analysed from the principles of capillary rise and, while the soil pores are not true capillaries, they can be considered as having an 'effective' or 'equivalent' capillary radius, r. The suction necessary to absorb water or matric potential, τ (dyne cm^{-2}) can then be expressed as

$$\tau = 2\gamma/r$$

where γ is the surface tension of water (73 dyne cm^{-1}). As the suction is increased, r becomes larger. The suction can be expressed in several ways:

$$1 \text{ bar} = 10^6 \text{ dyne cm}^{-2} = 0.987 \text{ atm} = 1022 \text{ cm } H_2O$$
$$= 75 \text{ cmHg} = 10^5 \text{ Pa}$$

An ill-defined term which is frequently used in connection with matric potential is field capacity; this is the water content of a soil, draining from saturation, when the rate of water loss due to gravity is small (*c.* -0.05 to -0.5 bar).

There are several methods of measuring matric potential. The simplest tensiometer uses the vacuum produced by the weight of a column of mercury to suck water from soil standing on sintered glass in a funnel. The matric potential is adjustable:

$$\tau = -h_w - h_m\,\rho_m \text{ (cm water)}$$

where ρ_m is the density of mercury and h_w and h_m are the respective barometric heights of water and mercury.

The osmotic potential of water, π, is a result of the presence of solutes. These decrease the entropy and increase the order, which can be expressed as follows:

$$\pi = -RT\rho\nu m\phi/10^9$$

where R is the gas constant (8.31×10^7), T, the temperature (K), ν, the ions per molecule (taken as 1 for non-ionic solutes); m, the molality; ϕ, the osmotic coefficient, and ρ, the density.

The moisture content varies depending on whether

water is being extracted or added to the system (see Fig. 3.1.5).

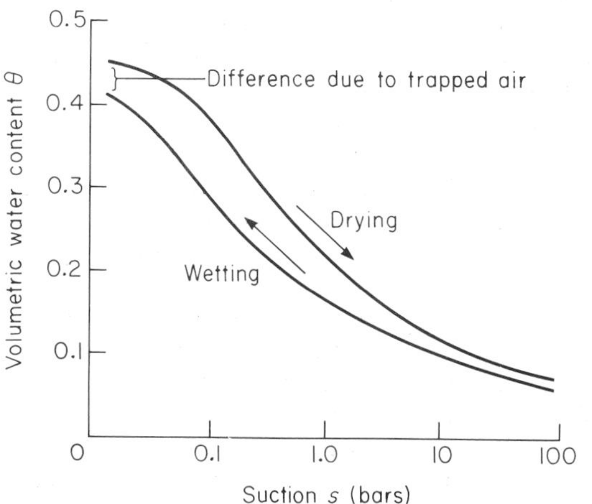

Fig. 3.1.5. Pore water content during wetting and drying. Redrawn with permission from White [98].

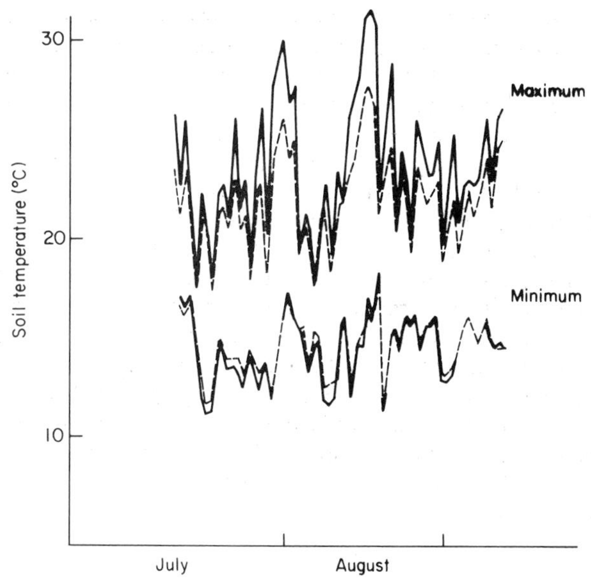

Fig. 3.1.6. Variation in soil temperatures at 75 cm depth at Camborne, Cornwall, 1972. ———, bare soil; ----, weed-covered soil. Courtesy of Dr A.A. Tompsett, Deputy Director, Rosewarne Experimental Horticulture Station.

Temperature

Temperature helps determine both the composition of the soil microflora and its activity. It varies with latitude and altitude and is cyclic both diurnally and seasonally. There is also a great variation with depth and this is influenced by the water content. The diurnal fluctuation can be as much as 35°C with minima below freezing and maxima about 60°C.

The presence of vegetation, plant residues or mulches reduces the variation in soil temperatures (see Fig. 3.1.6). Heat from the sun can also be reflected back by soil, particularly light-coloured soil.

Acidity and alkalinity

The composition of the soil microflora is highly dependent on soil pH. However, it is very difficult to measure the pH of a soil; this is largely a result of the diffuse double layer, which causes a hydrogen ion concentration gradient between the surface of soil particles and the surrounding soil solution. The highest concentration (lowest pH) is on the particle surface and the gradient is reduced if the electrolyte concentration in the soil solution is increased.

In order to overcome some of the problems in measurement, the pH of temperate soils is normally determined in a solution of 0.01 M calcium chloride. This gives the approximate ionic strength which is normally present in the soil solution. Determination of pH in such solutions results in values about 0.5–0.9 units lower than determinations made in water.

Metabolic activity, particularly that involving redox reactions, influences the pH of soil. For example the reduction of nitrate to nitrogen and sulphate to sulphide which takes place under anoxic conditions will cause the pH of the soil to become alkaline. However, the production of carbon dioxide during respiration causes the pH to become acidic. This change is approximately proportional to the logarithm of the partial pressure of carbon dioxide. The pH is also influenced by the time of year, the climate and the previous cropping history.

The soil pH is usually between 4.0 and 8.5. Rain leaches bases and this is one reason why most soils are on the acid side. The degradation of leaf litter can also cause a drop in pH. The result is often an acid pH on the surface layers of soil whereas the pH in lower horizons can be alkaline as a result of solution of calcium from mineral particles or from mollusc shell fragments [32]. Thus, while it is easy to measure the pH of the total soil solution, it is difficult to assess the pH in different locations within the soil, especially on particle surfaces where the micro-organisms and their extracellular enzymes may be present and active.

Soil atmosphere

In a well-drained soil many of the voids are filled with air. However, there will always be some nearly sealed spaces where microbial respiration reduces the oxygen concentration and increases the carbon dioxide and other gaseous metabolites (see Fig. 3.1.7). In wet soils there are more 'sealed' voids and consequently the mean concentration of oxygen throughout the soil is reduced, with some pockets becoming anoxic. This has a great influence on the soil microbial population because fungi and aerobic bacteria require oxygen and only rare instances have been reported of fungi other than yeasts acting as anaerobes. Although particle size is an important factor in soil drainage, even very small crumbs are likely to be heterogeneous. For example, the surfaces may be aerobic, allowing nitrification to occur, while the centres may be anaerobic, promoting denitrification with the formation of nitrous oxide and nitrogen.

The concentration of oxygen in the soil atmosphere can be expressed as its percentage or as its partial pressure, but account must be taken of diffusion. If soil crumbs are considered as respiring structures with an even distribution of micro-organisms, in a steady state the rate of oxygen uptake by the crumb is equivalent to the rate of oxygen diffusion to it. The diffusion depends on: (a) the difference in oxygen concentration between the soil gas phase and the water surrounding the soil crumb at a gas/water interface; (b) the geometry of the crumb structure, which can be affected by rainfall; and (c) the diffusion coefficients of the medium around the crumb. There is a limiting oxygen concentration at which no further reduction in oxygen uptake rate occurs; this is described by Michaelis–Menten kinetics. Oxygen is unlikely to be limiting at low matric potentials because most voids are filled with gas and the diffusion rate of oxygen is then 2.1 $\times$ 10^{-1} cm^2 s^{-1}. However, as the voids become filled with water there will be a limitation in oxygen availability because the diffusion rate is reduced to 2.6 $\times$ 10^{-5} cm^2 s^{-1}. The movement of oxygen in soil has been analysed [22, 36] in detail and can be summarized as:

$$r^2 = 6DC/M$$

where r is the maximum radius of a crumb in which oxygen can just reach the centre, D is the diffusion coefficient of oxygen in the crumb, C is the concentration of oxygen in the water on the outer surface of the crumb, and M is the rate at which oxygen is used up in root and microbial respiration. When reasonable figures are fitted to this equation, it is found that, in British soils under summer conditions, waterlogged crumbs larger than about 1 cm are likely to have cores which are anoxic but in winter this size is about 2 cm [78].

The concentrations of carbon dioxide in soil are generally between 0.002 and 0.02 atm (0.202 and 2.02 kPa) but values as high as 0.1 atm (10.1 kPa) have been recorded [65]. In a steady state, the carbon dioxide production by roots and micro-organisms must be equivalent to that lost in upward or downward movement. The production of the gas will be promoted by substrates for microbial growth and thus cropping, farmyard manure, crop residues and green manure increase production of carbon dioxide. However, these processes may also increase the air space in the soil, resulting in greater loss of the gaseous metabolite.

Ploughing is effective in aerating soil and might appear more effective than the modern practice of directly drilling seeds into soil without prior cultivation. However, there is no evidence which suggests that this is so, partly because in the latter method the number of earthworms are increased, because plant root residues are left in the soil surface layers and/or there is a reduction in injury to the worms from the plough.

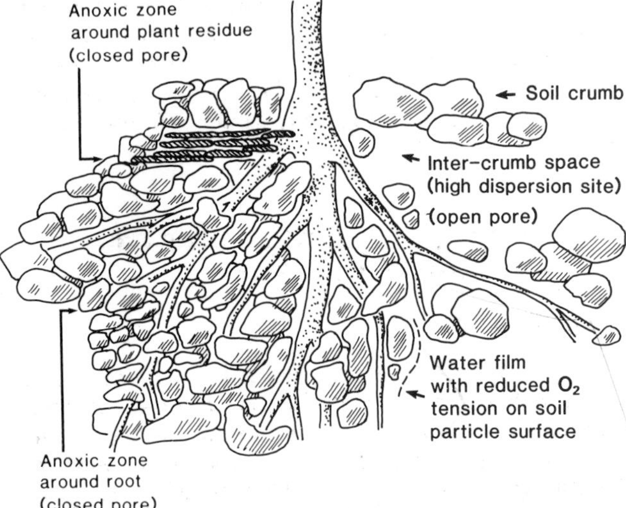

Fig. 3.1.7. Development of anaerobic zones in the soils as a consequence of microbial metabolism around roots and plant residues in closed pores.

When oxygen has disappeared from the soil atmosphere, microbial activity can of course be continued by those organisms capable of using alternative electron acceptors such as sulphate (see Chapter 2.5). This is activity governed by redox potential, which, like pH, is very difficult to determine in the field.

It is clear from the foregoing that, at any given time, there is a great heterogeneity in the chemical and physical nature of soil, even within a small area. This is further complicated by the fact that biological activity modifies soil conditions. Thus the soil must be treated as a collection of micro-environments which do not remain in a constant physical, chemical or biological state.

3.1.3 THE SOIL POPULATION

Types of organisms

Bacteria are the most numerous (10^6–10^9 viable cells cm^{-3}) organisms but owing to their small size (*c.* 1 μm) are not necessarily the major component of the soil biomass. Generally they are regarded as largely responsible for the cycling and transformation within the soil of carbon, nitrogen, phosphorus, iron and sulphur. The soil bacterial population is extremely diverse; a large proportion of the bacteria in Bergey's manual are soil-borne species and we can expect to find both new species and new genera in soils. The problems of identification are probably greater with bacteria from soil than with those from any other habitat and numerical taxonomy is perhaps the best approach.

Fungi, which are larger (spores *c.* 20 μm), are probably at least as important as bacteria in soil. Their tolerance of low pH makes them particularly significant in acid soils. They have major roles in plant residue decomposition and in plant pathogenesis. Soil fungi also exhibit great species diversity in the soil and identification is complicated by the fact that their life cycles in soil are often quite different from those in laboratory culture. Useful texts on the identification [25] and ecology [34] of soil fungi are available and a further text specifically describes the ecology of root-infecting fungi [30].

Classification of soil yeasts has been attempted by di Menna [67]. Their functions in soil are uncertain but they have been implicated in increasing soil structural stability through their ability to produce extracellular gums.

Other components of the soil microflora which have received relatively little attention are the actinomycetes and protozoa. *Streptomyces* spp. have been shown to be common producers of antibiotics. Protozoa occur in soil as flagellates, amoebae, ciliates and testacea — in the order of 10^3 (g wet soil)$^{-1}$ for each group. They may have a role in ecology as predators of soil bacteria [24]. Another group which has received relatively little attention as yet in soil microbial ecology is the viruses (see Chapter 2.2).

Soil animals such as nematodes, enchytraeid worms (potworms), lumbricid worms (earthworms), termites, springtails and mites also contribute to the soil biomass (see Chapter 3.3). It must be emphasized in this respect that these activities are not only the result of whole organisms but can also result from the activity of free enzymes in the soil. There is a great diversity of such activity in the soil [18].

Spatial and seasonal distribution

There have been many studies of the seasonal distribution of micro-organisms, particularly fungi [25], and, as might be expected, there is great variation. Similar studies have been made of species diversity down the soil profile but few have considered the correlation between the products of metabolism and the microbial population. One exception is the work on denitrification of nitrate added to the soil surface [26]. It was demonstrated that this process increased down the profile and was correlated with an overall increase in the number of denitrifiers.

Populations do not always correlate with activities and this was illustrated in studies [93] on the number of sulphate-reducing bacteria across a soil crumb in relation to the presence of sulphide (see Fig. 3.1.8). This demonstrated clearly the heterogeneity of the soil population even within a crumb.

3.1.4 SOURCES OF SUBSTRATES FOR MICROBIAL GROWTH

Inorganic sources

Soil micro-organisms are dependent on the minerals of soil in addition to any added fertilizers to satisfy their requirements for both major and trace elements. The chemolithotrophic population use these as a source of reducing equivalents, often coupled with light energy (see Section 2.5.4). Amongst the phototrophs are the blue-green algae, which occur in the surface horizons and can fix atmospheric nitrogen if there is an abundant source of light energy and surface water (see Section 2.5.5), and

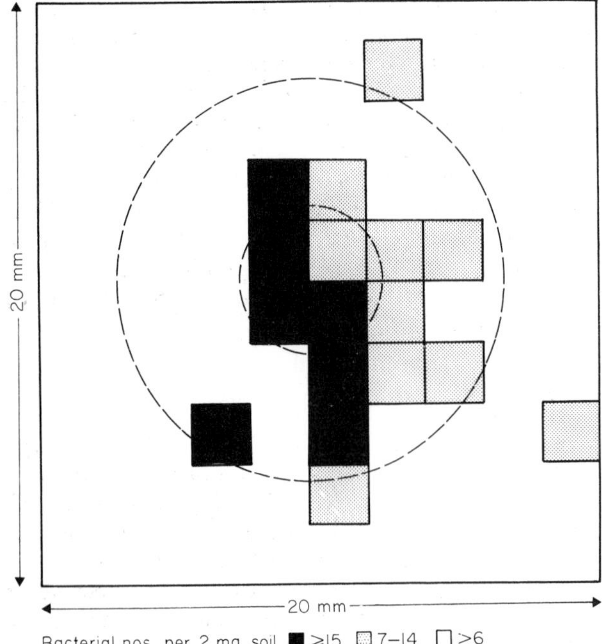

Fig. 3.1.8. Distribution pattern of sulphate-reducing bacteria when soil was incubated with starch. Starch (3 mg) was added to the soil; its location is indicated by the inner circle. The outer circle indicates the area where blackening occurred after incubation. After Wakao & Furusaka [93].

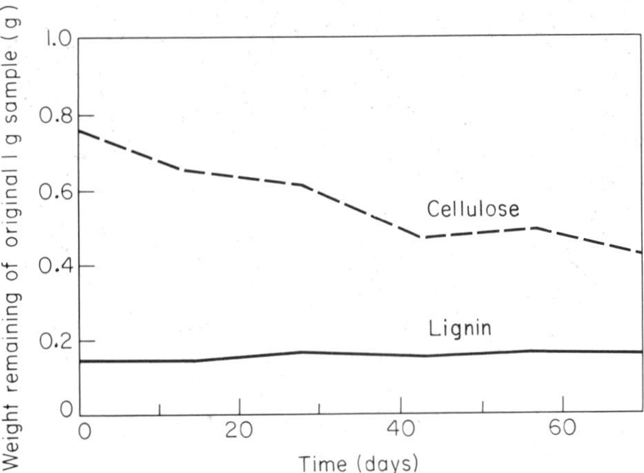

Fig. 3.1.9. The rate of breakdown of cellulose and lignin in wheat straw in clay soil as an index of secondary saprophytic activity. The mean soil temperature at 10 cm depth was 11°C during the experiment. After Harper & Lynch [41].

the nitrifying bacteria (see Sections 2.5.4 and 2.1.7). Plants can influence the availability of inorganic nutrients to micro-organisms and this is discussed later (see Section 3.1.9).

Crop residues and leaf litter

A major carbon substrate for saprophytes in forest ecosystems is leaf litter and the counterpart in agriculture is crop residues. The growth of heterotrophs in soil is usually considered to be limited by the availability of a suitable carbon source. However, there are exceptions such as microbial growth on straw (see Section 3.1.5) when nitrogen may be the limiting nutrient.

There is a consistent pattern of microbial succession on crop residues and leaf litter [44]; this is particularly well studied for fungi. The primary colonizers are those that can use the simple soluble materials, including monosaccharides and disaccharides, which are the first to be leached. Examples of these are the primary saprophytic sugar fungi, such as *Mucor* spp. and *Rhizopus* spp. They grow rapidly relative to later colonizers and it has been indicated that this may be characterized by a high specific rate of sugar utilization [61]. The secondary colonizers use the more complex material, such as polysaccharides, and commonly produce cellulases (see Fig. 3.1.9). Tertiary colonizers are capable of metabolizing some of the more complex polymers such as lignin, but as these are not readily assimilable energy sources the microbe's growth is quite slow. It should be recognized that such successional groupings usually overlap.

The colonization of plant residues by fungi is of particular agronomic significance if the colonizers are plant pathogens. Often the micro-organisms will only show their pathogenic properties if present in sufficient numbers. In this context Garrett [30] has defined inoculum potential as 'the energy of growth of a fungus available for colonization of a substrate at the surface of the substrate to be colonized'. Furthermore 'the share of substrate obtained by any particular fungal species will be determined partly by its intrinsic competitive saprophytic ability and partly by the balance between its inoculum potential and that of a competing species' (see also Section 3.5.5).

Rhizodeposition

The total carbon flow from roots is rhizodeposition. This flow is made up of several components. *Exudates* leak from living roots, *secretions* are actively pumped, and *lysates* are passively released from roots during autolysis.

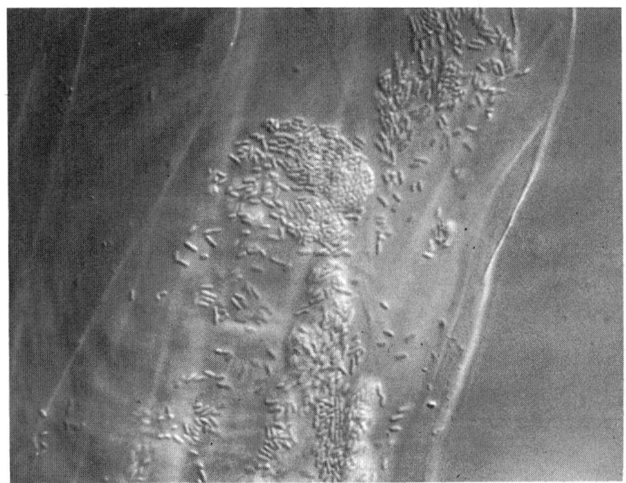

Fig. 3.1.10. Microcolony of bacteria spreading across wheat root epidermal cells viewed by Nomarski differential interference microscopy. Magnification ×800.

Mucigel arises from *plant mucilage* and is produced from epidermal and root cap cells. Microbial growth around roots leads to the production of *microbial mucilages*, which also form part of the mucigel. Fig. 3.1.10 shows the dense proliferation of micro-organisms around the root epidermis.

Investigations of these important substrates for microbial growth became possible when plants could be grown under aseptic conditions. These studies showed that root exudate material consisted usually of about 90% carbohydrate and 10% amino acids [6]. However, most studies were done in axenic solution culture, which does not subject the root to many of the physical stresses encountered in soil. Furthermore, the micro-organisms themselves could modify the pattern of exudation. It has also been shown that contact with solids, anaerobiosis, moisture stress, low temperature, herbicides and mechanical forces all increase exudation. Such problems were overcome recently when plants were grown in soil in the presence or absence of micro-organisms and with the shoots exposed to a gas phase containing $^{14}CO_2$ [8]. The radio-labelled photosynthate could be recovered from the roots and shoots of the plants, from the water soluble and insoluble material in the soil and from the CO_2 produced in root and microbial respiration. Up to 39% of the plant's dry matter production can be released by roots (see Table 3.1.1).

Soil organic matter

When biological residues decay in soil, they undergo humification [45]. Part of the process involves lignin, the more resistant woody part of decaying plant tissue left after hemicelluloses and celluloses have been metabolized. Lignin is composed mainly of phenolic polymers which eventually give rise to humic acid but neither lignin nor humic acid can be accurately defined. Humic acid is extracted from the soil with weak alkali and precipitated as a dark amorphous product with acid. The fraction of soil organic matter which cannot be readily extracted with cold alkali is termed humin whereas fulvic acid is the fraction which is soluble in both acid and alkali. There are three main aryl aldehydes in the polymers:

CHO, OH — *p*-Hydroxybenzaldehyde; CHO, OCH_2, OH — Vanillin; CHO, CH_3O, OCH_3, OH — Syringaldehyde

Table 3.1.1. ^{14}C loss from roots at 18°C day/14°C night and 16-hour day after growing plants in a chamber with a constant specific activity of $^{14}CO_2$ (70–90 mCi (g C)$^{-1}$) (after Whipps & Lynch [97])

Plant	(^{14}C lost/^{14}C fixed) × 100			
	Rhizosphere	Bulk soil	Respired CO_2	Total
Wheat (21 d) (*Triticum aestivum*)	5	12	23	40
Maize (28 d) (*Zea mays*)	4	4	10	18
Tomato (28 d) (*Lycopersicon esculentum*)	5	5	19	29
Pea (28 d) (*Pisum sativum*)	10	4	24	38

These substances in the polymeric form do not provide very favourable growth substrates for micro-organisms and this probably explains why some fractions of soil organic matter have been dated by ^{14}C as being at least 1000 years old [70, 72]. However, in the free state the aryl aldehydes can be degraded very slowly by basidiomycetes, ascomycetes, actinomycetes and some bacteria. The degradation is usually stepwise to give dihydroxyphenols followed by ring fissure. The oxidative coupling of phenols and amino acids by micro-organisms leads to the brown polymer pigment, melanin, although an additional energy source is usually required to produce this compound.

3.1.5 ENERGY FLOW

The principles of microbial energetics were outlined in Chapter 2.5 and the utilization of that energy for growth was described in Chapter 2.4. These considerations are relevant to the microbial cell growing in soil but most actual assessments have been made at a level no smaller than the microbial community. In fact, the total soil biomass is usually considered in ecological assessments of the energetics of soil micro-organisms. By comparison with laboratory studies of population dynamics, in field studies of soil micro-organisms the scale is much larger in terms of space (metres and kilometres against micrometres) and time (centuries and years against hours and days). The energetics of soil microbes are affected by resource quality, climate and soil conditions; the study of the biotic energetics and the chemical and physical factors affecting energetics make up 'production' ecology which quantifies animal, plant and microbial production in different ecosystems [85].

Primary productivity

Primary productivity (the net input of organic materials to soil) must be measured before energy budgeting can be attempted. Table 3.1.2 gives an indication of primary production and its turnover in the major biomes. Similar estimates can be made for agricultural soils, probably with more precision because the growth of individual plants can be more easily assessed.

Biomass measurements

The second stage in assessing the energy flow is to assess the microbial biomass which will potentially utilize the primary production. This can be done by direct microscopic observation of soil samples. In this approach, it is necessary to measure the number of bacterial cells or length of fungal mycelium, assess the cell density by growing organisms in pure culture and calculate the cell mass from:

$$\text{mass} = \text{density} \times \text{volume}$$

The biomass (total weight of living microbial cells) can be assessed from microscopic measurements. The volume can be measured from the cell dimensions and the density by suspending cells in solutions of salts (e.g. CsCl) with different densities, so that cells with a similar density to that of the suspending solution cannot be sedimented or centrifuged. Caution should be exercised, however, in the interpretation of such data because the osmoticum may extract water from the cell, giving an erroneously high result. Then:

$$\text{biomass} = \text{number of cells} \times \text{density} \times \text{volume}$$

A spherical bacterial cell of 1.0 μm diameter and density of 1.5 g cm^{-3} weighs *c.* 0.8×10^{-12} g. Fungal mycelium

Table 3.1.2. Primary production, organic matter standing crops, and turnover in major biomes (after Heal & Ineson [43])

Biome	Net primary production (t ha^{-1} yr^{-1})	Litter input (t ha^{-1} yr^{-1})	Standing dead (t ha^{-1})	Litter (t ha^{-1})	Soil organic matter (t ha^{-1})	K_L*	K_r*
Tundra	4.0	1.7	1.8	28.0	200.0	0.06	0.017
Boreal forest	8.0	5.8	1.3	35.0	150.0	0.17	0.043
Temperate forest	13.0	8.5	7.9	30.0	120.0	0.28	0.082
Temperate grassland	15.0	7.3	—	4.0	220.0	1.78	0.065
Desert	2.0	1.3	—	1.0	80.0	1.30	0.025
Tropical forest	17.0	15.8	13.5	7.5	85.0	2.11	0.160

* K_L = litter input/litter; K_r = net primary production/(standing dead + litter + soil organic matter).

can be assumed to be cylindrical (volume = $\pi r^2 l$) with diameter of *c.* 4 μm. The density is often quoted as about 1.5 g cm^{-3} but the basis for this is unclear and it seems rather heavy; possibly errors may originate during density gradient centrifugation.

Direct observation methods are extremely tedious and the first serious attempt to produce an indirect method was by Jenkinson and Powlson [48]. They killed the microbes by chloroform fumigation, allowed fresh micro-organisms to grow on the dead cells as substrates and measured the carbon dioxide respired in a 10-day period. It was also necessary to correct for the inherent respiration of the soil. To measure the percentage of cell carbon mineralized to carbon dioxide they added pure cultures of micro-organisms to soil, then fumigated them with chloroform and measured carbon dioxide produced in the succeeding 10 days. A mean figure (k) was obtained for the range of organisms and a value of 0.41 now seems to be most appropriate for incubations at 22°C, but this factor is temperature-dependent [1]. Then the biomass, B mg carbon (per 100 g dry soil), can be calculated from:

$$B = \frac{X - x}{k}$$

where X is the carbon as carbon dioxide produced by fumigated soil in 10 days and x is that produced by unfumigated soil. Others [15] have shown in their soils that:

$$B = \frac{0.673X - 3.53}{k}$$

which avoids the need to measure carbon dioxide from unfumigated cores; however, they have also indicated that anomalous results can be obtained if soils are sieved prior to the determination [63].

Another method is to measure the ATP content of soil (see Fig. 3.1.11); ATP is required for the biosynthetic and catabolic reactions of cells. The major variant of this technique, as used by different investigators, is the extractant used. With Na_2HPO_4 and paraquat dichloride as extractant, Jenkinson and Oades [47] related their estimates to those of the fumigation method by:

$$\text{biomass C in soil} = K\ (\text{ATP content of soil})$$

where K is 120, although recent evidence shows 160 to be a better value and that it is preferable to use purified luciferase with luciferin to generate light from extracted ATP [87]. The correlation may break down in soils that have recently received large additions of substrate, particularly when phosphate is in short supply.

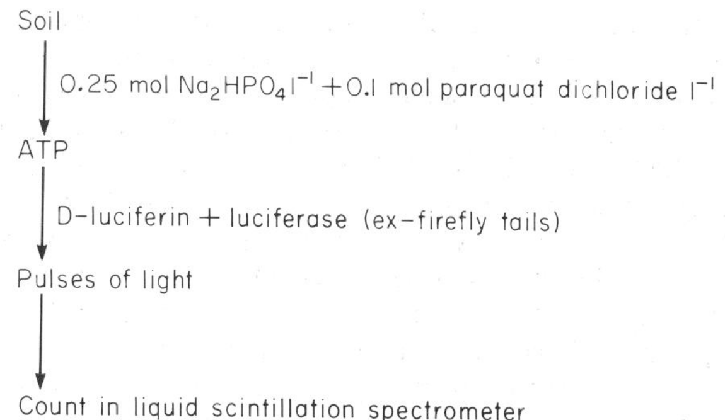

Fig. 3.1.11. ATP extraction procedure from cells. After Jenkinson & Oades [47].

Calculations of energy budgets

In recent years there have been several attempts to correlate the input of substrate energy into soil with the growth of micro-organisms. For continuous (or fed batch) culture:

$$\text{overall rate of energy consumption} = \text{consumption for growth} + \text{consumption for maintenance}$$

$$\frac{\mu x}{Y} = \frac{\mu x}{Y_g} + mx$$

where μ is the specific growth rate (h^{-1}), x is the biomass (g), Y is the observed growth yield (g dry weight per g substrate), Y_g is the true growth yield when no energy is used in maintenance (g dry weight per g substrate) and m is the maintenance coefficient (g substrate per g dry wt per h). The specific maintenance rate, a, can be expressed as mY_g.

In many natural systems the supply of energy substrate (ds/dt) can be calculated from a knowledge of the amounts of leaf litter, crop residues and root-derived materials which are added annually. The soil biomass can be calculated from counting the population by direct observation and by a knowledge of the mean cell or unit hyphal weight, as determined in the laboratory (see Section 3.1.7). It is most important that fungi are included with bacteria in estimates of biomass as many investigators have found these to be the major components of the soil biomass [2, 34, 80] (see Table 3.1.3).

The calculation of the growth rate or generation time of the soil population is heavily dependent on a knowledge of the 'average' maintenance energy. In laboratory cultures the maintenance energies of around 0.4 h^{-1} as found in the original calculations of Pirt [73] are probably

Table 3.1.3. Biomass estimates and annual litter production for Meathop wood soil (after Gray *et al.* [35] and Satchell [80])

Group	Dry matter biomass (kg ha^{-1})
Bacteria	36.9
Actinomycetes	0.2
Fungi	454.0
Protozoa	1.0
Nematodes	2.0
Earthworms	12.0
Enchytraeidae	4.0
Molluscs	5.0
Acari	1.0
Collembola	2.0
Diptera	3.0
Other arthropods	6.0
Total microflora	492.1
Total microfauna	36.0
Total biomass	528.1
Annual litter production	7640.0

only relevant to organisms growing in luxuriant nutrient media. Much lower figures have been suggested and there is now some experimental evidence for figures as low as 0.00003 h^{-1} for dormant organisms [3]. While it is almost certain that cells will have a finite maintenance energy in nature, there is a great need to find a method of evaluating this. One complication is that there are likely to be widely different values for different organisms depending, for example, on their ability to utilize energy from stored cell products. Seeds of plants can have extreme longevity (1000 years) by very slowly utilizing stored carbohydrates. In the broad distinction between *r* and *K* strategists (see Table 3.1.4) the *K* strategists might be considered to have very low maintenance energies. The ecological strategy adopted by an organism is dependent on the resource (substrate) quality and the climatic severity. For example, an organism normally regarded as an *r* strategist on root-derived carbon in summer might become a *K* strategist on lignocellulosic substrates in the cold of winter. Production ecology, therefore, only provides integrated energetic balances of *r* and *K* strategists (see Table 3.1.4).

Respiration

As respiration is often a good index of microbial activity, measures of the CO_2 output of soil should be a means of producing energy budgets. This has worked in some cases [32] but many investigators have failed to show that a good correlation exists between bacterial numbers and CO_2 output of soil.

If a bacterial cell weighs about 1.5×10^{-9} mg, then, as Gray and Williams [35] have calculated from a correla-

Table 3.1.4. Trends in microbial characteristics in early (exploitation) and late (interaction) stages of succession under favourable conditions and under conditions of continuous resource or discontinuous climatic adversity (after Heal & Ineson [43])

Characteristic	Exploitation (*r*)	Interaction (*K*)
Morphology	Small cells or diffuse mycelium	Large; often compact mycelium Cell walls resistant to animal enzymes
Physiology	Rapid growth rate, high yield efficiency, giving maximum colonization Use readily available substrates; nutrient-demanding Sensitive to plant defence compounds	Moderate growth rate and yield efficiency for substrate utilization Use of more resistant substrates; moderately nutrient demanding Production of defence compounds (antibiotics) against competitors and grazers
Life history	Simple; maximum production of wind- and water-dispersed propagules following short growth phase; sporing intermittent	Varied
Population dynamics	Explosive density independent, crash-through substrate depletion or opportunistic grazing	Relatively damped, density-dependent control by interspecific competition and selective grazing
Community structure	Diverse	Symbiotic relations extensive; very diverse

tion of soil respiration with bacterial numbers, 1 mg bacteria would evolve 0.121 mg CO_2 h^{-1}, i.e. 0.033 mg carbon. Assuming that this represents about 65% of the metabolized carbon, the remaining 35% will have contributed to an increase in biomass or excretion of metabolites. As the typical bacterial cell has about 50% carbon, the maximum increase in biomass would be 0.0333 mg h^{-1}, i.e. the doubling time of the population is *c.* 28.5 h. The authors indicated that unfortunately there are many sources of error in this calculation, leading to an overestimation of the growth rate. In particular, there is no consideration of the contribution of root respiration and it is erroneously assumed that respiration rates remain consistently high for the whole year.

^{14}C studies of plant residue breakdown

Autoradiography has been used to study the breakdown of ^{14}C-labelled plant parts in soil (see Fig. 3.1.12); after 6 months the label was mostly in the veins, which could be due to greater intensity of initial labelling or less decomposition of the lignin component or accumulation of label in decomposer fungal mycelium. A more quantitative investigation of the breakdown has been made [46] by adding ryegrass, uniformly labelled with ^{14}C, to soil (see Fig. 3.1.13). Initially the breakdown was rapid and only about one third of the ^{14}C label remained after 6 months; then breakdown slowed and about one fifth of the label was left after ten years. Over the same period the unlabelled carbon naturally present in the soil was also studied and it was found that this was much more resistant to degradation as only one tenth disappeared after 4 years. This shows that freshly incorporated organic matter is a much better substrate for microbial growth than that which has aged. Other similar studies [21] showed that, after 448 days incubation, 50% of the label was recovered as CO_2. The main sugars, glucose and xylose, were decomposed by 70% and there was little transformation to other sugars.

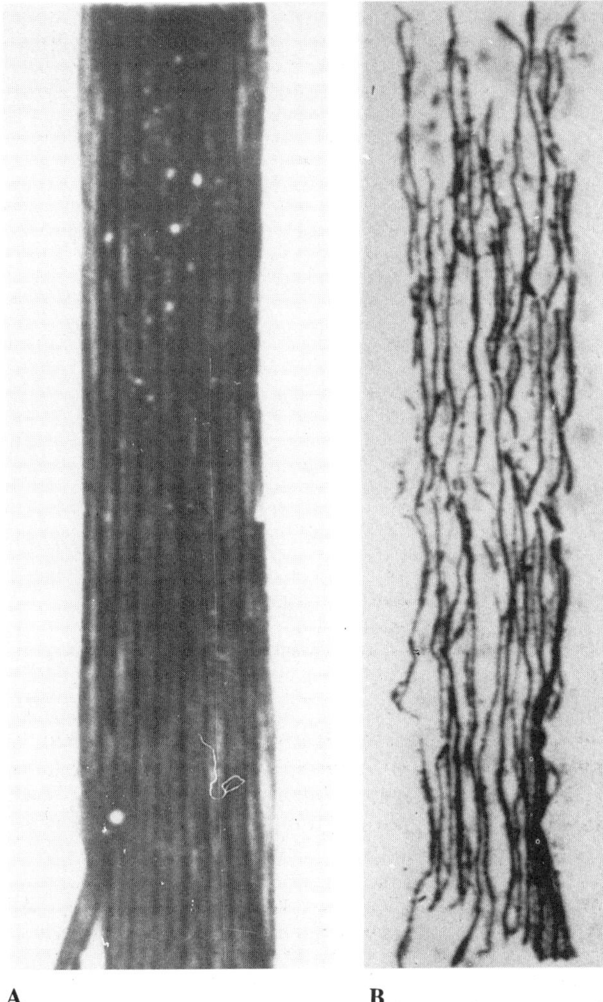

Fig. 3.1.12. Decomposition of ^{14}C-labelled leaves of ryegrass: (A) Zero time, (B) after 6 months. Photograph courtesy of Dr E. Grossbard. For further details see Grossbard [39].

Microbial growth on rhizodeposition products

The continuing flow of organic materials from roots sustains an active microbial population close by. Roots can release up to 39% of the plant's production so this is no trivial flow. A study of the amount of the flow and production of microbial biomass [7] found that the yields of bacteria were much greater than could have been derived from the quantities of carbohydrates exuded under aseptic conditions. This suggests that microorganisms on and around the roots enhanced the release of substrates. They possibly do this both directly, by making holes or by producing substances which stimulate the process, and indirectly, by utilizing the exudates and preventing their build-up in solution, thus increasing outward diffusion. However, this data must be interpreted with caution because some of the root-associated biomass may be growing oligotrophically on traces of carbon compounds in the air used to aerate the rooting medium and

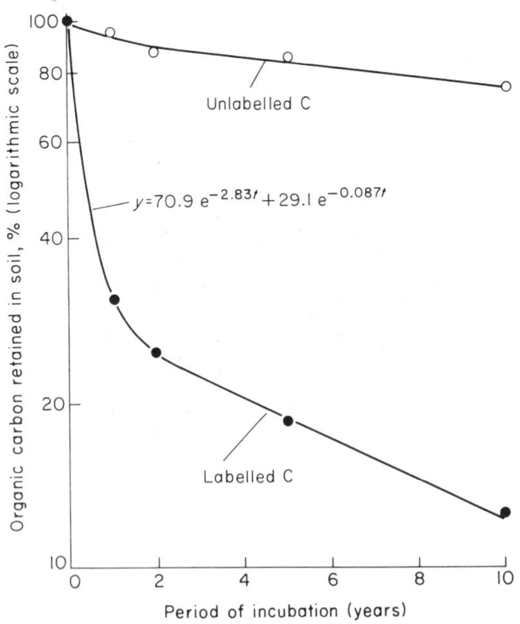

Fig. 3.1.13. Losses of labelled and unlabelled C from Rothamsted soils. Redrawn with permission from Jenkinson [46].

the cell size of bacteria in pure culture may be much greater than that on the roots.

Data on the growth of micro-organisms in the rhizosphere are useful in evaluating some of the processes which are of significance to cereals. For example, the efficiency of nitrogen fixation by free-living bacteria is only about 10–15 mg nitrogen per g sugar consumed (see Chapter 4.1) and, if the total exudation of about 0.2 mg sugar per mg plant dry weight were utilized in nitrogen fixation, only about 2–3 μg nitrogen per mg plant dry weight could be fixed, which is at most about 15% of the nitrogen content of the plant. The actual amount fixed would be much less, as this calculation assumes that all the rhizosphere population is able to fix nitrogen and in practice the proportion is likely to be very low, even when seeds have been inoculated. Furthermore, much of the nitrogen fixed may not be available to plants. The situation could be different for perennial species with slow growth rates but these calculations serve as a warning to temper some of the excitement [31] of the possibilities of non-symbiotic nitrogen fixation (see Chapter 4.1).

The carbon : nitrogen ratio

In laboratory studies of microbial growth on defined media, it is usually clear that either carbon or nitrogen is the growth-limiting nutrient. There is no such certainty in the natural environment, where organic materials can have widely different carbon : nitrogen (C : N) ratios (see Table 3.1.5). If the ratio is below 20, then the nitrogen is often adequate to satisfy requirements of the microflora which decompose the residues. However, ratios above this usually result in removal of nitrogen from the soil to the micro-organisms (immobilization), and this has to be replaced with fertilizer to avoid nitrogen deficiency. For example it can be assumed that 100 g straw contains about 40 g carbon and 0.5 g nitrogen. Assuming that about 66% of the carbon and nitrogen is available to the microflora in 6 months [46], that the mean C : N ratio of the microbial cells is about 5 : 1 and that 35% of the carbon (the remainder being respired or excreted) and 100% of the nitrogen is converted to biomass, an additional 1.5 g nitrogen per 100 g straw would have to be supplied from the soil. However, the immobilization of nitrogen is probably only temporary, except under anaerobic conditions when denitrification may occur, and thus when it is eventually mineralized the nitrogen may subsequently become available to plants.

Table 3.1.5. The carbon : nitrogen ratio of some organic materials (after Foth & Turk [28])

Material	C : N ratio
Soil humus	10
Sweet clover (young)	12
Barnyard manure (rotted)	20
Clover residues	23
Green rye	36
Cane trash	50
Corn stover	60
Straw	80
Timothy	80
Sawdust	400

More carbohydrate energy for nitrogen fixation is available from plant residues than from roots. Using an associative system where cellulolytic fungi growing on straw make available sugars for an anaerobic nitrogen-fixing bacterium, *Clostridium butyricum*, the nitrogen fixed could amount to 84 kg N ha^{-1} [62].

The place of mathematical models

DECOMPOSITION MODELS

Participants in the IBP Tundra Programme produced models of microbial decomposition with reference to the total microbial community in soil, i.e. synecology, rather than a consideration of individual species, i.e. autoecology. Initially, the studies were concerned with producing word models, which are essentially simple statements of some conceptual ideas. These may be of value to individuals in expressing ideas on paper, but their practical value is very limited.

These basic considerations have led to the development of mathematical models of the influences of abiotic variables [16] and chemical composition [15] on microbial respiration and substrate weight loss. The model for the former has been expressed as follows:

$$R\ (T,M) = \frac{M}{a_2 + M} \times \frac{a_2}{a_2 + M} \times a_3 \times a_4 \frac{(T - 10)}{10}$$

where $R(T,M)$ is the respiration rate (µl 0_2 g^{-1} h^{-1}) at temperature T (°C) and moisture level M (% water content on a dry weight basis), a_1 the % water content at which the substrate is half 'saturated' with water, a_2 the water content at which 50% of air spaces are water filled, a_3 the optimal respiration rate at 10°C when neither H_2O nor O_2 are limiting and a_4 the Q_{10} coefficient. The moisture level makes allowance for the oxygen availability. Simulations with the model were compared with measured rates of microbial respiration of the litter and generally produced good fits; an example is shown in Fig. 3.1.14. The authors admit that the forms of these evaluations are not rigorous but they do provide some prospects for future development in being able to predict seasonal patterns of weight loss under a variety of abiotic conditions.

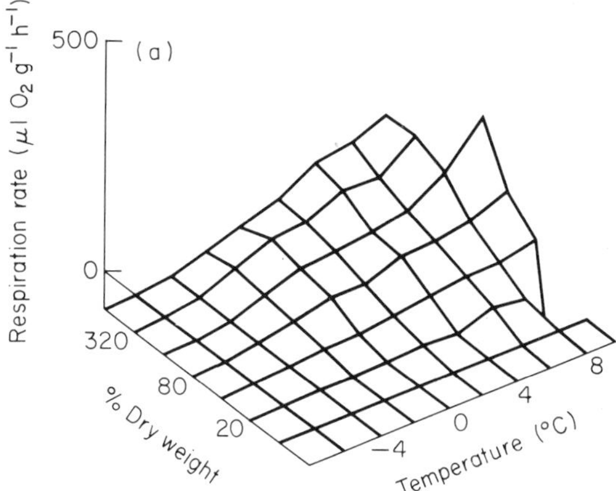

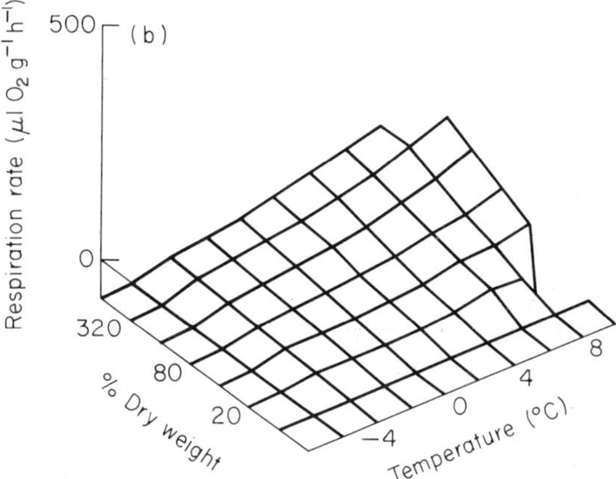

Fig. 3.1.14. Measured (a) and simulated (b) rates of microbial respiration from mixed graminoid litter from Barrow, Alaska. Redrawn with permission from Bunnell *et al.* [16].

NUTRIENT CYCLES

The flow of nutrients in the biosphere can be considered in three compartments: (a) the inorganic store which is mainly the soil itself; (b) the inanimate organic store formed by residues and excreta of organisms; and (c) the biomass store which comprises the living organisms. The principles of such cycling for the general case are outlined in Fig. 3.1.15.

The most extensively studied nutrient cycles are those of carbon and nitrogen. The carbon cycle has been considered at the cell level in Chapter 2.5 and the nitrogen cycle is considered in some detail in Chapter 4.1. However, all cycles are interdependent and Fig. 3.1.16 outlines a scheme linking carbon and nitrogen flow aimed at producing a mathematical model of decomposition in soil.

To be of practical value, it is essential that experimental data be fitted to such models. The data in Table 3.1.6 are based on one such model where experimental data had been fitted to the model. The information provided by these outputs should help in management decisions on cropping such as direct drilling (zero tillage), crop rotations and intercropping so that nutrients and energy will be used more effectively in crop production.

The phosphorus cycle has been studied particularly in relation to mycorrhiza (see Section 3.1.9). However, it should be recognized that all elements are cycled in soil. The sulphur cycle is significant to all soil organisms and its cycle is outlined in traditional terms in Fig. 3.1.17.

3.1.6 PHYSICAL EFFECTS

Moisture

Much of the interest in the effects of water potential on the soil microflora has been concerned with the survival,

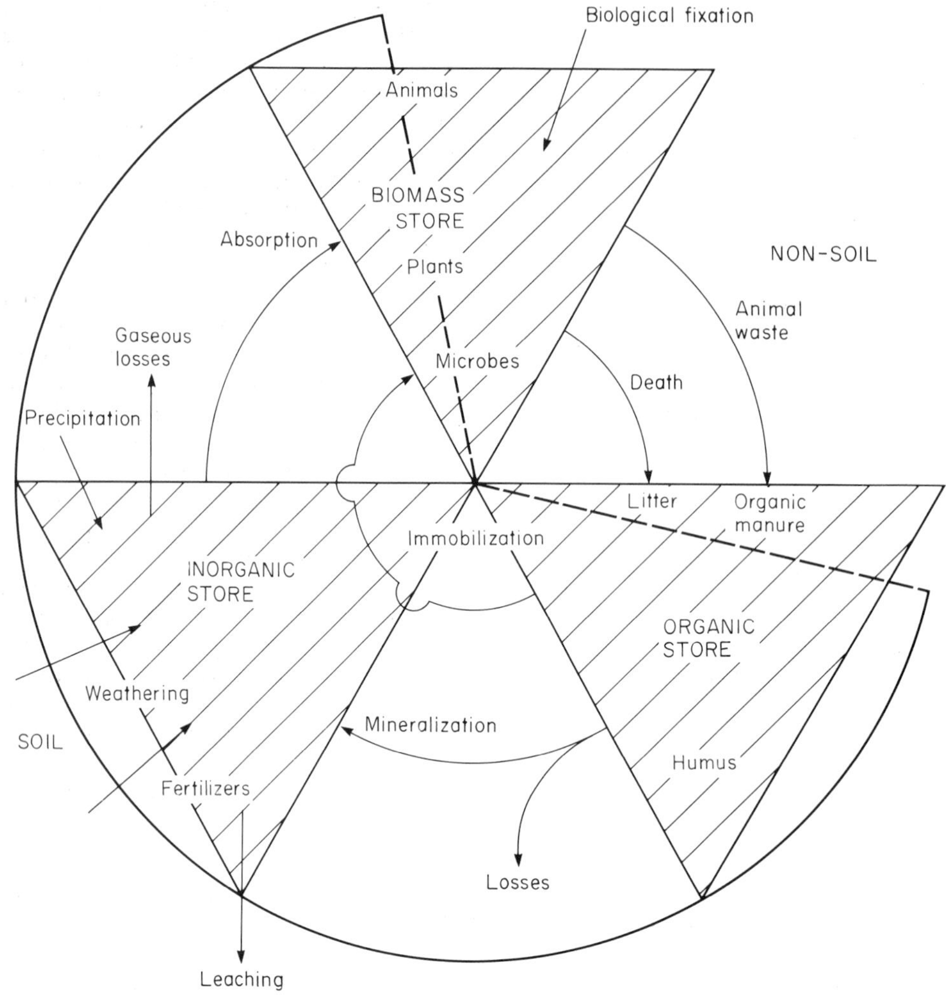

Fig. 3.1.15. Fundamentals of a nutrient cycle. Redrawn with permission from White [98].

Table 3.1.6. Carbon and nitrogen turnover in Saskatchewan wheat-fallow, Rothamsted continuous wheat, and Brazilian sugarcane soil–plant systems (reproduced with permission from Paul & Voroney [71])

Determination	Saskatoon	Rothamsted	Brazil
Soil weight (t ha^{-1})	2700	2200	2400
Organic C (t ha^{-1})	65	26	26
C inputs (t ha^{-1} yr^{-1})	1.6	1.2	13
Turnover of soil C (yr)	40	22	2.0
Microbial C (kg ha^{-1})	1600	570	460
Microbial N (kg ha^{-1})	360	95	84
Microbial turnover time (yr)	6.8	2.5	0.24
N flux through microbial biomass (kg ha^{-1} yr^{-1})	53	34	350
Crop removal of N (kg ha^{-1} yr^{-1})	40	24	220

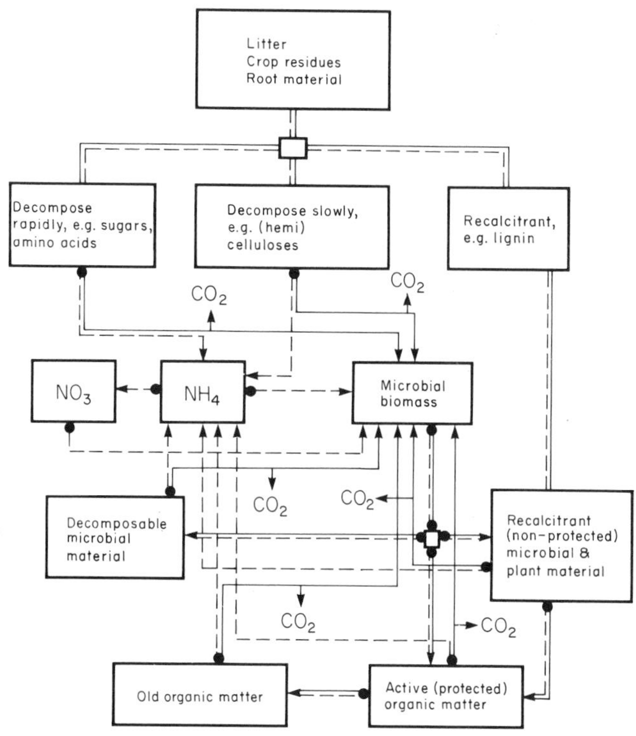

Fig. 3.1.16. Scheme of the N-mineralization–immobilization model. Redrawn with permission from van Veen *et al.* [92].

germination and growth of fungal plant pathogens. The effect of water potential on the rate of linear extension of some fungi is shown in Fig. 3.1.18; clearly there is considerable difference between species. Some bacteria can grow at potentials of −200 bar but most are limited to potentials between 0 and −100 bar. Amongst the activities affected by moisture are nitrification, which is maximal at matric potentials between −0.1 and −0.2 bar, and sulphur oxidation, which is maximal between −0.03 and −0.06 bar. Kouyeas [50] investigated the effect of water potential on the colonization of buried plant parts and found that there was a considerable variation between species: *Pythium* spp. and *Mortierella* spp. were isolated in the wettest conditions and *Mucor* spp., *Aspergillus* spp. and *Penicillium* spp. were isolated in the driest conditions; there was little effect on *Fusarium oxysporum, Trichoderma* spp. and *Gliocladium* spp. Such effects are of course important in determining the nature of the soil fungal population and in part explains why some plant diseases are restricted to particular regions. In Britain, waterlogging can be a serious problem and soils are often at field capacity (saturated) for much of the winter; fungi such as *Pythium* spp. may therefore produce more pathogenicity problems. This contrasts with dry conditions where, for example, it is known that *Aspergillus* spp. and *Penicillium* spp. can commonly attack seeds [37].

Soil water content indirectly affects the movement of

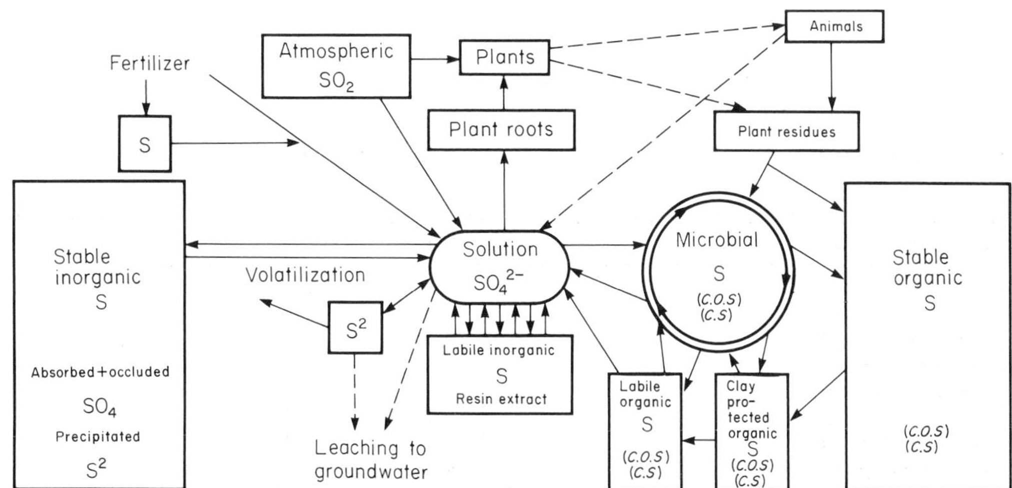

Fig. 3.1.17. Conceptual flow diagram of the main forms and transformations of sulphur in the soil–plant system. Redrawn with permission from Stewart [84].

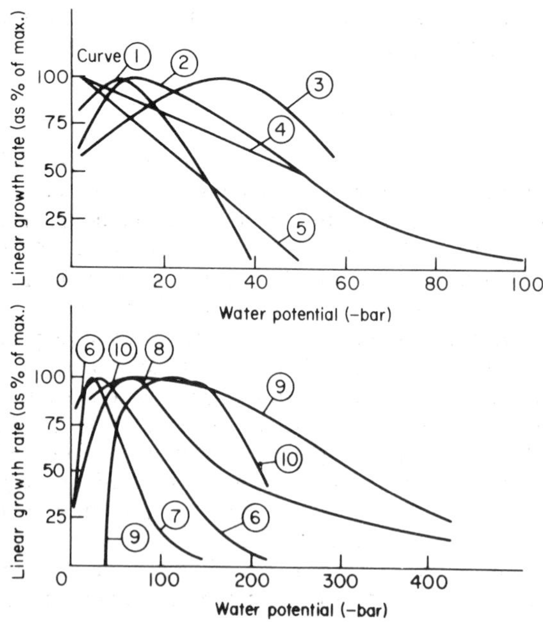

Fig. 3.1.18. Effect of water potential on the rate of linear extension of 11 fungi. 1, *Phycomyces nitens* and *Phytophthora cinnamomi*; 2, *Fusarium culmorum* (isolate from USA); 3, *F. culmorum* (isolate from Australia); 4, *Cochliobolus sativus*; 5, *Ophiobolus graminis* and *Rhizoctonia solani*; 6, *Aspergillus niger*; 7, *A. flavus*; 8, *A. amstelodami*; 9, *Xeromyces bisporus*; 10, *Stereum frustulosum*. Redrawn with permission from Griffin [38].

micro-organisms between pores, along roots and around plant residues. Bacteria only spread rapidly over surfaces when the thickness of the water film is comparable to the size of a bacterial cell. However, fungal mycelium can be anchored in the older parts of its length where water may be available and this allows water to be transported to the growing apex. Therefore, as the matric potential falls, the movement of bacteria is likely to be restricted more than that of fungi; the initial range in which this happens is around −0.5 to −5 bar. Therefore, in addition to the direct effect of water potential on fungi and bacteria, the reduction in movement, reducing the competitive ability to colonize substrates, provides a further basis for the observations that fungi are the major contributors to soil respiration (see Section 3.1.5), particularly in dry conditions.

It is often difficult to separate the effects of water and aeration. For example, it has been shown that the root and stalk rot of soybeans by *Phytophthora* sp. is increased when the bulk density at the base of the plough layer is high [29]; this could be the result of a high water potential or restricted aeration.

The effect of water has been investigated independently of aeration in the studies of denitrification in soil [66]. As denitrification requires anoxic conditions, it was possible to do the experiments in an atmosphere of argon and measure the respiratory losses of nitrogen gases. It was found that as the water content of the soil increased, denitrification also increased.

Aeration

It can be expected that, when the partial pressure of oxygen in the soil atmosphere is reduced as a result of increased water or compaction, the activity of anaerobic bacteria will ténd to become more significant than that of aerobic micro-organisms. In practice there have been few investigations of this hypothesis. Indeed, there would only be a limited value in such studies because, as has already been indicated (see Section 3.1.2), the existence of anaerobic micro-environments in otherwise aerobic soils would make the situation difficult to analyse. Similarly it might be thought that obligate anaerobes could not survive in aerobic soils but this consideration ignores the potential of spore formation and other possible means of 'protection' from oxygen, such as the dense layer of polysaccharide which forms around *Azotobacter* spp. to 'protect' the oxygen-sensitive enzyme nitrogenase. Therefore, many methods of isolation of the microbial populations will produce a false picture of the significance of the organisms because it is necessary to consider if the organisms are in an active vegetative state or in a non-growing or dormant state. It is perhaps of more value to consider the effects of oxygen on metabolic processes rather than on the populations themselves and in this respect the activities of anaerobic bacteria in soil have been reviewed [81, 88]. The following are two examples of agronomic significance.

APPLICATION OF COW SLURRY TO GRASSLAND

Studies on the effect of cow slurry on the composition of the soil atmosphere were made by inserting sampling probes at various depths in the soil profile [17]. The slurry provides a substrate for microbial growth and it was found that at 10 cm the concentration of oxygen was decreased while the concentrations of carbon dioxide, methane and nitrous oxide were increased; this effect persisted for about 6 months.

ETHYLENE IN THE SOIL ATMOSPHERE

In anaerobic soils ethylene is known to be formed at concentrations which could restrict the growth of cereal roots [83]. The microbiological processes involved were investigated and it was found that the substrate for the formation of the gas was methionine, in combination with an energy source, and that primary saprophytic sugar fungi were particularly important in this activity [52]. Further work revealed that increased oxygen tensions promoted the formation of the gas [60]. The apparent paradox with the observations for soil was rationalized by demonstrating that the limiting step for the formation of the gas was the mobilization of the substrate and that anaerobic conditions were required for this; thereafter oxygen promoted the formation of the gas. It was also shown that very low concentrations of oxygen reduced the growth rate of the fungus and the production of the gas increased exponentially as the growth rate decreased [61], thus providing an additional explanation of the effect of low oxygen concentrations in the soil. Furthermore, other studies have demonstrated that ethylene can be metabolized by the soil microflora [1], that this takes place most readily under aerated conditions and that the direct measurement of ethylene in the soil atmosphere is a balance between production and utilization:

$$\text{accumulation} = \text{production} - \text{utilization}$$

These observations justify clearly that there must be both field observations of soils and laboratory studies of pure cultures of micro-organisms and that these must be closely linked in order to provide explanations of microbial processes in an ecosystem.

Redox potential (E_h)

Redox potential (E_h) governs several microbial processes which are of significance to plants but it should be noted that it is very difficult to measure E_h in soil. Takai and Kamura [86] have outlined the effects of E_h on microbial activity (see Table 3.1.7).

SULPHATE REDUCTION

The activities of sulphate-reducing bacteria are of industrial importance because the sulphide produced is a major factor contributing to the corrosion of metal pipework and storage tanks. In most soils, any sulphide which is produced reacts with iron, is oxidized, or is converted to the non-toxic sulphydryl ion. In conditions where sulphide persists, particularly in wet acidic soils, it may exert a toxic effect on roots such as those in rice [91]. It is considered better to increase soil E_h by application of nitrate to prevent sulphide formation, rather than to promote the binding of the sulphide. However, an alternative may be to stimulate the catalase activity of rice roots; this promotes the activity of *Beggiatoa* spp., an organism which oxidizes sulphide [74].

Sulphate reduction is important in many other environments, e.g. the estuarine and marine ecosystems (see Chapter 4.4).

ORGANIC ACIDS

Short-chain organic acids, which are volatile in steam, are important phytotoxins. They are formed in wet soils particularly when plant residues are present [55, 78]. It appears that cellulose is the major substrate for the formation of the acids. At very low redox potentials, the acids themselves provide substrates for methanogenic bacteria. Volatile fatty acids also form in silage [101]. Here three fermentations are recognized: (a) lactic; (b) secondary (butyric or clostridial); and (c) aerobic or after-fermentation. In the first step lactic acid bacteria ferment water-soluble carbohydrates to produce acetic and lactic acids. Various chemical additives and bacterial inocula have been tested in attempts to stimulate the process and bring a rapid fall in pH of the silage. This

Table 3.1.7. Microbial metabolism in anaerobic soils (after Takai & Kamura [86])

	Organic matter decomposition	Metabolism	E_h volts	Organic acids
Stage 1 (rapid)	Aerobic/semi-anaerobic	O_2 respiration NO_3^- reduction Mn^{4+}, Fe^{3+} reduction	+0.6 to +0.3	Only accumulated if fresh organic matter present
Stage 2 (slow)	Anaerobic	SO_4^{2-} reduction CH_4 production	0 to −0.22	Rapid accumulation of early stages, rapid decrease in advanced stage

prevents further degradation by microbial action and thus conserves the fodder (see also Section 3.5.5).

Temperature

Laboratory experiments are usually carried out at temperatures which are quite unrelated to the natural environment. Furthermore, such experiments are usually at constant temperature and there are considerable diurnal, daily and seasonal fluctuations in the real world. Currie [23] showed that the consumption of oxygen in soil either bare of crops or under kale was reduced by a factor of 10 in the cold (3°C) weather of January compared with the warmer weather of July (17°C).

Hydrogen ion concentration

Studies on straw decomposition show that anaerobic fermentation results in the formation of organic acids, which can be responsible for a drop in pH. A similar phenomenon can occur with leaf litter, where aromatic and aliphatic acids can be leached when the material enters the soil and cause a drop in the pH in the immediate environment of the residue. The subsequent fermentation also can yield aliphatic acids, thereby making the conditions even more acidic. The soil usually provides a buffering action around the residues and, while the pH on the surface of the material could be very low, that of the bulk soil may only be affected slightly.

Streptomyces spp. seldom grow at pH values less than 5. However, it has been demonstrated [99] that they are active between pH 4 and 5 in some soils. The explanation provided was that they are present around fungal hyphae which lyse and liberate ammonia from the glucosamine constituents of chitin; this causes the localized pH in the microzone to increase towards neutrality. This type of activity may be important in the survival of many microbial species in soil.

Warcup [95] investigated the fungal populations in grassland soils of widely different pH values. There was a considerable species diversity over these soils and dilution plate counts showed that more colonies were isolated with decreasing pH.

3.1.7 MICROBIAL ACTIVITY AND ITS ELIMINATION

The progress of soil microbiology has in the past been impeded by lack of suitable methodology to assess microbial growth and activity in such a complex environment. However, major developments in methodology have been made recently and the reader is referred to recent texts [19, 90] as good source books.

Sterilization

In order to establish that micro-organisms are involved in particular processes, or to eliminate microbiological effects from the growth of plants, it is sometimes necessary to sterilize soil. Unfortunately all methods suffer the disadvantage of modifying the soil chemically or physically.

Steam sterilization can take a long time (e.g. 6 hours at 120°C) and it is common to do this in two cycles to increase the effectiveness of the kill. Chemical reactions during the process increase the concentration of soluble nutrients and organic matter which can be extracted from the soil. Dry heat is an alternative and sterilization can often be achieved by heating for about 16–18 h in an oven at 120°C. This process inevitably dries the soil and also liberates many substrates for microbial growth, resulting in a flush of microbial activity on re-wetting. Chemicals such as methyl bromide and ethylene oxide can also be used, but these are hazardous to use and can produce residual phytotoxicity. Chloroform has been used often as a so-called partial sterilant [48] and this has formed the basis for a method to estimate soil biomass (see Section 3.1.5). In this method the fumigated micro-organisms provide substrates for a fresh inoculum; the size of the flush of carbon dioxide is proportional to the biomass originally present. Soils can also be sterilized by gamma irradiation from a ^{60}Co source (5 Mrad). Unless this is available locally, it can be inconvenient and expensive and still suffers from the disadvantage of modifying the soil chemically. It does not usually, however, affect free enzymes.

3.1.8 INACTIVITY

Microbial growth in soil is either very slow indeed or periods of activity are followed by longer periods of inactivity (see Section 3.1.5.). During these inactive periods, microbes may be in resting structures which require very little or no energy [35]. Inactivity is often the result of stresses in the natural environment [33]. Most of the stresses can be considered as the inverse of the factors which promote activity as described above.

As early as 1909, Russell and Hutchinson [77] reported that untreated soil contained a factor which limited the development of soil bacteria. It is clear that soil bacteriostasis is at least partly biological in origin and quite common [14]. However, most attention to the inhibition of microbial growth in soil has been given to fungistasis, defined in a recent review [96] as follows: (a) viable fungal propagules not under the influence of endogenous or constitutive dormancy do not germinate in soil in conditions of temperature and moisture favourable for germination; or (b) growth of fungal hyphae is retarded or terminated by conditions of the soil environment other than temperature and moisture. The authors indicated that there were stages of induction, maintenance and release. Furthermore 'all environments of all organisms contain, in a general sense, stimulators and inhibitors of those organisms' and 'the evolutionary success of any organism depends on its ability to strike a balance between the stimulatory and inhibitory factors of its environment'. Probably both organic and inorganic chemicals give rise to the inhibition and stimulation.

Several investigators have suggested that volatiles are the most likely inhibitors. Smith and Cook [82] suggested that ethylene was the specific factor involved. With the vast array of potential substances which can be formed in soil, a single factor might be somewhat surprising and indeed the validity of their conclusions has been challenged [53].

Lockwood [51] has made a critical review of soil fungistasis and concluded that the most likely cause of no growth in soil is lack of available substrate; few inhibitory factors have been identified, ammonia being one. The phenomenon applies equally well to bacteria; relief of microbiostasis can be achieved by soil treatments which make carbon available from the native soil organic matter [56].

3.1.9 MICRO-ORGANISMS AND HIGHER PLANTS

Effects

Micro-organisms can have positive or negative influences on plant growth. Table 3.1.8 illustrates some effects of bacterial species in gnotobiotic culture with barley roots. The effect of the micro-organism will in part depend on the biomass of the particular microbial species that the root will support; in gnotobiotic culture this does not appear to be particularly sensitive to the inoculum size and the saturation biomass has been termed the colonization potential (see Fig. 3.1.19). In mixed microbial culture, the colonization potential of one species can be depressed by another. In Fig. 3.1.20 all species had

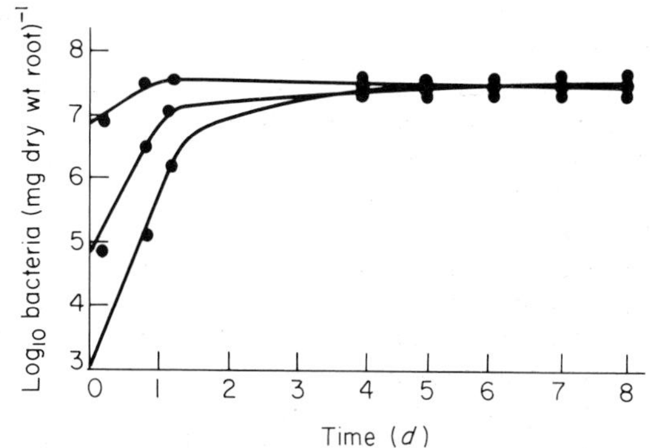

Fig. 3.1.19. Effect of original inoculum size on growth rate and final population of a fluorescent *Pseudomonas* sp. in the barley rhizosphere (mean of counts from the middle and older portions of the root). After Bennett & Lynch [10].

Table 3.1.8. Growth and activity of bacterial isolates from the rhizosphere of barley (*Hordeum vulgare*) (after Lynch & Clark [59])

				Seedling growth (% of axenic control)			
				Root		Shoot	
Isolate*	Cell length (μm)	Inoculum (10^7 per ml)	Colonization potential (10^6 per mg dry root)	Fresh length	Dry wt	Fresh length	Dry wt
S1	1.4	2.6	2.4	+18.0†	+27.0†	+19.6†	+23.7†
S3	1.6	4.6	3.1	+16.9†	+16.3†	+19.2†	+32.9†
S4	1.6	0.013	2.1	+3.0	−5.3	+7.8	+8.5†
S9	2.3	2.4	1.5	+1.1	+2.5	+10.9	+14.2†
Ag	2.0	7.8	5.1	−17.2†	−20.7†	−6.6†	−1.9

* Gram negative rods except Ag (*Arthrobacter globiformis*).
† Results significantly different from control at $P>0.05$.

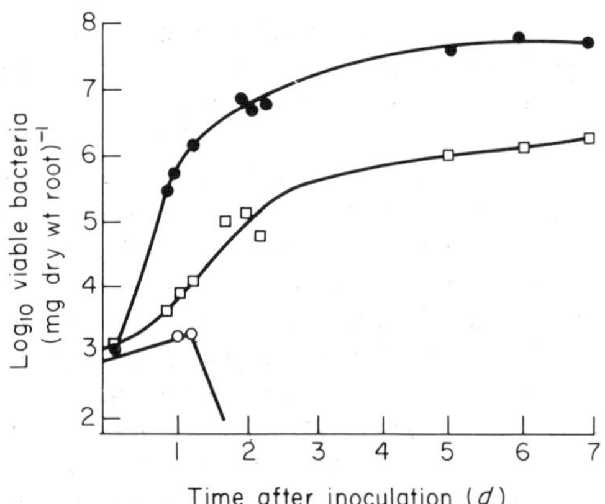

Fig. 3.1.20. Growth of a *Curtobacterium* sp. (○), a *Pseudomonas* sp. (●) and a *Mycoplana* sp. (□) when co-inoculated in the wheat rhizosphere. After Bennett & Lynch [9].

similar colonization potentials in pure culture but in mixed culture only *Pseudomonas* retained its potential.

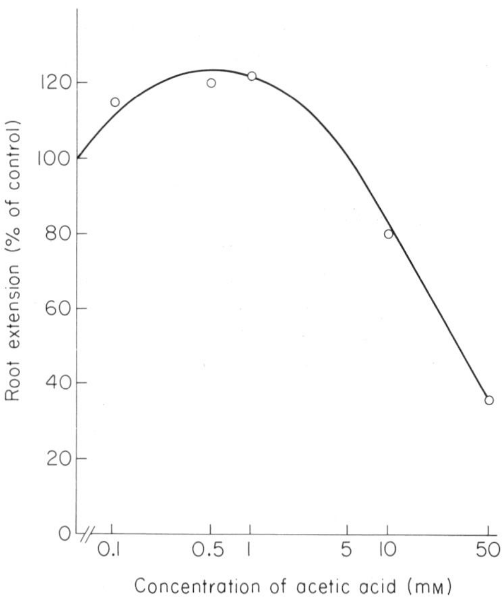

Fig. 3.1.21. Effect of acetic acid on root extensions of barley seedlings at pH 6.6. After Lynch [55].

Microbial products

With the diverse and complex substrates available to the soil microflora a great variety of organic compounds can be formed [54]. However, many of the substances will never occur at concentrations which can affect plant growth. Commonly, stimulatory effects can be obtained in low concentrations whereas higher concentrations are inhibitory (see Fig. 3.1.21).

PLANT GROWTH REGULATORS (HORMONES)

A plant growth regulator is an organic substance produced in a minute quantity in one part of a plant and transported to other parts where it exerts effects on activities such as cell elongation, root production, apical dominance and fruit production.

Most common plant growth regulators or their analogues can be produced by soil micro-organisms. However, there is little experimental evidence to show that they accumulate around roots or shoots in concentrations which are of significance to plants. For example, indolyl β-acetic acid is formed from tryptophan by numerous bacteria and fungi, but this precursor may not be readily available naturally.

At least 35 gibberellins, with the same central structure as gibberellic acid (GA_3) are known. The initial discovery of gibberellin activity was in Taiwan where it was found that infection of rice plants with the fungus *Gibberella fujikuroi* caused a foot rot but prior to this the plants became abnormally tall. The active principle was subsequently identified as GA_3.

PHYTOTOXINS

Throughout this century, there has been concern that the residues or exudates from one crop may harm a succeeding crop. This is often now described as 'allelopathy'. Many substances have been implicated in this phenomenon but it is not always clear if the substances are present in soil as a result of microbiological breakdown of residues or if they come from roots.

The production of volatile fatty acids has already been mentioned (see Section 3.1.6) and the evidence for the accumulation of antibiotics in soil was reviewed by Brian [13] (see also Section 3.1.8). It is still unclear whether these can form in phytotoxic concentrations in soil but evidence has been presented [64] which shows that quite large populations of the fungus, *Penicillium urticae*, which produces patulin, develop on straw in the stubble mulch, a farming practice in the USA, used mainly to conserve moisture and to prevent soil erosion.

Inorganic nutrients

Plants and soil micro-organisms can compete for a wide range of inorganic nutrients which are obtained from the soil minerals. Some nutrients, notably nitrogen and phosphorus, are required in fairly high concentrations whereas others are only required in traces [79].

Nitrogen is a constituent of proteins and it is essential for plant growth. The plant requires a very high nitrogen concentration for maximum growth and natural soil can seldom provide the amount necessary. As a result modern agriculture has become increasingly dependent on the fertilizer industry. However, as with any other components added to biological systems, overdoses can have distinctly harmful effects. Plants take up nitrogen either as nitrate or as ammonium ions. The former is largely taken up from solution whereas the latter is usually present in the soil as exchangeable cation. It is for this reason that nitrate fertilizer acts more rapidly than ammonium but is liable to leach from the soil and enter groundwater.

Phosphorus is necessary to the plant for the wide range of enzymic reactions which depend on phosphorylation. It is also commonly added as fertilizer.

Potassium is important in the synthesis of amino acids and proteins from ammonium ions but it has a much lesser role than nitrogen and phosphorus in plant metabolism. It has an essential function as a constituent of the plant fabric and is another of the major elements which are in sufficiently short supply in the soil to make it necessary to add regularly as fertilizer. A characteristic sign of a deficiency in this element is the premature death of leaves.

Sulphur, calcium, magnesium, sodium, silicon and chlorine are all essential in moderately high concentrations for optimal plant growth. Soil type and condition determines the availability of these ions and whether it is necessary to apply them as fertilizers. Other elements are needed in much smaller quantities; these are the trace or minor elements which include iron, manganese, zinc, copper, boron and molybdenum. It should be emphasized that deficiencies of these elements in the soil lead to deficiencies in the elements in the plants and subsequently this can lead to deficiency diseases in animals which are fed or grazed on the plants. For example, cobalt is needed in fairly large quantities by ruminants and, if the feed or pasture is deficient in this element, it must be given by mouth, such that it enters the rumen; injections into the body are not effective.

It is evident from the foregoing that the nutritional requirements of plants are similar to those of the soil microflora; thus micro-organisms have a profound influence on the inorganic nutrition of higher plants. However, micro-organisms do not necessarily compete for nutrients and indeed the microbial input of nitrogen into the soil has long been recognized as important in plant nutrition. The economic value of the nitrogen fixing bacteria will be discussed later (see Chapter 4.1.). The nitrifying species, *Nitrobacter* and *Nitrosomonas*, convert ammonium to nitrate. Most plants can utilize nitrogen in either form but ammonium ions are usually less readily leached from the soil. A problem of agricultural importance is the activity of a few microbial species which are able to reduce nitrate to gaseous nitrous oxide and dinitrogen; this can be of particular significance in the rhizosphere [100] (see Fig. 3.1.22).

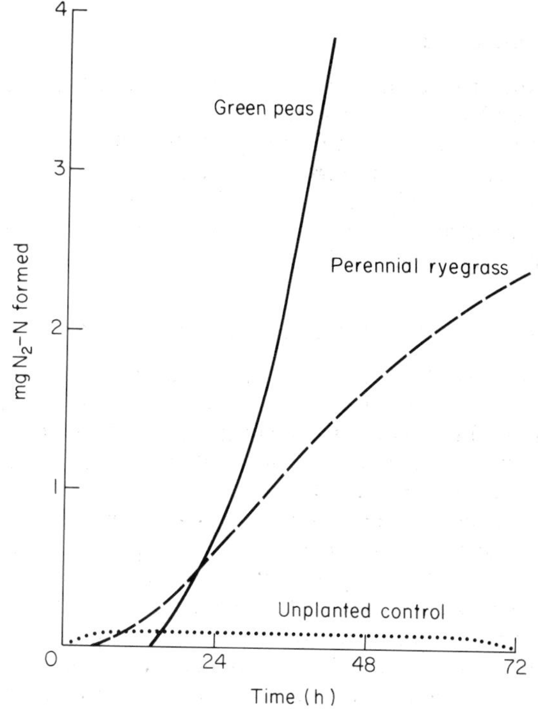

Fig. 3.1.22. Denitrification induced by the roots of peas and ryegrass. After Woldendorp [100].

Experimentally, it is very difficult to assess the effects of the soil microflora on plant nutrition in the complex environment of the soil. Plants, however, can be grown under gnotobiotic conditions in water culture and this provides a model in which to study the interaction. Bowen and Rovira [12] studied the uptake of 0.005 mM potassium phosphate over periods up to 30 min by 8-day-old tomato and clover seedlings. In complete unaerated nutrient solution the uptake was greater in the presence of soil micro-organisms than when seedlings were grown aseptically. In contrast Barber [4], using a different experimental procedure, found quite the reverse effect with barley. Micro-organisms increased the uptake and translocation by plants of phosphate in experiments lasting 30 min, but reduced the uptake in experiments lasting 24 hours [5]. Seedling age was also important; during 30 min periods, uptake and translocation were stimulated by micro-organisms more in 6-day-old than in 8-day-old seedlings, while for 12-day-old seedlings less phosphate was transferred to the shoots in the presence of micro-organisms.

The reduction of nutrient uptake by plants in the presence of micro-organisms is probably fairly easily explained as competition between the two for the available nutrients. The promotion of uptake, particularly phosphate, is a little more difficult to understand. However, it seems likely that the promotion of the uptake of trace elements such as manganese by micro-organisms is explained by the formation of microbial products which chelate the metal ion, increasing its solubility and uptake.

The above effects are generally considered in relation to the activity of the rhizosphere population but they should also be considered in relation to the endophytic symbionts of plants, notably the vesicular–arbuscular (v–a) mycorrhiza.

Symbionts

Rhizobial associations with legumes are covered in Chapter 4.1. The other symbiotic association of interest in soil is mycorrhiza.

Mycorrhizal fungi which live in symbiosis with the root surface of many plants (ectotrophic or sheathing) and those which enter the host (endotrophic or vesicular–arbuscular, v–a) have been discussed in a standard reference work [40]. Many of the endotrophs have not been classified into genera. In recent years it has been recognized that the v–a mycorrhiza probably have great significance in agriculture and forestry. They have not been grown in the absence of host plants and present a major challenge to the microbiologist. The fungi exist in soil as chlamydospores from which either germ tubes or hyphae emerge and invade the root, forming an appressorium which allows the fungi to penetrate between two cells. The intercellular hyphae, or arbuscles, branch dichotomously and end in a mass of fine hyphae; subsequently vesicles containing lipid globules may form (see Fig. 3.1.23). The occurrence of v–a mycorrhiza in plants is extremely widespread, although generally aquatic plants do not appear to be infected. A mathematical approach to the spread of infection has been outlined by Tinker [89]:

$$\frac{dL_i}{dt} = SL_i\left(1 - \frac{L_t}{nL_t}\right) \qquad (2)$$

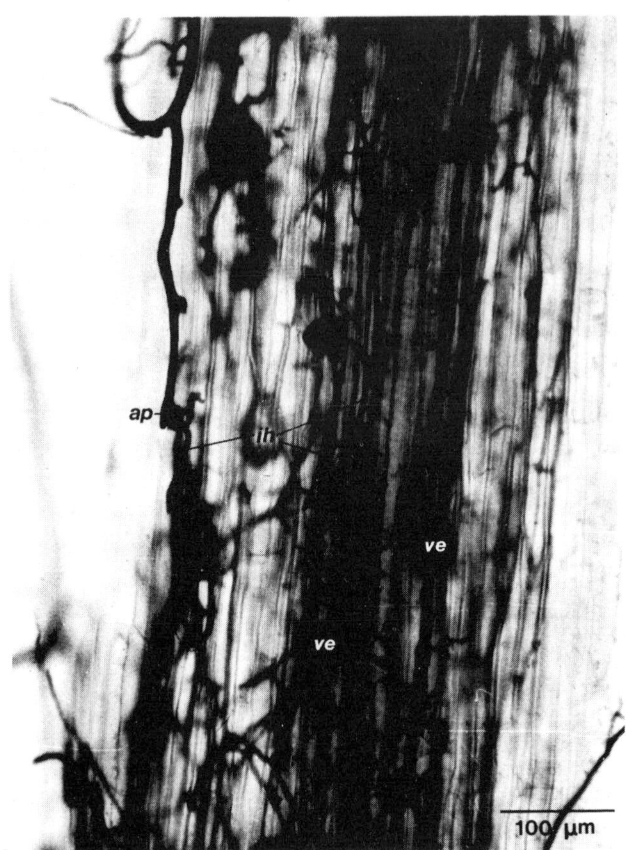

Fig. 3.1.23. Intercellular hyphae (ih) of yellow vacuolate end mycorrhiza in onion root, with entry point of external hyphal appressorium (ap) and vesicles (ve). Photograph courtesy of Dr F.E.T. Sanders, University of Leeds.

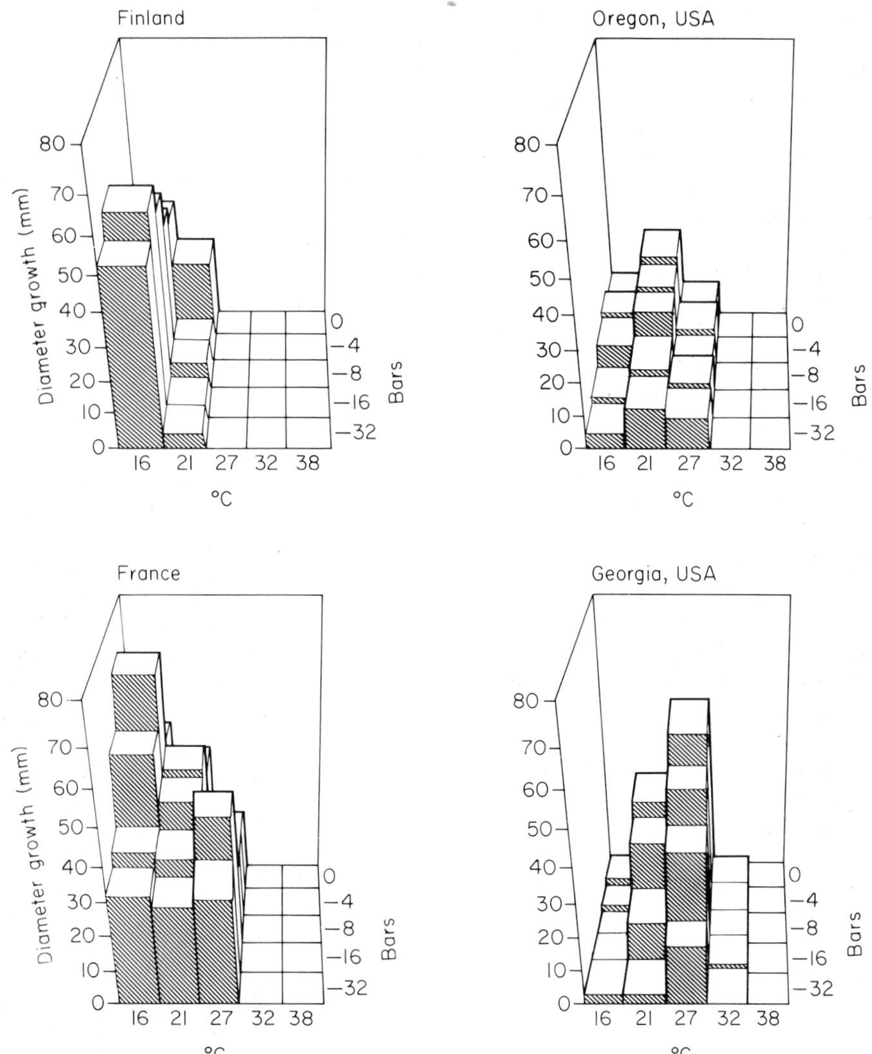

Fig. 3.1.24. Mycelial growth (y axis) of four distinct geographical isolates of *Cenococcum graniforme* in response to combinations of temperature (x axis) and osmotic water potential (z axis). Growth at each treatment combination represents three replicates. After Reid [75].

where L_t is the total root length at time t, L_i is the amount of infected root per unit soil volume, host, $\mathrm{d}L_i/\mathrm{d}t$ is the rate of formation of new infected root per unit volume of soil and S and n are constants. This model states that only the density of infected root is important until the fraction of the root not infected becomes significantly less than one. Experiments with spring wheat and white clover growing with different numbers of plants per plot in a soil−sand mixture were conducted and it was demonstrated that the model provided a good fit to the experimental data.

There are many environmental factors which influence spore formation and infection; these include temperature, light, pH, water regime and nutrient supply. Fig. 3.1.24 illustrates the effect of water potential on ectomycorrhizal fungi at different geographical locations. Clearly

Table 3.1.9. Stimulation of onion growth by inoculation with various fungal endophytes in a soil which had been sterilized by gamma irradiation (after Mosse [68])

	Soil 1 (pH 7)		Soil 2 (pH 4.6)	
Endophyte	Degree of infection	Plant dry wt (mg)	Degree of infection	Plant dry wt (mg)
Laminate	Moderate	737	Moderate	370
E_3	Abundant	708	Abundant	441
Honey-coloured	Moderate	689	Moderate	475
Small brown clusters	Moderate	649	Abundant	631
Yellow vacuolate	Slight	246	Moderate	410
Endogone microcarpa	Slight	151	Slight	134
Bulbous reticulate	Very slight	139	Very slight	153

water potential is critical to the colonization of roots but there may simultaneously be an effect on the plant because at low soil water potentials mycorrhiza may effectively extend the water-absorbing area of the root system. It is generally considered that v–a mycorrhiza stimulate the growth of their hosts by promoting the uptake of phosphorus. This has led to attempts to inoculate plants growing in phosphorus-deficient soils. Generally these have been successful and an example is given in Table 3.1.9. More recent practical approaches to mycorrhizal inoculation have been outlined by Hayman [42]. The mycorrhiza do not show a strong host specificity and they are frequently found in many crop plants and forest trees.

3.1.10 REFERENCES

[1] Abeles F.B., Cracker L.E., Forrence L.E. & Leather G.R. (1971) Fate of air pollutants: removal of ethylene, sulfur dioxide, and nitrogen dioxide by soil. *Science* **173**, 914–16.

[2] Anderson J.P.E. & Domsch K.H. (1973) Quantification of bacterial and fungal contributions to soil respiration. *Arch. Mikrobiol.* **93**, 113–27.

[3] Anderson T.-H. & Domsch K.H. (1985) Determination of ecophysiological maintenance carbon requirements of soil micro-organisms in a dormant state. *Biol. Fert. Soils* **1**, 81–9.

[4] Barber D.A. (1966) Effects of micro-organisms on nutrient absorption by plants. *Nature* **212**, 638–40.

[5] Barber D.A., Bowen G.D. & Rovira A.D. (1976) Effects of micro-organisms on absorption and distribution of phosphate in barley. *Aust. J. Plant Physiol.* **3**, 801–8.

[6] Barber D.A. & Gunn K.B. (1974) The effect of mechanical forces on the exudation of organic substances by the roots of cereal plants grown under sterile conditions. *New Phytol.* **73**, 39–45.

[7] Barber D.A. & Lynch, J.M. (1977) Microbial growth in the rhizosphere. *Soil Biol. Biochem.* **9**, 305–8.

[8] Barber D.A. & Martin J.K. (1976) The release of organic substances by cereal roots in soil. *New Phytol.* **76**, 69–80.

[9] Bennett R.A. & Lynch J.M. (1981) Bacterial growth and development in the rhizosphere of gnotobiotic cereal plants. *J. Gen. Microbiol.* **125**, 95–102.

[10] Bennett R.A. & Lynch J.M. (1981) Colonisation potential of bacteria in the rhizosphere. *Curr. Microbiol.* **6**, 137–8.

[11] Berkeley R.C.W., Lynch J.M., Melling J., Rutter P.R. & Vincent B. (eds) (1980) *Microbial Adhesion to Surfaces*. Ellis, Horwood, Chichester.

[12] Bowen G.D. & Rovira A.D. (1966) Microbial factor in short-term phosphate uptake studies by plant roots. *Nature* **211**, 665–6.

[13] Brian P.W. (1957) The ecological significance of antibiotic production. In *Microbial Ecology*, p.168 (ed. R.E.O. Williams & C.C. Spicer). Cambridge University Press, London.

[14] Brown M.E. (1973) Soil bacteriostasis limitation in growth of soil and rhizosphere bacteria. *Can. J. Microbiol.* **19**, 195–9.

[15] Bunnell F.L., Tait D.E.N. & Flanagan P.W. (1977) Microbial respiration and substrate weight loss. II. A model of the influences of chemical composition. *Soil Biol. Biochem.* **9**, 41–7.

[16] Bunnell F.L., Tait D.E.N., Flanagan P.W. & van Cleve K. (1977) Microbial respiration and substrate weight loss. I. A general model of the influences of abiotic variables. *Soil Biol. Biochem.* **9**, 33–40.

[17] Burford J.R. (1976) Effect of the application of cow slurry to grassland on the composition of the soil

atmosphere. *J. Sci. Fd. Agric.* **17**, 115–26.

[18] Burns R.C. (ed.) (1978) *Soil Enzymes*. Academic Press, London.

[19] Burns R.G. & Slater J.H. (eds.) (1982) *Experimental Microbial Ecology*. Blackwell Scientific Publications, Oxford.

[20] Campbell R. (1985) *Plant Microbiology*. Edward Arnold, London.

[21] Cheshire M.V., Mundie C.M. & Shepherd H. (1973) The origin of soil polysaccharide: transformation of sugars during the decomposition in soil of plant material labelled with ^{14}C. *J. Soil Sci.* **24**, 54–68.

[22] Currie J.A. (1961) Gaseous diffusion in the aeration of aggregated soils. *Soil Sci.* **92**, 40–5.

[23] Currie J.A. (1970) Movement of gases in soil respiration. In *Sorption and Transport Processes in Soil*, p.152. Soc. Chem. Ind. Monogr. No. 37. SCI, London.

[24] Darbyshire J.F. (1975) Soil protozoa-animalcules of the subterranean microenvironment. In *Soil Microbiology*, p.147 (ed. N. Walker). Butterworth, London.

[25] Domsch K.H. & Gams W. (eds.) (1972) *Fungi in Agricultural Soils*. Longman, London.

[26] Doner H.E., Volz M.G., Belser L.W. & Løken J.-P. (1975) Short term nitrate losses and associated microbial populations in soil columns. *Soil Biol. Biochem.* **7**, 261–3.

[27] Fitter A.H. (ed.) (1985) *Ecological Interactions in Soil. Plants, Microbes and Animals*. Blackwell Scientific Publications, Oxford.

[28] Foth H.D. & Turk L.M. (1972) *Fundamentals of Soil Science*, 5th edn. John Wiley, London.

[29] Fulton J.M., Mortimore C.G. & Hildebrand A. (1961) Note on the relation of soil bulk density to the incidence of *Phytophthora* root and stalk rot of soybeans. *Can. J. Soil Sci.* **41**, 247.

[30] Garrett S.D. (1970) *Pathogenic Root-infecting Fungi*. Cambridge University Press, Cambridge.

[31] Giller K.E. & Day J.M. (1985) Nitrogen fixation in the rhizosphere: significance in natural and agricultural systems. In *Ecological Interactions in Soil. Plants, Microbes and Animals,* p.127 (ed. A.H. Fitter). Blackwell Scientific Publications, Oxford.

[32] Gray P.H.H. & Wallace R.H. (1957) Correlation between bacterial numbers and carbon dioxide in a field soil. *Can. J. Microbiol.* **3**, 191–4.

[33] Gray T.R.G. (1976) Survival of vegetative microbes in soil. In *The Survival of Vegetative Microbes*, p.327 (ed. T.R.G. Gray & J.R. Postgate). Cambridge University Press, London.

[34] Gray T.R.G., Hissett R. & Duxbury T. (1974) Bacterial populations of litter and soil in a deciduous woodland. II. Numbers, biomass and growth rates. *Rev. Ecol. Biol. Soc.* **11**, 15–26.

[35] Gray T.R.G. & Williams S.T. (1971) Microbial productivity in soil. In *Microbes and Biological Productivity,* p.255 (ed. D.E. Hughes & A.H. Rose). Cambridge University Press, London.

[36] Greenwood D.J. & Berry G. (1962) Aerobic respiration in soil crumbs. *Nature* **195**, 161–3.

[37] Griffin D.M. (1966) Fungi attacking seeds in dry seed beds. *Proc. Linn. Soc. N.S.W.* **91**, 84–9.

[38] Griffin D.M. (1972) *Ecology of Soil Fungi*. Chapman & Hall, London.

[39] Grossbard E. (1969) A visual record of the decomposition of ^{14}C-labelled fragments of grasses and rye added to soil. *J. Soil Sci.* **20**, 38–51.

[40] Harley J.L. & Smith S.E. (1983) *Mycorrhizal Symbiosis*. Academic Press, London.

[41] Harper S.H.T. & Lynch J.M. (1981) The kinetics of straw decomposition in relation to its potential to produce the phytotoxin acetic acid. *J. Soil Sci.* **32**, 627–37.

[42] Hayman D.S. (1984) Methods for evaluating and manipulating vesicular–arbuscular mycorrhiza. In *Microbiological Methods for Environmental Biotechnology*, p.95 (ed. J.M. Grainger & J.M. Lynch). Academic Press, London.

[43] Heal O.W. & Ineson P. (1984) Carbon and energy flow in terrestrial ecosystems: relevance to microflora. In *Current Perspectives in Microbial Ecology*, p.394 (ed. M.J. Klug & C.A. Reddy). American Society for Microbiology, Washington, DC.

[44] Hudson H.J. (1968) The ecology of fungi on plant remains above the soil. *New Phytol.* **67**, 837–74.

[45] Hurst H.M. & Burges N.A. (1967) Lignin and humic acids. In *Soil Biochemistry*, p.260 (ed. A.D. McLaren & G.H. Peterson). Marcel Dekker, New York.

[46] Jenkinson D.S. (1977) Studies on the decomposition of plant material in soil. V. The effects of plant cover and soil type on the loss of carbon from ^{14}C labelled ryegrass decomposing under field conditions. *J. Soil Sci.* **28**, 424–34.

[47] Jenkinson D.S. & Oades J.M. (1979) A method for measuring adenosine triphosphate in soil. *Soil Biol. Biochem.* **11**, 193–9.

[48] Jenkinson D.S. & Powlson D.S. (1976) The effects of biocidal treatments on metabolism in soil. V. A method for measuring soil biomass. *Soil Biol. Biochem.* **8**, 209–13.

(49) Jenny H. (1980) *The Soil Resource*. Springer-Verlag, New York.

[50] Kouyeas V. (1964) An approach to the study of moisture relations of soil fungi. *Plant Soil* **20**, 351–63.

[51] Lockwood J.L. (1977) Fungistasis in soils. *Biol. Rev.* **52**, 1–43.

[52] Lynch J.M. (1972) Identification of substrates and

isolation of micro-organisms responsible for ethylene production in the soil. *Nature* **240**, 45–6.
[53] Lynch J.M. (1975) Ethylene in soil. *Nature* **256**, 576–7.
[54] Lynch J.M. (1976) Products of soil micro-organisms in relation to plant growth. *CRC Crit. Revs. Microbiol.* **5**, 67–107.
[55] Lynch J.M. (1977) Phytotoxicity of acetic acid produced in the anaerobic decomposition of wheat straw. *J. Appl. Bact.* **42**, 81–7.
[56] Lynch J.M. (1982) Limits to microbial growth in the soil. *J. Gen. Microbiol.* **128**, 405–10.
[57] Lynch J.M. (1983) *Soil Biotechnology. Microbiological Factors in Crop Productivity*. Blackwell Scientific Publications, Oxford.
[58] Lynch J.M. & Bragg E. (1984) Micro-organisms and soil aggregate stability. In *Advances in Soil Sciences Vol. 2*, p.133 (ed. B.A. Stewart). Springer-Verlag, New York.
[59] Lynch J.M. & Clark S.J. (1984) Effects of microbial colonization of barley (*Hordeum vulgare* L.) roots on seedling growth. *J. Appl. Bacteriol.* **56**, 47–52.
[60] Lynch J.M. & Harper S.H.T. (1974) Formation of ethylene by a soil fungus. *J. Gen. Microbiol.* **80**, 187–95.
[61] Lynch J.M. & Harper S.H.T. (1974) Fungal growth rate and the formation of ethylene in soil. *J. Gen. Microbiol.* **85**, 91–6.
[62] Lynch J.M. & Harper S.H.T. (1983) Straw as a substrate for co-operative nitrogen fixation. *J. Gen. Microbiol.* **129**, 251–3.
[63] Lynch J.M. & Panting L.M. (1981) Measurement of the microbial biomass in intact cores of soil. *Microb. Ecol.* **7**, 229–34.
[64] McCalla T.M. & Norstadt F.A. (1974) Toxicity problems in mulch tillage. *Agric. Environ.* **1**, 153–74.
[65] Macfadyen A. (1970) Simple methods for measuring and maintaining the proportion of carbon dioxide in air for use in ecological studies of soil respiration. *Soil Biol. Biochem.* **2**, 9–18.
[66] McGarity J.W. (1961) Denitrification studies on some South Australian soils. *Pl. Soil* **14**, 1–21.
[67] Menna M.E. di (1965) Yeasts in New Zealand soils. *N.Z.J. Bot.* **3**, 194–203.
[68] Mosse B. (1972) The influence of soil type and *Endogone* strain on the growth of mycorrhizal plants in phosphate deficient soils. *Rev. Ecol. Biol. Soc.* **9**, 529–37.
[69] Parr J.F., Gardner W.R. & Elliott L.F. (1981) *Water Potential Relations in Soil Microbiology*. Soil Science Society of America, Madison.
[70] Paul E.A., Campbell C.A. Rennie D.A. & McCallum K.J. (1964) Investigations of the dynamics of soil utilizing carbon dating techniques. *Trans. 8th Int. Congr. Soil Sci.* **3**, 201–8.
[71] Paul E.A. & Voroney R.P. (1984) Field interpretation of microbial biomass activity measurements. In *Current Perspectives in Microbial Ecology*, p.509 (ed. M.J. Klug & C.A. Reddy). American Society for Microbiology, Washington, D.C.
[72] Perrin R.M.S., Willis E.H. & Hodge C.A.H. (1964) Dating of humus podzols by residual radiocarbon activity. *Nature* **202**, 165–6.
[73] Pirt S.J. (1975) *Principles of Microbe and Cell Cultivation*. Blackwell Scientific Publications, Oxford.
[74] Pitts G., Allan A.I. & Hollis J.P. (1972) *Beggiatoa*: occurrence in the rice rhizosphere. *Science* **178**, 990–2.
[75] Reid C.P.P. (1984) Mycorrhizae: a root soil interface in plant nutrition. In *Microbial–Plant Interactions*, p.29 (ed. R.L. Todd & J.E. Giddens). Soil Science Society of America, Madison.
[76] Rose C.W. (1966) *Agricultural Physics*. Pergamon, *Oxford*.
[77] Russell E.J. & Hutchinson, H.B. (1909) The effect of partial sterilisation of soil on the production of plant food. *J. Agric. Sci.* **3**, 111–44.
[78] Russell E.W. (1973) *Soil Conditions and Plant Growth*, 10th edn. Longman, London.
[79] Russell R.S. (1977) *Plant Root Systems. Their Function and Interaction with the Soil*. McGraw-Hill, London.
[80] Satchell J.E. (1971) Feasibility study of an energy budget for Meathrop wood. In *Productivity of Forest Ecosystems*, Proc. Brussels Symposium 1969, p.619. Unesco, Paris.
[81] Skinner F.A. (1975) Anaerobic bacteria and their activities in soil. In *Soil Microbiology*, p.1 (ed. N. Walker). Butterworth, London.
[82] Smith A.M. & Cook R.J. (1974) Implications of ethylene production by bacteria for biological balance of soil. *Nature* **252**, 703–5.
[83] Smith K.A. & Russell R.S. (1969) Occurrence of ethylene and its significance in soil. *Nature* **222**, 769–71.
[84] Stewart J.W.B. (1984) Interrelation of carbon, nitrogen, sulfur and phosphorus cycles during decomposition processes in soil. In *Current Perspectives in Microbial Ecology*, p.442 (ed. M.J. Klug & C.A. Reddy). American Society for Microbiology, Washington D.C.
[85] Swift M.J., Heal O.W. & Anderson J.M. (1979) *Decomposition in Terrestrial Ecosystems*. Blackwell Scientific Publications, Oxford.
[86] Takai Y. & Kamura T. (1966) The mechanism of reduction in waterlogged paddy soil. *Folia Microbiol.* **11**, 304–13.

[87] Tate K.R. & Jenkinson D.S. (1982) Adenosine triphosphate measurement in soil: an improved method. *Soil Biol. Biochem.* **14**, 331–5.

[88] Tiedje J.M., Sextone A.J., Parkin T.B., Revsbech N.P. & Shelton D.R. (1984) Anaerobic processes in soil. *Pl. Soil* **76**, 197–212.

[89] Tinker P.B.G. (1982) Mycorrhizas: the present position. *Trans. 12th Int. Contr. Soil Sci. New Delhi* **5**, 150–64.

[90] Tinsley J. & Darbyshire J.F. (eds.) (1984) *Biological Processes and Soil Fertility*. Martinus Nijhoff/Dr W. Junk, The Hague.

[91] Vamos R. (1959) 'Brusone' disease of rice in Hungary. *Pl. Soil* **11**, 65–77.

[92] Van Veen J.A., Ladd J.N. & Frissel M.J. (1984) Modelling C and N turnover through the microbial biomass in soil. *Pl. Soil* **76**, 257.

[93] Wakao N. & Furusaka C. (1976) Influence of organic matter on the distribution of sulphate-reducing bacteria in a paddy-field soil. *Soil Sci. Plant Nutr.* **22**, 203–5.

[94] Waksman S.A. & Martin J.P. (1939) The role of micro-organisms in the conservation of the soil. *Science* **90**, 304–5.

[95] Warcup J.H. (1951) The ecology of soil fungi. *Trans. Brit. Mycol. Soc.* **34**, 376–99.

[96] Watson A.G. & Ford E.J. (1972) Soil fungistasis — a re-appraisal. *Ann. Rev. Phytopath.* **10**, 327–48.

[97] Whipps J.M. & Lynch J.M. (1986) Energy losses by the plant in rhizodeposition. *Ann. Proc. Phytochem. Soc. Eur.* **26**, 59–71.

[98] White R.E. (1979) *Introduction to the Principles and Practice of Soil Science*. Blackwell Scientific Publications, Oxford.

[99] Williams S.T. & Mayfield C.I. (1971) Studies on the ecology of actinomycetes in soil. III. The behaviour of neutrophilic streptomycetes in acid soils. *Soil Biol. Biochem.* **3**, 197–208.

[100] Woldendorp J.W. (1963) The influence of living plants on denitrification. *Meded. Landbouwhogeschool Wageningen* **63**, 1–100.

[101] Woolford M.K. (1984) *The Silage Fermentation*. Marcel Dekker, Maidenhead.

3.2 The aquatic environment

3.2.1 AQUATIC ENVIRONMENTS AND MICRO-ORGANISMS

Micro-organisms always live in water, whether it be in the fluids within animals or the film of water around a soil particle. Some are adapted to live in the earth's natural water bodies, which range from small temporary ponds to the world's oceans. Most of the world's natural water bodies are salt: freshwater lakes and rivers contain only 0.01% of the earth's water, groundwater accounts for 0.3–0.8% and oceans contain around 97.5%. Micro-organisms inhabit an immense variety of aquatic environments, existing at temperatures from −1.9°C to more than 90°C, at pHs from 1 to 11, at salt concentrations from those of distilled water to 5.0 M NaCl, and at pressures of more than 600 atm (60.6 Pa) (extreme environments are reviewed in Chapter 3.4).

Although the extremes provide fascinating examples of microbial adaptivity and versatility, this chapter will focus on a typical micro-organism living in the average temperate lake, estuary or ocean and the factors and interrelationships controlling microbial survival and growth. The microscopic algae are included since these micro-organisms are the dominant primary producers of organic matter in most aquatic environments.

Salt concentration is the fundamental property that separates oceans from other water bodies; oceans have more than 30 parts 1000^{-1}, estuaries 0.5 to 30, and fresh water less than 0.5. Throughout the world's oceans the proportions of various ions making up the salt are essentially constant, although the total quantity may vary slightly because of dilution with river water or excess evaporation. Sea water is predominantly a sodium chloride solution (see Table 3.2.1) but fresh water varies according to the geology of its drainage basin. Some lakes may be sodium chloride solutions while others are dominated by bicarbonate or magnesium sulphate. Despite the 1000-fold range of salt concentrations between oceans and fresh water, the salinity is usually not an important factor affecting the ecology of micro-organisms.

Some species of marine bacteria have a sodium requirement [60], but bacteria at least do not need to resist the osmotic inflow of water. Instead, freshwater bacteria evidently develop high turgor pressure and can withstand distilled water. For example, some *Pseudomonas* species have remained viable for more than 20 years in distilled water [40]. Algae have also adapted to waters of various salinities, and salt content is not an important ecological factor of concern. Different species are found in salt and fresh water but the same general types are present in

Table 3.2.1. The concentrations (mmol l^{-1}) of the main ionic components of standard sea water and of low alkalinity river water (after Dryssen & Wedborg [15])

	Sea water	River water
Sodium	479	0.300
Potassium	10.4	0.065
Magnesium	54.4	0.030
Calcium	10.5	0.087
Chloride	559	0.20
Sulphate	28.9	0.15
Bicarbonate	1.86	0.1
Silicon	0.002–0.16	0.218
pH	8.12	7.30
Salinity ‰	35	0.041

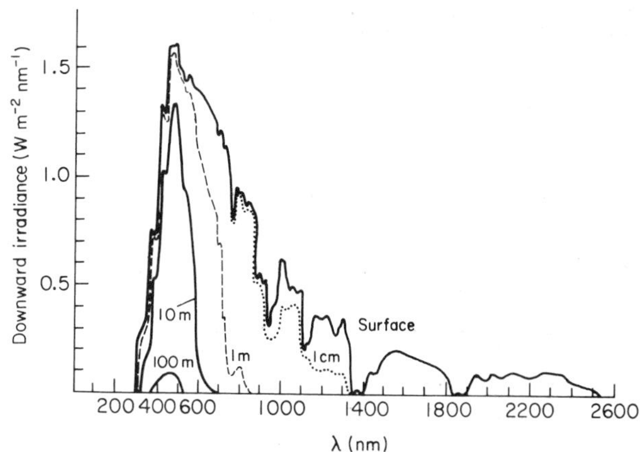

Fig. 3.2.1. Spectral distribution of light energy at the sea surface and at depths of 1 cm, 1 m, 10 m and 100 m. Redrawn with permission from Jerlov [43].

both. As a result, the full complement of microbial processes occur in both fresh water and sea water.

Within aquatic environments, there are a number of different microbial habitats. These include the water column itself and a range of diverse habitats located in the bottom sediments, on submerged solid surfaces, or in association with plants or animals. In the water column of lakes and oceans, most of the bacteria and all of the algae are free-floating, i.e. they are *planktonic* organisms, and their movement in the bulk water is determined by turbulence and water currents. Generally, a relatively small proportion of micro-organisms in the water column are attached to particles. This contrasts with the *benthic* environment, i.e. the bottom sediments and rocks of oceans, lakes and streams. Here bacteria (and algae where there is light) are attached to particles, confined to interstitial water between sediment particles, or embedded in the organic films that cover all submerged surfaces. Where the water is shallow, micro-organisms attached to surfaces such as submerged rocks or vegetation comprise a significant portion of the microbial population.

3.2.2 ENVIRONMENTAL FACTORS AFFECTING MICRO-ORGANISMS

Light, temperature and inorganic nutrients

In natural waters, light is both absorbed and scattered. In clear oceans, light penetrates to about 100 m (see Fig. 3.2.1) but in most lakes, where dissolved and particulate organic matter are abundant, light penetrates to only a few metres. Red and infrared wavelengths, which contain most of the energy of sunlight, are absorbed in the top metre or so while blue wavelengths penetrate to great depths. In clear waters, 53% of the total light energy is transformed into heat in the first metre [105]. As a consequence of the relatively shallow penetration of light, algal photosynthesis is restricted to the upper 100 m of oceans and the upper 10 m of most lakes (see Fig. 3.2.2a). As the average depth of the oceans is 3900 m, the photic zone makes up a tiny part of the oceans' volume.

The absorption of light energy heats the water. The restriction of the heating to the upper layers of bodies of water causes stratification because the heated water is less dense than the cooler, underlying waters. Winds mix the warmer waters close to the surface but not the heavier deep waters and set up a two-layer stratification (see Fig. 3.2.2a). In temperate lakes, the deep waters mix with the surface waters only in the spring and autumn when the temperatures of the upper and lower layers are the same. Most regions of the oceans have both seasonal stratification of the upper layers and a permanent deeper stratification which restricts vertical mixing to the upper several hundred metres.

Because of stratification, algae remain for months in the upper, lighted layers of lakes or oceans where they can grow. In fact, as clearly explained in Parsons *et al.* [68], algae do not begin to grow in the spring until the depth of mixing decreases enough for them to be able to photosynthesize more than they respire each day. However, when algae die and sink into deeper waters, the nutrients released by decomposition accumulate in the deep waters (see Fig. 3.2.2b) and are inaccessible to

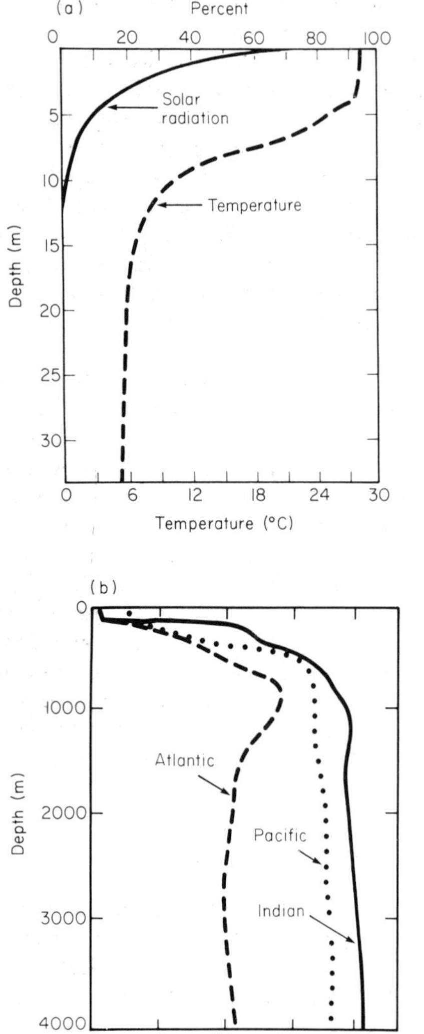

Fig. 3.2.2. Temperature, light and nutrients. (a) Temperature (°C) and light (% of surface solar radiation) in a temperate lake, July. After Wetzel [105]. (b) Nitrate (μmoles l^{-1}) in the Atlantic Ocean. After Valiela [98].

photosynthesizers. In lakes, the mixing of the waters in the spring and autumn makes these nutrients accessible again to algae. In contrast, in the oceans the trapped nutrients do not reach the surface for 500 to 1000 years, the average circulation time of the deep water. This circulation is speedier in regions of upwelling where the combination of winds, shallows, and currents move deep, nutrient-rich water to the surface (e.g. around islands and near the west side of continents such as the Peruvian coast). When the upwelled water reaches the surface, the nutrients are in the right ratio for explosive photosynthesis and algal growth. In Fig. 3.2.2b, only nitrate concentrations are shown, but as A.C. Redfield pointed out in 1934 (reviewed in [75]) phosphate and silicate are also in the correct proportions (the Redfield ration is C : N : P = 106 : 16 : 1 as moles). This finding is not surprising given that these nutrients have accumulated because of the decomposition of algae; it is a dramatic example of the way micro-organisms control some of the chemistry of the sea.

Dissolved gases

The amount of oxygen dissolved in water depends mainly upon temperature and pressure. At normal pressure and 20°C, about 9 mg O_2 l^{-1} is present in fresh water; sea water has 20% less. Colder water contains more oxygen. Since oxygen diffuses rapidly across the air–water interface, the surface waters of lakes and of oceans are usually in equilibrium with the atmosphere; additional oxygen is added to water during photosynthesis.

Although the process of decomposition continually removes oxygen, this removal is usually balanced by diffusion from the atmosphere and by photosynthesis. When thermal stratification is present, however, the water mass is isolated from the atmosphere and decomposition begins to reduce the amount of oxygen. In oceans and oligotrophic (nutrient-poor) lakes, there is always enough oxygen to last until the next period of mixing. But in estuaries and eutrophic (nutrient-rich) lakes there is so much organic matter present that microbial respiration may use up all of the oxygen in the deeper waters. As a result, low oxygen or even anaerobic conditions are often found in lakes at the end of summer stratification, in estuaries and in stratified layers of coastal oceans. In the sediments, oxygen concentrations depend upon the amount of organic matter added each year. If only a little is added, the upper centimetre of sediments will be oxic; if a lot of organic matter is added, then all but the top millimetre of sediment will be anoxic. When oxygen disappears microbial communities change and such processes as bacterial photosynthesis, sulphate reduction and methanogenesis become important.

Carbon dioxide is another important gas in natural waters. It not only dissolves in water but also takes part in chemical reactions and forms carbonic acid [105].

Because this weak acid dissociates, the total inorganic carbon of natural waters includes bicarbonate, carbonate and dissolved CO_2 gas. Additional inorganic carbon comes from weathering reactions, which produce cations such as calcium, magnesium, sodium, potassium and bicarbonates. Concentrations of inorganic carbon in sea water are around 2 mmol l^{-1} and in fresh water are 0.05–10 mmol l^{-1}. In all waters, the CO_2 equilibrium system acts as a reservoir for carbon for photosynthesis, and bicarbonate is an important buffer in both fresh and sea water. Although CO_2 is used in photosynthesis, lack of CO_2 rarely limits rates in natural waters because it readily diffuses into surface waters from the atmosphere.

Organic matter

The dissolved and particulate organic matter in lakes, rivers and oceans provides the energy for bacteria and fungi. This material has two sources: part is derived from external processes (allochthonous), that is, organic matter entering streams and eventually lakes and oceans from soils and terrestrial vegetation; and part is derived from internal processes (autochthonous), that is, from excretion and decomposition of algae, rooted aquatic plants and aquatic animals. The organic matter of aquatic systems, mostly in the ocean and mostly dissolved, totals $1-2 \times 10^{18}$ g C, an amount equal to the organic matter on land [98]. However, in water the distinction between dissolved, colloidal and particulate forms is difficult to make [82] for organic matter exists in a continuum of sizes from dissolved to colloidal (>0.001 μm) to particulate (>0.5 μm). In most research, the colloidal organic matter is combined with the dissolved organic matter (DOM) and the DOM is defined as the organic matter passing through a 0.5 μm filter.

The total amount of DOM ranges from 0.4 mg C l^{-1} in the deep ocean to 1.5 mg C in the surface ocean and oligotrophic lakes and 20 mg C or more in forest lakes dark with humic compounds. In most waters, the ratio of organic carbon in dissolved form to that in particulate form to that in living organisms is about 100 : 10 : 2, indicating the overwhelming quantitative importance of DOM. But the DOM is abundant because it consists mostly of high molecular weight polymers, such as humic and fulvic acids, lignins and proteins, mostly resistant to microbial breakdown [105]. The rate of breakdown is so slow that the age of DOM determined by ^{14}C dating in the deep sea is 3400 years [109]. Eventually even the most resistant molecules of the DOM are broken down or otherwise removed from solution — if this were not so there would be much more and much older DOM in the ocean.

A small amount of DOM, usually about 100 μg C l^{-1}, is made up of small molecular weight organic compounds that are either rapidly assimilated (sugars, amino acids, small chain fatty acids, etc.) or broken down (peptides, sugar polymers, etc.) by micro-organisms. New compounds are found in the DOM with each advance in measurement technology. Everything is present from dissolved DNA and ATP to all possible amino acids and sugars but the amount of each individual compound is low, usually 1–10 μg l^{-1} for glucose, sucrose, alanine, acetate, etc.

Micro-organisms take up and oxidize these compounds as fast as they are formed. This removal rate, measured by incubation of water samples with ng l^{-1} concentrations of radiolabelled organic compounds, can be as rapid as 5 min or as long as thousands of hours [68, 105, 112]. In all cases, the concentrations of these small molecular weight compounds are set by the ability of the microbes to take them up (see Section 3.4.2 on oligotrophy). This control must be exerted by micro-organisms with multiple transport and growth systems, all with low half-saturation constants. Although micro-organisms cannot grow in a chemostat at these low concentrations of single substrates, they can grow if a number of substrates are present in the same low concentrations as natural water [2, 27, 68].

The surface micro-environment

All aquatic environments contain solid surfaces in some form, for example, the particles in the bottom sediment or submerged rocks, suspended detritus or inorganic particles, or man-made structure such as ships or platforms. Physicochemical, nutritional and hydrodynamic conditions at such solid–liquid interfaces are quite unlike those in the bulk phase, and thus surface micro-environments tend to harbour quite different populations from those found in the water column. Similarly, conditions at the air–water interface are different, and a different population, called the *neuston*, colonizes this habitat. The neuston frequently has a higher proportion of prosthecate or pigmented bacteria, and the surfaces of these bacteria tend to be more hydrophobic than those of bacteria maintained in the water column [53]. The relative importance of surface-attached or neuston micro-organisms in an aquatic environment depends on the relative proportion of water mass to surface, and becomes more important in shallow waters or water with a large suspended load. For example, attached micro-organisms will be considerably more important in shallow streams flowing over a rocky bed than in the deep ocean. Similarly, in waters with a large amount of suspended material, particle-associated bacteria can make an appreciable contribution to the total microbial activity within the system [30].

Surface micro-environments are particularly important in flowing waters where water is constantly moving over the submerged surface. Such surfaces are commonly coated with a thick 'slime layer' or biofilm, which is composed of attached bacteria, algae and usually invertebrates, which are embedded in an intercellular polymeric matrix (see Fig. 3.2.3). This matrix is produced by the bacteria, and by the algae when present. It consists largely of highly hydrated polysaccharides, which through their gel diffusion properties or through constituent anionic groups protect the embedded micro-organisms from potentially harmful or toxic factors in the adjacent water. For example, toxic metals such as copper may be complexed by negatively-charged polysaccharides [95], or biofilm polysaccharides may prevent the extensive penetration of biocides or toxins.

Surface micro-environments are not only a protective microhabitat, but in many cases they also appear to offer resident micro-organisms a nutritional advantage for the following reasons. First, in a flowing system, the organism attached to a surface is exposed to a larger flux of nutrients than is the micro-organism which is 'locked' into its suspending water mass. The planktonic microbe will move with the currents and turbulence. Thus, these factors do little to assist the delivery of fresh nutrients to the organism, which is continually removing the nutrients from its immediately surrounding water. Replenishment of these nutrients may well become diffusion-limited. In contrast, where water is flowing over a surface, nutrients are constantly being delivered to the surface and replacing nutrients moved into the biofilm by diffusion or bacterial uptake.

A second way in which the surface micro-environment is nutritionally advantageous is that some nutrients are concentrated at interfaces through adsorption at the surface. Nutrients such as charged substrates (e.g. charged sugars, amino acids, ions), surface-active compounds (e.g. fatty acids), and organic macromolecules or detritus are adsorbed on surfaces, and thus their concentration is increased compared with the water column. They are therefore generally more available to micro-organisms which attach to those surfaces. Similarly, surface-active substances adsorb at the air–water interface and are available to the neuston. By contrast, in some cases, the adsorption of substrates on surfaces may make them less available; for example, the adsorption of amino acids to certain clays, which are also highly charged, can prevent their utilization [93]. However, with most surfaces in natural environments, attached micro-organisms appear to be able to use much of the adsorbed nutrient and benefit from it.

A third way in which the surface micro-environment can be nutritionally favourable is that the surface may also be the substrate. Thus, comparatively large concentrations of micro-organisms are usually found on particles of organic detritus or on dead or senescent eukaryotic organisms [67].

Finally, biofilm communities can be particularly rich environments because of the diversity and complexity of the resident members. Frequently, the micro-organisms embedded within the polymeric matrix comprise diverse groups which interact functionally and supply each other with nutrients and a means of getting rid of metabolic products (see Fig. 3.2.3). For example, in mature, thick biofilms, the spatial segregation of different functional types parallels that found in sediments (see p.151). In such mature films, aerobic bacteria in the surface layers may use oxygen at a faster rate than it can diffuse to deeper parts of the film. Hence, the deeper layers become anoxic and support anaerobic bacteria, such as sulphate-reducing bacteria [28]. If light is available, microalgae, such as diatoms, will be present in the biofilm and can provide heterotrophic bacteria with much, if not all, of their carbon and energy requirements. This was demonstrated in an oligotrophic stream, where carbon fixed by phototrophs in a biofilm was channelled into the bacterial heterotrophs, and the sizes of the two types of population changed together in a seasonal cycle [26]. Thus, the biofilm community may be to a large extent self-sufficient, but it is also a dynamic community changing in time. New cells colonize the biofilm surface, through either cell replication or recruitment from the liquid phase, whereas biofilm turnover is assisted by cell death, invertebrate grazing or natural sloughing of the film.

3.2.3 ALGAL POPULATIONS

Types of algae

The microscopic algae of aquatic environments come from a diversity of groups: diatoms (Bacillariophyceae),

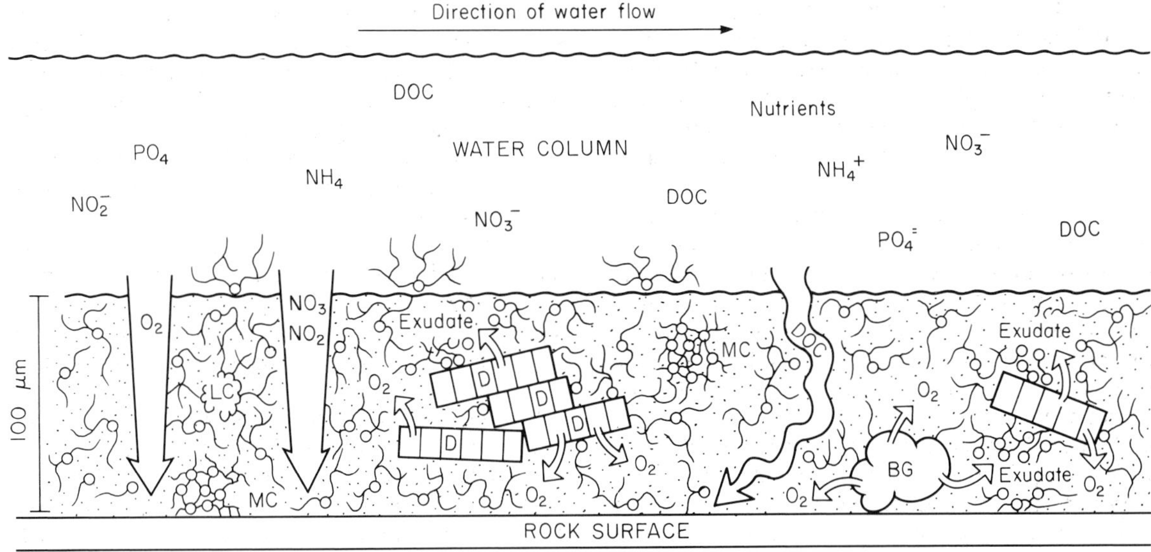

Fig. 3.2.3. Diagrammatic representation of a mature biofilm in which bacteria (○) live within a continuous matrix of exopolysaccharide produced by themselves and by algae. The diagram illustrates possible processes within the biofilm, where diatoms and cyanobacteria are physiologically integrated with the biofilm bacteria. BG, blue-green bacteria (cyanobacteria); D, diatoms; DOC, dissolved organic carbon; LC, lysed cyanobacteria; MC, microcolony. After Costerton *et al.* [12].

dinoflagellates (Dinophyceae), coccolithophorids in the sea (Haptophyceae), Cyanophyceae (blue-green algae or blue-green bacteria), Cryptophyceae and Chrysophyceae [87]. For planktonic environments, ideas about the importance of the various types have changed completely in recent years as techniques of collecting and examination have improved. The first technique was the plankton net, which indicated that the microalgae, 20–200 μm in size, were dominant (see Fig. 3.2.4a). In the last few decades the use of the inverted microscope and settling chambers revealed that most of the time the nanoplankton (2–20 μm) are more abundant both in number and in total mass (see Fig. 3.2.4b). Finally, in the last few years the use of polycarbonate filters and the epifluorescent microscope has shown that the blue-green bacteria of the picoplankton (0.2–2 μm) are not only almost as numerous as the nanoplankton but often are responsible for most of the photosynthesis. Picoplankton are mainly unicellular, nonflagellated species of *Synechococcus*, 0.6–2 μm in size.

Microscopic algae, especially diatoms, are also abundant on the exposed sediments of marshes and attached to surfaces of rocks, sand, mud and plants in the shallow-water zone of lakes, oceans, estuaries and streams [105].

Interactions with their environment

Microalgae have adapted to a wide range of temperatures. They are found in hot springs and colonize the ice crystals immediately below the ice cover of the Arctic and Antarctic oceans. Particularly at low temperatures, their photosynthesis is relatively independent of temperature (see Fig. 3.2.5; see also Fig. 3.4.6. for algae in and on rock surfaces of the Antarctic). In their adaptation to low temperatures, they are different from the bacteria which are relatively inactive at temperatures below 5°C. Algae also have a kind of succession of dominance in planktonic systems. The causes of this succession are difficult to sort out among temperature, light and nutrients, but Wetzel [105] points out that diatoms are active at low temperatures while green and blue-green algae grow well above 15°C.

The amount of light in natural systems is important in several ways. First, the absence of light in the nutrient-rich deep waters of oceans and lakes halts all photosynthesis. Second, with increasing depth a decrease in algal photosynthesis parallels the decrease in light (see Fig. 3.2.6). Third, photosynthesis is inhibited by too much light (note the top of the water column in Fig. 3.2.6).

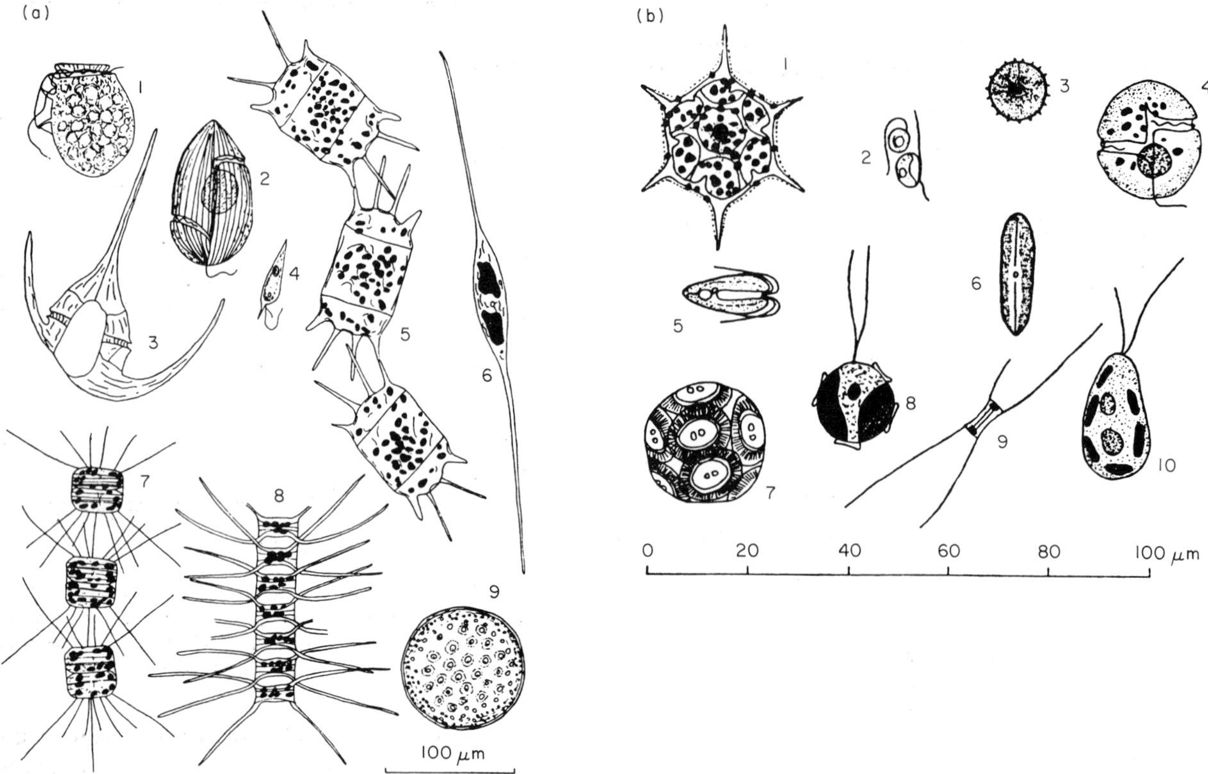

Fig. 3.2.4. (a) Examples of microphytoplankton (scale is 100 μm): dinoflagellates (1–4) and diatoms (5–9). (b) Examples of nanophytoplankton (note expanded scale): flagellates (1, 2, 4, 5), coccolithophores (7) and diatoms (3, 6, 9). Redrawn with permission from Parsons *et al.* [68].

Does light, however, really limit algal photosynthesis? Would more light increase the photosynthesis and production of most ecosystems? The answer is: probably not, for algae appear to be most limited by the rate of supply of nutrients.

In fresh water, the limiting nutrient for algal growth is usually phosphorus while in sea water it is nitrogen. The reasons for this are not well understood. In most cases identifying the limiting nutrient is unimportant, for both nitrogen and phosphorus are needed for algal growth and usually the supply of both is close to the ratio needed. Fertilization of a lake with phosphorus, for example, soon leads to nitrogen limitation. In lakes and oceans, events such as upwelling or pollution with sewage add both nitrogen and phosphorus and the response of the natural populations of algae is usually in proportion to the nutrient added. At least in lakes, the relationship between nutrient loading and algal growth and biomass is so clear (see Fig. 3.2.7) that it is now easy to predict what the effect on a lake will be of adding 100, 500 or 5000 persons km^{-2} to a drainage basin in Europe, Canada or the USA.

In unpolluted lakes and oceans, algal species are well adapted to taking up the low levels of nutrients and may well compete in this way. For example, the kinetics of uptake of nitrate are not only different between species but also between different clones of the same species [8]. Clones of two diatoms, *Thalassiosira pseudonana* and *Fragilaria pinnata*, had K values (half-saturation for uptake) of less than 0.75 $\mu mol\ l^{-1}$ when isolated from low nutrient oceanic water but K values greater than 1.5 $\mu mol\ l^{-1}$ when isolated from an estuary. These values reflect the natural substrate levels (nitrate is usually found at less than 0.5 $\mu mol\ l^{-1}$ in the ocean [98]).

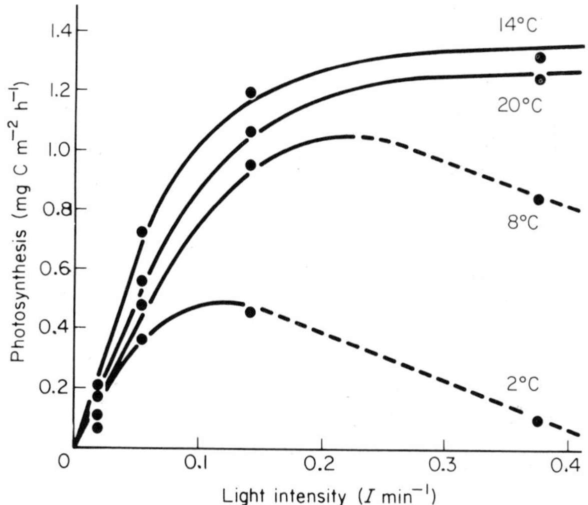

Fig. 3.2.5. Photosynthesis of phytoplankton at different temperatures and at different light levels. After Stanley & Daley [92].

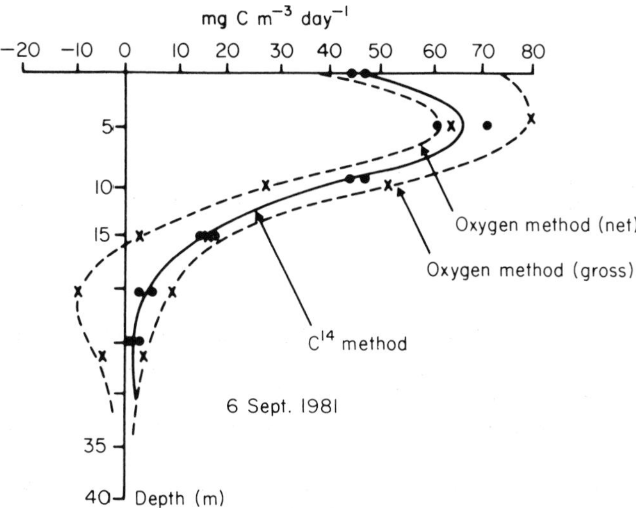

Fig. 3.2.6. Photosynthesis of algae at different depths of the water column as measured by changes in oxygen in light and dark bottles and by incorporation of $^{14}CO_2$. Redrawn with permission from Gieskes & Kraay [23].

Production

Algal production is the base of aquatic food chains and is directly linked to the amount of man's harvest of fish. All of the world's best fisheries occur in areas of high primary

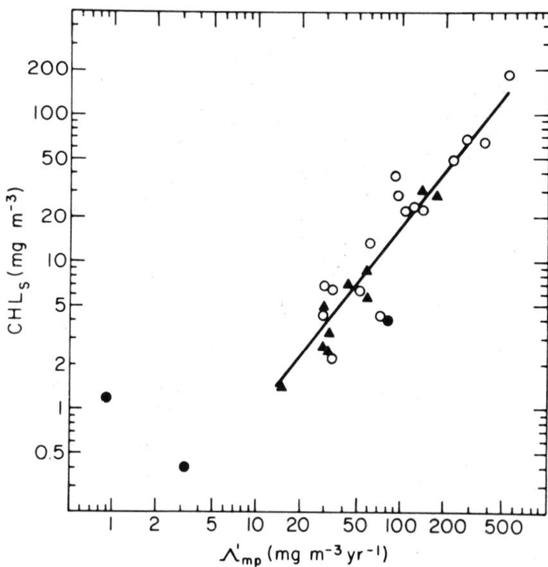

Fig. 3.2.7. Algal mass in various lakes, exemplified by the amount of chlorophyll *a* m^{-3}, plotted against the amount of phosphate m^{-3} entering that lake each year from the drainage basin. Redrawn with permission from Oglesby [66].

production such as the Grand Banks of Newfoundland and the upwelling areas off Peru and off Southwest Africa. This is illustrated by Ryther's [80] synthesis of ocean data (see Table 3.2.2). The best harvest comes from the upwelling areas where fish, such as anchovy, are able to consume phytoplankton directly. The efficiency of harvest decreases drastically in the open ocean where the fixed carbon passes through as many as five trophic levels as it moves from algae to fish. The table also illustrates the variation in annual primary production to be expected within ocean environments. Within freshwater lakes, the range is even greater (an arctic lake produced 4 g C m^{-2} yr^{-1} [49] while Lake Victoria in Africa produced 650 g C m^{-2} yr^{-1} [96]).

How do algae interact with the nutrient supply rate, the most important factor controlling algal production in natural waters? Algae, like other micro-organisms, can grow rapidly; in laboratory cultures they may double five times day^{-1}. In natural water bodies they have 0.5 to more than 1 doublings day^{-1} in nutrient rich waters and 0.2–0.5 doublings day^{-1} in nutrient-poor waters [68]. These growth rates are rapid enough to deplete nutrient supplies in a few days or weeks. When nutrients are added suddenly to the surface layers of natural bodies of water, as occurs in upwelling regions of the ocean or in

Table 3.2.2. Primary production and fish production in oceanic communities (after Ryther [80])

Environment	Primary production ($g\ C\ m^{-2}\ yr^{-1}$)	Trophic levels	Fish production ($g\ C\ m^{-2}\ yr^{-1}$)
Oceanic	50	5	0.0005
Shelf	200	3	0.34
Upwelling	300	1.5	36

spring and autumn mixing of lakes, there is rapid growth of algae and even so-called luxury consumption of nutrients. For example, a diatom cell may store enough phosphorus for eight divisions [78]. In lakes, nutrients are rapidly depleted by the spring bloom of algae; the subsequent summer growth depends upon nutrients from streams, upon the regeneration of nutrients during decomposition, and upon release of nutrients by zooplankton. In the ocean, Eppley and Peterson [17] estimated that in inshore and coastal waters 54–70% of the primary production is based upon regenerated nitrogen while in the open ocean the value is 82–94%. This regeneration step is one of the tight links between bacteria and algae for in many systems much of this regeneration is carried out by very small cells: in coastal waters off California, 40% of the regeneration of ammonium was carried out by cells less than 1 μm in size [29].

Animals graze the algae, on the rocks of streams as well as in the water columns of lakes and oceans, but the effect on the production of the algae is still the subject of much debate. There is no doubt that in many cases zooplankton and herbivores consume 40–95% of the primary production in aquatic ecosystems [98], and that this contrasts with deciduous forests or old fields where 10% or less of the primary production is consumed by herbivores. If animals removed 90% of the primary production of a forest or of a field or of attached macro-algae on a rocky shore, the total photosynthetic production would be drastically reduced. But what happens when 50% of the micro-algae are removed in one day, yet they have the potential to double every day? The answer seems to be that high rates of algal growth and even of photosynthesis are often found when the biomass of algae is relatively low because of grazing. Possible explanations for the high rates include the release of nutrients by the grazers or the selection of fast-growing species of algae. The complete answers remain to be worked out, but the more we learn about the picoplankton, the nanoplankton and their grazers, the more it appears that most of the production in planktonic systems is quickly grazed. When large amounts of algal biomass build up it is because the grazers are not abundant; in these circumstances the algae are likely to be less active per cell, and the total production per unit area is much less than the maximum possible.

Compared with microsites of forests or soil, a cubic metre of water from a lake or ocean is homogeneous. If the algal species of the plankton in this cubic metre responded in the same way as micro-organisms in a chemostat under constant conditions, we would expect that a single species would soon become dominant. Yet this does not occur as shown by data from three stations close to the California coast where 147 taxa of phytoplankton were found over several months of sampling [25]. This was certainly an underestimate as many of these taxa were groupings of small flagellates that contained a number of undescribed species. Also, in a several-year study of a subarctic lake, more than 400 species of planktonic algae were found [36]. Why then are so many species of algae present? This unexpected presence of assemblages of many species is Hutchinson's [39] 'paradox of the plankton'. The explanation [76, 98] lies in the continual changes occurring in the water column from day to day and week to week. Light, temperature and nutrients continually change, grazer numbers wax and wane. These changes may occur in patches so that over time the planktonic ecosystem is not homogeneous and is not constant for long enough periods for one species to achieve a competitive advantage. Blooms of one or several species do occur but their duration is short enough for other species to survive. In other words, unlike a chemostat, the physics, chemistry and biology of any body of water are never constant; the algae of a planktonic ecosystem never come into equilibrium.

3.2.4 BACTERIAL AND FUNGAL POPULATIONS

The microbial flora

All types of microbes have been found in natural waters, but some of these originated from the soils of a watershed or the sewage outfalls of cities. These visitors may remain viable for some time and cause disease in man through transmission in drinking water or shellfish (*Salmonella*, hepatitis virus), but they do not reproduce and therefore they gradually disappear [74]. Some pathogens do grow in natural waters. There is good evidence that

Legionella pneumophila is more abundant in lakes where algae are abundant [20] and that *Vibrio cholerae* multiplies on the exoskeleton of copepods and perhaps other crustaceans [10].

Micro-organisms from natural waters have been cultured and described for the past century, and an imposing list of species of bacteria and fungi has been amassed. Rheinheimer [74] remarks that more than 60 genera and 16 of the 19 groups of bacteria from Bergey's Manual have been found in aquatic systems. These have almost all been identified from cultures grown on agar plates or with MPN methods. These identifications have provided a description of many microbially mediated processes in nature and are adequate to answer many questions. We now know a great deal about the sequence of transformations and of microbial species involved in methanogenesis, nitrogen cycling, fermentation, sulphate reduction, etc. For the past 20 years, however, there has been a growing realization that the plate count and MPN techniques are not quantitative when used in aquatic ecosystems. And quantitative procedures are needed to take the next steps in microbial ecology of measurements of rates of microbial processess and controls of populations and biomasses.

The culture approach to microbial ecology relies on the unstated belief that identification of a species will be the key to the ecology of that species. Yet the biochemical attributes of species of bacteria often change in culture. The exact amount of difference that distinguishes a species is not clear. For example, K.-P. Witzel (personal communication) has isolated more than 1500 strains of heterotrophic bacteria (e.g. *Pseudomonas*) from two lakes in northern Germany. On the basis of some 180 tests of each strain, all appear to be different. We must conclude that for heterotrophic bacteria, at least, the identification of species in the laboratory tells us little about the ecology of the species in nature. Of course, if the species were a specialized form such as a nitrifying bacteria or an obligate sulphate-reducer, then identification would give a very good idea of what the organism was doing in nature although it would not answer the questions of how fast the process was occurring or of what limits the organism.

The picture becomes even cloudier with the realization that only a small fraction of the total number of bacteria present in nature can be cultured in the laboratory. For example, in the plankton of the eutrophic lakes of northern Germany, Witzel (personal communication) found that only about 1% of the bacteria could be cultured out of the several million per millilitre he found by direct counts with the epifluorescent microscope; similar results were found in sea water [74]. Kuznetsov [54] notes that a higher percentage of the total will grow on agar plates when the waters are very rich in organic compounds (e.g. polluted waters) than when the waters are unpolluted and low in organic compounds.

For the future, better methods must be developed for culturing aquatic microbes or they must be identified in water samples without culturing. One promising but difficult technique for identification is immunofluorescence. This has been used to count the ammonium-oxidizing bacteria *Nitrosococcus* and *Nitrosomonas* at concentrations of 10–30 cells ml^{-1} in the ocean [101, 102].

The abundance and size of bacteria

The first attempts to count bacteria directly were in fresh water because the bacteria had to be concentrated by evaporation or chemical precipitation (both difficult in sea water) and counted with oil immersion microscopy. In the 1920s and 1930s, Henrici and Kuznetzov (described in [59]) found 10^4–10^6 bacteria ml^{-1} while the plate count numbers were only 10^1–10^3. Since then, each improvement in technique, such as concentrating on membrane filters, gave increased numbers, but the improvement 'plateau' was apparently reached with the development of the epifluorescent microscope to count bacteria concentrated on polycarbonate filters [32]. In this technique, used also for algae less than 2 μm and for small flagellated protozoans, the nuclear material of the micro-organisms is stained with nuclear dyes such as acridine orange [32] or DAPI [72]. The micro-organisms appear bright against a dark background (see Fig. 3.2.8) when viewed by oil immersion through a microscope in which the excitation light passes down through the microscope lens.

The new direct count methods (reviewed in [99]) have been tested against the scanning electron microscope methods and there is no doubt that even the small objects in Fig. 3.2.8A are bacteria. These bacteria are also alive, since they have been shown by autoradiography to take up radioactive glucose and amino acids (see Fig. 3.4.3).

Bacteria are found in all natural waters and sediments. In the plankton, there may be as few as 40×10^3 ml^{-1} and as many as 12×10^6 ml^{-1} (see Table 3.2.3). The numbers depend upon the amount of usable organic matter present; they are highest in the rich algal soup of estuaries and enriched lakes and lowest in the depths of the oceans and in the bottom of deep wells. It is safe to

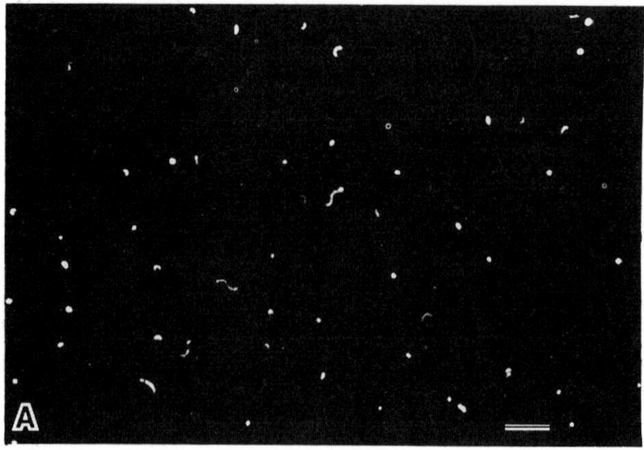

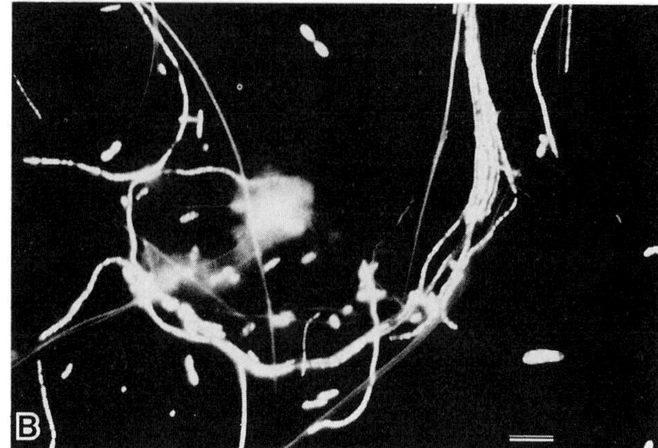

Fig. 3.2.8. Natural populations of bacteria from the Sargasso Sea stained with DAPI and viewed with epifluorescence (marker bars are 5.0 μm). (A) Small bacteria typical of free-living plank-tonic forms. (B) Large bacteria that colonize particulate material called marine snow. Reproduced with permission from Sieburth [88].

Table 3.2.3. The numbers of bacteria, estimated by direct count, in the plankton and sediments of various aquatic systems

	Bacteria	Reference
Plankton	10^6 ml^{-1}	
Oligotrophic lakes	0.5–2	—
Estuaries	5–12	Van Es & Meyer-Reil [99]
Tidal inlets, coastal waters	1–5	Van Es & Meyer-Reil [99]
Offshore	0.05–1	Van Es & Meyer-Reil [99]
Sargasso Sea	0.6	Hobbie *et al.* [34]
Deep ocean		
10 m	1.4	Watson & Hobbie [103]
100 m	0.5	Watson & Hobbie [103]
1000 m	0.1	Watson & Hobbie [103]
1800 m	0.04	Watson & Hobbie [103]
Sediments (marine)	10^9 (g dry wt)$^{-1}$	
Sandy, offshore	0.01	Hobbie *et al.* [34]
Coastal, intertidal, subtidal	4–17	Rublee [79]
Saltmarsh		
surface	5–14	Rublee [79]
5 cm	3–7	Rublee [79]
10–20 cm	1–4	Rublee [79]

assume that in most natural waters (rivers, lakes, estuaries, oceans) there are around 1×10^6 ml^{-1} bacteria.

The sediments of lakes and oceans resemble soils in that there are around 1×10^9 bacteria (g dry weight)$^{-1}$ (see Table 3.2.3) although the numbers range from 0.01 in sandy sediments to 14 in the surface of a salt marsh. Rublee [79] pointed out that in sediments the number of bacteria is directly related to the organic carbon content. Sediments with 0.1–1% w/w organic carbon had 0.1–1 $\times 10^9$ cells (g dry wt)$^{-1}$, sediments with 1–10% had 1–10 $\times 10^9$ cells, and those with 10–50% organic matter had 10–100 $\times 10^9$ cells. Expressed as carbon, the bacterial cells are a little less than 1% w/w of the total organic carbon.

There have been only a few quantitative studies of fungi in aquatic systems. They are apparently only important where plant materials accumulate in such places as marsh sediments and streams. In one salt marsh (see

Table 3.2.4. Estimates of biomass ($\mu g\ C\ cm^{-3}$) of microbial groups in the sediments of a saltmarsh (after Rublee [79]) and in litter of the salt marsh grass *Spartina alterniflora* incubated in the field and in the laboratory (after Marinucci *et al.* [61])

	Bacterial biomass	Fungal biomass	Algal biomass	Protozoan biomass
Sediments				
0–1 cm	200	130	1200	4–46
5–6 cm	125		21	
20–21 cm	50		3	
Litter				
Lab incubated *	216	1864		
Field incubated *		1000		

* Data from Marinucci *et al.* [61].

Table 3.2.4), fungal and bacterial biomasses were equal at the surface of the sediment, but the fungi were restricted to the upper centimetre by anaerobic conditions below 2 cm. The amount of fungi in grass litter incubated in the field was seven times that found in the sediments; the same litter incubated in the laboratory had 14 times more fungal biomass than the sediments.

Bacteria in the plankton of natural waters are small compared with the bacteria grown in laboratory cultures (see Fig. 3.2.8). In this figure, most of the planktonic bacteria are 0.4–0.6 μm in largest dimension, but some are as small as 0.2 μm. These planktonic Sargasso Sea bacteria resemble those from the Irish Sea where the average volume was 0.04 μm^3 [97] (see also discussion in Section 3.4.2). Larger bacteria (see Fig. 3.2.8B) are found associated with particles in the ocean or in polluted waters.

Bacteria in aquatic sediments are slightly larger than planktonic forms. In a study in a salt marsh, Rublee [79] found that most of the bacteria were in the 0.05 and 0.10 μm^3 size range (average 0.2), but that some bacteria were as large as 2.0 μm^3. Bacteria attached to particles have also been found to be larger than planktonic forms, with cell lengths of about 1–2 μm and about 0.2 μm, respectively [35].

Biomass of micro-organisms

For ecological studies of production and food webs, the biomass of micro-organisms must also be determined. One way to do this is to calculate biomass in amount of carbon per area or per volume from direct count data. In this calculation, the numbers of bacteria or kilometres of fungal hyphae are multiplied by factors for dry weight, density, organic carbon content and cell volume. None of these factors can be measured very accurately for populations from the field because cultured micro-organisms are much larger than those seen in the field, because there is often shrinkage of micro-organisms during preservation, and because converting wet weight to dry weight is difficult. As a result of these problems, all biomass calculated in this way has to be taken as ±50%.

Biomass may also be determined indirectly by using chemical measurements of some component of micro-organisms. This approach is attractive because of the tediousness of the direct count methods and because of the ease and reproducibility of a chemical measurement. A number of techniques have been proposed such as the determination of ATP [51], lipid and fatty acids [106], and the cell wall constituents muramic acid [62] and lipopolysaccharide [103]. These methods all have the problem that concentrations must be multiplied by a conversion factor to get to biomass. This factor varies. For example, the ratio of ATP to carbon may vary from 43 to 9500, but a value of 250 is standard [51]. Also, a standard value cannot be used for the conversion of muramic acid to bacterial biomass because Gram-positive bacteria contain much more muramic acid than do Gram-negative bacteria. Another problem is that some of the constituents are not specific to micro-organisms (ATP is found in algae, bacteria, fungi and animals, lipopolysaccharide is found in blue-greens, etc.).

3.2.5 MICROBIAL ACTIVITY AND PRODUCTION

Field measurements of activity and production of micro-organisms

MEASUREMENTS IN NATURE

There are two ways to study the rates of microbial activity in nature: one can measure chemical changes directly in the environment or measure changes in radioisotopes or compounds in a sample of water or sediment enclosed in a container. An ideal sitaution for using chemical change methods is a stratified lake (see Fig. 3.2.2a). Because the deep waters are cut off from contact with the atmosphere for months, rates of O_2 decrease and CO_2 increase (respiration, decomposition), H_2S increase (sulphate reduction) and CH_4 increase close to the sediment (methanogenesis) may be calculated from measurements of concentration made each week [105].

In some situations it is not possible to obtain a number of samples over time yet rates may still be calculated from a single sample. In sediments from lakes or oceans, for example, the assumption may be made that the concentration of SO_4^{2-} in the porewater is at steady state. The concentration is set by the rate sulphate is consumed by sulphate reduction and the rate of diffusion of SO_4^{2-} into the sediment. The rate of sulphate reduction may then be calculated from a diffusion constant and the concentration gradient of the sulphate as measured in the porewater at different depths [4].

A single sample may also be useful for calculations of respiration in the subsurface water masses of oceans (reviewed in [42]). There are a number of methods for doing this but all rely upon the assumption that the water mass was at one time at the surface of the ocean and at that time its oxygen concentration was in equilibrium with the atmosphere. Later, a few months to decades, the water mass has moved deeper in the ocean and is isolated from the surface waters by temperature or salinity stratification. Because the temperature of this water mass is unchanged, the initial concentration of oxygen may be estimated. A water sample from this water mass can be analysed to give the change in oxygen during the time the water mass has been isolated from the atmosphere. The final number needed for the rate calculation is the time since isolation; this age or velocity of the water mass is derived from either calculations based upon the physics of the ocean system (e.g. [77]) or determinations that use radioactive or stable isotope tracers as clocks (e.g. [13]). Tritium (^3H) produced by nuclear weapon tests in the 1950s and 1960s is especially useful for the months-to-decade period. Tritium measures are enhanced by measurements of ^3He, its stable isotope daughter which builds up in the water mass after its isolation from the atmosphere. Figure 3.2.9 illustrates the power of these methods to measure the low rates of microbial activity in the sea. The highest rates shown, close to 1 ml O_2 l^{-1} yr^{-1}, are hundreds of times slower than rates in a eutrophic lake and the techniques are valid for rates two orders of magnitude slower as well.

MEASUREMENTS IN CONTAINERS

The measurement of rates of processes such as growth or respiration through changes in enclosed samples has many pitfalls for the unwary. This measure assumes that the rate measured in the bottle or tube is the same as that in nature. Therefore, during the incubation there must not be any appreciable change in the concentration of the substrates, of the concentration of oxygen, or of the microbial populations. The first problem that arises is the insensitivity of most chemical methods. The rates of microbial activity in samples from nature are almost always slow — so slow that chemical changes take days to build up to a measurable level. Because long incubation times may induce changes in the rates, the chemical change technique is not usually applicable. A second problem is that even 1-day incubations may induce growth in some populations of bacteria and cause death in others (note the changes over 24 hours in the size distribution of bacteria from the Irish Sea shown in Fig. 3.4.2).

Many of the problems are avoided if radioisotopes are used. The incubations can be short, 10 min to a few hours, and the great sensitivity of the method allows

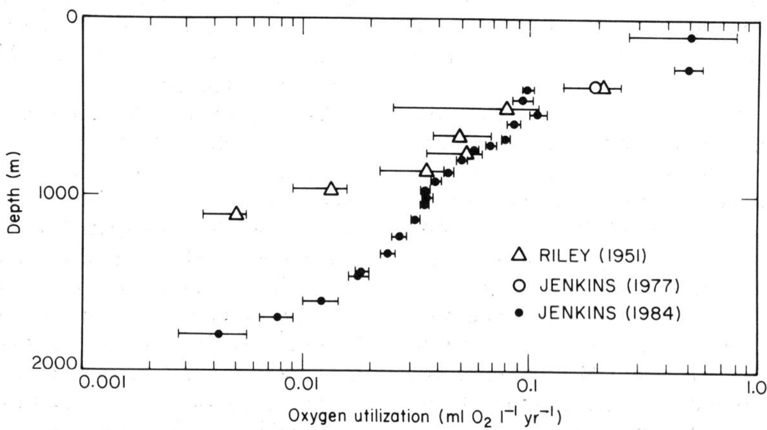

Fig. 3.2.9. Oxygen utilization rates for various depths in the Sargasso Sea as calculated from physical measures (Riley [77]) and from ^3H–^3He measures. After Jenkins [41, 42].

time-course measures of the change in form or location of the isotope. If the rate does not change over the entire time-course measurement, then the assumption is made that enclosing the sample did not introduce any errors and that the rate is the same as in nature. Some of the radio-labelled compounds used are $^{14}CO_2$ (incorporation measures photosynthesis, chemosynthesis), $^{32}PO_4^{3-}$ (incorporation measures rate of transfer), and $^{35}SO_4^{2-}$ (change to $H_2{}^{35}S$ measures sulphate reduction).

While radioisotope experiments solve some of the rate-measurement problems, other pitfalls open up. For example, concentrations of organic substrates are so low in natural waters that ^{14}C-labelled substrates added in an experiment may increase the concentrations fivefold. This increase could cause the rates of uptake to increase. Tritium-labelled substrates are better for this type of experiment. Another problem is the equilibrium of the radioisotope among various molecules. When $^{35}SO_4^{2-}$ is reduced to $H_2{}^{35}S$, for example, the ^{35}S ends up in pyrite, elemental sulphur, FeS or dissolved H_2S. While the total rate of sulphate reduction can be measured by summing the label in all the reduced forms, the rate of formation of the individual compounds cannot be measured because the isotope equilibrates among the reduced forms.

It is difficult to measure exactly what heterotrophic micro-organisms are doing in nature because the community, and perhaps individuals as well, use many organic substrates simultaneously. If, for example, the incorporation of ^3H-glutamic acid is measured, then it is likely that this is only a small fraction of the total uptake of organic compounds, which also includes other amino acids, fatty acids and sugars. To calculate the production of heterotrophic microbes from uptake experiments, many radio-labelled compounds would have to be used and the concentrations of many compounds would have to be determined.

One way to circumvent this problem is to measure the rate of DNA or RNA synthesis and convert this to microbial production. To make this measurement, plankton samples are incubated for short periods, usually less than 1 hour, with radioactive precursors of DNA and RNA. Most microbes can utilize external supplies of these precursors, such as adenine and thymidine, as a supplement to *de novo* synthesis. After incubation, the microbes are removed from the water by filtration and the radioactivity incorporated into DNA and RNA is measured. Karl [52] found that ^3H-adenine incorporation labels the production of the entire microbial population. However, as algae are also a part of the microbial population, Fuhrman and Azam [21] suggested that the production of the bacteria alone could be measured by incubating water samples with 5 nM ^3H-thymidine; bacteria incorporate the thymidine into DNA by a pathway apparently lacking in most algae and fungi [63]. Multiplication by the appropriate conversion factor (1 mole of thymidine incorporation represents the production of 2×10^{18} bacterial cells) gives the actual production rate of micro-organisms. This technique is simple and rapid but has the same problem as other production measures in that a conversion factor is used; this factor is derived from laboratory measures and it is not known how well it applies to bacteria in nature.

Bacterial autotrophs

Most bacteria and fungi are heterotrophic organisms that consume, not produce, organic matter. However, a small group of photosynthetic and chemoautotrophic bacteria fix carbon dioxide and are therefore classified as autotrophs. These organisms are rare in nature, and their share in the total production of organic matter is small [74]. The main reason for this is that several special environmental conditions are required, and that these are seldom present simultaneously.

The photosynthetic bacteria have two types of photosynthesis. In the cyanobacteria the photosynthesis is similar to that of plants, and oxygen is produced [19]. The cyanobacteria are often abundant in the plankton of rivers and lakes when nutrients are plentiful (some species fix nitrogen and thereby take advantage of high phosphorus concentrations), but only a few forms have adapted to oceanic life. The reasons for the lack of cyanobacteria in the sea are still unknown. One theory [38] is that SO_4^{2-} is so abundant in seawater that it competitively inhibits the uptake of the similarly-sized MO_4^{2-} molecule needed in nitrogen fixation.

Bacteria other than cyanobacteria do not evolve oxygen during photosynthesis and are able to use a whole range of electron donors for reducing power [19] such as H_2S or organic acids. They are obligate anaerobes or microaerophiles. The special conditions of abundant H_2S, low oxygen, and light seldom coincide in natural waters except in the subsurface waters of some lakes or in the top few millimetres of marsh or estuarine sediments. One situation in which all the conditions did occur was in Lake Belovod in the Soviet Union [54]. In this lake, the upper waters were oxygenated to a depth of 13 m; H_2S occurred below 13 m, and light penetrated to 16.5 m. The layer of water at 13.5–14.0 m was pink with purple sulphur bacteria (*Chromatium*) whose total photosynthesis was six times that of the phytoplankton in the surface waters. Most lakes, however, do not have abundant H_2S in the bottom waters and do not have enough light reaching the interface of the H_2S and oxygen to support bacterial photosynthesis.

Another group of autotrophic micro-organisms is the chemosynthetic bacteria. These satisfy their energy requirements by oxidizing reduced inorganic substrates such as compounds of inorganic nitrogen, sulphur and iron or elements such as sulphur and hydrogen (see Chapter 2.5). Most chemosynthetic bacteria require free oxygen as the electron acceptor, but some (*Thiobacillus denitrificans, Desulfovibrio desulfricans*) can use bound oxygen derived from nitrate or sulphate [68]. Their physiology

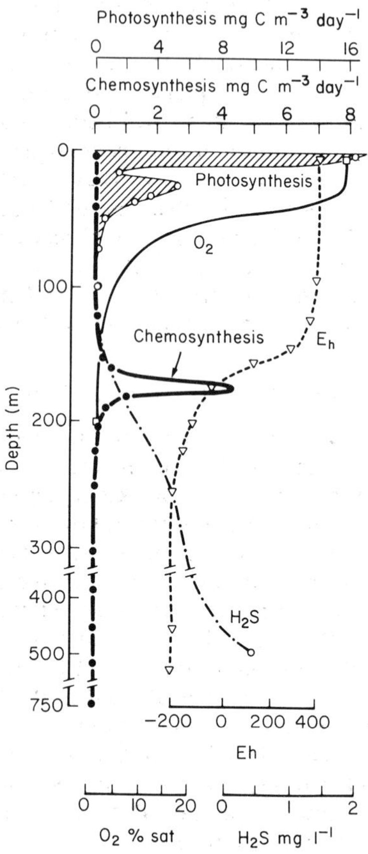

Fig. 3.2.10. Rates of chemosynthesis and photosynthesis and concentrations of oxygen and hydrogen sulfide on one day in October in the Black Sea. After Sorokin [91].

has long interested microbiologists and is well known (see for, example, Section 3.4.3 on *Thiobacillus* and *Sulfolobus*). Their ecology is less well known, but they may play key roles in cycles of several important elements.

The reduced substrates of these bacteria are almost all derived from the decay of organic matter. Consequently, chemosynthetic bacteria are not considered to be primary producers in the sense of a food or energy food chain. Exceptions to this rule are the bacteria that utilize the reduced sulphur compounds from volcanic activity or other geological processes (for bacteria in hot springs see Section 3.4.7, bacteria in deep sea vents Section 3.3.5). Another consequence is that these bacteria sometimes flourish at the interface of aerobic and anaerobic environments such as in sediments or marine fjords. In the Black Sea, for example, the waters below 150 m are anaerobic and contain H_2S (see Fig. 3.2.10). At the interface between the aerobic and anaerobic zones, at a depth of about 180 m, chemosynthesis by *Thiobacillus* spp. reached a peak rate of 9 mg C m^{-3} day^{-1}, nearly the same as the maximum rate of photosynthesis. However, the total photosynthesis per unit of area was about 350 mg C m^{-2} day^{-1} while the total chemosynthesis was about 200 mg C m^{-2} day^{-1} [91].

Chemosynthetic bacteria are easiest to study in rare situations like the Black Sea where, for example, all of the uptake of $^{14}CO_2$ in experiments carried out below the photic zone is due to chemosynthesis. But there are similar interfaces between aerobic and anaerobic zones in the many microhabitats of sediments, of faecal pellets of zooplankton, and perhaps of organic particles in the plankton. When the organisms are not abundant, however, the processes are difficult to study. We do know that large chemical changes occur; chemosynthesizers are likely to be quite important in natural element and nutrient cycles.

One technique that has been applied at the microscale is the combination of immunofluorescence and autoradiography [101,102]. Samples from the north-east Pacific Ocean were incubated for 24 hours with $^{14}CO_2$ and then filtered on to cellulose nitrate filters. The filters were then dipped into nuclear track emulsion and exposed for 3 weeks in the dark. After this the autoradiographs were developed and then stained with antiserum against ammonium-oxidizing bacteria (*Nitrosococcus oceanus, Nitrosomonas marina*). Finally, the filters were stained with fluorescent antibody. The numbers of ammonium oxidizers found, 10–20 ml^{-1}, are very low — these are typical concentrations for chemosynthetic bacteria. As described later, there is other evidence that ammonium oxidation and many of the other transformations carried out by chemosynthetic bacteria are very important in element cycling and involve large amounts of material. This raises the question of whether the low numbers of these specialists that are found in nature can be responsible for all of the transformations. Is it possible that facultative oxidizers are doing most of the processing?

Microbial heterotrophy

ORGANISMS

Most of the micro-organisms living in rivers, lakes and oceans are heterotrophic bacteria; bacteria are much more important heterotrophs in aquatic habitats than fungi and algae. Certainly fungi are present in low numbers in most aquatic habitats but they only become important in streams or marshes where large amounts of litter from terrestrial plants are present. When fungi are present, they apparently carry out the same processes of breakdown of polymers as they do in terrestrial soils. As seen in Table 3.2.4, fungi are not restricted to fresh water but are also found in abundance in salt marsh surface soils and in the grass litter on the marsh.

Heterotrophic algae are also known but their importance is not clear. Some non-pigmented forms of diatoms exist in sediments and these undoubtedly exist heterotrophically. A.E. Linkens (personal communication) has shown that at least some of these diatoms release an extracellular cellulase and take up the dissolved products. However, there has also been a great deal of interest in chlorophyll-containing algae as heterotrophs. Most research has centred upon their ability to grow on various dissolved organic substrates in the laboratory [55]. In spite of a vast physiological literature, there is little evidence that in nature any pigmented algae actually exist on dissolved organic matter [44, 89].

Compared with bacteria, algae do not have as effective transport systems for compounds such as sugars and amino acids and do not utilize such a wide range of compounds. Bacterial transport systems are so effective that they are half-saturated at substrate concentrations of a few nanomoles per litre. Bacteria are able to use these substrates as fast as they are formed from larger molecules and keep the concentrations very low (close to the half-saturation values).

Algal transport systems for substrates have half-saturation values at least ten times higher than those of the bacteria. The present state of knowledge is that algae do have transport systems for dissolved organic substrates and can exist heterotrophically when given high concentrations of substrates in the laboratory. The presence of these transport systems is often used as proof that algae in nature must be heterotrophic. However, proof should also include evidence that in nature the total amount of dissolved organic matter incorporated into an algal cell is adequate for the cell's energy needs. Because algal cells are orders of magnitude larger in volume than bacterial cells and because substrate concentrations in nature are orders of magnitude smaller than laboratory concentrations, it is doubtful that algae can take up enough organic substrate to exist as heterotrophs. It is possible that taking up a little substrate might aid survival (e.g. in lakes under ice or in deep water beneath the photic zone) or that there are some microhabitats (at the surface of sediments, attached to aquatic plants) where relatively high substrate concentrations allow algal heterotrophy.

Recently some chlorophyll-containing algae have been proved to be phagotrophs, that is, they ingest bacteria, yeasts and cyanobacteria [5, 81]. The finding is not surprising given the similar morphology of many pigmented and non-pigmented flagellates. Techniques are still being developed to determine whether this form of heterotrophy is of primary or secondary importance to the algae.

SOURCES OF ENERGY FOR HETEROTROPHS

The organic substrates used by heterotrophic organisms originate from plants. In the oceans the greater portion of organic matter comes from planktonic algae except in coastal regions where the organic carbon in river runoff can be locally important. There is some disagreement about the total amounts and about the importance of macrophyte production in the sea. One estimate [110] is of 23 gigatons (Gt) of carbon fixed annually in photosynthesis by phytoplankton and 1.7 Gt by macrophytes (a Gt = 10^{15} g). Another estimate [98] is of 20–30 Gt C for phytoplankton and 0.3 Gt C entering the ocean from river runoff. Much of this fixed carbon is ultimately mineralized by heterotrophic micro-organisms, but there are a variety of pathways for the DOM as it moves from algae to other micro-organisms.

A small fraction of the photosynthetically fixed carbon is immediately released as DOM by algae and macrophytes. This is probably less than 15% of the total fixed [83] and consists of a variety of amino acids, short-chain acids, glycerol, carbohydrates and polysaccharides [68]. Zooplankton also release organic carbon originally fixed by algae. Some 15–20% of that ingested is released as DOM when algal cells are eaten ('sloppy feeding', [11]); some DOM is also excreted by zooplankton [107]. Another fraction of the algal carbon, at least 30%, is released as DOM at the time of death of the cell. From these numbers, it is obvious that most of the photosynthetically fixed carbon is not lost as DOM but persists as particulate matter that must be enzymatically decomposed before use by micro-organisms. This particulate matter, including algae, zooplankton and larger organisms, is almost completely decomposed in oceans and lakes. In the open ocean, where sediments accumulate at 0.001–0.006 mm yr^{-1}, about 0.01–0.04% of the primary production is preserved [64]. In highly productive lakes and ocean regions (Baltic, Peru upwelling), up to 1 mm of sediments accumulates each year, and 10–20% of primary production is preserved. This high rate of preservation is rare and only occurs where the water column is shallow and anaerobic conditions prevail throughout the sediment.

The sources of energy for heterotrophs in rivers, lakes and estuaries are not only algae and macrophytes, but also DOM and particulate organic matter (POM) from the watershed. By the time this organic matter enters streams, the DOM and POM have lost their most microbially available fraction through leaching and perhaps some microbial decomposition. In streams the fungi dominate the microbial biomass (and presumably the decomposition) during the initial stages of processing of leaves. After the leaf surface has been partially broken down and the cells invaded by hyphae, bacteria become more important and dominate the last stages of processing [94].

UTILIZATION OF DISSOLVED ORGANIC MATTER

As noted earlier, there are many species of heterotrophic bacteria present in any aquatic system, and each must obtain energy from the pool of substrates. Each substrate is present at concentrations of 10–100 nmol l^{-1}. Each species of bacteria has membrane transport systems for a variety of simple molecules such as sugars, amino acids, organic acids, fatty acids and even peptides up to six amino acids [14, 112, R. Coffin, personal communication]. One possible strategy is specialization, in which each species is adapted to one or a few substrates. Indeed, if water samples from nature are placed in a chemostat, one species outcompetes the rest.

There is little evidence for this strategy of specialization in nature. The operative strategy seems to be generalization; all species are able to take up many substrates simultaneously. Evidence for this strategy comes from measurements of the kinetics of uptake of natural communities. A single set of parameters is found at concentrations close to the natural substrate levels as if all species had similar uptake kinetics. Other evidence comes from autoradiographs; many cells take up substrate when low levels of ^{3}H-glucose or ^{3}H-amino acid are incubated with water samples. In some cases, up to 60% of the cells become labelled [37].

The question remains, if all the heterotrophic species appear to be carrying out the same processes, why has not one species outcompeted the rest? The answer will be more complicated than differences in the transport constants and may lie in the different efficiencies of growth or in the different exoenzymes produced. The answer may even be the same as it is for the algae: a constantly changing environment.

Whatever their strategy, heterotrophic bacteria are certainly effective at removing organic substrates from natural waters. Substrate concentration levels are nearly always low and change little even when the measured uptake rate changes by orders of magnitude. For example, removal times for individual substrates are less than an hour in highly productive estuaries [14], about 1 day in estuaries, 1–10 days in coastal waters, and 10–100 days in oceanic waters [107, 111]. The conclusion follows that aquatic bacteria are always limited by the amount of carbon substrate and that they take up substrate as fast as it appears.

UTILIZATION OF PARTICULATE MATTER

About half of the photosynthetically fixed organic carbon is released as DOM during algal exudation, leaching from dead cells, and 'sloppy feeding' of zooplankton. The rest remains as particles that are both consumed by aquatic animals, from protozoa to fish, and eventually decomposed by bacteria. When organic particles are abundant — in salt marshes and algal blooms — size fractionation measurements have shown that most of the microbial activity can be attributed to attached bacteria; when particles are rare — the usual case — free-living bacteria are more active than attached forms [100].

There is little information about the rate of decomposition of particulate matter in planktonic systems. The only study is by Cole *et al.* [9] in Mirror Lake, New Hampshire. For a substrate, the researchers used ^{14}C-labelled phytoplankton harvested on to a glass fibre filter. The filter was attached to a nylon fishing line and incubated in a flask of lake water. Each day the water in the flask was changed and the amount of $^{14}CO_2$ measured. The rates of conversion of algal particulate matter to CO_2 were 6% day^{-1} in summer and 0.5% day^{-1} in winter. The conversion was fastest in the upper layers of the lake and temperature proved to be the single most important factor controlling the rate. After one year, 70% of the algal particulate matter was decomposed to CO_2. In the upper layers of the lake, of the photo-synthetically fixed organic matter oxidized 25% came from algal exudates and 75% came from particulate matter.

3.2.6 MICROBIAL CYCLING OF CARBON, NITROGEN AND SULPHUR

The cycling of the nutrient elements in aquatic ecosystems, i.e. carbon, nitrogen, phosphorus and sulphur, is biologically regulated by two main processes: assimilation of inorganic nutrients by photosynthetic organisms and the subsequent mineralization by heterotrophs [46]. The photosynthetic processes have been discussed earlier (Sections 3.2.2 and 3.2.3); this section deals with the cycling by heterotrophs.

Carbon cycle

The decomposition of organic matter in natural waters can be either an aerobic or an anaerobic process or a combination. In the open ocean, the water is so deep (average 3900 m) and contains so much oxygen, that most of the algal-formed POC decomposes aerobically before it sinks to the bottom. Sediment trap data summarized in Valiela [98] show that only about 2% of the organic matter from primary production reaches a depth of 3500 m. In shallow waters, coastal oceans, estuaries and lakes, 25–60% of the organic carbon produced in photosynthesis may settle out of the upper layers and be decomposed anaerobically in the bottom waters or in the sediments. Given that 70% of the earth is ocean and that most of that is deep ocean, most of the decomposition in natural waters must be aerobic.

AEROBIC PROCESSES

The stepwise breakdown of organic matter involves a large number of different organisms and a complex food web. At least some of this complex cycle may be conceptually simplified in oxic environments because planktonic algae all have a similar chemical composition and because all of the decomposing organisms have the same respiratory metabolism. Thus, both for an individual organism and for an ecosystem, the decomposition or mineralization parts of the process can be described in terms of the overall stoichiometry. The history and applications of the concepts have been reviewed by Richards [75].

The ratios of elements used and released in the following equation, known as the Redfield ratio (after A.C. Redfield), are used in many of the calculations and predictions of biogeochemistry of sea water:

$$(CH_2O)_{106}(NH_3)_{16}(H_3PO_4) + 138\ O_2 = 106\ CO_2 + 122\ H_2O + 16\ HNO_3 + H_3PO_4$$

Another way to examine cycling by micro-organisms is to measure the mass of organic carbon in the various components of the food web and the fluxes of organic carbon in and out of the components. A good example of how these all fit together in an aquatic ecosystem [56] comes from studies of Mirror Lake, a small, stratified, 10 m deep lake in New Hampshire (see Fig. 3.2.11). Here, oxygen disappears completely from the deepest water for about a month at the end of the summer. The volume of water deoxygenated, which lies at a depth of

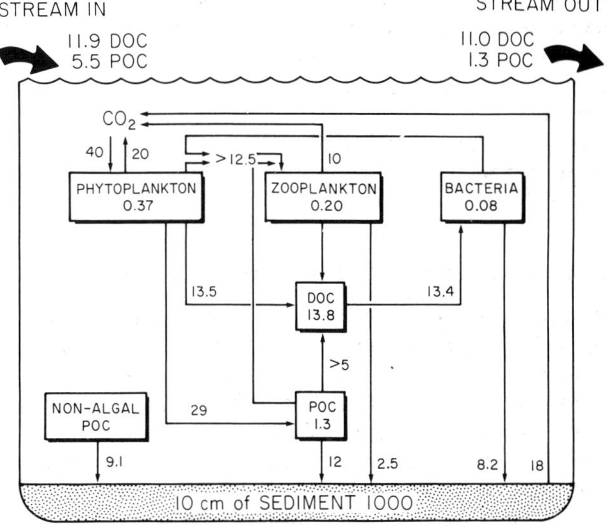

Fig. 3.2.11. The average quantity of carbon (g C m^{-2}, within boxes) and the annual transfer between components (g C m^{-2} yr^{-1}, arrows) in Mirror Lake, New Hampshire. Data from Likens [56].

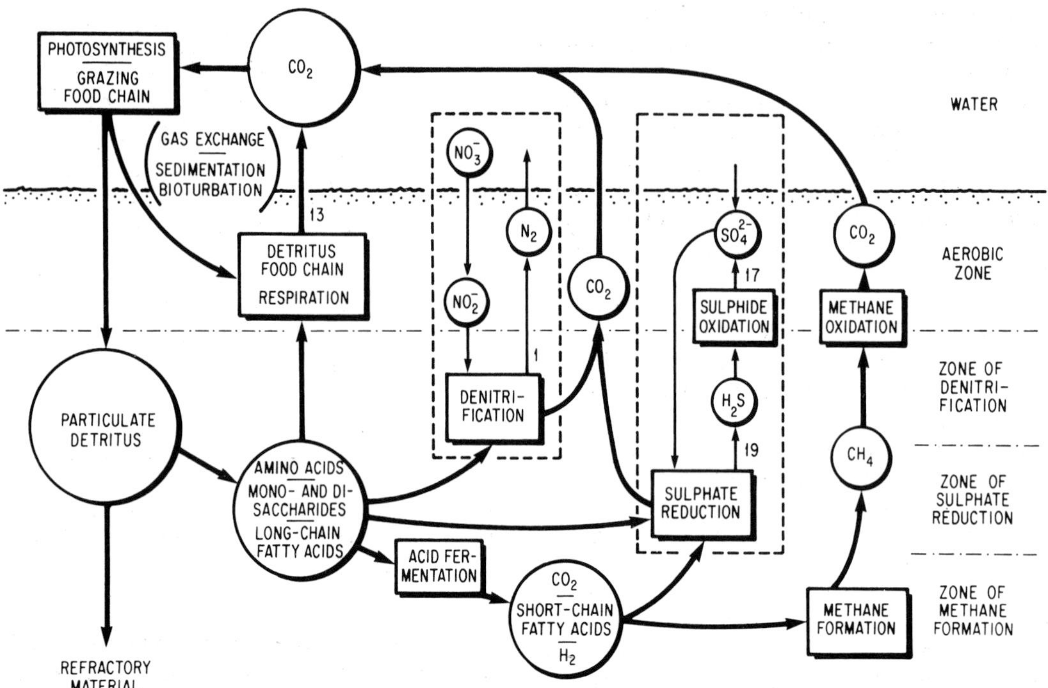

Fig. 3.2.12. Transformations of organic carbon during anaerobic decomposition of a marine sediment and some aspects of the nitrogen and sulphur cycles. (Redrawn with permission from Jørgensen [47].) The rates (mmol m^{-2} day^{-1}) are calculated as carbon oxidation equivalents and come from Jørgensen [46].

between 9 and 10 m, is only 1% of the volume of the lake. Thus the water and the upper layers of the sediment are mostly aerobic.

About 80% of the annual income of organic carbon to Mirror Lake is from phytoplankton photosynthesis (40 g C m^{-2}); the remainder comes from stream inputs of DOC and POC and from the photosynthesis of attached water plants and algae. The mass of the phytoplankton (0.37 g C m^{-2}) is small, yet each year 100 times this mass is produced in net photosynthesis (40 g C m^{-2}). The 2×10^6 bacteria ml^{-1} in the water column have an even smaller mass (0.08 g C), about 10% that of the POC and 1% that of the DOC. Despite this small mass, they process about 1/3 of the primary production. Another 1/3 is consumed and respired by zooplankton and the remainder is decomposed by sediment bacteria (18 g C). In this lake, about 13 g C is permanently buried in the sediment each year; much of this may come from the non-algal inputs to the lake.

These studies did not include the microflagellates, an important grazer of bacteria and perhaps a prey of zooplankton. In spite of this, the conclusions of the study are valid: most (here 88%) of the primary production is decomposed each year, bacteria of the plankton cycle about 1/3 of this carbon despite their small mass, and decomposition in the sediment is about equal to that in the plankton.

ANAEROBIC PROCESSES

There are important contrasts between aerobic and anaerobic processes in aquatic ecosystems. In aerobic processes, all organisms carry out the same general metabolic process but specialize in other ways, such as in their enzymes for breaking apart different complex organic compounds or linkages. A complete mineralization takes place within one organism. In anaerobic processes, organic carbon can move along a number of different pathways because nitrate, sulphate and carbon dioxide function as electron acceptors for different types of respiring organisms. In addition, the energy does not stay in organic compounds but is often transferred to reduced inorganic compounds such as H_2S or CH_4. These compounds may be transported to aerobic zones and later oxidized by other organisms. Thus complete mineralization takes place only within associations of physiologically different types of organisms [46]. The overall scheme is shown in Fig. 3.2.12.

Many of the processes, especially sulphate reduction and methane formation, are dependent upon fermentation as a first step. As Fenchel and Blackburn [19] point out, the process in lake or marine sediments appears to be similar to, but slower than, ruminant and sludge fermentation. Reeburgh [73], in his review of this topic, shows how rapidly the short-chain substrates are cycled (minutes to hours) but the absolute rates are still in doubt since substrates like acetate may well be present largely in a biologically unavailable form.

In sediments the various processes are segregated by depth. The first segregation is caused by the availability of oxygen, nitrate and sulphate which diffuse from the water column into the sediments and are used up as they diffuse farther down into the sediments. A second segregation is caused by the energy yield from the various types of respiration; this yield decreases in the order oxygen, nitrate, sulphate and carbon dioxide. As a consequence of these two types of segregation, the processes are distributed with depth. First all of the oxygen is used, then below this the nitrate is used, then sulphate and then CO_2 (in methane formation). In marine sediments the oxygen is used up in the top few millimetres, the nitrate in the top few centimetres, and the sulphate is often reduced in the top few metres. Sulphate is seldom used up in marine sediments because it is present at 29 mM concentrations in sea water. The sequence of respiration processes with depth is illustrated in Fig. 3.2.13. Carbon dioxide is reduced and methane is often produced at greater depths than those shown in the figure; methane formation correlates with the depletion of sulphate.

In a shallow marine embayment, Limfjord in Denmark (see Fig. 3.2.12), more than half of the organic matter was mineralized by an anaerobic process, the fermentation–sulphate reduction pathway (19 mmol m^{-2} day^{-1}), while 13 mmol was mineralized aerobically [46, 90]. In a different shallow system, a freshwater reservoir in the Netherlands [1], methane production was the most important decomposition process in the sediment (see Table 3.2.5). This predominance of methanogenesis occurs only when sulphate is very low in concentration, as in fresh waters or in sediments with such high rates of sulphate reduction near the sediment surface that no sulphate diffuses deeper in the sediment. One reason for the separation of these two processes is that methanogenic bacteria cannot compete with the sulphate reducers for substrates [47, 58]. In the case of this reservoir, the two processes occurred at

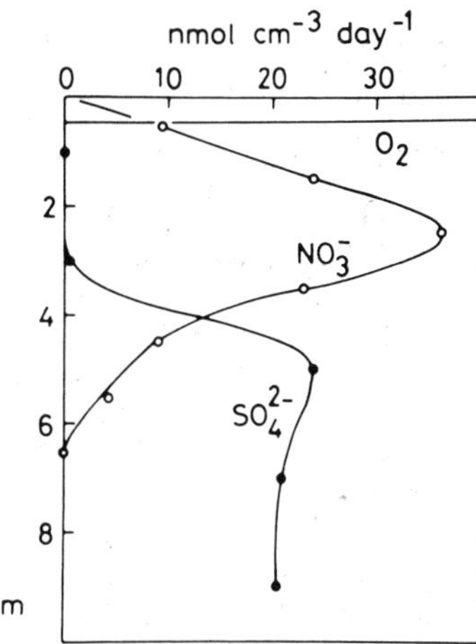

Fig. 3.2.13. Oxygen, nitrate and sulphate respiration rates in a Danish marine sediment. Redrawn with permission from Jørgensen [46].

Table 3.2.5. The carbon budget of a shallow Netherlands reservoir (after Adams & van Eck [1])

Process	Rate (g C m^{-2} yr^{-1})	Depth (cm)
Primary productivity	300	
Sedimentation	80	
autochthonous	60	
allochthonous	20	
Burial	20	
Oxidation in sediment	60	
aerobic	50	0–0.4
anaerobic	10	0–30
denitrification	0.5	0–4
sulfate reduction	4	0–10
methanogenesis	5.5	10–30

different depths but no methane escaped from the sediment surface because methane was oxidized by bacteria in the presence of sulphate [48].

The overall control of the rates of anaerobic decomposition lies in the amount and availability of the organic matter [104]. The type of respiration that occurs, however, is controlled by the availability of electron acceptors, and availability is controlled by the diffusion rates through sediment compared with the overall reaction rates and other environmental factors.

Nitrogen cycle

More than in any other major element cycle, micro-organisms control the nitrogen cycle. Abiotic transformations are extremely slow while microbial oxidation and reductions are rapid. The result is that micro-organisms are responsible for fixation, nitrification, denitrification, etc. Not only do these transformations from one form to another fuel microbial life but the form of nitrogen is often important to nitrogen availability to plants, to nitrogen movements within sediments or soil, and to nitrogen loss from estuaries.

In lakes, rivers, estuaries and coastal regions, a large part of the annual budget of nitrogen comes from the land (e.g. Narragansett Bay in Fig. 3.2.14). In this type of open cycle, the primary production and cycling rates are driven by the rate of input from outside the ecosystem. In contrast, in the open ocean and in large lakes the cycle is almost closed; some nitrate is mixed from deeper waters but most of the primary production is driven by ammonia regenerated from dead algae [17].

NITROGEN FIXATION

Although the most abundant form of nitrogen is N_2 gas, the main cycling of this element is back and forth from organic forms to inorganic forms. The small losses to the gas form can be matched by nitrogen fixation [19]. In planktonic systems, nitrogen fixation is almost entirely carried out by photosynthetic cyanobacteria containing heterocysts. As mentioned in Section 3.2.5, the planktonic nitrogen fixers live mainly in fresh water (perhaps because of their molybdenum needs). The actual amounts contributed by nitrogen fixation to the total nitrogen cycling in an ecosystem varies tremendously from $<1\%$ (the marine Narragansett Bay in Fig. 3.2.14) in oceanic and oligotrophic fresh waters to more than 43% of the total nitrogen input in the eutrophic Clear Lake, California (freshwater results summarized in [105]). The present understanding is that most of the nitrogen fixation in fresh waters is planktonic, but that in the marine systems

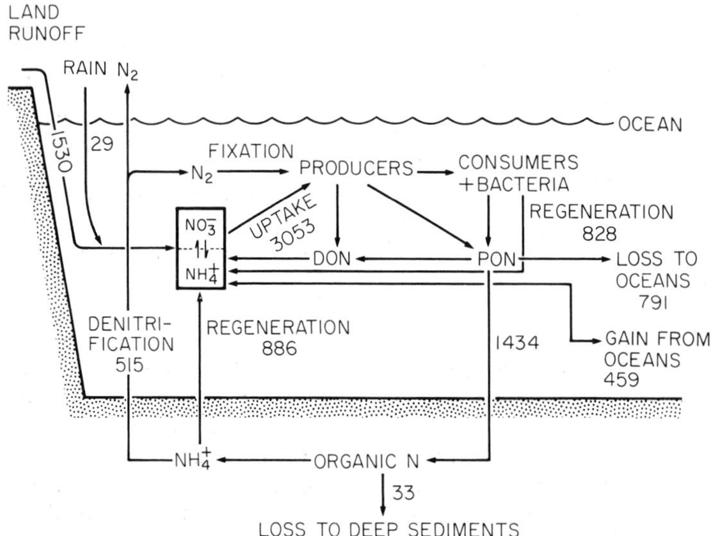

Fig. 3.2.14. Major transformations and transport of nitrogen in the aquatic environment. The numbers next to arrows (mmol N m^{-2} yr^{-1}) are the annual rates for Narrangansett Bay, USA, as reported by Nixon & Pilson [65].

75% of the fixation is by benthic micro-organisms (on 2% of the benthic area) and only 25% by plankton [6, 7]. In some marine systems, such as seagrass beds, coral reefs, salt marshes and mangroves, nitrogen fixation can be as high as 500–2000 mmol m^{-2} yr^{-1}, nearly enough to provide all of the nitrogen needs for algal production in a rich estuary (see Fig. 3.2.14).

AMMONIA PRODUCTION AND USE

When cell protein is decomposed under aerobic and anaerobic conditions by micro-organisms and animals, ammonia is liberated. The yearly flow in aquatic ecosystems is almost the same as the amount of nitrogen in the total primary production (the Redfield ratio is C : N = 106 : 16). The process of liberation, called regeneration, is responsible for supplying 54–94% of the nitrogen for oceanic primary production [17] (see Section 3.2.3). The organisms responsible for regeneration and their relative importance vary from system to system. For example, in one planktonic system 90% was regenerated by organisms smaller than 35 μm and 40% by organisms smaller than 1 μm [29]. Valiela [98] summarizes data showing that, when zooplankton are abundant, they regenerate nearly all of the ammonia. When ammonia is regenerated in sediments, it diffuses upwards and into the water column, as in Narragansett Bay (see Fig. 3.2.14).

In the euphotic zone of the open ocean, ammonia is taken up by algae as fast as it is produced by regeneration [24] so that concentrations in the water are <0.5 μmol l^{-1}. In coastal regions and in estuaries the concentrations may be 1–10 μmol l^{-1} and in polluted lakes and estuaries, where algal uptake does not keep pace with the regeneration, the concentrations are often >100 μmol l^{-1} [98]. However, another process, nitrification, also competes for ammonia in aquatic systems and is dominant when there is no photosysnthesis (e.g. deep in the ocean).

NITRIFICATION

This process, in which ammonia is oxidized to nitrate with oxygen as the electron acceptor, is carried out mainly by autotrophic bacteria (see Section 3.2.5). It is not a trivial process for it has formed all of the tremendous pool of nitrate existing in the deep waters of the ocean (see Fig. 3.2.2). Even more important is the nitrification in the euphotic zone (see Fig. 3.2.14). Kaplan [50] estimates an annual amount of 3×10^9 Mt N for nitrification in the ocean.

Nitrification occurs at all levels of oceans and lakes but will be most prevalent where intense decomposition is occurring or where ammonia is diffusing from an anaerobic zone into waters containing some oxygen. For example, in the Black Sea (see Fig. 3.2.10) nitrification appears to be occurring at depths of 100–200 m. As noted earlier in the section on autotrophy (see Section 3.2.5), the numbers of autotrophic nitrifiers in the ocean appear to be too low to account for all of the transformation unless one makes the unlikely assumption that generation times in the laboratory are longer than those in the sea [50].

The implications of nitrification for microbial ecology are [19]: (a) the process supports a portion of the microbial community (but only $1–5 \times 10^4$ cells l^{-1} [101] out of $1–10 \times 10^8$ cells l^{-1}); (b) algae and bacteria must assimilate nitrate instead of ammonia, and this takes more energy; and (c) an oxidized form of nitrogen is produced that participates in the denitrification process (an important loss of fixed nitrogen from ecosystems).

DENITRIFICATION

In this process (see Fig. 3.2.12), nitrate is reduced but is not incorporated into cellular material; the intermediate product is nitrite and the final products are the gases N_2 or nitrous oxide (N_2O). Reduced carbon or sulphur compounds are also necessary and most of the organisms carrying out this process are heterotrophs occupying anoxic or nearly anoxic environments. The specialized conditions required, low oxygen and nitrate, are contradictory and usually exist where several processes are occurring simultaneously. In the Black Sea (see Fig. 3.2.10), the nitrite found at 50 m and 75 m is an indication of denitrification. In sediments, where the nitrate concentration limits the process, the denitrification zone may be restricted to the few millimetres at the top of the sediments where there is enough oxygen for nitrate formation [19].

Denitrification also occurs in restricted areas of the oceanic water column where oxygen concentrations are low. In these areas, the abundance of organic matter may well limit the rates [31]. If there is little exchange of materials in this restricted water mass, then the amount of denitrification that has occurred can be calculated from the observed loss of oxygen and the observed concentrations of nitrate. The calculations are based on the Redfield ratio; that is, based on the formula previously given, a certain amount of loss of oxygen should have produced a certain amount of nitrate. Any difference between expected nitrate and measured nitrate content in the water body is caused by denitrification [31].

In the ocean, denitrification is the principal process that balances the input of fixed nitrogen to the sea (inputs = river runoff, precipitation, nitrogen fixation; outputs = burial in sediments, denitrification). As described for Narragansett Bay (see Fig. 3.2.14), the denitrification was 800 and the burial was 33 mmol m^{-2} [65]. Another consequence of high rates of denitrification in coastal marine systems could be to alter ratios of nitrogen and phosphorus. The theory (explained in [65] and [98]) is that the mineralization of organic matter causes all of the organic phosphorus to be mineralized, but that some of the organic nitrogen is eventually lost through denitrification to N_2. The resulting inorganic nutrients are lower in nitrogen than the needs of the algae according to the Redfield ratio, and algal growth is reduced because of nitrogen limitation.

Sulphur cycle

The main processes of the sulphur cycle were discussed under the topic of anaerobic cycling of carbon, since sulphate reduction is the main pathway in marine sediments. For example, in Limfjord more than half of the photosynthetically produced organic carbon was decomposed by the fermentation–sulphate reduction pathway (see Fig. 3.2.12). One should remember, however, that only in productive oceanic ecosystems does anaerobic cycling play such an important role in decomposition. While it is important in shallow coastal waters and on the shelf and slope of continents, it is not important in oligotrophic open ocean and deep sea environments.

Part of the H_2S produced during sulphate reduction in the sediment diffuses up to the oxic zone and part combines with iron to form pyrite [19]. In subtidal sediments of Limfjord, Jørgensen [45] showed that 65% of the H_2S was produced in the top 10 cm of sediment; 90% of all the sulphide was oxidized at the sediment surface and 10% remained in the sediment as pyrite (see Fig. 3.3.12). In intertidal sediments, such as in salt marshes, pyrite formation and oxidation is much more rapid than previously suspected [22].

While the process of oxidation of reduced sulphur compounds is extremely important (see Figs. 3.2.10 and 3.2.12), the quantitative contribution of chemosynthetic bacteria to sulphur cycling is difficult to determine because H_2S and other reduced sulphur compounds are spontaneously oxidized by oxygen without bacterial mediation [19]. Unlike the case of ammonia oxidation, the micro-organisms oxidizing reduced sulphur must compete

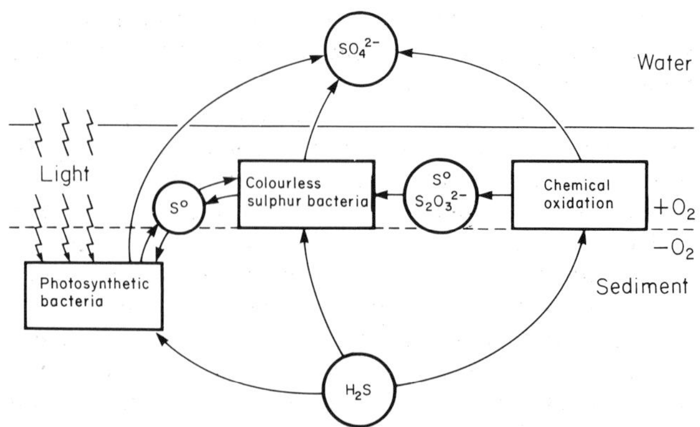

Fig. 3.2.15. Pathways of H_2S oxidation in sediments. Redrawn with permission from Jørgensen [48].

against rapid chemical oxidation. Certainly such forms as *Thiobacillus* derive energy from the oxidation of reduced sulphur compounds in the extreme environments of hot springs (see Section 3.4.3) and are the base of an entire food chain at the recently discovered thermal vents of the ocean floor (see Section 3.3.5). In sediments, *Beggiatoa* can form dense white (with sulphur inclusions) mats on the sediment surface and are abundant in the rhizosphere where oxygen is introduced into the sulphide zone via roots [48]. Some of the pathways of H_2S oxidation are shown in Fig. 3.2.15. But these colourless sulphur bacteria mentioned above are not the only sulphide oxidizers, for pigmented photosynthetic bacteria, such as *Chromatium*, are also found in the top layers of sediments. They are restricted, however, to water depths less than about 10 m and are only rarely important because they require light plus anaerobic conditions.

3.2.7 MICRO-ORGANISMS AND AQUATIC FOOD WEBS

Only in the last decade has there been any realization that micro-organisms are at all important in the pathways of movement of energy and carbon in aquatic ecosystems. These pathways are called food webs (see, for example, Fig. 3.2.16). Earlier it was believed that microbial biomass, as determined by plate counts, was too small to account for any significant pathway of energy flux although their dominance of the remineralization–regeneration process was recognized. The traditional picture of a planktonic food web in the ocean began with algae as the producers and traced the flux of carbon and energy directly to the small zooplankton and eventually to macrozooplankton and fish (in Fig. 3.2.16 the classic food web is the upper right half of the diagram).

The recent view, first stated by Pomeroy [70], is that a microbial food web exists that undoubtedly cycles a great deal of the organic matter in all aquatic ecosystems. The organisms in this microbial food web (in the lower left half of Fig. 3.2.16) and their trophic relationships are reasonably well known. Bacteria and picoplanktonic algae are fed upon by protozoan microflagellates and ciliates, and these in turn are food for copepods, etc. The unanswered questions are: (a) how much of the total carbon fixed in photosynthesis is processed by the microbial food web; and (b) how much carbon and energy moves from the microbial food web into copepods, salps, larval fish and other organisms leading to the top of the food web?

The presence of the microbial food web raises the additional question of control. Does predation by microflagellates and ciliates control bacterial numbers and productivity? Although all of the processes discussed here are also at work in rivers, lakes and estuaries, these habitats are mentioned only briefly in this discussion and in the examples. For these questions the ocean is by far the best studied system.

Food web quantification

One approach to carbon flow studies is to construct a conceptual model that brings together ideas and understanding. At the current level of knowledge, these are at best guidelines for future research. Peterson [69] summarized for the mixed layer of the open ocean the information on carbon stocks and flows. Bacteria process 33% of the primary production, and the microflagellates and ciliates graze half of this (all of the net bacterial production). Twenty per cent of the primary production is grazed directly by larger zooplankton, and an additional 10% of the primary production reaches the zooplankton

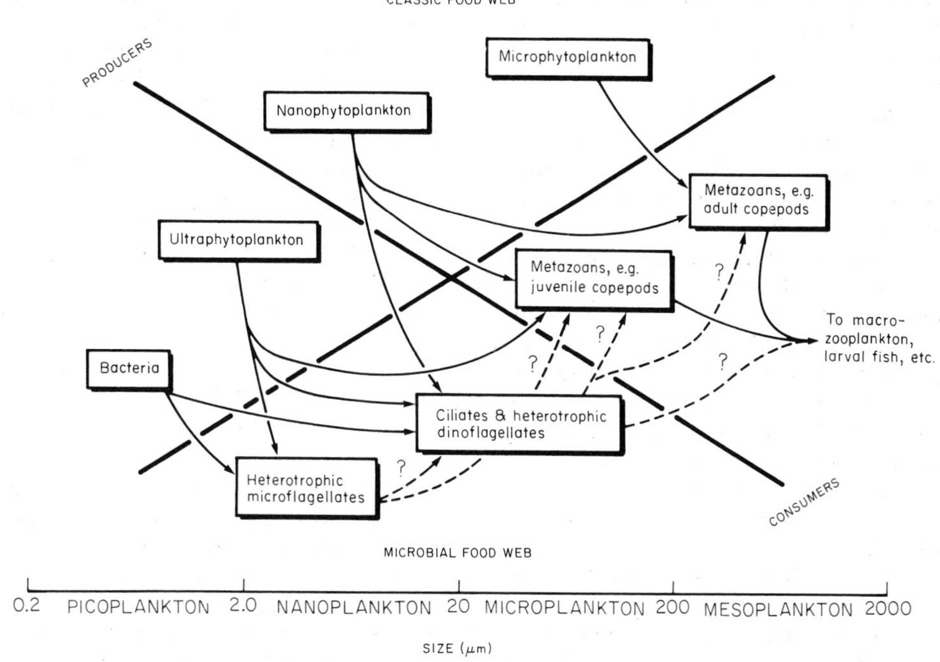

Fig. 3.2.16. A diagram of the marine pelagic food web (after Sherr *et al.* [86]). The diagram is divided by a diagonal (running from the lower left to the upper right) into producers (above the line) and consumers (below the line) of organic matter. The other diagonal (from upper left to lower right) separates the classic food web (above) from the microbial food web (below). Broken lines indicate trophic links that are postulated but not yet demonstrated.

via their grazing of the microflagellates and ciliates. Williams [108] also summarized current understanding but estimated that 56% of the primary production is processed by bacteria and none reaches the higher levels of the classic food web.

The importance of temperature for the functioning of the microbial food web is the subject of a recent theory by Pomeroy and Deibel [71]. They point out that, in the tropics and subtropics, bacterial metabolism and growth are so rapid that the carbon from primary production turns over every 1–3 days. The microbial food web is so active that there is little energy and carbon left for the metazoans (larger consumers). In temperate regions, on the other hand, low temperatures hold the bacterial production and respiration at low rates during the spring bloom of phytoplankton. This allows more primary production to be consumed by the metazoan grazers, both benthic and planktonic. The authors postulate that this difference in the food webs in seasonally very cold water could enhance production of the organisms of the classic food web. They state that 'it may be crucial in sustaining the major fisheries of the Grand Banks and Bering Sea'.

Another approach to the question of carbon flow to higher trophic levels is experimental. Rather than attempting to dissect the systems into its component parts, these studies introduce an isotope label into the microbes and then determine how much of the label reaches the zooplankton or fish. One experiment [16] was conducted in a 5 m diameter, 15 m deep plastic cylinder in a Scottish sea loch. The cylinder was inoculated with 6 mCi of ^{14}C-glucose and the isotope amount followed for 50 days as it moved into different size particles (<1, >1, >10, >100 μm). Most of the label appeared to pass directly from DOM into <1 μm particles (presumably bacteria) and from there into $^{14}CO_2$ and DOM. After 13 days, 20% of the label was in particulate matter but only 2% of the label that was fixed into the bacteria was in larger organisms. The authors concluded that bacteria were important principally as regenerators rather than as food for larger organisms. Because there are only three or four studies of this type, few conclusions can be drawn at this time. For example, the extrapolations from this experiment have been criticized because a sea loch is atypical of the world ocean [84].

The experiment does remind us that carbon transfer from one trophic level to another is not necessarily a very efficient process (see Table 3.2.2). Two transfers would at best result in only a few percent of the label accumulating in larger organisms. The amount accumulating is also dependent upon the efficiency of bacterial transformation of DOM into particulate matter. Laboratory studies of cultures growing at various rates on various compounds give growth yields (proportion of total uptake going into particulate matter) of 20–80% [19]. A simple compound such as glucose may give yields of 40%. Yields from bacteria growing on natural detritus and DOM are reviewed by Linley and Newell [57], who concluded that it is the ratio of carbon to nitrogen in the DOM that is the most important factor. When the C:N is close to 5 (the value for bacteria), yields can approach 90%; when phytoplankton were the source of carbon (C:N = 5–8), yields were 40–60%; when structural materials from salt marsh or sea grasses were the source, yields were close to 15% but lignin may also play a role.

Predation and control by protozoans

Techniques used in studies of small flagellates and ciliates of the microbial food web are based mainly on epifluorescent direct counts of samples on polycarbonate filters [3, 85]. The most difficult process to measure is the feeding rate of protozoans on bacteria in nature. The difficulty lies in the methods. What tracer can be added to samples of plankton that will be ingested by the protozoans at the same rate as natural bacteria are ingested? Latex beads with fluorescent dyes are ingested but are artificial. Laboratory bacteria with a radioisotope label are ingested but they are usually ten times larger in volume than the natural bacteria so the rates are suspect. Two promising techniques are the use of mini-bacteria or of natural bacteria stained and preserved by lyophilization (M. Pace and B. Sherr, personal communication). Mini-bacteria are non-reproducing buds from *E. coli*. These small bacterioids can be labelled with radioisotopes or dyes.

Colourless nanoflagellates [87, 88] are present in natural waters at about the same concentrations as algae, that is, hundreds to thousands per millilitre (see Table 3.2.6). These same numbers are found in fresh water [33]. Only 1–10 ciliates are found per millilitre. Are flagellates and ciliates abundant enough to control the bacteria?

In the laboratory small flagellates consume bacteria, ciliates consume flagellates, and the numbers of these populations oscillate in classic predator–prey relationships (see, for example, [57]). Conditions in a test-tube are very different from those in nature, however, because the bacteria in a test-tube are 10 times larger than those in natural waters and are about 100 times more abundant.

For these reasons, experiments have been carried out in large freshwater chambers (60 m^3) in a northern lake

Table 3.2.6. The numbers (ml^{-1}) and the wet weight (μl^{-1}) of the photosynthetic and heterotrophic micro-organisms of the shelf and open ocean regions of the North Atlantic (from Sieburth [88]), of the shelf region of southeastern USA (Sherr *et al.* [85]), and of Limfjord (Denmark) (Andersen & Sørensen [3])

	Shelf		Open ocean	
	no. (ml^{-1})	wet wt ($\mu g\ l^{-1}$)	no. (ml^{-1})	wet wt ($\mu g\ l^{-1}$)
Picoplankton (0.2–2 μm)				
cyanobacteria	60 000	12	12 000	1
bacteria	1 000 000	115	600 000	20
Nanoplankton (2–20 μm)				
photosynthetic algae	3000	74	500	14
flagellates	3000	74	500	14
flagellates *	300–800	12–110		
flagellates†	200–17 000			
ciliates	0.4–6	2–12		
ciliates†	1.5–160			

* Sherr *et al.* [85], southeastern USA estuary (C × 20 = wet wt).
† Andersen & Sorensen [3], Limfjord, Denmark.

[33]. The lake in which the chambers were located was oligotrophic and the large changes in bacterial biomass and the subsequent growth of microflagellates were only seen when plankton algae became abundant after fertilization with nitrogen and phosphate (see Fig. 3.2.17). From these data (plus growth measurements of the bacteria), it was calculated that 20–30% of the water of the entire chamber was cleared each day by the microflagellates except that when the microflagellates reached their peak 200% was cleared. In chambers in a shallow Danish fjord [3] it was found that a range of 5–365% was cleared per day by microflagellates (average 45%). In both of these studies, the means of the clearance rate per cell (0.6×10^{-5} ml hr^{-1} and 0.24×10^{-5} ml hr^{-1} for the enclosure and the fjord, respectively) were similar to the $0.2–4 \times 10^{-5}$ ml hr^{-1} found for six species of flagellates in pure culture [18]. The conclusion is that the predation rate of microflagellates in natural waters is high enough to control the numbers of bacteria.

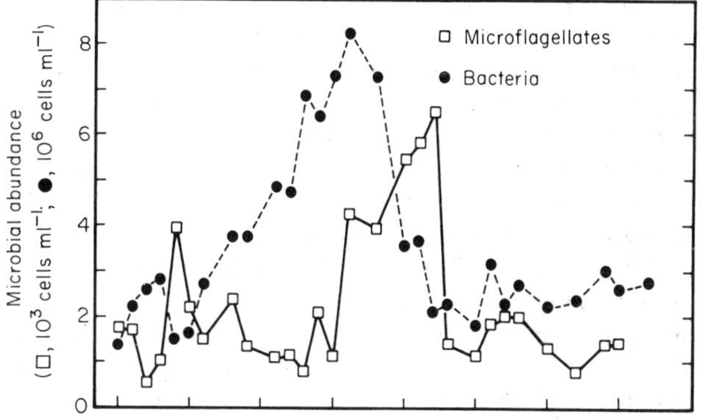

Fig. 3.2.17. Numbers of bacteria and colourless microflagellates in a 60 m^3 enclosure in an arctic lake. The enclosure was fertilized daily with N and P. The units are 10^3 microflagellate cells ml^{-1} and 10^6 bacterial cells ml^{-1}. After Hobbie & Helfrich [33].

Conclusions from test-tube cultures and from experimental enclosures, however large, must be tested in nature. To investigate the microbial food web, it is necessary to sample several times a week and to sample the same water mass each time. No one knows exactly what the 'same water mass' means for microbial processes, but there is at least one site where the cycles of bacteria, colourless flagellates and ciliates are in synchrony over a large area (tens of kilometres) of slowly moving water. Apparently the same water mass is sampled for several months at a time. This is Limfjord, a shallow marine bay in Denmark [3]. In this fjord, populations showed strong, successive fluctuations throughout the summer with a total of 8 bacterial and 7 predator (flagellate and ciliate) population peaks (see Fig. 3.2.18). Each bacterial peak was followed by a flagellate peak 3–8 days later. Each flagellate peak was followed by a ciliate peak within 4–6 days. These data are strong evidence for the control of bacterial numbers by the grazing of microflagellates. They also indicate that flagellates, too, may be controlled by grazing by ciliates (see also [85]).

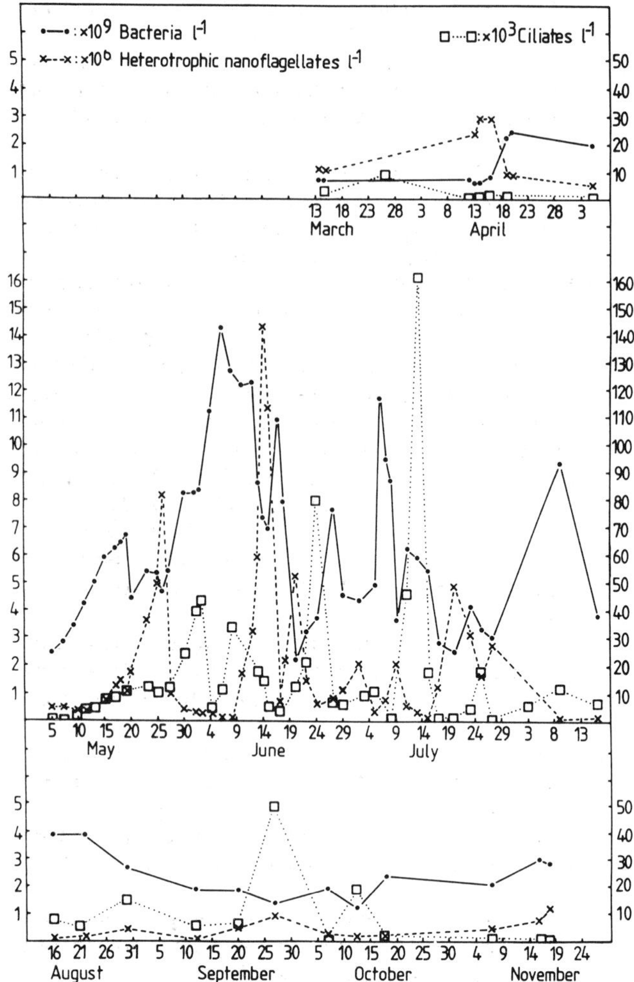

Fig. 3.2.18. Bacteria, microflagellates and ciliates (mean concentrations at 1 and 2 m) in Limjford, Denmark. Redrawn with permission from Andersen & Sørensen [3].

There can be no doubt that the microbial food web exists in both freshwater and marine systems. It is highly dynamic and consists of several tightly coupled processes encompassing producers, consumers and mineralizers.

3.2.8 REFERENCES

[1] Adams D.D. & van Eck G.Th.M. (in press) Biogeochemical cycling of organic carbon in the sediments of Grote Rug reservoir. In Proceedings of a Workshop on Measurement of Microbial Activity in the Carbon Cycle of Aquatic Ecosystems (ed. T. Cappenberg & C.L.M. Steenbergen). *Ergebnisse der Limnol., Arch. Hydrobiol.*

[2] Ammerman J.W., Fuhrman J.A., Hagström Å. & Azam F. (1984) Bacterioplankton growth in seawater: I. Growth kinetics and cellular characteristics in seawater cultures. *Mar. Ecol. Prog. Ser.* **18**, 31–9.

[3] Andersen P. & Sørensen H.M. (1986) Population dynamics and trophic coupling in pelagic microorganisms in eutrophic coastal waters. *Mar. Ecol. Prog. Ser.* **33**, 99–109.

[4] Berner R.A. (1974) Kinetic models for the early diagenesis of nitrogen, sulfur, phosphorus, and silicon in anoxic marine sediments. In *The Sea*, Vol. 5, p.427 (ed. D. Goldberg). John Wiley, New York.

[5] Bird D.F. & Kalff J. (1986) Bacterial grazing by planktonic lake algae. *Science* **231**, 493–5.

[6] Capone D.G. (1983) Benthic nitrogen fixation. In *Nitrogen in the Marine Environment*, p.105 (ed. E.J. Carpenter & D.G. Capone). Academic Press, New York, London.

[7] Carpenter E.J. (1983) Nitrogen fixation by marine *Oscillatoria (Trichodesmium)* in the world's oceans. In *Nitrogen in the Marine Environment*, p.65 (ed. E.J. Carpenter and D.G. Capone). Academic Press, New York, London.

[8] Carpenter E.J. & Guillard R.R.L. (1971) Intraspecific differences in nitrate half-saturation constants for three species of marine phytoplankton. *Ecology* **52**, 183–5.

[9] Cole J.J., Likens G.E. & Hobbie J.E. (1984) Decomposition of planktonic algae in an oligotrophic lake. *Oikos* **42**, 257–66.

[10] Colwell R.R. (ed) (1984) *Vibrios in the Environment*. John Wiley, New York.

[11] Conover R.J. & Huntley M.E. (1980) General rules of grazing in pelagic ecosystems. In *Primary Productivity in the Sea*, p.461 (ed. P.G. Falkowski). Plenum Press, New York.

[12] Costerton J.W., Marrie T.J. & Cheng K.-J. (1985) Phenomena of bacterial adhesion. In *Bacterial Adhesion*, p.3 (ed. D.C. Savage & M. Fletcher). Plenum Press, New York.

[13] Craig H. (1971) The deep metabolism: oxygen consumption in abyssal ocean water. *J. Geophys. Res.* **76**, 5078–91.

[14] Crawford C.C., Hobbie J.E. & Webb K.L. (1974) The utilization of dissolved free amino acids by estuarine microorganisms. *Ecology* **55**, 551–63.

[15] Dryssen D. & Wedborg M. (1980) Major and minor elements, chemical speciation in estuarine waters. In

Chemistry and Biogeochemistry of Estuaries, p.71 (ed. E. Olausson and I. Cato.) John Wiley, Chichester, New York.

[16] Ducklow H.W., Purdie D.A., Williams P.J.leB. & Davies J.M. (1986) Bacterioplankton: a sink for carbon in a coastal marine plankton community. *Science* **232**, 865–7.

[17] Eppley R.W. & Peterson B.J. (1979) Particulate organic flux and planktonic new production in the deep ocean. *Nature* **282**, 677–80.

[18] Fenchel, T. (1982) Ecology of heterotrophic microflagellates II. Bioenergetics and growth. *Mar. Ecol. Prog. Ser.* **8**, 225–31.

[19] Fenchel T. & Blackburn T.H. (1979) *Bacteria and Mineral Cycling*. Academic Press, London, New York.

[20] Fliermans C.B., Cherry W.B., Orrison L.H., Smith S.J., Tison D.L. & Pope D.H. (1981) Ecological distribution of *Legionella pneumophila*. *Appl. Environ. Microbiol.* **41**, 9–16.

[21] Fuhrman J. & Azam F. (1982) Thymidine incorporation as a measure of heterotrophic bacterioplankton production in marine surface waters: evaluation and field results. *Mar. Biol.* **66**, 109–20.

[22] Giblin A.E. & Howarth R.W. (1984) Pore water evidence for a dynamic sedimentary iron cycle in salt marshes. *Limnol. Oceanogr.* **29**, 47–63.

[23] Gieskes W.W. & Kraay G.W. (1984) State-of-the-art in the measurement of primary production. In *Flows of Energy and Materials in Marine Ecosystems*, p.171 (ed. M.J.R. Fasham). NATO Conf. Series IV-13. Plenum Press, New York, London.

[24] Goldman J.C. & Glibert P.M. (1983) Kinetics of inorganic nitrogen uptake by phytoplankton. In *Nitrogen in the Marine Environment*, p.233 (ed. E.J. Carpenter & D.G. Capone), Academic Press, New York, London.

[25] Goodman D., Eppley R.W. & Reid F.M.H. (1984) Summer phytoplankton assemblages and their environmental correlates in the Southern California Bight. *Jour. Mar. Res.* **42**, 1019–49.

[26] Haack T.K. & McFeters G.A. (1982) Nutritional relationships among microorganisms in an epilithic biofilm community. *Microb. Ecol.* **8**, 115–26.

[27] Hagström Å., Ammerman J.W., Henrichs S. & Azam F. (1984) Bacterioplankton growth in seawater: II. Organic matter utilization during steady-state growth in seawater cultures. *Mar. Ecol. Prog. Ser.* **18**, 41–8.

[28] Hamilton W.A. (1987) Biofilms: microbial interactions and metabolic activities. In *Ecology of Microbial Communities*, p.361 (ed. M. Fletcher, T.R.G. Gray & J.G. Jones). Cambridge University Press, Cambridge, UK.

[29] Harrison W.G. (1978) Experimental measurements of nitrogen remineralization in coastal waters. *Limnol. Oceanogr.* **23**, 684–94.

[30] Harvey R.W. & Young L.Y. (1980) Enumeration of particle-bound and unattached respiring bacteria in the salt marsh environment. *Appl. Environ. Microbiol.* **40**, 165.

[31] Hattori A. (1983) Denitrification and dissimilatory nitrate reduction. In *Nitrogen in the Marine Environment*, p.191 (ed. E.J. Carpenter and D.G. Capone). Academic Press, New York, London.

[32] Hobbie J.E., Daley R.J. & Jasper S. (1977) Use of Nuclepore filters for counting bacteria by fluorescence microscopy. *Appl. Environ. Microbiol.* **33**, 1225–8.

[33] Hobbie J.E. & Helfrich J.V.K. (in press) The effect of grazing by microprotozoans on production of bacteria. In *Proceedings of a Workshop on Measurement of Microbial Activity in the Carbon Cycle of Aquatic Ecosystems* (ed. T. Cappenberg & C.L.M. Steenbergen). *Ergebnisse der Limnol., Arch. Hydrobiol.*

[34] Hobbie J.E., Novitsky T.J., Rublee P. A., Ferguson R.L. & Palumbo A.V. (in press) Microbiology. In *Georges Bank* (ed. R. Backus). MIT Press, Cambridge, Mass.

[35] Hodson R.E., MacCubbin A.E. & Pomeroy L.R. (1981) Dissolved adenosine triphosphate utilization by free-living and attached bacterioplankton. *Marine Biol.* **64**, 43–51.

[36] Holmgren S. (1968) *Phytoplankton Production in a Lake North of the Arctic Circle*. Fil. Lic. Thesis, Institute of Limnology, Uppsala University.

[37] Hoppe H.-G. (1978) Relations between active bacteria and heterotrophic potential in the sea. *Neth. J. Sea Res.* 12, 78–98.

[38] Howarth R.W. & Cole J.J. (1985) Molybdenum availability, nitrogen limitation and phytoplankton growth in natural waters. *Science* **29**, 653–5.

[39] Hutchinson G.E. (1961) The paradox of the plankton. *Amer. Nat.* **95**, 137–45.

[40] Iacobellis N.S. & Devay J.E. (1986) Long-term storage of plant-pathogenic bacteria in sterile distilled water. *Appl. Environ. Microbiol.* **52**, 388–9.

[41] Jenkins, W.J. (1977) Tritium helium dating in the Sargasso Sea: a measurement of oxygen utilization rates. *Science* **196**, 291–2.

[42] Jenkins W.J. (1984) The use of tracers and water masses to estimate rates of respiration. In *Heterotrophic Activity in the Sea*, p.391 (ed. J.E. Hobbie & P.J. leB. Williams). Plenum Press, New York, London.

[43] Jerlov N.G. (1976) *Marine Optics*. Elsevier, Amsterdam.

[44] Jones A.K. (1982) The interactions of algae and bacteria. In *Microbial Interactions*, Vol. 1, p.189 (ed.

A.T. Bull & J.H. Slater). Academic Press, London, New York.

[45] Jørgensen B.B. (1977) The sulfur cycle of a coastal marine sediment (Limfjorden, Denmark). *Limnol. Oceanogr.* **22**, 814–32.

[46] Jørgensen B.B. (1980) Mineralization and the bacterial cycling of carbon, nitrogen and sulfur in marine sediments. In *Contemporary Microbial Ecology*, p.239 (ed. D.C. Ellwood, M.J. Latham, J.N. Hedger, J.M. Lynch & J.H. Slater). Academic Press, London, New York.

[47] Jørgensen B.B. (1983a) Processes at the sediment–water interface. In *The Major Biogeochemical Cycles and Their Interactions*, Scope 21, p.477 (ed. B. Bolin & R.B. Cook). John Wiley, Chichester, New York.

[48] Jørgensen B.B. (1983b) The microbial sulphur cycle. In *Microbial Geochemistry*, p.91 (ed. W.E. Krumbein). Blackwell Scientific Publications, Oxford,

[49] Kalff J. & Welch H.E. (1974) Phytoplankton production in Char Lake, a natural polar lake, and in Meretta Lake, a polluted polar lake, Cornwallis Island, Northwest Territories. *J. Fish. Res. Bd. Canada* **31**, 621–36.

[50] Kaplan W. (1983) Nitrification. In *Nitrogen in the Marine Environment*, p.139 (ed. E.J. Carpenter & D.G. Capone). Academic Press, New York, London.

[51] Karl D.M. (1980) Cellular nucleotide measurements and applications in microbial ecology. *Microbiol. Rev.* **44**, 739–96.

[52] Karl D.M. (1981) Simultaneous rates of ribonucleic and deoxyribonucleic acid syntheses for estimating growth and cell division of aquatic microbial communities. *Appl. Environ. Microbiol.* **42,** 802–10.

[53] Kjelleberg S. (1985) Mechanisms of bacterial adhesion at gas–liquid interfaces. In *Bacterial Adhesion*, p.163 (ed. D.C. Savage and M. Fletcher). Plenum Press, New York.

[54] Kuznetsov, S.I. (1973) *The Microflora of Lakes and its Geochemical Activity*. A translation edited by C.H. Oppenheimer. University of Texas Press, Austin, London.

[55] Lewin J.C. & Lewin R.A. (1960) Autotrophy and heterotrophy in marine littoral diatoms. *Can. J. Microbiol.* **6**, 127–34.

[56] Likens G.E. (ed.) (1985) *An Ecosystem Approach to Aquatic Ecology*. Springer-Verlag, New York, Berlin.

[57] Linley E.A.S. & Newell R.C. (1984) Estimates of bacterial growth yields based on plant detritus. *Bull. Mar. Sci.* **35**, 409–25.

[58] Lovley D.R. & Klug M.J. (1986) Model for the distribution of sulfate reduction and methanogenesis in freshwater sediments. *Geochim. Cosmochim. Acta* 50, 11–18.

[59] McCoy E. & Sarles W.B. (1969) Bacteria in lakes: populations and functional relations. In *Eutrophication: Causes, Consequences, Correctives*, p.331. National Academy of Sciences, Washington, DC.

[60] Macleod R.A. (1965) The question of the existence of specific marine bacteria. *Bact. Rev.* **29**, 9–23.

[61] Marinucci A.C., Hobbie J.E. & Helfrich J.V.K. (1983) Effect of litter nitrogen on decomposition and microbial biomass in *Spartina alterniflora*. *Microb. Ecol.* 9, 27–40.

[62] Moriarty D.J.W. (1980) Measurement of bacterial biomass in sandy sediments. In *Biogeochemistry of Ancient and Modern Environments*, p.131 (ed. P.A. Trudinger, M.R. Walter & B.J. Ralph). Australian Academy of Sciences, Canberra, and Springer-Verlag, Berlin.

[63] Moriarty D.J.W. (1984) Measurements of bacterial growth rates in some marine systems using the incorporation of tritiated thymidine in DNA. In *Heterotrophic Activity in the Sea* p.217 (ed. J.E. Hobbie & P.J. leB. Williams). Plenum Press, New York, London.

[64] Muller P.J. & Suess E. (1979) Productivity, sedimentation rate, and sedimentary organic matter in the oceans. I. Organic carbon preservation. *Deep-Sea Res.* **26**, 1347–62.

[65] Nixon S.W. & Pilson M.E.Q. (1984) Estuarine total system metabolism and organic exchange calculated from nutrient ratios: an example from Narragansett Bay. In *The Estuary as a Filter*, p.261 (ed. V.S. Kennedy). Academic Press, New York, London.

[66] Oglesby R.T. (1977) Phytoplankton summer standing crop and annual productivity as functions of phosphorus loading and various physical factors. *J. Fish. Res. Bd. Canada* **34**, 2255–70.

[67] Paerl H.W. (1985) Influence of attachment on microbial metabolism and growth in aquatic ecosystems. In *Bacterial Adhesion*, p.363 (ed. D.C. Savage & M. Fletcher). Plenum Press, New York.

[68] Parsons T.R., Takahashi M. & Hargrave B. (1977) *Biological Oceanographic Processes*. Pergamon Press, Oxford, New York.

[69] Peterson B.J. (1984) Synthesis of carbon stocks and flows in the open ocean mixed layer. In *Heterotrophic Activity in the Sea*, p.547 (ed. J.E. Hobbie & P.J.leB. Williams). Plenum Press, New York, London.

[70] Pomeroy L.R. (1974) The ocean's food web, a changing paradigm. *BioScience* **24**, 499–504.

[71] Pomeroy L.R. & Deibel D. (1986) Temperature regulation of bacterial activity during the spring bloom in Newfoundland coastal waters. *Science* 233, 359–61.

[72] Porter K. & Feig Y.S. (1980) The use of DAPI for

identifying and counting the aquatic microflora. *Limnol. Oceanogr.* **25,** 943–8.

[73] Reeburgh W.S. (1983) Rates of biogeochemical processes in anoxic sediments. *Ann. Rev. Earth Planet. Sci.* **11**, 269–98.

[74] Rheinheimer G. (1980) *Aquatic Microbiology*, 2nd edn. John Wiley, New York.

[75] Richards F.A. (1984) Nutrient interactions and microbes. In *Heterotrophic Activity in the Sea*, p.289 (ed. J.E. Hobbie & P.J.leB. Williams). Plenum Press, New York, London.

[76] Richerson P., Armstrong R. & Goldman C.R. (1970) Contemporaneous disequilibrium, a new hypothesis to explain the 'paradox of the plankton'. *Proc. Nat. Acad. Sci.* **67**, 1710–14.

[77] Riley G.A. (1951) Oxygen phosphate and nitrate in the Atlantic Ocean. *Bull. Bingham Oceanogr. Coll.* 13, 1–126.

[78] Rodhe W. (1948) Environmental requirements of freshwater plankton algae. Experimental studies in the ecology of phytoplankton. *Symbol. Bot. Upsalien.* **10**, 1–149.

[79] Rublee P.A. (1982) Bacteria and microbial distribution in estuarine sediments. In *Estuarine Comparisons*, p.159 (ed. V.S. Kennedy). Academic Press, New York, London.

[80] Ryther J.H. (1969) Photosynthesis and fish production in the sea. The production of organic matter and its conversion to higher forms of life vary throughout the world ocean. *Science* **166**, 72–6.

[81] Sanders R.W. & Porter K.G. (in press) Phagotrophic phytoflagellates. *Advances in Microbial Ecology*.

[82] Sharp J.H. (1973) Size classes of organic carbon in seawater. *Limnol. Oceanogr.* **18**, 441–56.

[83] Sharp J.H. (1984) Inputs into microbial food chains. In *Heterotrophic Activity in the Sea*, p.101 (ed. J.E. Hobbie & P.J.leB. Williams). Plenum Press, New York, London.

[84] Sherr E.B., Sherr B.F. & Albright L.J. (1987) Bacteria: link or sink? *Science* **235**, 88.

[85] Sherr E.B., Sherr B.F., Fallon R.D. & Newall S.Y. (1986) Small, aloricate ciliates as a major component of the marine heterotrophic nanoplankton. *Limnol. Oceanogr.* 31, 177–83.

[86] Sherr E.B., Sherr B.F. & Paffenhofer G.A. (1986) Phagotrophic protozoa as food for metazoans: a 'missing' trophic link in marine pelagic food webs. *Mar. Microb. Food Webs* **1**, 2–79.

[87] Sieburth J.McN. (1979) *Sea Microbes*. Oxford University Press, New York.

[88] Sieburth J.McN. (1984) Protozoan bacterivory in pelagic marine waters. In *Heterotrophic Activity in the Sea*, p.405 (ed. J.E. Hobbie & P.J.leB. Williams). Plenum Press, New York, London.

[89] Smith A.J. (1983) Modes of cyanobacterial carbon metabolism. *Ann. Microbiol.* **134B**, 93–113.

[90] Sørensen J., Jørgensen B.B. & Revsbech N.P. (1979) A comparison of oxygen, nitrate and sulphate respiration in coastal marine sediments. *Microb. Ecol.* 5, 105–15.

[91] Sorokin Y.I. (1964) On the primary production and bacterial activities in the Black Sea. *J. Cons. Int. Explor. Mer* **29**, 41–60.

[92] Stanley D.W. & Daley R.J. (1977) Environmental control of primary productivity in Alaskan tundra ponds. *Ecology* **57**, 1025–33.

[93] Stotzky G. & Burns R.G. (1982) The soil environment: clay–humus–microbe interactions. In *Experimental Microbial Ecology*, p.105 (ed. R.G. Burns & J.H. Slater). Blackwell Scientific Publications, Oxford, UK.

[94] Suberkropp K. & Klug M.J. (1976) Fungi and bacteria associated with leaves during processing in a woodland stream. *Ecology* **57**, 707–19.

[95] Sutherland I.W. (1983) Microbial exopolysaccharides — their role in microbial adhesion in aqueous systems. *CRC Crit. Rev. Microbiol.* **10**, 173–201.

[96] Talling J.F. (1965) The photosynthetic activity of phytoplankton in East African lakes. *Int. Rev. Ges. Hydrobiol.* **50**, 1–32.

[97] Turley C. & Lochte K. (1986) Diel changes in the specific growth rate and mean cell volume of natural bacterial communities in two different water masses in the Irish Sea. *Microb. Ecol.* 12, 271–82.

[98] Valiela I. (1984) *Marine Ecological Processes*. Springer-Verlag, New York, Berlin.

[99] Van Es F.B. & Meyer-Reil L.-A. (1982) Biomass and metabolic activity of heterotrophic marine bacteria. In *Advances in Microbial Ecology*, Vol. 6, p.111 (ed. K.C. Marshall). Plenum Press, New York, London.

[100] Wangersky P.J. (1984) Organic particles and bacteria in the ocean. In *Heterotrophic Activity in the Sea*, p.263 (ed. J.E. Hobbie & P.J.leB. Williams). Plenum Press, New York, London.

[101] Ward B.B. (1984) Combined autoradiography and immunofluorescence for estimation of single cell activity by ammonium-oxidizing bacteria. *Limnol. Oceanogr.* **29**, 402–10.

[102] Ward B.B. & Perry M.J. (1980) Immunofluorescent assay for the marine ammonium-oxidizing bacterium *Nitrosococcus oceanus*. *Appl. Environ. Microbiol.* **39**, 913–18.

[103] Watson S.W. & Hobbie J.E. (1979) Measurement of bacterial biomass as lipopolysaccharide. In *Native*

Aquatic Bacteria: Enumeration, Activity, and Ecology, p.82 (ed. J.W. Costerton & R.R. Colwell). ASTM Special Technical Pub. 695, Amer. Soc. Test. Materials, Philadelphia.

[104] Westrich J.T. & Berner R.A. (1984) The role of sedimentary organic matter in bacterial sulfate reduction: the G model tested. *Limnol. Oceanogr.* **29**, 236–49.

[105] Wetzel R.G. (1975) *Limnology*. W.B. Saunders, Philadelphia, London.

[106] White D.C., Bobbie R.J., King J.D., Nickels J. & Amoe P. (1979) Lipid analysis of sediments for microbial biomass and community structure. In *Methodology for Biomass Determinations and Microbial Activities in Sediments*, p.87 (ed. C.D. Litchfield & P.L. Seyfried). ASTM STP 673, American Society for Testing and Materials, Philadelphia.

[107] Williams P.J.leB (1975) Biological and chemical dissolved organic material in sea water. In *Chemical. Oceanography*, Vol. 2, 2nd edn, p.301 (ed. J.P. Riley & G. Skirow). Academic Press, New York.

[108] Williams P.J. leB. (1984) Bacterial production in the marine food chains: the emperor's new suit of clothes? In *Flows of Energy and Materials in Marine Ecosystems*, p.271 (ed. M.J.R. Fasham). NATO IV:13. Plenum Press, New York, London.

[109] Williams P.M., Oeschger M.H. & Kinney P. (1969) Natural radiocarbon activity of the dissolved organic carbon in the Northeast Pacific Ocean. *Nature* **224**, 256–8.

[110] Woodwell G.M., Whittaker R.H., Reiners W.A., Likens G.E., Delwiche C.C. & Botkin D.B. (1978) The biota and the world carbon budget. *Science* **199**, 141–6.

[111] Wright R.T. (1984) Dynamics of pools of dissolved organic carbon. In *Heterotrophic Activity in the Sea*, p.121 (ed. J.E. Hobbie & P.J.leB. Williams). Plenum Press, New York, London.

[112] Wright R.T. & Hobbie J.E. (1966) The use of glucose and acetate by bacteria and algae in aquatic ecosystems. *Ecology* **47**, 447–64.

3.3 The animal environment

3.3.1 INTRODUCTION

All animals are continually exposed to micro-organisms present in the atmosphere, in the food and in the environment. The external surfaces of the animal, the alimentary tract, and in higher animals the eyes, nose and urinogenital regions offer a wide range of potential colonization sites, but they do not all offer the correct conditions for microbial survival and replication. Indeed, a variety of physical, chemical and ecological conditions and host defences may mitigate against the survival of the micro-organisms.

Exposure to environmental micro-organisms commences during birth in viviparous animals, or after hatching from the egg in birds, reptiles and most invertebrates. Successive invasions of micro-organisms occur in each habitat offered on or in the animal. The newly arrived micro-organisms integrate with, or displace, previously successful species until eventually, under stable environmental conditions, a climax population is established in each of the habitats. Species which occur in this climax population in all members of the animal community, and which in all probability were present during the evolution of the animal, are defined as indigenous or autochthonous species; other species which may occur in the same habitat, but do not normally establish themselves there in the absence of perturbations of the system, are termed allochthonous species [136]. Many allochthonous species are present in most animal habitats at all times, but some autochthonous species are unique to a single type of animal habitat, and some may be restricted to a single species of animal.

The indigenous organisms associated with external surfaces of the animal often apparently offer little benefit to the host, although they may offer some protection against pathogens. Conversely, indigenous micro-organisms within the alimentary tract have been exploited by many herbivorous animals for the digestion of plant structural carbohydrates and other beneficial functions such as the utilization of non-protein nitrogen, detoxication of dietary components and provision of vitamins. With the possible exception of a few invertebrates, no animals can digest cellulose, hemicellulose or pectin without the aid of these symbiotic micro-organisms. Because the rate of enzymic hydrolysis of cellulose and hemicellulose is slow, the rate of passage of plant tissues through the alimentary tract of herbivorous mammals and some insects (for example, wood-eating termites) is slowed by the provision of relatively large chambers in which the microbial hydrolysis of plant polysaccharides occurs. Such chambers are either pregastric complex stomachs as found in ruminants, kangaroos, colobid monkeys, camels, llamas and some whales, or are post-gastric chambers, such as the large intestine and caecum occurring in primates, pigs, horses, many rodents, birds and some reptiles. Many mammals with a pregastric chamber also possess a well-developed large intestine and caecum where additional fermentation of dietary components occurs.

One of the most surprising aspects of the animal–

micro-organism system is the enormous numbers of micro-organisms that are involved. It has been calculated [81] that the human being may be colonized by as many as 10^{14} indigenous microbial cells and that, of the total human–micro-organism community, only 10% of the cells were human [132]. In the sheep rumen, viable bacterial populations as high as 2×10^{10} cells g^{-1} of digesta are often recorded. If each organism were only 1 μm long, those present in a rumen with a fluid volume of 5 litres would stretch, end to end, 2.5 times round the earth.

The types of habitat provided by the animal for micro-organisms vary from fully aerobic sites, such as those found on external surfaces, to highly anaerobic sites of low redox potential in the alimentary tract. Sites of intermediate redox potential occur in the oral cavity and lower reproductive tract. In external environments the habitat may be relatively dry with a limited nutrient supply; cells are often removed by abrasion and sloughing of the external layer of skin or cuticle. Internal environments are usually moist, with abundant nutrients. The temperature of the habitat varies from ambient on the external surfaces of invertebrates and cold-blooded animals to slightly above core body temperature in the gastrointestinal tract of mammals.

Many of the bacteria in a wide variety of aqueous systems, including the gastrointestinal tract, grow in microcolonies attached to surfaces surrounded by a glycocalyx through which molecules and small particles pass [46, 47]. The glycocalyx consists principally of exopolysaccharide which binds to both the bacterial surface and the substratum; it probably protects the bacteria from bacteriophage or colicin attack, traps nutrients by its ion exchange capacity and minimizes loss by abrasion. The surface of some animal cells lining the alimentary tract is enclosed in a polysaccharide glycocalyx [37] presenting a surface similar to that of the bacterial glycocalyx [127]; this surface can supply nutrients to adherent bacteria in the form of polysaccharides and glycoproteins. In the rat, the mucus layer lining the mid-colon may also prevent adhesion of luminal bacteria to the colon epithelium [19].

Some autochthonous micro-organisms perform only a limited range of functions in their natural environment and it is easy to allocate a niche to these species. In contrast, others have a broader range of functions. For example, micro-organisms with some flexibility of substrate utilization may change their functions according to the carbon sources present. Thus no specific niche may be allocated to these species. More intimate associations of micro-organisms with animals occur where the organisms are enclosed within the animal cells in direct contact with the cytoplasm of the host. These endosymbionts are usually found in particular organs or specific anatomical regions such as, for example, those associated with the gastrointestinal tract of some insects.

The role of micro-organisms in many animal–microbe systems is incompletely understood. Much research has been conducted on microbial communities in ruminants, man and invertebrate pest species, but many of the principles operating undoubtedly apply in others.

3.3.2 MICROBIOLOGY OF EXTERNAL SURFACES

The skin and lower genital tract of man

The skin is sterile before birth, but during birth it is contaminated by the vaginal flora. In normal healthy individuals, the skin is highly resistant to invasion by many species of bacteria. The keratinized epithelium provides a physical barrier, and its dryness, low pH (5.5), the production of antibacterial compounds such as long chain unsaturated fatty acids in sebaceous secretions of glands, and interference between bacterial populations all exert some control over which species dominate the population of the skin [139, 163].

Bacteria inhabiting the skin belong either to a resident population or to a transient population. The transient population represents species from the environment or species such as *Staphylococcus aureus* more commonly found in the nose which in some individuals periodically occur in large numbers on the skin [99]. The resident flora consists of species normally found on the skin, but the species composition varies considerably with the anatomical site. The major anatomical regions can be divided into three groups: the exposed surfaces (face, neck, hands); the moister regions (armpits, perineum, toewebs) and the remainder (trunk, the remainder of arms and legs). Higher bacterial population densities occur on the exposed surfaces, where populations are dominated by corynebacteria, micrococci and sometimes *Staph. epidermitidis*. In the moister regions, Gram-negative bacteria including *Pseudomonas* spp. are more common but rarely dominant, except in some pathological conditions.

Counts of bacteria present on the skin vary according to the methods used; surface samples indicate a bacterial population density of up to 8×10^4 cm^{-2} for the armpit and 2.3×10^4 cm^{-2} on the forearm. However, skin from cadavers which was homogenized and treated with Triton X-100 yielded population densities of 9.2×10^5 cm^{-2} and 1.3×10^6 cm^{-2} for the armpit and scalp, respectively [160]. Electron microscopy has revealed bacterial colonies both on the skin surface and in the stratum corneum [90]. The different values obtained by the two methods is explained by this observation. Undoubtedly the highest bacterial populations occur on the head, armpit, groin, perineum, hands and feet; numbers greatly increase after hydration of the skin by washing. For this reason bacterial colony counts from female skin are generally higher than counts from male skin, because of more frequent washing.

The eyes are well protected from bacterial invasion by lachrymal secretions which wash them free of micro-organisms; the secretions drain through tear ducts to the nose. These secretions contain lysozyme, which kills susceptible bacterial cells and helps prevent bacterial invasion of the eye socket.

Although much is known of pathogenic microbial species inhabiting the female lower genital tract, only a little is known of the normal microbial flora [67]. One microbial habitat is the walls of the vagina; its inner surface consists of stratified epithelium lubricated by secretions of the cervix and is clearly favourable for the growth of anaerobes. Glycogen in the vaginal mucosa is hydrolysed (probably by enzymes in the epithelial cells) and provides substrates which are fermented by micro-organisms. Acids are produced which maintain the vaginal pH at 4–5 in sexually mature women. The posterior fornix of the vagina is therefore inhabited by species tolerating a low pH including streptococci, yeasts (particularly *Candida albicans*), anaerobic lactobacilli, coliforms and mycoplasmas, many of which are of potential harm to both the newborn baby and the mother.

Another habitat, the skin of the labia majora, contains sebaceous glands which secrete bacterial inhibitors and sweat glands which provide ammonia and other substrates for microbial growth. The microbial population here is dominated by *Corynebacteria, Staph. epidermitidis* (common skin bacteria), lactobacilli, faecal streptococci and micrococci.

Microbiology of invertebrate integuments

The micro-organisms present on invertebrate integuments have a wide spectrum of relationships with the host, from acting as pathogens, through the benign or commensal relationships of most invertebrate–microbial interactions, to the close mutualistic symbioses illustrated by the carriage of fungal inocula in special integument structures by ambrosia beetles. In all these cases the species complement and abundance of the epibiota are affected by a combination of integument characteristics, the habitat in which the animal lives and the associated environmental conditions. These variables operate in a hierarchical series: the gross physicochemical environmental conditions determine the form, abundance and identity of the micro-organisms potentially available as colonists; the occurrence of an animal in a habitat associated with particular microclimatic conditions favours closer contact with a more specific component of the microbial community and the integument characteristics of the animal then operate at the finest level of selection.

The behaviour of the animal is an important modifying factor and aggregation, dispersal, feeding activities and even grooming influence the species complement and abundance of attached micro-organisms.

These factors do not have equal weighting and in particular circumstances one or another may be dominant, but this holistic concept serves to emphasize that, in searching for ultimate factors to account for particular ecological phenomena, it is easy to introduce artefacts or draw false conclusions as a consequence of isolating the animal from its natural environment. Unfortunately, in many studies this supporting detail is incomplete. The ecology of attached micro-organisms will therefore have to be assembled from discrete details starting with the nature of the surface layers — integuments — of invertebrates.

BODY SURFACE CHARACTERISTICS

The basic structure of the invertebrate integument is elaborated in different ways in different groups of invertebrates but in terms of microbial attachment there appears to be a fundamental division between soft-bodied animals, where the superficial layers of the integument are living structures, and the arthropods with exoskeletons where the integument surface is more or less inert.

In the coelenterates (including the stony corals), free-living annelids and most groups of molluscs, secretions of mucus and protein play a large part in protecting the epidermis. The mucopolysaccharides are rapidly degraded by micro-organisms but the dynamic processes of more or less continuous secretion reduce microbial attachment. In addition bacteriolytic agents have been identified in earthworm mucus [151] and may be found in the epidermal secretions of other groups. The echinoderms are coelomates and have an internal calcareous skeleton, but the body surface is similar to that of the corals in having a sheet of ciliated epidermal cells covering the test. The cilia clean the surface of the animal and prevent bacterial or algal growths. A powerful bacteriocide is also present in the coelomic fluid, which may influence bacterial attachment.

The opposite extreme is found in bivalve and gastropod molluscs where an inert calcareous shell encloses the body. Algal growths may be observed on the shells of sessile intertidal mussels and limpets but the majority of gastropods appear free of epibionts. This may be due to the inhibition of microbial colonization by the periostracum, a thin horny layer of conchiolin covering the shell; the abrasion of this layer may facilitate attachment.

The procuticle in arthropods is soft and flexible when first secreted by the epidermis but becomes hardened with tanned proteins. In terrestrial arthropods, and particularly in insects, layers of wax are deposited on the surface of the tanned procuticle, waterproofing the integument and providing a hydrophobic surface layer. The wax may be further protected by a hard layer of tanned protein. The detailed structure and functioning of arthropod and insect cuticles varies widely between groups and is reviewed in [80] and [98].

MECHANISMS OF MICROBIAL ATTACHMENT

There is much information on micro-organisms isolated from the external surfaces of invertebrates, considered in the following section, but the mechanisms of attachment have not generally been investigated.

Many fungi produce sticky spores which adhere to the integument [70]; this is the principal means of dispersal in some species such as *Ceratocystis* spp., which include Dutch elm disease. On the other hand it has been shown [129] that the strongly hydrophobic spores of *Streptomyces griseus* are attracted to arthropod cuticles and hence this could be true of other micro-organisms.

Studies of the mechanism of bacterial adherence to mammalian tissue culture cells have shown that the degree of hydrophobicity of the bacterial surface is correlated with the ability of the organisms to adhere to the cells. The hydrophobic molecules allow the bacteria to approach the negatively charged epithelial cell and thereby enable binding molecules on each of the cell surfaces to interact and form specific bonds [106].

Little work has been carried out on the specificity of microbial attachment to invertebrate surfaces except for pathogenic fungi, where the stimulus for research has been the development of biological control measures against insect nematode pests. Entomopathogenic fungi normally invade through the surface integument and so a spore must remain in contact with the cuticle long enough for germination and invasion to take place. For this purpose *Entomophthora* spp. produce sticky spores, which germinate and penetrate the thin cuticles of flies, mosquitoes and aphids [84] at any point. In larger insects, penetration of *Entomophthora* is through the non-sclerotized membranes between the segments, in physically protected, humid sites. The setae and setal sockets are covered by a cuticle which is much thinner than that of the integument and some Laboubeniales germinate and penetrate at these sites [149]. If the thallus cannot penetrate at the point of germination, a haustorium is extended to a more susceptible area.

Some entomopathogenic fungi, however, appear to have highly specific attachment sites. An example is *Coelomomyces psorophorae* where the distribution of cysts on the integument of mosquito larvae indicates that sites are identified by differences in cuticle texture [35]. The final product of attachment, host invasion, may be a result of enzymatic degradation of the cuticle but mechanical penetration by haustoria appears to be the main mode of entry through the host integument.

More than 150 species of fungi are known to attack nematodes or their eggs [8] and these species can be classified into predatory and endoparasitic forms. The predatory fungi capture nematodes by adhesive devices (hyphae, knobs, branches, nets) or non-adhesive snares and hyphal loops. After capture the nematode cuticle is penetrated and the fungal hypha enters the body.

The endoparasitic fungi have no hyphal development outside the body of the host and infection is generally by adhesive spores. In *Meria coriospora* the spores are tear-drop shaped with a small adhesive bud at the distal end. They can attack at any point on the surface of the nematode but do so mostly in the region of the mouth and on sensor organs. Nordbring-Hertz and colleagues have carried out a series of elegant experiments, reviewed in [101], demonstrating that site recognition by *Meria*, and nematode capture by some trapping species with adhesive knobs and hyphal networks, is initiated by a lectin–carbohydrate interaction between the nematode and fungal surfaces.

MICRO-ORGANISMS ASSOCIATED WITH INVERTEBRATE INTEGUMENTS

Most hard-bodied invertebrates have an extensive surface microbiota which broadly reflects the species complement of fungi and bacteria characteristic of the habitat where the animals are found. Mobile invertebrates will therefore act as passive carriage and dispersal agents for micro-organisms. In other cases, almost exclusively involving fungi, dispersal is primarily effected by the 'deliberate' or active carriage of inocula by insect vectors. In passive dispersal propagules may be found on any part of the integument, but in active carriage propagules may be confined to specialized morphological structures. Propagules may also be dispersed actively or passively in animal guts.

It would seem reasonable to assume that the species complement of bacteria associated with invertebrate integuments would reflect the animal's habitat. The integument of the marine wood-boring isopod *Limnoria lignorum* was dominated by *Pseudomonas* and *Vibrio* species, which are common in the water, as well as *Aeromonas hydrophila*, which was associated with the wood from which the animals were collected, but was not isolated from water [21]. Similarly *Pseudomonas* and *Bacillus* species might be expected to be prevalent on the exoskeleton of terrestrial arthropods but, paradoxically, only *Streptomyces* (Actinomycetes) were isolated from the exoskeleton of woodland millipedes [145]. No Eubacteria were recorded on the integument, even though Enterobacteriaceae were common in the faeces. These results imply that the attachment of *Streptomyces* spores is more specific than the electrostatic mechanism previously suggested [129]. Differential survival of the bacterial inocula would not seem to be implicated, given the moist conditions in which the millipedes live. In contrast, various species of flies have been found to carry a wide range of human pathogenic bacteria, including *Shigella dysenteriae* (bacterial dysentery), *Salmonella typhi* (typhoid) and virulent strains of *Escherichia coli*. The relatively few bacteria on a fly's feet, picked up from food or faeces, soon dry up and are killed. Gut carriage and transmission by faeces is a far more important mechanism of passive dispersal of pathogens.

The fungi associated with soil invertebrates reflect most of the common saprophyte genera (*Mortierella*, *Cladonia*, *Penicillium*, *Aspergillus*, etc.), which sporulate abundantly in dead organic matter [15, 117]. Spores of invertebrate parasites (*Beauvaria* and *Paecilomyces*) and plant pathogens (e.g. *Fusarium*) are also found. Most studies on Collembola and mites suggest that these animals carry inocula of about 20 species [153] but the number of genera appears to be related to body surface area; microarthropods had an average of 0.5 species per individual while beetles and worms had one or two species [117]. A total of 120 taxa of fungi were isolated from tracts and whole-body squashes (which include fungi in gut contents) of species of woodland collembola. The species present on the integument were generally similar to those present in the soil and litter horizons from which the animals were extracted. Two exceptions were the parasite *Beauvaria bassiana*, which was more abundant on collembola than in soil, and *Trichoderma* species which were never isolated from the animals. Soil invertebrates may therefore be important passive dispersal agents for fungal spores in soil and litter systems. The carriage of spores may even be an essential dispersal mechanism for many wood-rotting basidiomycetes which form sporophores on the underside of branches and logs on the forest floor, where air currents are unlikely to be sufficient for aerial dispersal [147].

Above ground, wind and rain are generally considered to be more important for the dispersal of fungi. However, the physiological characteristics of *Ceratocystis* spores make it unlikely that these species are air-dispersed [50]. Transmission takes place when *Scolytus* spp. beetles disperse from their breeding material in spring and fly to the tops of healthy elm trees to feed on young sappy bark in twig crotches. Some of the wounds are contaminated with spores of *C. ulmi* and the fungus spreads through the xylem vessels of the tree, producing the characteristic wilt symptons. Once the tree is dead or dying, the female beetle excavates a breeding gallery in the bark where the larvae develop, surrounded by fungus in and around the chambers. Pupation occurs in the autumn and the beetles emerge bearing loads of 250–2500 viable spores for *Scolytus scolytus* with a range from 1 to 20 000 spores [156]. Although three spore types are produced by the fungus, the spore load on the beetles is dominated by one type, conidiospores, because of the synchrony between beetle emergence and the phenological characteristics of the fungus (see Fig. 3.3.1).

Active (epizootic) dispersal of fungal propagules carried in body pouches (mycangia or mycetangia) is characteristic of a number of insect groups, notably the ambrosia beetles, woodwasps and some termites, which exploit living or dead wood as a food resource. The winged females of leaf-cutting ants, which maintain fungal monocultures in fungus gardens through a series of physical and biochemical controls [143], also transport pellets of hyphae from the garden in a mycangium located below the mouth [157].

There are more than 1500 species of ambrosia beetles (Scolytidae), mostly tropical species in the tribe Xyloborini and members of the genus *Xyloborus*. Each species is associated with one or more species of Ascomycetes or Deuteromycetes with *Fusarium* the major genus. Many

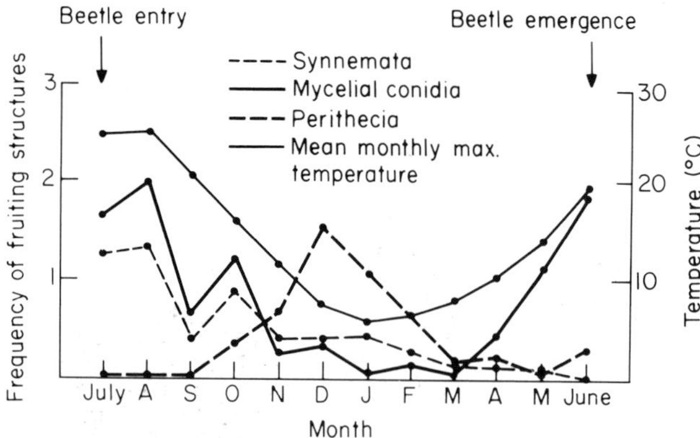

Fig. 3.3.1. Changes in the frequency of the three spore stages of *Ceratocystis ulmi* in beetle breeding galleries between beetle entry (left) and beetle emergence (right). (Copyright Forestry Commission. After Webber & Brazier [156].)

species attack unhealthy or newly felled logs, and are often restricted to a single family of trees. Ambrosia beetles have been studied for over 150 years because of their economic importance and reviews have been published on this group and their fungal associations [77, 102, 103].

The female beetles carry the fungal propagules in small pouches (mycangia) on the exoskeleton; these may be membranous pouches at the base of the mandibles, or sclerotized pouches at the base of the elytra or between the thoracic segments. The fungus is carried as a yeast-like form which multiplies while in the mycangium. When the beetle starts boring to construct the brood galleries, glands associated with the mycangium produce a copious oily secretion which flushes out the fungal inoculum. This inoculum germinates on the tunnel walls to produce a new growth of ambrosia: a tightly packed mass of conidiophores which form the adult and larval food.

The ambrosia beetles have co-evolved with the fungal symbionts to the extent that initiation of reproduction in the female is dependent upon the high concentrations of lysine, methionine, arginine and histidine synthesized by the fungus. It has also been demonstrated that the nutritional basis of the symbiosis for the growth of the larvae of one species is the provision of ergosterol by the fungus [77].

Syricid woodwasps infest a wide range of hard and soft woods in the northern hemisphere. The adult female drills through the bark of the tree into the xylem where the eggs are deposited together with fungal spores from a pair of pouches located at the base of the ovipositor. The larvae also have paired mycangia on the first abdominal segment where spores are held in a dormant condition embedded in wax platelets. The larval integument is shed at pupation but the wax platelets in the exuvium are shattered by the movements of the emerging adult. The spores then germinate and the fungus grows rapidly to invade and inoculate the mycangium of the female before she emerges from the tree [85]. Fungi associated with the woodwasp, including the basidiomycetes *Stereum sanguinolentum*, *Stereum chailletii*, *Amylostereum areolata* and *Daedalea unicolor*, appear to play a complex role in the ecology of the insect in that they may facilitate host invasion by weakening the tree, as well as having nutritional significance for larval development.

LUMINOUS BACTERIA

Bioluminescence on land is found in fungi, bacteria and some insect groups, notably Coleoptera (fireflies and glow-worms). Numerous records exist of bioluminescent invertebrates, but these mostly represent contamination of the integument by luminous bacteria from the moist habitats in which these animals are found. Some animals, however, are infected with pathogenic luminous bacteria by parasitic nematodes [118]. The bacteria infect the haemocoele of caterpillars and other insects when the nematode has burrowed through the gut wall. The bacteria multiply, killing the host, and the progeny of the nematodes are reinfected as they complete their life cycle in the insect's haemocoele. The functional significance of this symbiosis is unknown.

Bioluminescent bacteria have been most extensively studied in the marine environment where they occur free-living, as commensals on the integument of many animals and in specific organ-bound associations. Recent studies [11] indicate that there are two main taxonomic groups: the genus *Photobacterium*, found mainly in symbiotic association with cephalopods, tunicates and fish, and *Beneckia* (*Vibrio*), which are free-living.

The symbiotic bacteria are located in special organs in a variety of shapes and locations, often accompanied by morphological adaptations which act as lenses, reflectors or shutters to control light emission (see Fig. 3.3.2). Two fish which live in the coastal Indonesian seas, *Anomalops katoptron* and *Photoblepharon palpebratus*, have large bean-shaped luminous organs under each eye which can be covered with a membrane so that it flashes periodically [88]. Many deep-sea fish have luminous organs but

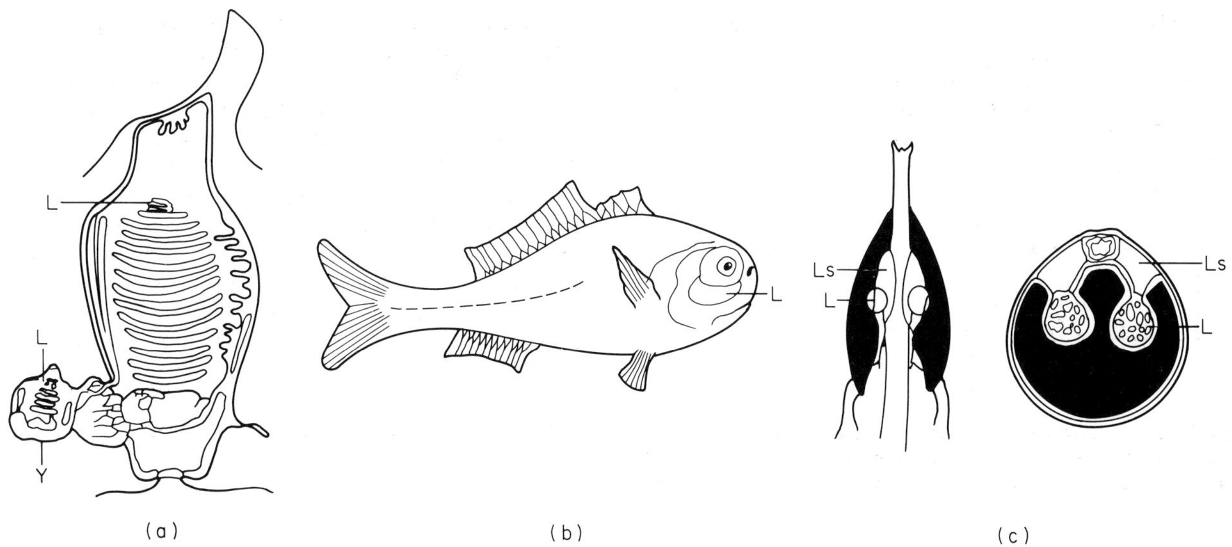

Fig. 3.3.2. Luminous organs which contain symbiotic luminous bacteria. (a) *Pyrosoma* sp. (a tunicate). An individual of a colony in side view with a luminous organ (L) above the gill-intestine. A young animal (Y), developed on the stolon, also has a luminous organ. (b) *Anomalops katoptron*. The bean-shaped luminous organ (L) is situated below the eye. (c) *Loligo edulis* (squid). A male animal with spindle-shaped lenses (Ls) and round luminous organs (L), and a cross-section through the trunk showing the luminous organs embedded in the ink sac and the lenses which flank the rectum. After Buchner [24].

in most species these are chemoluminescent and light generation does not involve bacteria. Similarly, of the 19 families of luminous squids, only two families have genera which use luminous bacteria as their source of light [60].

The mutualistic basis of the symbiosis accrues from the provision of nutrients and a protected environment for the bacteria while the light can be used to attract prey, to assist in escaping or diverting predators, and for communication [91] by the host animal.

Quantitatively, however, the most important habitats of luminous bacteria are the guts of marine animals including mussels, scallops, crabs and fish. Many marine fish carry these bacteria in their intestinal tract at population densities of 0.5 to 5 $\times$ 10^7 cells ml^{-1} of gut material. These bacteria may be important in the digestion of chitin [128].

3.3.3 VERTEBRATE GUT SYSTEMS

The vertebrate gut as a habitat

Unlike the external surfaces of animals, the alimentary tract offers many habitats highly conducive to microbial growth. A generalized diagram of the alimentary tract is shown in Fig. 3.3.3, and the relative capacities of different regions of the tract are given in Table 3.3.1. Numerous differences in the anatomy of the tract occur in the animal kingdom [9, 82], each related to the diet of the animal. Each segment of the gut provides different environmental conditions for microbial growth; these conditions frequently change in some way as the animal matures. In some sites definite successions of micro-organisms have been recognized, commencing at or soon after birth [95, 109]. Three interacting populations of micro-organisms may be found in each region of the gut: those free in the aqueous phase of the gut lumen, those attached to food particles, and those adherent to the epithelium lining the tract [36]. In some regions of the tract, such as the abomasum of cattle and the ileum of the mouse, a continuous viscous stream of mucus flows over the epithelium surface and this viscous stream contains most of the bacteria and protozoa. Despite earlier ideas, the number of micro-organisms actually attached to the epithelium in these sites is small [47]. Population densities for lumen micro-organisms and environmental conditions at different sites in the alimentary tract are given in Table 3.3.2. In addition, the structure and

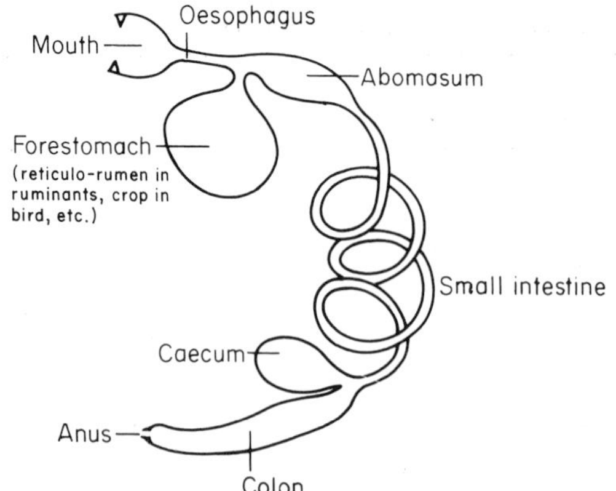

Fig. 3.3.3 Generalized diagram of the alimentary tract of animals.

Table 3.3.1. Parts of alimentary tracts of mammals, expressed as % of total tract

	Pig	Cow	Horse	Rabbit
Stomach	20	71	9	34
Small intestine	33	19	30	11
Caecum	6	3	16	49
Colon	32	8	45	6

Table 3.3.2. Microbial populations ($\log_{10}$ per g of organ contents), redox potential and pH of regions of the human alimentary tract*

	Stomach	Jejeunum	Ileum	Colon
Aerobes and facultative anaerobes	0–3	0–4	3–8	6–10
Anaerobes	0	0	2–7	8–12
Total	0–3	0–4	5–8	10–12
E_h (mV)	+150	−50	−150	−200
pH	3.0	6–7	7.5	6.7–7.3

* Condensed from [49].

physiology of sections of the tract may change with the age of the animal. This is best exemplified by the young ruminant, which does not have a fully functioning rumen until after 3–4 months.

Chewing of the food is an important first step in digestion, as the reduction in particle size increases the surface area of food available for subsequent attack by digestive enzymes. At the same time, as it is being chewed the food is mixed with saliva, which buffers the pH, generally contains some digestive enzymes, and helps to lubricate the passage of food through the oesophagus.

The mouth

The mouth of a mammal presents a variety of sites for colonization by ingested bacteria, and may be regarded as an ideal microbial incubator. Not only is the temperature maintained near core body temperature (at 35–36°C in man), but the abundant water and food provides a rich variety of microbial substrates, as well as habitats for both aerobes and anaerobes. The mouth normally harbours a complex microflora, with each species having differing growth rates, nutrient requirements and preferred sites for growth [100]. Bacterial population densities of 10^{11} (g wet wt)$^{-1}$ in plaque and 10^9 ml^{-1} of saliva have been recorded [57].

The microbial population mostly occurs attached to the teeth, the cheeks and the tongue and in gingival crevices (see Table 3.3.3). The species composition of populations differs from site to site; for example, species occurring on the aerobic crown of the tooth differ from those of the anaerobic gingival crevices, and cheek populations differ from tongue populations. The microbial population of the saliva represents mostly attached cells which have become dislodged by abrasion or washed off by saliva. The upper limit of the microbial population is controlled by the net effect of washout by salivary flow, abrasion during eating, and washout of cells from gingival crevices by tissue fluids.

In man, the mouth of the child is quickly contaminated after birth by a variety of micro-organisms, principally streptococci, staphylococci, coliform bacteria and Gram-positive rods originating from the mother and local environment. The initial population is mostly of aerobes and facultative anaerobes, but changes occur in a more or less definite succession because of alterations in oral conditions with the age of the child. This is best demonstrated by the succession of *Streptococcus* spp. During the first 48 hours after birth, the oral mucosa does not bind streptococci but *S. salivarius* is normally present in the saliva [107]. This species is probably transferred directly from the mother [27]. After the teeth have erupted, the gingival crevices are invaded by facultative anaerobes, including *S. sanguis*, which attaches more strongly to tooth surfaces than *S. salivarius* [28]. In the gingival cre-

Table 3.3.3. Approximate proportional distribution (%) of more common bacteria on oral surfaces and in saliva*

	Gingival crevice	Coronal	Plaque	Tongue	Buccal	Dorsum	Saliva	Mucosa
Str. salivarius	<1	<1		20	11		20	
Str. mitis	8	15		8	60		20	
Str. sanguis	8	15		4	11		8	
Str. mutans	—	1		<1	<1		<1	
Gram-positive filaments	35	42		20			15	
Veillonella sp.	10	2		12	1		10	
Bacteroides oralis	5	5		4				
Bacteroides melaninogenicus	6	<1		<1	<1		<1	
Enterococci	0–10	<0.1		<0.1	<0.1		<0.1	
Others, unidentified	>27–19		>17		>29	>14		>24

* Modified from Gibbons & van Houte [57].

vices, the redox potential decreases and the habitat becomes suitable for true anaerobes such as *Bacteroides melaninogenicus*, *Haemophilus* spp. and *Spirillum* spp. After about 1 year, *Streptococcus mutans* is found. Both *S. sanguis* and *S. mutans* are cariogenic plaque species and of considerable significance in oral health. In contrast, *S. mitis* is the dominant species on the buccal mucosa.

Many oral bacteria produce extracelluar dextrans, laevans and other polymers which form a glycocalyx surrounding the cells, binding colonies of different organisms together to form dental plaque [26]. Many of the glycocalyx components can themselves be hydrolysed by mouth bacteria, but mutan (an α-1, 3-glucan) produced by *S. mutans*, a common cariogenic plaque species, is water-insoluble and less easily hydrolysed by bacteria. Mutan is probably of major significance in maintaining the structural integrity of the plaque, and acidic conditions (pH 6.0) generated in carious lesions help maintain populations of the acid-tolerant plaque species. Plaque formation is not restricted to man, but is widespread in the animal kingdom [48].

Although continually exposed to reinoculation by environmental micro-organisms, the mouth has defence mechanisms which restrict invasion and growth. These mechanisms of anatomical and physiological barriers include mucous membranes, squamous epithelium, the flow of saliva, the anatomy and chemical composition of the teeth, and microbial products inhibitory to growth, phagocytosis by macrophages and leukocytes, humoral antibodies and cell-mediated responses in gingival crevices, and salivary lysozyme of immunoglobulins [100].

Less is known of the oral microbiology of animals, but many similarities undoubtedly exist between microbial populations in warm-blooded animals and man. In ruminants, regurgitation of food results in a large number of rumen organisms entering the mouth but rumen anaerobes probably do not survive there. Populations of the filamentous bacteria *Alysiella* spp. and *Simonsiella* spp. occur on the tongue [83]; these species are tenaciously adherent, even in this highly abraded environment.

In monogastric animals, food leaving the mouth passes down the oesophagus to the stomach. In species with a pregastric fermentation chamber such as the ruminant, the food passes first into the forestomach before passing on to the omasum, abomasum and remainder of the alimentary tract.

Forestomach fermentation systems

Considerably more is known of the rumen microbial communities than other forestomach fermentation systems. The young suckling ruminant is, however, essentially a monogastric animal. Ingestion of liquids stimulates the reflex closure of a structure known as the oesophageal groove, which channels the food directly to the abomasum and bypasses the undeveloped rumen [30]. When the young animal starts to eat small amounts of forage, the oesophageal groove does not close and forage is allowed to enter the immature rumen. The rumen increases in size and, with an increased forage intake and modification of its flora, becomes functional. The rumen microbial population initially consists of facultative anaerobes, principally lactic streptococci and lactobacilli, which ferment the small quantities of milk which accidentally enter the rumen from the oesophageal groove. These organisms also decrease the redox potential, which allows a succession of increasingly oxygen-sensitive organisms to proliferate and ferment the initially small quantities of ingested forage. Inoculation is direct from the mother, by aerosol or by eating contaminated feed, and typical anaerobic rumen bacteria occur in low numbers within the first days of the animal's life. In sheep the climax population may not be reached for 3–4 months when the rumen becomes fully functional [66]; the microbial flora of the caecum, however, is functional within 15 days of birth [109].

The food is prepared for microbial metabolism by an initial chewing and mixing with saliva in the buccal cavity to form a bolus. Saliva is produced copiously and continuously, and an adult cow produces about 60–100 l day^{-1} The saliva is rich in bicarbonate, phosphate, sodium, calcium and potassium ions, contains mucoproteins and urea and usually buffers the rumen contents to pH 6.4–6.8 [66]. The bolus is swallowed and is moved down the oesophagus into the rumen. Here the plant fragments are invaded and colonized by anaerobic and facultatively anaerobic rumen micro-organisms. The invasion process is a mixture of active invasion involving chemotaxis in some species (ciliate protozoa, anaerobic chytridiomycete zoospores and motile bacteria) and adherence by non-motile bacteria. One of the first of the ciliate species to invade freshly ingested plant tissue is *Isotricha intestinalis* (see Fig. 3.3.4A), followed by the ophryoscolecid ciliates, chytridiomycete zoospores and bacteria (see Fig. 3.3.4B). *Isotricha intestinalis* is a large, highly motile species; it has been observed that 90% of the population of this species in the rumen fluid transfers to plant tissues within 5 min of their ingestion [112]; the same species also migrates to and from the rumen epithelium.

The animal mixes the rumen contents by regular contractions of the rumen wall. This mixing is of crucial importance to the efficiency of rumen fermentation since it serves to help inoculation with rumen micro-organisms, spreads the saliva throughout the rumen, enhances the absorption of fermentation products by replenishing the rumen liquor near the rumen epithelium, reduces the tendency of the plant fragments to float, and assists the mechanical breakdown of the fragments. Despite this, stratification of digesta does occur to some extent. The larger, less dense, little-digested fragments form an upper digesta layer, and the smaller, more dense, more digested fragments form successively lower layers. The least dense fragments can be returned to the mouth by eructation

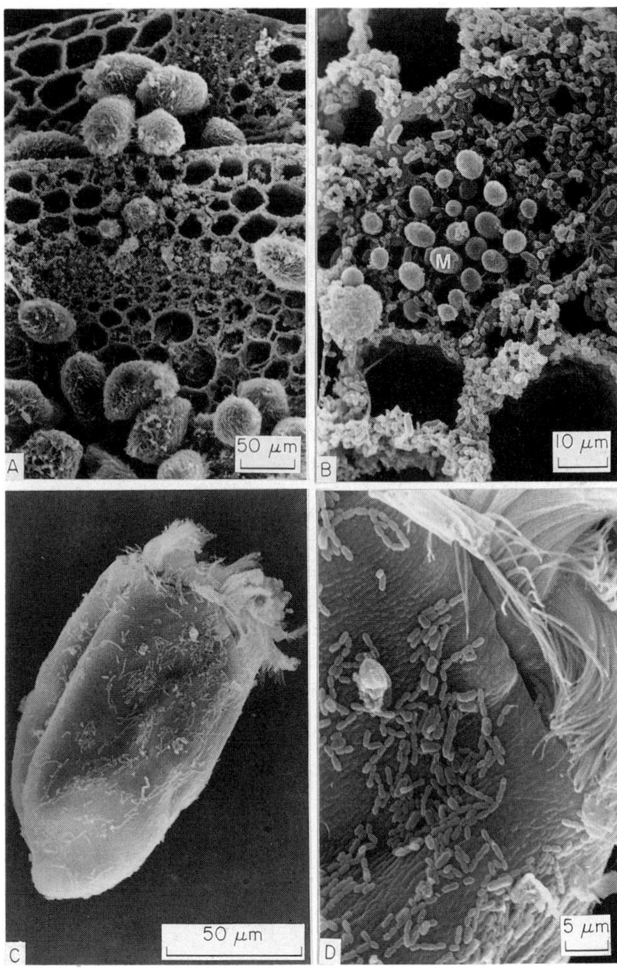

Fig. 3.3.4. Rumen micro-organisms. The 'invasion' of plant tissue by the protozoon *Isotricha intestinalis* (A) and bacteria, including what are probably cells of the large bacterium *Magnoovum eadii* (B), prior to digestion. The ophryoscolecid ciliates often carry bacteria attached to their external cuticle; this cell of *Eudiplodinium maggii* (C) has large numbers of rod-shaped bacteria, attached singly and in chains (D).

where they are fragmented by further chewing and grinding (rumination) and swallowed again. This action produces new fracture sites on the plant cell walls for microbial colonization. The smallest, most dense particles eventually leave the reticulorumen via the reticulo-omasal orifice with rumen liquor containing suspended micro-organisms. Turnover times for the liquid and solid phases of bovine rumen contents are 4–30 hours and 20–55 hours, respectively. The turnover time, which varies with the animal species, is influenced by the composition, rate of intake and particle size of the diet.

In the rumen, the micro-organisms produce essential nutrients for the host animal (see Fig. 3.3.5). The plant components are fermented principally to short-chain fatty acids, mostly acetate, propionate and *n*-butyrate, with the generation of carbon dioxide, methane and microbial cells. A number of other types of animal also have foregut fermentation; concentrations of volatile fatty acids are given in Table 3.3.4. In cattle, daily production has been calculated to be 3.7 kg of acetic acid, 1.1. kg of propionic acid and 0.6 kg of *n*-butyric acid, which yield 399 g of bacterial and 315 g of protozoon cells (dry wt) [162]. The fatty acids are absorbed directly through the rumen epithelium to serve as carbon and energy sources for the animal; the gases are lost by eructation. The microbial cells pass out of the rumen to the abomasum and remainder of the alimentary tract where they themselves are digested. Plant tissues passing from the rumen undergo secondary fermentation in the caecum and large intestine. In the caecum the secondary fermentation provides 8–30% of the total energy removed from the ingested feed. In contrast, ruminal fermentation provides 60–65% [52].

The extent of microbial digestion in the rumen is determined by the balance between the rate of fermentation and the rate of passage of the rumen contents. Rates of passage also affect the composition of the rumen microbial population since organisms with generation times greater than the turnover time wash out.

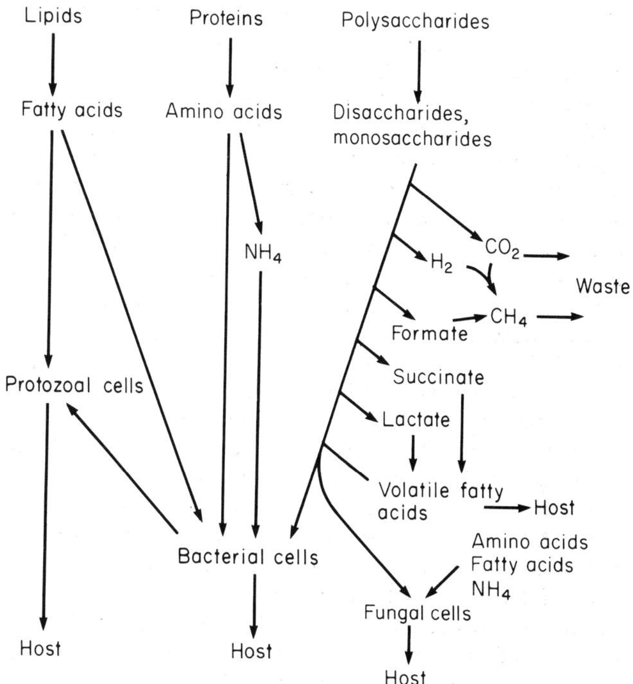

Fig. 3.3.5. Major metabolic activities occurring in the rumen.

THE MICROBIOLOGY OF THE RUMEN

The rumen microbial population is very complex, consisting of many different types of interacting eukaryotic and prokaryotic micro-organisms. The population is dominated by bacteria (10^{10}–10^{11} ml^{-1}), ciliate protozoa (10^4–10^6 ml^{-1}) and chytridiomycete fungi; flagellate protozoa, amoebae and bacteriophages also occur. In addition, there are many types of micro-organisms which have been observed in the light microscope and electron microscope that have yet to be identified. Reviews of the biochemistry and ecology of rumen micro-organisms are available in [62, 63, 66, 108, 121, 122].

Table 3.3.4. Volatile fatty acid concentrations and pH in forestomachs of some animals with pregastric fermentation chambers*

	Volatile fatty acids, mM				
	Total	Acetate	Propionate	*n*-butyrate	pH
Colobid monkeys	103–219	50–64	22–28	6–14	5.5–7.0
Cattle	56–224	45–70	9–30	7–30	5.5–7.0
Sheep	32–187	38–69	15–30	9–17	5.5–7.0
Sloth	35– 95	—	—	—	5.2–6.7
Hippopotamus	60–180	—	—	—	5.0–5.7
Macropod marsupials	101–137	71–85	11	17	4.6–8.0

* Modified from Bauchop [9] and Hungate [66].

The rumen micro-organisms live in three distinct but interacting habitats — the rumen liquor, the digesta fragments and the rumen wall. Many of the micro-organisms are obligate anaerobes found only in the rumen or gastrointestinal tracts of mammals, but facultatively anaerobic bacteria are also present. Some of these bacteria, particularly *Streptococcus faecium*, may play a significant role in ureolysis on the rumen epithelium. Possibly the major role of facultative anaerobes in the rumen is to scavenge oxygen and initially to aid the establishment of the rumen flora; subsequently they help maintain a low redox potential in the system. Some of the most common rumen micro-organisms, their major substrates and their fermentation products are given in Table 3.3.5.

CELLULOLYSIS

The bulk of the diet of ruminants consists of cellulose (30–60%) and hemicellulose (30–50%), and it is on the hydrolysis of these polymers that most of the microbial growth is dependent, either directly or indirectly. As in all other ecological niches, cellulose decomposition or digestion is generally carried out by a 'cellulase complex' consisting of three classes of hydrolytic enzymes: endoglucanases (C_x-cellulases), cellobiohydrolases (C_1-cellulases) and cellobiases (β-glucosidases). The endoglucanases attack amorphous (partially hydrated) regions on the surface of crystalline cellulose and create 'nick' sites. This releases cellobiose, exposes additional sites for attack

Table 3.3.5. Physiological niches, substrates utilized and fermentation products of more abundant rumen micro-organisms*

Organism/niche	Substrates fermented	Major end products of fermentation
Cellulose degradation		
Bacteria		
Bacteroides succinogenes	Cellulose, cellobiose, glucose	Succinate, acetate, formate
Ruminococcus spp.	Cellulose, cellobiose, xylose	Succinate, lactate, acetate, formate
Butyrivibrio fibrisolvens	Cellulose, widely adapted	Butyrate, lactate, formate
Protozoa		
Epidinium ecaudatum	Cellulose, hemicellulose, starch	Acetate, butyrate
Ophryoscolex sp.	Cellulose, hemicellulose, starch	Acetate, butyrate
Fungi		
Neocallimastix frontalis	Cellulose, hemicellulose	Acetate, lactate, formate, ethanol
Pectin degradation		
Bacteria		
Lachnospira multiparus	Pectin, cellobiose, glucose	Formate, lactate, acetate
Succinivibrio dextrinosolvens	Pectin, maltose, xylose	Acetate, succinate, lactate
Starch degradation		
Bacteria		
Streptococcus bovis	Starch, soluble sugars	Lactate
Bacteroides amylophilus	Starch, maltose	Succinate, acetate, lactate
Protozoa		
Entodinium caudatum	Starch	Acetate, butyrate
Methane production		
Bacterium		
Methanobacterium ruminantium	$H_2 + CO_2$, formate	Methane
Miscellaneous		
Bacteria		
Selenomonas ruminantium	Soluble sugars	Acetate, propionate
Megasphera elsdenii	Lactate, some sugars	Range of volatile fatty acids
Bacteroides ruminicola	Range of sugars	Succinate, acetate, formate
Protozoa		
Isotricha intestinalis	Sugars	Lactate, butyrate, acetate

* Modified from Hungate [66] and Ogimoto & Imai [108].

by endoglucanases and generally disrupts the structure of the crystalline cellulose matrix. The cellobiose is hydrolysed to glucose by cellobiase. The endoglucanases and cellobiase (as well as other α-glucosidases) are widespread among animal groups and many claims of cellulolytic activity in guts are based on assays using carboxymethyl cellulose (CMC) or other forms of hydrated cellulose. However, efficient cellulose digestion must involve cellobiohydrolases to denature crystalline cellulose, though comminution of the plant material will serve to disrupt the cellulose matrix to some extent. Other enzymes, for example pectinase, may also be important in exposing cellulose fibrils to enzymic attack. All these classes of enzymes may occur in the same microbe, for a cellulosome complex, providing complete 'cellulase' activity, has been isolated from *Clostridium thermocellum* [12].

Relatively few of the rumen bacteria are truly cellulolytic. The most common are *Bacteroides succinogenes*, *Ruminococcus albus* and *R. flavefaciens*. Some strains of *Butyrivibrio fibrisolvens* are also cellulolytic, particularly in ruminants living at nutritional extremes, such as the high-arctic Svalbard reindeer [114] and starving zebu cattle fed poor forage [86]. In *Bacteroides succinogenes* and the cellulolytic *Ruminococcus* spp. attachment by glycocalyx to the cellulosic substratum is a prerequisite to cellulolysis by cell-wall bound bacterial enzymes. In *B. fibrisolvens* attachment of this type does not occur, and this species forms satellite microcolonies around plant particles outside the glycocalyx formed by the attached cellulolytic species [36]. It is likely that these species release cellodextrins and cellobiose produced during cellulolysis into the rumen liquor, where they may be fermented by a variety of other bacteria including *Selenomonas ruminantium*, non-cellulolytic *Bacteroides* spp. and *Streptococcus bovis* [130]. This fermentation probably supports continuing cellulolysis by the removal of compounds that inhibit the activity of cellulases by end-product inhibition.

The anaerobic chytridiomycetes are a recently recognized group of micro-organisms, so far found only in the alimentary tracts of herbivores. The life cycle of rumen species consists of an alternation of generations between the motile flagellated zoospore stage, free in the rumen liquor, and the vegetative, reproductive stage (see Fig. 3.3.6), which grows on the digesta fragments [10, 110, 111, 113]. All the species so far isolated from the rumen are cellulolytic, and their populations are higher in animals fed high-fibre diets than in those fed concentrates or at pasture. Growth of the fungi on plant fragments weakens and disrupts the tissues, which probably allows

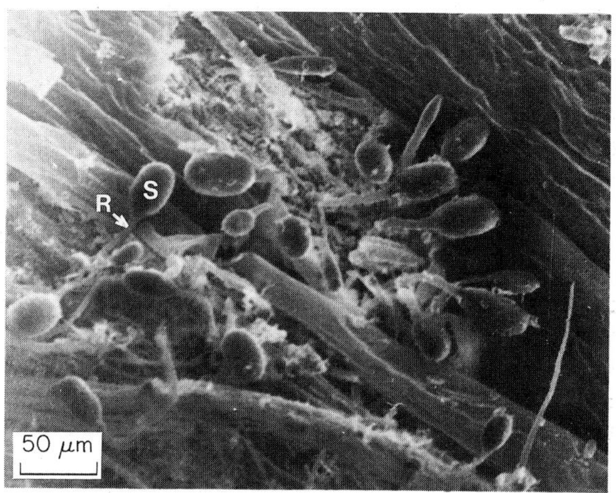

Fig. 3.3.6 The anaerobic rumen fungus *Neocallimastix patriciarum* growing on plant tissue *in vitro*. The rhizoids (R) and sporangia (S) are clearly visible.

the rumen bacteria greater access to the interior of the fragment and thereby accelerates the digestion.

Many species of ophryoscolecid ciliates ingest fibrous plant tissues. Since these ciliates always contain ingested bacteria, and bacteria may live within the cell and on the external cuticle [41, 68, 154] (see Figs. 3.3.4C, D), there has been some dispute as to how much of the cellulolytic activity associated with the ciliates is truly of protozoal origin. It is probable that the protozoa produce some cellulase, since in *Eudiplodinium maggii* (see Fig. 3.3.4C) at least 70% of the cellulase was found not to be of bacterial origin [42]. However, the most abundant protozoan genus, *Entodinium*, is not cellulolytic.

Bacteria are often attached to the external cuticle of ophryoscolecid ciliates (see Figs. 3.3.4C, D). Two bacterial types, methanogens [154] and *Ruminococcus albus* [68], have been shown by immunofluorescence to attach to ciliates, but other morphological types are common. The significance of this interaction has yet to be demonstrated, but the methanogens may stimulate the growth of the ciliate by removing hydrogen evolved during fermentation.

Hindgut fermentation systems

In the Equidae (horses and related species) microbial fermentation for their energy supply takes place in the large intestine (the horse has a 25–30 l caecum and a 55–70 l colon). The process is similar to that in the rumen and volatile fatty acids formed are absorbed through the caecum wall.

Two main features distinguish hindgut fermentation from rumen-type fermentation systems. First, in non-ruminants, the microbial substrates have already been exposed to the digestive enzymes of the host animal prior to fermentation, and, second, the microbial cells formed in the hindgut cannot be directly utilized by the animal and are excreted in the faeces. The result of the loss of the microbial cells is that herbivores relying on caecal or colonic fermentation generally have higher dietary requirements for preformed amino acids and vitamins than ruminants. Losses of nitrogen and vitamins are reduced by some mammals (e.g. rabbits) which practise coprophagy of special soft faecal pellets [64]. These pellets accumulate in the stomach where they inoculate freshly ingested food prior to passage to the caecum.

Numerous other mammals, birds, reptiles and possibly some fish have pouches in their large intestine or caecum which house large microbial populations [82, 125]. The significance of these populations in many species has yet to be determined. Some birds, however, use the volatile fatty acids generated by caecal microbes in response to urea which passes through the caecal wall [92].

Host–micro-organism interactions

The epithelium lining the rumen, caecum and large intestine in many animals is the site of regulation of the uptake of volatile fatty acids and the release of urea. A part of this process is due to the bacterial populations which are attached to the epithelium [36]. Some populations are very host specific; for example, filamentous adherent micro-organisms colonizing the ileum of mice will not colonize the ileum of rats and vice versa [148]. In the rumen some of the adherent bacteria are ureolytic facultative anaerobes which may regulate urea flux across the rumen epithelium [38]. Micro-organisms adherent to the wall of the bird and horse caeca are believed to have a similar function. In addition, the adherent facultatively anaerobic population aids in the maintenance of rumen anaerobiosis by utilizing blood oxygen after it diffuses through the epithelium. Another type of adherent cell digests the protein of dead epithelial cells. Although the adherent bacterial population is in tight contact with the epithelial cells of the rumen, antibody production to these bacteria occurs not in the rumen wall but in the caecum wall, and possibly the wall of the colon and small intestine [138]. The rumen micro-organisms *per se* do not appear to be pathogenic to the host, and they may play a role in protecting the host against pathogens by stimulating the production of cross-reacting antigens.

Adherent populations sometimes affect the structure and topography of the epithelium cells [133, 134] or modify chemical receptors on the surface, which controls the selection of species [56].

Intermicrobial relationships

The microbial attack and digestion of plant particles in the rumen is by a consortium of species. The cellulolytic species adhere to the particles in close contact with the cell wall polymers; satellite microcolonies of non-adherent, non-cellulolytic and cellulolytic species occur nearby [36]. Fermentation products, such as hydrogen, formate, succinate, ethanol and carbon dioxide which are produced by the adherent polysaccharide fermenters, may be utilized by the satellite species in addition to hexoses and cellodextrins which may be released by the cellulolytic species. When these ideas were tested in the laboratory, it was found that growth in co-culture could increase bacterial growth yield and change the pattern of fermentation products [161]. For example, succinate generated by *Bacteroides succinogenes* during growth on cellulose was used as an energy source by *Selenomonas ruminantium*, and hydrogen generated by *Ruminococcus flavefaciens* was used in methanogenesis by *Methanobacterium ruminantium* [79].

Antagonistic relationships occur, particularly involving the predation of bacteria by ciliate protozoa. The oligotrich protozoa of the genera *Entodinium*, *Polyplastron*, *Ophryoscolex* and *Epidinium* are the most active, and, in an average bovine rumen, may be responsible for the ingestion of 180 g of bacterial biomass per day [41]. Much of the bacterial nitrogen is released into the rumen fluid as free amino acids.

Interprotozoal predation also occurs as large species, such as *Polyplastron multivesiculatum*, ingest *Epidinium* spp. and *Entodinium bursa* ingests smaller *Entodinium* spp. An interesting result of the latter behaviour is the development of defensive caudal spines by the small entodinia in the presence of *E. bursa* [43].

3.3.4 INVERTEBRATE GUT SYSTEMS

There have been few studies on the role of micro-organisms in invertebrate guts in which all components have been functionally integrated at a level of understanding comparable to that of the rumen. Termites and cockroaches are exceptions and with certain of these species extensive research has been stimulated by applied problems of their pest status and control. The role of micro-organisms in the digestion of plant polysaccharides and the nitrogen balance of termites is therefore used as a framework for considering other invertebrate systems.

The invertebrate gut as a habitat

The food, feeding mechanisms, gut structure, gut physiological environment and digestive processes exhibit enormous differences both within and between classes, orders and even families of invertebrates. At the same time the gut microbiota range from symbionts with specific nutritional roles, through commensal gut bacteria, to ingested micro-organisms which survive gut transit to varying degrees. Ingested micro-organisms consist of free-living micro-organisms which may be selected as food or be taken in with other food materials. In both instances they encounter a sequence of events during gut transit which protects the indigenous micro-organisms from competitive displacement, determines the status of the animal as a carriage and dispersal agent for fungal and bacterial inocula and influences the susceptibility of the host to infection by pathogens.

Feeding

The first factor affecting the ingestion of microbial cells is the habitat occupied by the animal. Soil and litter animals, for instance, show a food preference related to the availablity of items in microhabitats occupied at any instant in time [116, 153]. Thus individual mites of the same species may have totally different gut contents even though they occupy soil or litter microhabitats separated by only a few millimetres [1].

The proportion of bacteria, fungi and other food materials ingested by invertebrates is also a function of size. Protozoa, for example, are able to selectively feed on bacterial cells or individual fungal spores or hyphae, while, at the other extreme, large litter-feeding invertebrates ingest fungi and bacteria in the proportions in which they occur at the feeding site. On the other hand, some fairly large invertebrates, such as certain beetle and fly larvae, are entirely mycophagous as they feed on or in large fungal fruiting bodies.

Structure and physiological environment of the gut

Although the gut of invertebrates differs markedly in structure and function between groups, three regions are evident on morphological and histological criteria: the foregut (stomodaeum), midgut (mesenteron) and hindgut (protodaeum). The relative proportions of these regions varies with the species, but the hindgut is often particularly large and elaborate in saprovores.

In arthropods the foregut and hindgut are ectodermal in origin and are lined with a cuticle which is shed on ecdysis. The gut lining in these regions offers potential attachment sites for micro-organisms and the hindgut walls are often heavily colonized by bacteria; spines or cuticular folds may prevent washout of symbionts (Fig. 3.3.7). Permeability of the foregut and hindgut walls is restricted by the cuticle, but invertebrates with basic nutritional dependence upon hindgut symbionts have patches or invaginations with thin epicuticle to facilitate the absorption of microbial fermentation products [17].

The midgut is endodermal in origin, is uncuticularized and is the main digestive and absorptive surface. In the majority of arthropods the midgut is lined with a delicate chitinous peritrophic membrane which protects the cells of the gut wall from abrasion by gut contents. The peritrophic membrane is continually being renewed and is thus an ephemeral surface for microbial attachment.

As in vertebrates, the chemical environment of the gut is affected by the nature of the ingested food material, the activity of the indigenous microbiota and substances secreted into the lumen. In most invertebrates the gut pH falls within the physiological optima (pH 6–9, see Table 3.3.6) for bacteria [17] with more acidic conditions prevailing in foreguts; midguts and hindguts are generally more alkaline. The survival of ingested bacteria may be affected by extreme conditions; for example, *Escherichia coli* is inactivated by the acid midgut (pH 3.0–3.5) of the blowfly larva *Lucilia* and bacteria present in this region are limited to a few species of spore-forming bacilli [58].

Little information is available on the redox conditions in invertebrate guts except for wood-eating insects. In nine species of termites both the foreguts and midguts were aerobic with an E_h in excess of +100 mV whereas the paunch and colon, where fermentation occurs, were anaerobic with E_h as low as −270 mV [152]. In the larva of the moth *Tineola* feeding on keratinous materials (wool, hair, feathers), the formation of cysteine and hydrogen sulphide decreases the redox potential in the gut to −200 mV [58] compared to +200 to +300 mV for many other insects [17].

In addition to enzymes and sloughed cells, the wide spectrum of organic and inorganic compounds secreted into invertebrate guts provide favourable conditions for

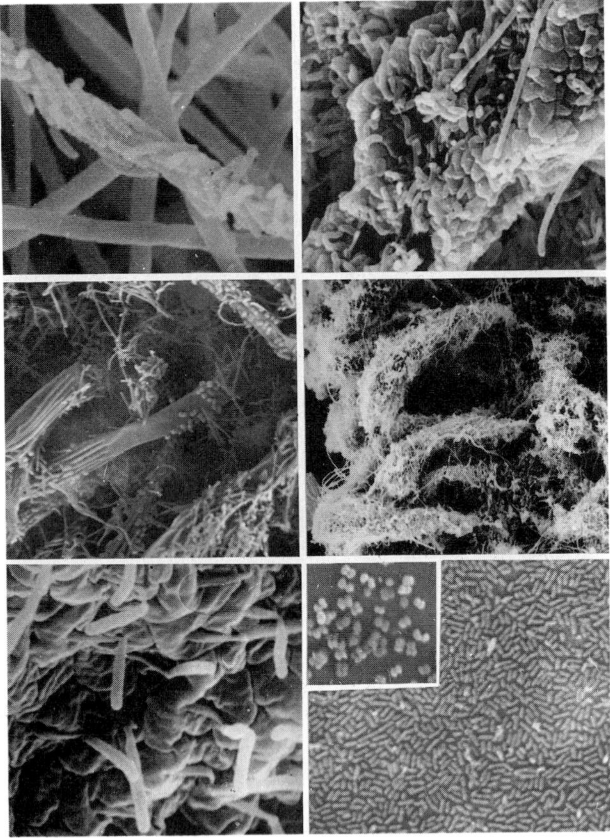

Fig. 3.3.7. Patterns of microbial attachment to cuticle. (A) On flat, relatively permeable surfaces a monolayer of rods or coccobacilli is most commonly encountered. *Schistocerca gregaria* (locust), colon. The inset shows coccoid forms which are occasionally encountered in the foregut. (B) End-on attachment is sometimes employed on less permeable cuticles, for example crustacean hindgut. It would be a suitable orientation for spore-formers or serve to remove parts of the cell from non-stirred layers adjacent to the surface. *Oniscus asellus* (woodlouse), hindgut. (C) Elongated spines may be furnished by the host as attachment surfaces for symbiont organisms, ensuring that the microbes come into intimate contact with food material in transit. Posterior colon of *Procubitermes aburiensis* (soil-feeding termite) showing filamentous prokaryotic organisms adhering to spines. (D) An alternative form of spine, branched distally, serving as an attachment site for long rods and coccoid couples. Ileum of *Acheta domestica* (house cricket). (E) Filaments may also attach to flat cuticular surfaces although a clearly differentiated hold-fast is not apparent. Posterior hindgut of *Cylindroiulus* sp. (millipede). (F) Epibionts may be found in some locations, suggesting secondary symbiotic relationships amongst the constituents of a gut flora. Hindgut of *Tachypodiulus* sp. (millipede). After Bignell [17].

bacterial growth. The main ionic fluxes are K^+, Na^+ and Cl^- associated with fluid inputs from Malpighian tubes and other excretory organs. Nitrogenous compounds excreted into the gut include uric acid, allantoin, urea, ammonia, xanthine and proteinaceous materials [40]. Microbial exploitation of these compounds may facilitate nitrogen conservation by the animal as well as make up any dietary deficiencies of essential amino acids [23]. Finally, mucus secreted into the gut of molluscs and earthworms, and from the Malpighian tubes of arthropods, forms a readily metabolizable carbon source for gut micro-organisms. The addition of this readily metabolized material to the gut contents of tropical earthworms may prime the depolymerization of recalcitrant soil organic matter fractions by gut bacteria [7].

Digestive processes

In general terms the enzyme complements of invertebrate guts reflect the diet of the animal; predators show high protease and lipase activity while herbivores and saprovores show higher levels of carbohydrase activity. The assimilation efficiency of animals feeding on plant materials is generally inversely related to the lignin and cellulose content of the food. However, many fungi and some bacteria are very effective in utilizing cellulose and, to a lesser extent, lignin, as energy sources [165]. Some insect groups, notably termites [23], cockroaches [16] and the larvae of dung- and wood-feeding beetles [13, 14] have symbiotic associations with micro-organisms which help digest these structural compounds.

The lower termites are predominantly wood-eaters and the assimilation efficiencies for cellulose and hemicelluloses (54–93%) are high in comparison with most other saprovores. There is some debate over the extent to which lower termites produce cellulases in addition to the gut protozoa, but quantitatively there is little doubt about the role of the symbionts. The hindgut paunch is the main site of polysaccharide digestion by the flagellate protozoa, which can constitute up to a third of the termite biomass. Wood fragments engulfed by the protozoa are fermented anaerobically to form acetic acid, hydrogen and carbon dioxide. About 80% of the acetate is derived from cellulose and 20% from hemicellulose [105]. While the protozoa do possess some cellulolytic activity not derived from endosymbiotic bacteria, they require bacteria as a source of nutrients or growth factors [164]; the hindgut bacteria therefore appear to be essential for the maintenance of the protozoa. The products of bacterial

Table 3.3.6. pH of the alimentary canal in representative arthropods [17]

Species	Foregut	Midgut	Hindgut
Insecta			
Periplaneta americana	4.8–6.8	6.1–6.6	6.3–6.8
Melanoplus bivattatus	5.5	6.7	6.4
Zootermopsis nevadensis	6.8	7.1	7.1
(Alkaline gut) (highest pH recorded)			
Culex pipiens and other mosquito larvae	—	10.5	7.0
Oryctes nasicornis	8.5	10.4	6.7
Cubitermes severus	6.7	7.5	10.4
Various Lepidoptera	—	10.3	8.9
Tipula abdominalis (larva)	—	11.6	—
(Acid gut) (lowest pH recorded)			
Lucilia cuprina	7.3	3.3	7.8
Various Diptera	5.2	2.8	7.3
Crustacea			
Calanus finmarchicus	—	6.0–8.0	—
Daphnia magna	—	6.0–6.2	—
Ligia oceanica	—	6.4	—
Astacus fluviatilis	4.2–6.6	—	—
Cancer pagurus	5.8–6.0	—	—
Acari			
Phthiracarus sp.	—	5.4–6.6	—

fermentation (lactate, butyrate and acetate) are absorbed by the termites (see Fig. 3.3.8) for respiration and biosynthesis. Assimilative carbon dioxide reduction is also proposed as a sink for carbon dioxide produced by fermentation [23] together with the production of methane by methanogenic bacteria as in the rumen. Rates of methane emission from termites under laboratory conditions may, however, exceed those of ruminants on a weight-specific basis [23], and it has been estimated that termites could be a major biogenic source of atmospheric methane [166]; this, however, is disputed [45].

Far less is known about the role of gut micro-organisms in the digestive processes of higher termites. It was shown [104] that 19% of the cellulase activity in *Nasutitermes exitiosus* was located in the foregut, 58% in the midgut, 14% in the mixed segment and 8% in the hindgut. Removal of the gut flora with tetracycline did not affect the location or activity of the enzymes, suggesting that the termite did not depend upon the gut flora for cellulose digestion. This type of experiment is, however, complicated by the phenomenon of acquired microbial enzymes, as demonstrated in woodlice [59] and other insect–fungus associations [89].

In the fungus-growing Macrotermitinae the basidiomycete *Termitomyces* sp. produces small spherical conidiophores on the surface of the fungus comb and these are selectively consumed by the termites together with comb material. In the midgut of *Macrotermes natalensis* all three classes of cellulases are present, while the anterior hindgut, where there are high numbers of bacteria, contains mainly endoglucanase and cellobiase. Extracts of the salivary glands and midgut tissues actively degraded carboxymethylcellulose but not microcrystalline cellulose, showing that the termites themselves could produce the endoglucanases but not cellobiohydrolases. In contrast, there was high activity of all types of cellulase in the nodules on the fungus comb. The midgut of the termite, an active site of cellulose digestion, contains the enzyme complex acquired from the fungus as well as low levels of animal cellulases [89].

NITROGEN

A general feature of mycophagy and bacteriophagy is that animals are able to acquire a high-quality food derived from resources deficient in nitrogen. This is particu-

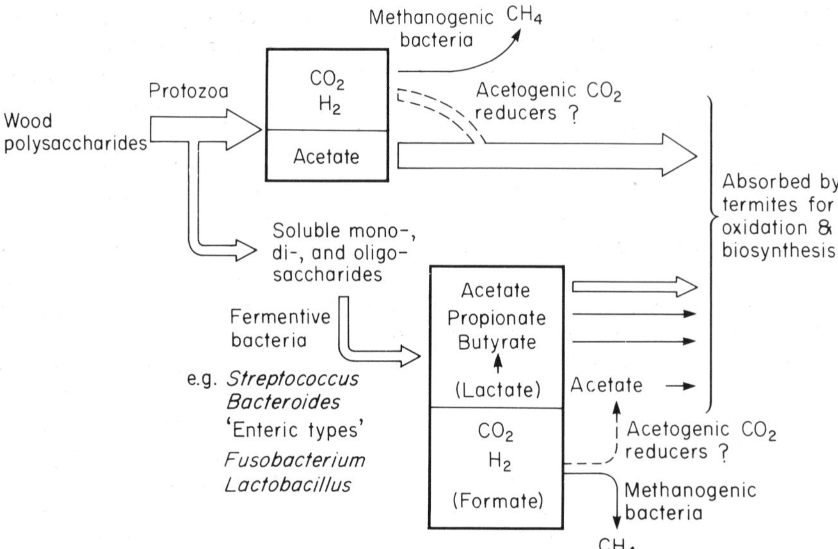

Fig. 3.3.8. Proposed working model for wood polysaccharide degradation in the hindgut of lower termites. Thickness of arrows represents approximate relative quantitative contribution of the respective reactions. Major products of the hindgut fermentation are indicated in boldface letters. After Breznak [23].

larly important in animals exploiting wood as food [144] and in the fungus-growing termites [44]. The woody materials collected by the termites contain less than 1% nitrogen and the nitrogen concentration in the fungal comb is 1.4%; the conidiophores, however, contain up to 38% protein and provide all the amino acids required for termite growth and reproduction [126]. Excretory nitrogen from the termites deposited on the combs as uric acid is also recycled by the fungus.

Uric acid is the major nitrogenous excretory product of terrestrial invertebrates because it can be voided in a crystalline form and thus minimizes water loss. In the cockroach *Periplaneta americana* uric acid accumulated in fat bodies when the animals were fed a high nitrogen diet and this store was mobilized when nitrogen was scarce [96]. The mechanism of nitrogen mobilization was not determined but, since insects lack uricase, a microbial intermediary was implicated. This has been definitively demonstrated [119] for the lower termite *Reticulitermes flavipes* where uricolysis occurs as an anaerobic process carried out in the hindgut by *Streptococcus* sp., *Bacteroides termitidis* and *Citrobacter* sp., the uric acid being fermented to ammonia, carbon dioxide and acetate. Uric acid was also transported in Malpighian tubules [119] from its site of synthesis and storage in fat bodies to the gut; microbial uricolysis *in situ* also released nitrogen that was re-used by the termites for biosynthesis. The same processes may occur in the millipede *Glomeris marginata* [4] and the endosymbionts of anobiid beetles may also utilize uric acid [5].

Dinitrogen fixation in invertebrate guts has been demonstrated in a wide variety of animal groups from many habitats, particularly in animals feeding on nitrogen deficient material, such as wood. Most determinations have been carried out on termites, and results show enormous variability in rates of dinitrogen fixation. These rates have been expressed in terms of the 'time to double nitrogen' (TDN) in the body of the termite [23] and it has been concluded that TDN values range from 6 months to more than 2000 years. Feed selection by the animals is probably more important in providing dietary nitrogen. The extreme variability of the results may also be due to disturbance factors or variable demands during growth, for example, in marine shipworms [29].

The gut microflora

Much information is available on the qualitative and quantitative characteristics of bacteria in the guts of aquatic animals [e.g. 76, 135]. Most of these studies show that aquatic invertebrates are similar to terrestrial invertebrates and contain a range of unicellular and filamentous bacterial morphotypes, occurring in similar attachment sites in the gut. Few examples of fungal commensals or dispersal of fungal inocula have been recorded in freshwater or marine environments, though algal symbiosis is common [141]. The following account is therefore based on terrestrial examples which illustrate the full spectrum of these invertebrate–microbial relationships.

Termites are well known for their nutritional dependence upon symbiotic protozoa but the majority of termites belong to the family Termitidae or 'higher termites', which harbour bacteria in their guts. Even in the 'lower termites' with protozoan symbionts, key metabolic processes are mediated by bacteria [22]. The paunch (anterior proctodaeum), which contains the protozoa, also holds a heterogeneous assemblage of bacterial cells. Many of these have hold-fast elements to secure them to the gut wall or an elongated growth form which may prevent washout.

The soil-feeding termites have a gut microbiota of unparalleled complexity with many novel features. For instance, in *Cubitermes severus* and *Procubitermes aburiensis* the midgut is heavily colonized by a diverse assemblage of bacteria including spirochaetes inserted between the microvilli of the mesenteric tissues. The interface between the midgut and the hindgut forms a 'mixed segment' which was found to have a carpet of spirochaetes of such uniform morphology as to suggest a pure culture [18]. These bacteria may contribute to the low redox conditions (−104 mV) in the large hindgut by removing oxygen from the digesta [10]. Spirochaetes, free-living as well as attached, appear to be of general occurrence in termite guts [22] and form spectacular 'motility assemblages' in the termite *Mastotermes darwiniensis* where co-ordinated undulations of spirochaetes attached to the protozoon *Mixotricha paradoxa* provide propulsion for the cell [39].

The hindgut of *Procubitermes* and *Cubitermes* is densely colonized by bacteria but in addition the posterior hindgut has dense wefts of actinomycetes, containing spirochaetes and other bacteria, attached to paddle-shaped cuticular spines (see Fig. 3.3.7). These associations are not unique, for scanning electron microscopy has revealed filamentous bacteria, probably actinomycetes, to be a dominant feature in the guts of many terrestrial invertebrates (see Fig. 3.3.7).

The protection of the gut microbiota, and indeed the organism itself, from invasion by ingested micro-organisms may be maintained by physical or biological mechanisms. The earlier example of *Lucilia* larvae, which maintain a highly acid foregut, is analogous to the acid stomach conditions in many mammals which lyse ingested bacteria. In the larvae of the wax moth *Galleria mellonella*, *Streptococcus faecalis* is the only bacterium present in high populations in the gut. *Streptococcus faecalis* suppresses the growth of bacteria ingested with food by the production of a bacteriocin as well as a bacteriolytic enzyme [73]. Such tight regulation of the gut microbiota may not be a general phenomenon in invertebrates, but may occur in animals with specialized diets where the bacteria have specific nutritional roles.

The faeces of soil animals such as millipedes, woodlice, fly larvae and earthworms may contain more than 500 times more viable bacteria per gram than the food materials on which the animals are feeding [69]. However, there are few invertebrates where the gut microbiota has been differentiated into free-living and indigenous bacteria.

Several studies of earthworm gut bacteria suggest that essentially the same groups are present as in the soil in which the animals are living [115, 131]. On the other hand, in earthworms feeding on rotten wood, it was found that 73% of 473 bacterial strains isolated from the gut contents belonged to the genus *Vibrio* (including pathogenic strains such as *V. cholerae*, *V. parhaemolyticus*, and *V. alginolyticus*) [87].

A similar increase in counts of viable bacteria during gut transit in arthropods has been recorded for many invertebrate groups [78] including woodlice [69, 123], fly larvae [146] and millipedes [2, 6, 145]. The calculated generation times for bacteria in the guts of the litter-feeding millipede *Glomeris marginata* are about 4 hours [69]. Actual generation times may be less than this because there are regions of the gut where a significant proportion of the ingested bacteria are lysed: approximately 53%, 53% and 29% of *Pseudomonas syringiae*, *Erwinia herbicola* and *Escherichia coli*, respectively, were assimilated by the millipede when fed to the animals on leaf discs [3]. Analysis of isolates from the food, guts and faeces of the millipede revealed that, although there was some overlap, the bacterial floras of the food litter and guts were distinct [69]. Similar results were obtained for the woodlouse, refuting other work [20] suggesting that the guts of these animals are sterile (see also Fig. 3.3.7). The majority of isolates from the food litters of *Oniscus* and *Glomeris* were *Pseudomonas* spp. The most predominant member of the gut population of *Oniscus* was *Klebsiella pneumoniae* and this organism was also a common isolate from the faeces of both animals [69]. The ability to degrade uric acid or urea is a characteristic of bacteria isolated from insect guts and was also found in *Glomeris* and *Oniscus* gut isolates. While this characteristic may be selected by the excretory products of *Glomeris* [4] it is not selected by those of *Oniscus*.

In studies on the gut flora of another millipede, *K. pneumoniae* was identified [6], together with members of the genera *Sarcina*, *Bacillus* and *Corynebacterium*; *K. pneumoniae* was also associated with the *Vibrio* complex of *Eiseria lucens* [87]. If the occurrence of *K. pneumoniae* is widespread in the guts of saprotrophic animals, this could resolve the anomalous occurrence of this coliform in rotten wood, sawdust and living trees [69]. It is not thought, however, that invertebrates are often vectors of pathogens of medical importance. There are few diseases where transmission of free-living bacterial pathogens to man by invertebrates is a significant facet of the epidemiology of the disease, reviewed in [25]. Notable exceptions are the carriage of the enterobacteria *Shigella* (bacterial dysentery) and *Salmonella typhi* (typhoid) by flies. The bacteria pass rapidly through the guts but persist in the lumen for up to 7 days. Transmission occurs by 'vomit drops' or frequent defaecation.

The most notorious insect-transmitted diseases are typhus and bubonic plague, which are parasites of specific vectors. Typhus is caused by *Rickettsia prowazeki*, which multiplies in the gut of the louse after being ingested with a blood meal from an infected host. The bacteria remain virulent in dry louse faeces for at least three months. Infection occurs when this material is inhaled, rubbed into mucous membranes or scratched into the skin. In bubonic plague, the bacterium *Yersinia pestis* proliferates in the gut of the flea and more or less blocks it. The flea is unable to take in the next blood meal and as blood is sucked up, it spurts back into the host, carrying with it detached clumps of bacteria. The infection is generally fatal to both host and vector.

The occurrence of fungal spores in the guts of soil animals has been comprehensively reviewed [153]. A very diverse group of invertebrates and fungi are involved, though there are few cases where dependence on invertebrate vectors for dispersal has been demonstrated. Spores of many wood-decomposing fungi were present in the guts of woodlice [147] but spore viability of *Merulius lacrymans* was reduced after egestion. In contrast it has been found that it was essential for one spore type of a basidiomycete to pass through the gut of a fly larva before it would germinate [103].

The gasteromycete fly *Phallus impudicus* is a good example of a vector for spore transmission. These flies are attracted to the odour of the sticky spore slime on the fungal cap. The spores pass through the gut of the flies unharmed. Several types of fungi have evolved dispersal mechanisms which involve attracting flies or other insects to feed on sugary secretions containing spores. In *Claviceps purpurea* (which produces ergots on rye and other members of the plant family Graminaceae), primary infection of the host flowers is by wind dispersal of ascospores but the ovary of the plant is converted into a mass of fungal conidiophores which secrete a noxious-smelling, sugary secretion containing conidia which are dispersed to healthy plants through insect guts [70].

Most instances of transmission of fungal spores in invertebrate guts involve passive mechanisms, but the fungus-growing termites carry inocula of *Termitomyces* between one colony and the next [74]. Reproductive stages of *Macrotermes* and five species of *Microtermes* carry *Termitomyces* conidia in their guts during the nuptial flight to start a fungus comb in the new colony. The winged stages of *Macrotermes sublyalinus*, *Ancistrotermes* spp. and *Odontotermes* spp. do not do this, however, and the combs appear to originate from basidiospores collected by the workers.

The entomopathogenic fungi provide interesting insights into the gut as an environment for micro-organisms [38, 159]. The buccal cavity, foregut and hindgut are the main sites of fungal penetration. Gut conditions, however, are generally unfavourable for spore germination. Conidia of *Metarhizium anisopliae* fail to germinate in the gut of the beetles *Oryctes rhinoceros* and *Hylobus pales* although they pass through the gut in a viable state. Conidia of *Beauvaria bassiana* were inhibited from germinating *in vitro* by gut contents of several insect species, a phenomenon attributed to high gut pH or nutritional unsuitability [35]. These conditions did not account for the failure of *M. anisophliae* conidia to germinate in the gut of locusts, where a heat-labile toxin, possibly produced by the gut bacterial flora, was implicated.

The Trichomycetes, a diverse but ecologically defined group of lower fungi, are obligate commensals of a very wide range of invertebrates [93]. The majority of species are found in the gut lumen and have fairly specific attachment sites in the foregut, in the hindgut and on the peritrophic membrane. All the species are characterized by a specialized hold-fast for thallus attachment and a life cycle adapted to reinfecting the host after ecdysis. Attachment of spores is rapid and the trichospores of *Smittium culisetae* become attached to the hindgut cuticle of mosquito larvae within 30 minutes of ingestion. There is little evidence of harm or benefit to the host from the association.

3.3.5 ORGAN-BOUND ASSOCIATIONS

Endosymbiosis, in which cells of the host contain micro-organisms, is widespread among insect groups with specialized diets such as blood, plant sap and wood. Nearly all anobiid and cerambycid beetles have yeast-like endosymbionts contained in gut caecae [75] and endosymbiotic bacteria, particularly rickettsiae, are found in at least 10% of all insects [158]. The endosymbionts of Homoptera show a particularly wide range of morphotypes [65] from typical rods to the bizarre 'c' and 't' symbionts of leafhoppers. Other notable bacterial associations occur in annelids and pogonophora, and micro-algae are found in many coelenterates and protozoa.

Endosymbiotic bacteria are found in the cells of the gut wall in some cerambycid larvae where they are constantly voided into the gut lumen. Generally, though, they are contained in specialized cells termed bacteriocytes. These may be aggregated (as in the midgut wall of tsetse flies, lice and ants of the genus *Camponotus*); located in modified Malpighian tubules (as in some plant-feeding weevils and crysomelid beetles); or scattered through fat bodies (cockroaches, most homopteran bugs, bloodsucking bugs and many wood-feeding beetles). The relationship between the endosymbionts and the host is highly specific and often monospecific. Consequently transfer between generations often occurs within the ovary. In the bed-bug *Cimex* (Heteroptera) the bacteria invade the nurse cells and hence are transferred to the eggs while in *Rhizopertha* (Coleoptera) the bacteria pass with sperm from the male to the female and enter the micropyle of the egg during fertilization [24]. In other species infection is via the gut. The eggs of anobiid beetles are smeared with secretions containing yeast from the glands around the ovipositor and the hatching larvae are infected when they eat part of the eggshell. The tsetse fly (*Glossina*) larva undergoes full development within the body of the female and the bacteria are transferred from 'milk glands' to the larval gut. Unlike ectosymbiotic associations, endosymbionts are not lost by ecdysis.

Studies on the functional significance of endosymbionts to the host have been carried out by eliminating the symbionts or by the maintenance *in vitro* of isolated symbionts and mycetocytes [34, 65]. It has been demonstrated that the symbionts can correct nutritional deficiencies in insect diets by producing various amino acids, sterols and B vitamins. For example, when larvae of the anobiid beetle *Stegobium* were devoid of symbionts, they were only able to develop with the addition to the diet of thiamin, riboflavin, pyrimidine, biotin, nicotinic acid, pantothenic acid, folic acid, choline and cholesterol [54, 55]. Symbionts of *Lasioderma*, however, could supply all the ten essential amino acids, but only lysine and phenylalanine, limiting in the diet, were in sufficient quantities for optimum growth [75].

Another role for endosymbionts is found in the leech *Hirundo medicinalis*, which depends upon the bacterium *Pseudomonas hirudinis* to lyse ingested erythrocytes in the gut caeca [124].

Despite the extensive surveys which have been carried out on bacterial endosymbionts, new and unusual functional relationships are still being reported. One of the most spectacular recent discoveries was the finding of chemoautotrophic bacteria in the Pogonophora. The Pogonophora are worm-like animals (deuterostomes) which do not possess a digestive tract (see Fig. 3.3.9). They are mainly found in the deep seas, consequently they were not discovered until the beginning of the 20th century.

In the late 1970s a remarkable discovery was made of 'giant', plumed pogonophora, subsequently named *Riftia pachyptila*, together with large clams and other fauna, clustering around hydrothermal vents at sites 2.5 km deep in the Galapagos Rift and the East Pacific Rise [72]. Normally benthic populations of animals are sparse at this depth because of the remoteness of photosynthetically productive surface waters. In this community, it has been shown that the primary nutrient and energy source is the chemoautotrophic production of bacterial biomass utilizing hydrogen sulphide emitted from volcanic vents [71]. *Riftia*, which are up to 1.5 m in length and 38 mm in diameter, are much larger than pogonophora from the north-western Pacific (10–35 mm) but the basic body structure is similar. Chemoautotrophic symbionts are found in bacteriocytes which line a fluid-filled cylinder at the core of the body [31, 32]. Transfer of metabolites is believed to be by bacterial excretion and uptake by host membranes, or possibly by lysis of bacteria. Evidence for the chemoautotrophic nature of the bacteria comes from electron micrographic studies which show the abundance of the bacteriocytes and the endosymbiotic habit. Detailed studies of European *Schlerolinum* and *Siboglinum*, which live in rotten wood and organic sediments with high sulphide content, have revealed similar bacterial structures (see Fig 3.3.10) as well as carbon dioxide-fixing ribulosebisphosphatecarboxylase, which also points

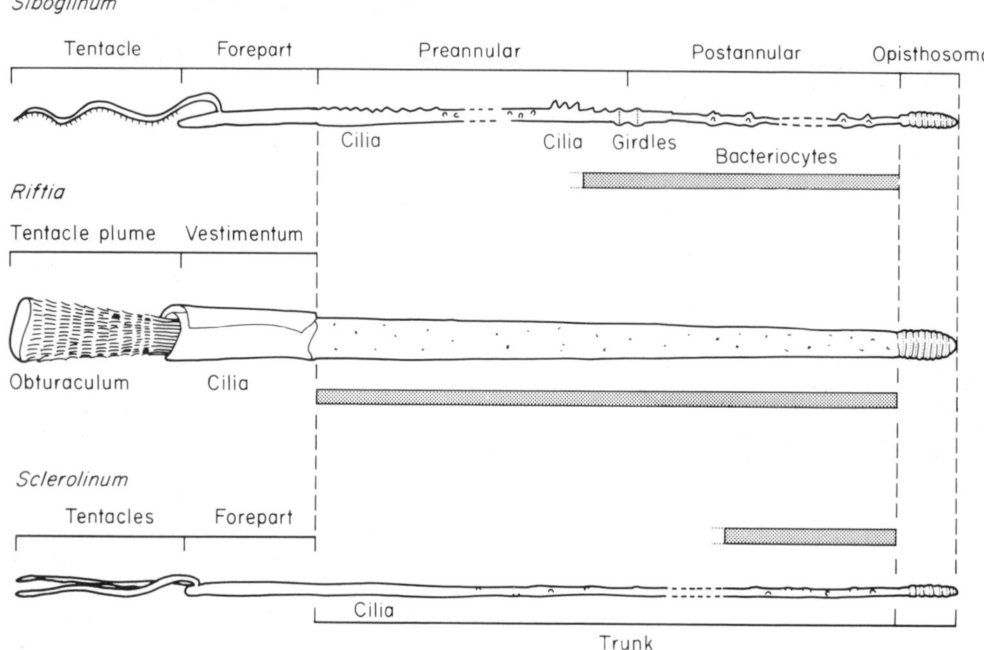

Fig. 3.3.9. Diagrams comparing body divisions in Pogonophora — much shortened. *Siboglinum*, pattern found in most Perviata; *Riftia*, pattern found in Obturata (Vestimentifera); *Sclerolinum*, pattern found in one family of Perviata (Sclerolinidae). From Southward [142].

to chemoautotrophic processes [142]. On the other hand the gutless clam, *Solemya reidi*, from sulphide-rich habitats, has gills containing symbiotic, chemolithotrophic bacteria, which were assumed to oxidize sulphide to provide the major energy source for the symbiosis. However, it was later found that the initial steps in sulphide oxidation occur in specific organelles in the gill cytoplasm and not in the bacteria [120]. This study illustrates the dangers of assuming functional roles of animal–microbial associations.

Another recent discovery is the occurrence of symbiotic bacteria in the gland of Deshayes in several species of shipworm. Shipworms are wood-boring marine bivalve molluscs in which the valves are reduced in size and used for boring tunnels in submerged wood. The body of the mollusc completely fills the tunnel and the animal feeds by ingesting wood fragments produced by the rasping action of the valves. The gland of Deshayes contains aerobic, nitrogen-fixing cellulolytic bacteria; these are believed to secrete cellulase into the gut of the mollusc to enable it to digest cellulose in the wood fragments. This gland is the only known example of an association between a cellulolytic bacterium and an organ that is not part of the digestive tract [155].

ALGAL SYMBIOSIS

A variety of aquatic invertebrates posses symbiotic algae. Coelenterates provide many examples such as the green hydra (*Hydra viridissima*) and reef-forming corals. Algal symbionts are also found in protozoa, platyhelminths, molluscs and freshwater sponges; altogether, about 150 invertebrate groups are known to be involved [141]. The algae, which are usually coccoid forms, are from three classes: Cyanophyceae, Chlorophyceae and Dinophyceae. Various types of associations between algae and invertebrates are summarized in Table 3.3.7.

The nutritional role of zoochlorellae in *Hydra viridissima* has been extensively investigated. The algae synthesize maltose, which is available to the *Hydra* for metabolism and growth [97].

The protozoon *Paramecium bursaria* typically has several hundred zoochlorellae per cell and again there is

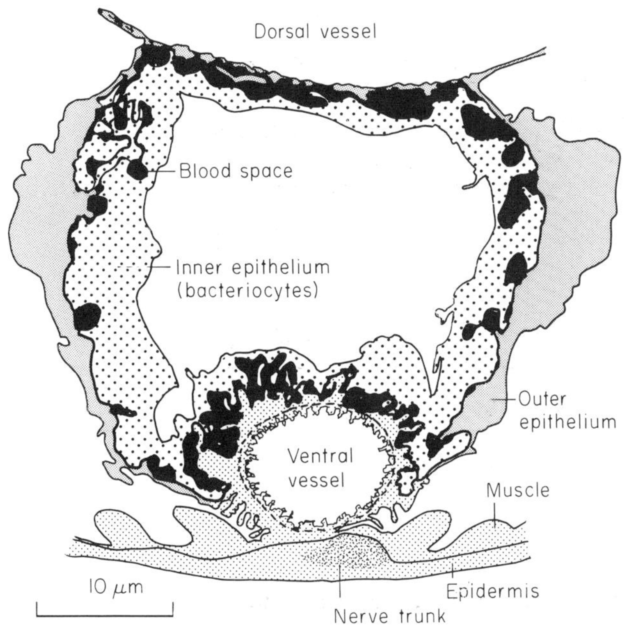

Fig. 3.3.10. Part of a transverse section of the postannular region of *Siboglinum fiordicum* showing the bacterial cylinder between the dorsal and ventral longitudinal blood vessels. Blood in the blood space shown black. From Southward [142].

algal strain recognition and carbon transfer as maltose. The photosynthetic activity of the algae facilitates the colonization of sediments with low oxygen tensions.

The zoochlorellae of corals are also important as oxygen suppliers since reef corals only do well in shallow waters where photosynthesis exceeds respiration [51]. It has often been suggested that the symbiotic algae assist in the carbon metabolism and calcification of corals. Tropical sea water is generally saturated with calcium bicarbonate and, as bicarbonate ions are removed from the water by algal photosynthesis, calcium carbonate is precipitated.

3.3.6 APPLICATIONS

Knowledge gained from understanding the microbial processes operating in the alimentary tracts of animals has far-reaching applications in the fields of nutrition, health and biotechnology. Fermentative digestion is only advantageous to an animal for utilizing substances that it is unable to digest using its own enzymes or which otherwise have poor nutritive value. Fermentation of protein, soluble carbohydrates and starch from the food, and hexoses and other products generated during plant cell wall digestion, inevitably leads to a loss of carbon, nitrogen and energy. The efficiency of animal production may be optimized by properly balancing the fermentation against the animal's own digestive capacity and requirements.

Ruminal fermentation of compounds that the animal itself can utilize in its lower alimentary tract can be controlled by protecting these components from the action of rumen micro-organisms and by controlling the rumen microbial species present or their activities. For example, protection of proteins from rumen fermentation has been achieved by treatment with heat, tannins and aldehydes [53]. Heat treatment, however, results in decreased digestibility of other dietary components, but treatment with uncondensed tannins causes cross-linking between proteins and other macromolecules which reduces microbial hydrolysis in the rumen.

Treatment of forages with formaldehyde protects protein from ruminal digestion without rendering it indigestible in the small intestine. Many chemical reactions occur when protein is treated with formaldehyde, but the major reactions — the formation of methylol groups on the terminal amino acids and the amino group of lysine, followed by condensations — are reversed under the acid conditions in the small intestine; this exposes the protein to enzymic digestion.

The amino acids methionine, cysteine, lysine and threonine are essential to the host animal, but are all subject to deamination in the rumen. Thus they may be growth-limiting when the animal is fed a poor-quality diet. Encapsulation of these animo acids in styrene copolymer matrices [53] reduces ruminal losses and allows significant quantities of these amino acids to be carried to the small intestine where some are absorbed. An alternative encapsulation method, coating nutrients with a layer of formaldehyde-treated protein, has proved successful for protecting unsaturated lipids from ruminal hydrogenation [137]. This method may be used to transport a variety of compounds (for example vitamins, hormones and pharmacological agents) to the small intestine.

Chemicals which inhibit methanogenesis have also been well studied and it was found that the ionophores monensin and lasalocid [33] and amicloral are partially successful. Monensin is currently used as a feed additive for cattle. The precise mode of action of monensin on methanogenesis is not known, but it is likely that it affects the generation of H_2 by non-methanogenic organisms, rather than the methanogens themselves. Monensin

Table 3.3.7. Summary of the different types of associations between algae and invertebrates found in sea water and fresh water [140]

	Sea water			Fresh water		
	Alga	Animal	Comments	Alga	Animal	Comments
Zoochlorellae	*Platymonas convolutae* (Pyramimonadales) ? Chlorophyceae	*Convoluta roscoffensis* (platyhelminth) *Anthopleura* sp. (sea anemone)	Zoochlorellae are uncommon in marine associations	*Chlorella* sp. (with some reports of *Scenedesmus*)	Protozoans Sponges *Hydra* Platyhelminths	The commonest type of freshwater association
Zooxanthellae	Dinoflagellates Cryptomonads Diatom	Sponges Coelenterates Molluscs *Cyclotrichium* (ciliate) *Convoluta convoluta*	Dinoflagellates are very common, especially in tropical oceans			None reported
Cyanellae	Unidentified blue-green algae	Echiuroid worms Sponges	Uncommon	*Synechococcus*	Protozoans	Uncommon
Chloroplasts	Siphonaceous seaweeds (*Codium, Caulerpa, Bryopsis*) *Cladophora* ? Red algae ?	Molluscs (*Elysia, Tridachia, Tridachiella, Placobranchus*) Molluscs (*Limapontia, Acteonia*) Mollusc (*Hermaea*) Rotifers	Previously described as zoochlorellae			None reported

treatment causes an increase in propionate and a decrease in acetate and *n*-butyrate concentrations in the rumen. This is advantageous for growing animals, which use the propionate for glucogenesis, but not for lactating animals, which rely on high ruminal acetate production for maximum lipogenesis in the mammary gland. It is also true that complete inhibition of methanogenesis is undesirable because many rumen organisms gain energy from the metabolic pathways involved in H_2 production; if there were no methanogens, the high ruminal H_2 concentrations would inhibit these pathways.

In the large intestine of man, microbial metabolism of ammonia and other nitrogenous compounds may be a cause of bowel cancer. To decrease the risk it has been recommended that diets contain a high level of fibre. The rationale is that a high fibre intake causes greater microbial activity and thus greater microbial nitrogen requirements in the large intestine. This results in low residual levels of free nitrogenous compounds in the bowel and also in relatively high levels of production of *n*-butyrate, an anti-tumour substance in animals.

In biotechnology, anaerobic fermentations have the major advantage that, unlike aerobic processes, they do not need expensive aeration and agitation. Understanding the reactions of gut anaerobes, both in pure culture and in stable co-cultures, has indicated a potential for the development of systems yielding products such as hydrogen, methane and ethanol from high-fibre agricultural wastes. For example, the rumen fungus *Neocallimastix frontalis* ferments straw and other fibrous plant tissues to a mixture of acetate, lactate, formate and ethanol. In co-culture with methanogenic bacteria, the major products are acetate and methane [94]. Thus, ethanol and methane can be produced from straw in a single culture. Similarly, *Methanobacterium ruminantium* may be co-cultured with the cellulolytic *Ruminococcus flavefaciens*, yielding methane, acetate and succinate from cellulose [79]. While these systems have not been used commercially, it is likely that, with improved strains of micro-organisms, large-scale plants could be operated digesting a variety of fibrous agricultural wastes. Similarly, genes isolated from gut anaerobes which degrade plant cell walls and cloned into organisms more suitable for industrial fermentation may be used for the controlled digestion of plant wastes for the production of chemical feedstocks [61, 140, 150].

Mixed rumen organisms have been used as starter inocula for digesters fermenting agricultural wastes such as manure. The climax population in fermentation systems varies considerably from the inoculum, but, in appropriately run fermenters, methane is a major product.

One of the most exploitable areas of studies of insect–micro-organism interactions is the use of entopathogenic bacteria, fungi and viruses as biocontrol agents against pests (see Chapter 4.2).

3.3.7 REFERENCES

[1] Anderson J.M. (1978) Competition between two unrelated species of soil Cryptostigmata (Acari) in experimental microcosms. *J. Anim. Ecol.* **47**, 787–803.

[2] Anderson J.M. & Bignell D.E. (1980) Bacteria in the food, gut contents and faeces of the litter-feeding millipede, *Glomeris marginata* (Villers). *Soil Biol. Biochem.* **12**, 251–4.

[3] Anderson J.M. & Bignell D.E. (1982) Assimilation of ^{14}C-labelled leaf fibre by the millipede *Glomeris marginata* (Diplopoda, Glomeridae). *Pedobiologia* **23**, 120–5.

[4] Anderson J.M. & Ineson P. (1983) Interaction between microorganisms and soil invertebrates in nutrient flux pathways of forest ecosystems. In *Invertebrate–Microbial Interactions*, p. 59 (ed. J.M. Anderson, A.D.M. Rayner & D. Walton). Cambridge University Press, Cambridge.

[5] Baker J.M., Laidlow R.A. & Smith H.C.A. (1970) Wood breakdown and nitrogen utilization by *Anobium punctatum* Deg. feeding on Scots pine sapwood. *Holzforschung* **24**, 46–54.

[6] Baleux B. & Vivares, C.P. (1974) Étude préliminaire de la flore bactérienne intestinale de *Schizophyllum sabulosum* var. *ribupes* Lat. (Myriapoda, Diplopoda). *Bull. Soc. Z. Franc.* **99**, 771–9.

[7] Barois I. & Lavelle P. (1987) Changes in respiration rate and some physicochemical properties of a tropical soil during transit through *Pontoscolex corethrurus* (Glossoscolecidae, Oligochaeta). *Soil Biol. Biochem.* **18**, 539–42.

[8] Barron, G.L. (1982) Nematode destroying fungi. In *Experimental Microbial Ecology*, p. 533 (ed. R.G. Burns & J.H. Slater). Blackwell Scientific Publications, Oxford.

[9] Bauchop T. (1977) Foregut fermentation. In *Microbial Ecology of the Gut*, p. 223 (ed. R.T.J. Clarke & T. Bauchop). Academic Press, London.

[10] Bauchop T. (1979) Rumen anaerobic fungi of cattle and sheep. *Appl. Env. Microbiol.* **38**, 148–58.

[11] Baumann P. & Baumann L. (1977) Biology of the marine Entobacteria: genera *Beneckea* and *Photobacterium. Ann. Rev. Microbiol.* **31**, 39–61.

[12] Bayer E.A., Setter E. & Lamed R. (1985) Organization and distribution of the cellulosome in *Clostridium thermocellum. J. Bact.* **163**, 552–9.

[13] Bayon G. (1980) Volatile fatty-acids and methane production in relation to anaerobic carbohydrate fermentation in *Oryctes nasicornis* larvae (Coleoptera: Scarabeidae). *J. Insect Physiol.* **26**, 819–28.

[14] Bayon C. & Mathelin J. (1980) Carbohydrate fermentation and by-product absorption studied with labelled cellulose in *Oryctes nasicornis* larvae (Coleoptera: Scarabaeidae). *J. Insect Physiol.* **26**, 833–40.

[15] Behan V.M. & Hill S.B. (1978) Feeding habits and spore dispersal of Oribatid mites in the North American arctic. *Rev. Ecol. Biol. Sol* **15**, 497–516.

[16] Bignell D.E. (1981) Nutrition and digestion. In *The American Cockroach*, p. 57 (ed. W.J. Bell & K.G. Adixodi). Chapman & Hall, London.

[17] Bignell D.E. (1984) The arthropod gut as an environment for micro-organisms. In *Invertebrate–Microbial Interactions*, p. 205 (ed. J.M. Anderson, A.D.M. Rayner & D. Walton). Cambridge University Press, Cambridge.

[18] Bignell D.E., Oskarsson H., Anderson J.M. & Ineson P. (1983) Structure, microbial associations and function of the so-called 'mixed segment' of the gut of two soil-feeding termites, *Procubitermes aburiensis* and *Cubitermes severus* (Termitidae, Termitiune). *J. Zool. London* **201**, 445–80.

[19] Bollard J.E., Vanderwee M.A., Smith G.W., Tasman-Jones C., Gavin J.B. & Lee S.P. (1986) Location of bacteria in the mid-colon of the rat. *Appl. Env. Microbiol.* **51**, 604–8.

[20] Boyle P.J. & Mitchell R. (1978) Absence of micro-

organisms in crustacean digestive tracts. *Science* **200**, 1157–9.

[21] Boyle P.J. & Mitchell R. (1981) Bacterial microflora of a marine wood-boring isopod. *Appl. Env. Microbiol.* **42**, 720–9.

[22] Breznak J.A. (1982) Intestinal microbiota of termites and other xylophagous insects. *Annual Review of Microbiology* **36**, 323–43.

[23] Breznak J.A. (1984) Biochemical aspects of symbiosis between termites and their intestinal microbiota. In *Invertebrate–Microbial Interactions*, p. 173. (ed. J.M. Anderson, A.D.M. Rayner & D.W.H. Walton). Cambridge University Press, Cambridge.

[24] Buchner P. (1965) *Endosymbiosis of Animals with Plant Micro-organisms*. John Wiley, New York.

[25] Busvine J.R. (1985) *Arthropod Vectors of Disease*. Studies in Biology No. 5. Edward Arnold, London.

[26] Carlsson J. (1967) Dental plaque as a source of salivary streptococci. *Odont. Revs.* **18**, 173–80.

[27] Carlsson J. (1970) The early establishment of *Streptococcus salivarius* in the mouth of infants. *J. Dental Res.* **49**, 415–22.

[28] Carlsson J., Grahnen H., Jonsson G. & Wikner S. (1970) Establishment of *Streptococcus sanguis* in the mouth of infants. *Arch. Oral Biol.* **15**, 1143–8.

[29] Carpenter E.J. & Culliney J.L. (1975) Nitrogen fixation in marine shipworms. *Science* **187**, 551–2.

[30] Carr D.H., McLeay L.M. & Titchen D.A. (1970) Factors affecting the reflex responses of the ruminant stomach. In *Physiology of Digestion and Metabolism in the Ruminant*, p. 35 (ed. A.T. Phillipson). Oriel Press, Newcastle upon Tyne.

[31] Cavanaugh C.M. (1983) Symbiotic chemotrophic bacteria in marine invertebrates from sulfide-rich habitats. *Nature, Lond.* **302**, 58–61.

[32] Cavanaugh C.M., Gardiner S.L., Jones M.L., Jannasch H.W. & Waterbury J.B. (1981) Procaryotic cells in the hydrothermal vent tube worm *Riftia pachyptila* Jones: possible chemoantotrophic symbionts. *Science* **213**, 340–1.

[33] Chalupa W. (1980) Chemical control of rumen microbial metabolism. In *Digestive Physiology and Metabolism in Ruminants*, p. 325 (ed. Y. Ruckenbusch & P. Thivend). M.T.P. Press, Lancaster.

[34] Chang K.P. (1981) *Nutritional Roles of Intracellular Symbiotes in Insects and Protozoa*. CRC Nutrition and Food Series, CRC Press, Boca Raton, Florida.

[35] Charnley A.K. (1984) Physiological aspects of destructive pathogenesis in insects by fungi: a speculative review. In *Invertebrate–Microbial Interactions*, p. 229 (ed. J.M. Anderson. A.D.M. Rayner & D.W.H. Walton). Cambridge University Press, Cambridge.

[36] Cheng K.-J. & Costerton J.W. (1980) Adherent rumen bacteria — their role in the digestion of plant material, urea and epithelial cells. In *Digestive Physiology and Metabolism in Ruminants*, p. 227 (ed. Y. Ruckenbusch & P. Thivend). M.T.P. Press, Lancaster.

[37] Cheng K.-J., Irvin R.T. & Costerton J.W. (1981) Autochthonous and pathogenic colonization of animal tissues by bacteria. *Can. J. Microbiol.* **27**, 461–90

[38] Cheng K.-J. & Wallace A.J. (1979) The mechanism of the passage of endogenous urea through the rumen wall and the role of ureolytic bacteria in urea flux. *Br. J. Nut.* **42**, 553–9.

[39] Cleveland C.R. & Grimstone A.V. (1964) The fine structure of the flagellate *Mixotricha paradoxa* and its associated micro-organisms. *Proc. R. Soc. Lond. B* **159**, 668–86.

[40] Cochran D.G. (1975) Excretion in insects. In *Insect Biochemistry and Function*, p. 177 (ed. D.J. Cardy & J.P. Reinecke). Chapman & Hall, London.

[41] Coleman G.S. (1975) The interrelationship between rumen ciliate protozoa and bacteria. In *Digestion and Metabolism in the Ruminant*, p. 149 (ed. I.W. McDonald and A.C.I. Warner). University of New England Publishing Unit, Armidale, NSW, Australia.

[42] Coleman G.S. (1978) The metabolism of cellulose, glucose and starch by the rumen ciliate protozoon *Eudiplodinium maggii*. *J. Gen. Microbiol.* **107**, 359–66.

[43] Coleman G.S., Laurie J.I. & Bailey J.E. (1977) The cultivation of the rumen ciliate *Entodinium bursa* in the presence of *E. caudatum*. *J. Gen. Microbiol.* **101**, 253–8.

[44] Collins N.M. (1983) The utilization of nitrogen resources by termites (Isoptera). In *Nitrogen as an Ecological Factor*, p.381 (ed. J.A. Lee, S. McNeil & I.A. Rorison). Blackwell Scientific Publications, Oxford.

[45] Collins N.M. & Wood T.G. (1984) Termites and atmospheric gas production. *Science* **224**, 84–5.

[46] Costerton J.W., Gleesey G.G. & Cheng K.-J. (1978) How bacteria stick. *Sci. Am.* **238**, 86–95.

[47] Costerton J.W., Ingram J.M. & Cheng K.-J. (1974) Structure and function of the cell envelope of gram-negative bacteria. *Bact. Revs.* **38**, 87–110.

[48] Dent V.E., Hardie J.W. & Bowden G.H. (1976) A preliminary study of dental plaque on animal teeth. *International Association of Dentistry Research Abstract* **85**, *J. Dental. Res.* **55**, Special Issue.

[49] Donaldson R.M. & Toskes P.P. (1983) The relation of enteric bacterial populations to gastrointestinal function and disease. In *Gastrointestinal Disease*, Vol. 1, p.44 (ed. M.H. Sleisenger & J.H. Fordtran). W.B. Saunders, Philadelphia.

[50] Dowding P. (1984) The evolution of insect–fungus

relationships in the primary invasion of timber. In *Invertebrate–Microbial Interactions*, p.133 (ed. J.M. Anderson, A.D.M. Rayner & D.W. Walton). Cambridge University Press, Cambridge.

[51] Droop M.R. (1963) Algae and invertebrates in symbiosis. In *Symbiotic Associations*, p.171 (ed. P.S. Nutman & B. Mosse). Cambridge University Press, Cambridge.

[52] Faichney G.J. (1975) The use of markers to partition digestion in the gastrointestinal tract. In *Digestion and Metabolism in the Ruminant*, p.277 (ed. I.W. McDonald & A.C.I. Warner). University of New England Publishing Unit, Armidale, NSW.

[53] Ferguson K.A. (1975) The protection of dietary proteins and amino acids against microbial fermentation in sheep. In *Digestion and Metabolism in the Ruminant*, p.448 (ed. I.W. McDonald & A.C.I. Warner). University of New England Publishing Unit, Armidale, NSW.

[54] Fraenkel G. & Blewett M. (1943a) The vitamin B-complex requirements of several insects. *Biochem. J.* **37**, 686–92.

[55] Fraenkel G. & Blewett M. (1943b) The sterol requirements of several insects. *Biochem. J.* **37**, 692–5.

[56] Gibbons R.J. (1984) Adherence of the oral flora. In *Attachment* of *Organisms to the Gut Mucosa*, Vol. I, p.11 (ed. E.C. Boedeker). CRC Press, Boca Raton, Florida.

[57] Gibbons R.J. & van Houte J. (1975) Bacterial adherence in oral microbial ecology. *Ann. Rev. Microbiol.* **29**, 19–44.

[58] Gilmour D. (1964) *The Metabolism of Insects*. Oliver & Boyd, Edinburgh.

[59] Hassall M. & Jennings J.B. (1975) Adaptive features of gut structure and digestive physiology in the terrestrial isopod *Philoscia muscorum* (Scop). *Biol. Bull.* **149**, 348–64.

[60] Hastings J.W. & Nealson K.H. (1981) The symbiotic luminous bacteria. In *The Prokaryotes*, p.1332 (ed. M.P. Starr). Springer-Verlag, Berlin.

[61] Hazlewood G.P., Mann S.P., Orpin C.G. & Romaniec M.P.M. (1986) Prospects for the genetic manipulation of rumen microorganisms. In *Recent Advances in Anaerobic Microbiology*, p. 162 (ed. S.P. Borellio & J.M. Hardie). Martinus Nijhoff, Dordrecht.

[62] Hobson P.N. & Wallace A.J. (1982a) Microbial ecology and activities in the rumen: Part 1. *CRC Critical Reviews in Microbiology*, April, 165–225.

[63] Hobson P.N. & Wallace A.J. (1982b) Microbial ecology and activities in the rumen: Part 2. *CRC Critical Reviews in Microbiology*, May, 253–320

[64] Hornicke H. & Bjornhag G. (1979) Coprophagy and related strategies for digestia utilization. In *Digestive Physiology and Metabolism in Ruminants*, p.707 (ed. Y. Ruckenbusch & P. Thivend). MTP Press, Lancaster.

[65] Houk E.J. & Griffiths G.W. (1980) Intracellular symbionts of the Homoptera. *Ann. Rev. Ent.* **25**, 161–87.

[66] Hungate R.E. (1966) *The Rumen and its Microbes*. Academic Press, London.

[67] Hurley R., Stanley V.C., Leask B.G.S. & Louvois J. (1974) Microflora of the vagina during pregnancy. In *The Normal Microflora of Man*, p.155 (ed. F.A. Skinner & J.G. Carr). Academic Press, London.

[68] Imai S. & Ogimoto K. (1978) Scanning electron and fluorescent microscopic studies on the attachment of spherical bacteria to ciliate protozoa in the rumen. *Jap. J. Vet. Sci.* **40**, 9–19.

[69] Ineson P. & Anderson J.M. (1986) Aerobically isolated bacteria associated with the gut and faeces of the litter-feeding macroarthropods *Oniscus asellus* and *Glomeris marginata*. *Soil Biol. Biochem.* **17**, 843–9.

[70] Ingold C.T. (1971) *Fungal Spores: Their Liberation and Dispersal*. Clarendon Press, Oxford.

[71] Jannasch H.W. & Nelson D.C. (1984) Recent progress in the microbiology of hydrothermal vents. In *Current Perspectives in Microbial Ecology*, p.170 (ed. M.J. Klug & C.A. Reddy). American Society for Microbiology, Washington.

[72] Jannasch H.W. & Wirsen C.O. (1979) Chemosynthetic primary production at East Pacific sea floor spreading centers. *Bioscience* **29**, 492–8.

[73] Jarosz J. (1979) Gut flora of *Galleria mellonella* suppressing ingested bacteria. *J. Invert. Path.* **34**, 192–8.

[74] Johnson R.A., Thomas R.J., Wood T.G. & Swift M.J. (1981) The inoculation of the fungus comb in newly founded colonies of some species of Macrotermitinae (Isoptera) from Nigeria. *J. Nat. Hist.* **5**, 751–6.

[75] Jurzitza G. (1979) The fungi symbiotic with anobiid beetles. In *Insect–Fungus Symbiosis: Nutrition, Mutualism and Commensalism*, p.65 (ed. L.R. Batra). Halsted Press, John Wiley, New York.

[76] Klug M.J. & Kotarski S. (1980) Bacteria associated with the gut tract of larval stages of the aquatic crane fly *Tipula abdominalis* (Diptera; Tipulidae). *Appl. Env. Microbiol.* **40**, 408–16.

[77] Kok L.T. (1979) Lipids of ambrosia fungi and the life of mutualistic beetles. In *Insect–Fungus Symbiosis*, p.33 (ed. L.R. Batra). Halsted Press, John Wiley, New York.

[78] Kozlovskaja L.S. (1970) Der Einfluss der Wirbellosen auf die Tätigkeit der Mikroorganismen in Torfboden. In *IV Colloquium Pedobiologiae*, p.81. INRA Publ. 71–7. Institut National de la Recherche Agronomique,

Paris.
[79] Latham M.J. & Wolin M.J. (1977) Fermentation of cellulose by *Ruminococcus flavefaciens* in the presence and absence of *Methanobacterium ruminantium. Appl. Env. Microbiol.* **34**, 297–301.
[80] Locke M. (1974) The structure and formation of integument of insects. In *The Physiology of Insecta*, 2nd edn, Vol. VI, p.123 (ed. M. Rockstein). Academic Press, New York.
[81] Luckey T.D. (1972) Introduction to intestinal microecology. *Am. J. Clin. Nut.* **25**, 1292–5.
[82] McBee R.H. (1977) Fermentation in the hindgut. In *Microbial Ecology of the Gut*, p.185 (ed. R.T.J. Clarke & T. Bauchop). Academic Press, London.
[83] McCowan R.P., Cheng K.-J. & Costerton J.W. (1979) Colonisation of a portion of the bovine tongue by unusual filamentous bacteria. *Appl. Env. Microbiol.* **37**, 1224–9.
[84] Macleod D.M. (1963) Entomopthorales infections. In *Insect Pathology: An Advanced Treatise*, Vol. 2, p.233 (ed. E.A. Steinhaus). Academic Press, New York.
[85] Madden J.L. & Coutts M.P. (1979) The role of fungi in the biology and ecology of woodwasps (Hymenoptera: Siricidae). In *Insect–Fungus Symbiosis*, p.165 (ed. L.R. Batra). Halsted Press, John Wiley, New York.
[86] Margherita S.S. & Hungate R.E. (1963). Serological analysis of *Butyrivibrio* from the bovine rumen. *J. Bact.* **86**, 853–60.
[87] Marialigeti K. (1979) On the community structure of the gut-microbiota of *Eiseria lucens* (Annelida, Oligochaeta). *Pedobiologia* **19**, 213–20.
[88] Marshall N.B. (1965) *The Life of Fishes*. Weidenfeld & Nicholson, London.
[89] Martin M.M. (1984) The role of ingested enzymes in the digestive processes of insects. In *Invertebrate–Microbial Interactions*, p.155 (ed. J.M. Anderson, A.D.M. Rayner & D.W. Walton). Cambridge University Press, Cambridge.
[90] Montes L.F. & Wilborn W.H. (1969) Location of bacterial skin flora. *Br. J. Derm.* **81**, Suppl. 1, 23–32.
[91] Morin J.G., Harrington A., Nealson K.H., Krieber N., Baldwin T.O. & Hastings J.W. (1975) Light for all reasons: versatility in the behavioral repertoire of the flashlight fish. *Science* **190**, 74–6.
[92] Mortensen A. (1985) Importance of microbial nitrogen metabolism in the caeca of birds. In *Current Perspectives in Microbial Ecology*, p.273 (ed. M.J. Klug & C.A. Reddy). American Microbiological Society, Washington, USA.
[93] Moss S.T. (1969) Commensalism of the Trichomycetes. In *Insect–Fungus Symbiosis*, p.175 (ed. L.R. Batra). Halsted Press, John Wiley, New York.
[94] Mountfort D.O., Asher R.A. & Bauchop T. (1982) Fermentation of cellulose to methane and carbon dioxide by a rumen anaerobic fungus in a triculture with *Methanobrevibacter* sp. RAI and *Methanosarcina barkeri. Appl. Env. Microbiol.* **44**, 128–34.
[95] Mueller R.E., Asplund J.M. & Ianotti E.L. (1984) Successive changes in the epimural bacterial community of young lambs as revealed by scanning electron microscopy. *Appl. Env. Microbiol.* **47**, 715–23.
[96] Mullins D.E. & Cochran D.G. (1975) Nitrogen metabolism in the American cockroach. II. An examination of negative nitrogen balance with respect to mobilization of uric acid stores. *Comp. Biochem. Physiol.* **50A**, 501–10.
[97] Muscatine L. & Lenhoff H.M. (1965) Symbiosis of hydra and algae. II. Effects of limited food and starvation on growth of symbiotic and aposymbiotic hydra. *Biol. Bull.* **129**, 316–28.
[98] Neville A.C. (1975) *Biology of the Arthropod Cuticle*. Springer-Verlag, Berlin.
[99] Noble W.C. & Somerville D.A. (1974) *Microbiology of the Human Skin*. Saunders, London.
[100] Nolte W.A. (1982) *Oral Microbiology*. Mosby, London.
[101] Nordbring-Hertz B. & Jansson H.-B. (1984) Fungal development predacity and recognition of prey in nematode-destroying fungi. In *Current Perspectives in Microbial Ecology*, p.327 (ed. M.J. Klug & C.A. Reddy). American Society for Microbiology, Washington.
[102] Norris D.M. (1979) The mutualistic fungi of *Xyloborini* beetles. In *Insect–Fungus Symbiosis*, p.53 (ed. L.R. Batra). Halsted Press, John Wiley, New York.
[103] Nuss I. (1982) Die Bedeutung der Proterosporen: Schlussfolgerungen aus Untersuchungen an *Ganoderma* (Basidiomycetes). *Plant Syst. Ev.* **141**, 53–79.
[104] O'Brien G.W., Veivers P.C., McEwen S.E., Scaytor M. & O'Brien R.W. (1979) The origin and distribution of cellulase in the termites *Nasutitermes exitiosus* and *Coptotermes lactens. Insect Biochem.* **9**, 619–25.
[105] Odelson D.A. & Breznak J.A. (1983) Volatile fatty acid production by the hindgut microbiota of xylophagous termites. *Appl. Env. Microbiol.* **45**, 1602–13.
[106] Ofek I. & Beachey E.H. (1980) General concepts and principles of bacterial adherence in animals and man. In *Bacteria Adherence*. p.1 (ed. E.H. Beachey). Chapman & Hall, London.
[107] Ofek I., Beachey E.H., Eyal F. & Morrison J.C. (1977) Postnatal development of binding of streptococci and lipoteichoic acid by oral mucosal cells of humans. *J. Infect. Dis.* **135**, 267–74.
[108] Ogimoto K. & Imai S. (1981) *Atlas of Rumen Microbiology*. Japan Scientific Societies Press, Tokyo.
[109] Oh J.H., Hume I.D. & Torrell D.T. (1972). Devel-

opment of microbial activity in the alimentary tract of lambs. *J. Anim. Sci.* **35**, 450–9.
[110] Orpin C.G. (1977a) Invasion of plant tissues in the rumen by the flagellate *Neocallimastix frontalis*. *J. Gen. Microbiol.* **98**, 423–30.
[111] Orpin C.G. (1977b) The rumen flagellate *Piromonas communis*: its life-history and invasion of plant material in the rumen. *J. Gen. Microbiol.* **99**, 107–17.
[112] Orpin C.G. (1985) Association of rumen ciliate populations with plant particles *in vitro*. *Microbiol. Ecol.* **11**, 56–9.
[113] Orpin C.G. & Letcher A.J. (1979) Utilization of cellulose, starch, xylan and other hemicelluloses for growth by the rumen phycomycete *Neocallimastix frontalis*. *Curr. Microbiol.* **3**, 121–4.
[114] Orpin C.G., Mathiesen S.D., Greenwood Y. & Blix A.S. (1985) Seasonal changes in the ruminal microflora of the High-arctic Svalbard reindeer *Rangifer tarandus platyrhynchus*. *Appl. Env. Microbiol.* **50**, 144–51.
[115] Parle I.N. (1963) A microbiological study of earthworm casts. *J. Gen. Microbiol.* **31**, 13–22.
[116] Peterson H. (1971) The nutritional biology of Collembola and its ecological significance. *Ento. Medd.* **39**, 97–118.
[117] Pherson D.A. & Beattie A.J. (1979) Fungal loads of invertebrates in beech leaf litter. *Rev. Ecol. Biol. Soil* **18**, 291–303.
[118] Poinar G.O., Thomas G., Haygood M. & Nealson K.H. (1980) Growth and luminescence of the specific bacterium associated with *Heterorhabditis bacteriophora*. *Nematologica* **23**, 97–102.
[119] Potrikus C.J. & Breznak J.A. (1980) Uric acid-degrading bacteria in guts of termites (*Reticulitermes flavipes* (Kollar)). *Appl. Env. Microbiol.* **40**, 117–24.
[120] Powell M.A. & Somero G.N. (1985) Sulfide oxidation occurs in the animal tissue of the gutless clam, *Solemya reidi*. *Biol. Bull.* **169**, 164–81.
[121] Prins R.A. (1977) Biochemical activities of gut micro-organisms. In *Microbial Ecology of the Gut*, p.73 (ed. R.T.J. Clarke & T. Bauchop). Academic Press, London.
[122] Prins R.A. & Clarke R.T.J. (1980) Microbial ecology of the rumen. In *Digestion Physiology and Metabolism in Ruminants*, p.179 (ed. R.Y. Ruckenbusch & P. Thivend). M.T.P Press, Lancaster.
[123] Reyes V.G. & Tiedje J.M. (1976) Ecology of the gut microbiota of *Tracheoniscus rathkei* (Crustacea, Isopoda). *Pedobiologia* **16**, 67–74.
[124] Rheinheimer G. (1974) *Aquatic Microbiology*. Wiley Interscience, London.
[125] Robinson I.M., Whipp S.C., Bucklin J.A. & Allison M.J. (1984) Characterization of predominant bacteria from the colons of normal and dysenteric pigs. *Appl. Env. Microbiol.* **48**, 964–9.
[126] Rohrman G.F. & Rossman A.Y. (1980) Nutrient strategies of *Macrotermes ukuzii* (Isoptera, Termitidae). *Pedobiologia* **20**, 61–73.
[127] Rozee K.R., Cooper D., Lam K. & Costerton J.W. (1982) Microbial flora of the mouse ileum mucous layer and epithelial surface. *Appl. Environ. Microbiol.* **43**, 1451–63.
[128] Ruby E.G. & Morin J.G. (1979) Luminous enteric bacteria of marine fishes: a study of their distribution, densities and dispersion. *Appl. Env. Microbiol.* **38**, 406–11.
[129] Ruddick S.M. & Williams S.T. (1972) Studies on the ecology of actinomycetes in soil. V. Some factors influencing the dispersal and adsorption of spores in soil. *Soil Biol. Biochem.* **4**, 93–103.
[130] Russell J.B. (1985) Fermentation of cellodextrins by cellulolytic and non-cellulolytic rumen bacteria. *Appl. Env. Microbiol.* **49**, 572–6.
[131] Satchell J.E. (1984) Earthworm microbiology. In *Earthworm Ecology: from Darwin to Vermiculture*, p. 351 (ed. J.E. Satchell). Chapman & Hall, London.
[132] Savage D.C. (1979) Microbial ecology of the gastrointestinal tract. *Ann. Rev. Microbiol.* **31**, 107.
[133] Savage D.C. (1980) Adherence of normal flora to mucosal surfaces. In *Bacterial Adherence*, p.31 (ed E.H. Beachey). Chapman & Hall, London.
[134] Savage D.C. (1984) Adherence of the normal flora. In *Attachment of the Oral Flora to Gut Mucosa*, Vol. I, p.3 (ed. E.C. Broedeker). CRC Press, Boca Raton, Florida.
[135] Savage D.C. & Blumershine R.H.V. (1974) Surface–surface associations in microbial communities populating epithelial habitats in the murine gastrointestinal ecosystem. *Inf. Imm.* **10**, 240–50.
[136] Savage D.C., Dubos R. & Schaedler R.W. (1968) The gastrointestinal epithelium and its autochthonous bacterial flora. *J. Exp. Med.* **127**, 67–75.
[137] Scott T.W. & Cook L.J. (1975) Effect of dietary fat on lipid metabolism in ruminants. In *Digestion and Metabolism in the Ruminant*, p.510 (ed. I.W. McDonald & A.C.I. Warner). University of New England Publishing Unit, Armidale, NSW.
[138] Sharpe M.E., Latham M.J. & Reiter B. (1975) The immune response of the host animal to bacteria in the rumen and caecum. In *Digestion and Metabolism in the Ruminant*, p.193 (ed. I.W. McDonald and A.C.I. Warner). University of New England Publishing Unit, Armidale, NSW.
[139] Shinefield H.R., Ribble J.C., Eichenwald H.F., Boris M. & Sutherland J.H. (1965) Bacterial interference. In *Skin Bacteria and their Role in Infection*, p.235 (ed.

H.I. Maibach).

[140] Smith C.J. & Hespell R.B. (1983) Prospects for the development and use of recombinant deoxynucleic acid techniques with ruminal bacteria. *J. Dairy Sci.* **66**, 1536–46.

[141] Smith D.C. (1973) *Symbiosis of Algae with Invertebrates*. Oxford Biology Readers No. 43, Oxford University Press, Oxford.

[142] Southward E.C. (1982) Bacterial symbionts in Pogonophora. *J. Mar. Biol. Ass. U.K.* **62**, 889–906.

[143] Stradling D.J. (1977) Food and feeding habits of ants. In *Production Ecology of Ants and Termites*, p.81 (ed. M.V. Bran). International Biological Programme 13, Cambridge University Press, Cambridge.

[144] Swift M.J. & Boddy L. (1984) Animal–microbial interactions in wood decomposition. In *Invertebrate–Microbial Interactions*, p.89 (ed. J.M. Anderson, A.D.M. Rayner & D.W. Walton). Cambridge University Press, Cambridge.

[145] Szabo I.M., Jagler K., Contreras E. Marialigeti K., Dzingov A., Barabas G.Y. & Pobozsny M. (1983) Composition and properties of the external and internal microflora of millipedes (Riplopoda). In *New Trends in Soil Biology*, p.197 (ed. P.L. Lebrun, H.M. Andre, A. De Medts, C. Gregoire-Wibo & G. Wanthy). Dieu-Brichart, Louvain-la-Neuve.

[146] Szabo I.M., Marton M. & Buti I. (1969) Intestinal microflora of the larvae of St. Marks fly. IV. Studies on the intestinal bacterial flora of the larval population. *Acta Microbiol. Acad. Sci. Hung.* **16**, 381–97.

[147] Talbot P.H.B. (1952) Disperal of fungus spores by small animals inhabiting wood and bark. *Trans. Br. Myc. Soc.* **35**, 123–8.

[148] Tannock G.W., Miller J.R. & Savage D.C. (1984) Host specificity of filamentous, segmented microorganisms adherent to the small bowel epithelium in mice and rats. *Appl. Env. Microbiol.* **47**, 441–2.

[149] Tavares I.I. (1979) the Laboubeniales and their arthropod hosts. In *Insect–Fungus Symbiosis*, p.229 (ed. L.R. Batra). Halsted Press, John Wiley, New York.

[150] Teather R.M. (1985) Application of gene manipulation to rumen microflora. *Can. J. Anim. Sci.* **65**, 563–74.

[151] Vallenbois P., Roch P., Lassegues M. & Dauant N. (1982) Bacteriostatic activity of a chloragen cell secretion. *Pedobiologia* **24**, 191–6.

[152] Veivers P.C., O'Brien R.W. & Slaytor M. (1980) The redox state of the gut of termites. *J. Insect Physiol.* **26**, 75–7.

[153] Visser S. (1985) Role of soil invertebrates in determining the composition of soil microbial communities. In *Ecological Interactions in Soil*, p.297 (ed. A.H. Fitter, D. Atkinson, D.J. Read & M.B. Usher). Blackwell Scientific Publications, Oxford.

[154] Vogels G.D., Hoppe W.F. & Stumm C.K. (1980) Association of methanogenic bacteria with rumen ciliates. *Appl. Env. Microbiol.* **40**, 608–12.

[155] Waterbury J.B., Calloway C.B. & Turner R.D. (1983) A cellulolytic nitrogen-fixing bacterium cultured from the gland of Deshayes in Shipworms (Bivalvia: Teredinidae). *Science* **221**, 1401–2.

[156] Webber J.F. & Brazier C.M. (1985) The transmission of Dutch elm disease: a study of the processes involved. In *Invertebrate–Microbial Interactions*, p.271 (ed. J.M. Anderson, A.D.M. Rayner & D.W. Walton). Cambridge University Press, Cambridge.

[157] Weber M.A. (1979) Fungus culturing by ants. In *Insect–Fungus Symbiosis*, p.77 (ed. L.R. Batra) Halsted Press, John Wiley, New York.

[158] Weiss E. (1982) The biology of rickettsiae. *Ann. Rev. Microbiol.* **36**, 345–70.

[159] Whistler H.C. (1979) The fungi versus the arthropods. In *Insect–Fungus Symbiosis*, p.1 (ed. L.R. Batra). Halsted Press, John Wiley, New York.

[160] Williamson P. (1965) Quantitative estimation of cutaneous bacteria. In *Skin Bacteria and their Role in Infection*, p.42 (ed. H.I. Maibach & G. Hildick-Smith). McGraw-Hill, New York.

[161] Wolin M.J. (1975) Interactions between the bacterial species of the rumen. In *Digestion and Metabolism in the Ruminant*, p.134 (ed. I.W. McDonald and A.C.I. Warner). University of New England Publishing Unit, Armidale, NSW.

[162] Wolin M.J. (1981) Fermentation in the rumen and human large intestine. *Science* **112**, 1463–8.

[163] Woodroffe R.C.S. & Shaw D.A. (1974) Natural control and ecology of microbial populations on skin and hair. In *The Normal Microflora of Man*, p.13 (ed. F.A. Skinner & J.G. Carr). Academic Press, London.

[164] Yamin M.A. (1980) Cellulose metabolism by the termite flagellate *Trichomitopsis termopsidus. Appl. Env. Microbiol.* **39**, 859–63.

[165] Zeikus J.G. (1981) Lignin metabolism and the carbon cycle: polymer biosynthesis, biodegradation, and environmental recalcitrance. *Adv. Microbiol. Ecol.* **5**, 211–43.

[166] Zimmerman P.R., Greenberg J.P., Wandiga S.O. & Crutzen P.J. (1982) Termites: a potentially large source of atmospheric methane, CO_2 and molecular H_2. *Science* **218**, 563–5.

3.4 Microbial adaptations to extreme environments

3.4.1 INTRODUCTION

Micro-organisms live in most of the environments of the earth, from the depths of the sea to mountain tops, from frigid Antarctic soils to near boiling hot springs. Their ubiquity raises the immediate question of what is normal and what is extreme from the microbial point of view. It is perhaps best to think of gradients of temperature, salinity, nutrients and pressure and define extreme conditions as the ends of these environmental gradients. Organisms in nature often exist at the extremes of several gradients, such as the bacteria living in the hot water escaping from vents in the bottom of the ocean.

The study of micro-organisms in extreme environments, both natural and man-made, has been amply reviewed [6, 16, 27, 37, 40].

3.4.2 OLIGOTROPHS

Micro-organisms that are adapted to live in environments with low rates of supply of nutrients are called oligotrophs. These low rates of supply occur because the habitat is a long way from the source of organic matter, such as in the deep sea or in deep groundwaters, or because the rate of photosynthesis by algae or by land plants is low. Photosynthesis is low usually because of low rates of supply of inorganic nutrients, especially nitrogen and phosphorus. It is the rates of supply that are important because bog and other anaerobic environments may contain large amounts of organic matter yet be oligotrophic. With this and similar exceptions, most oligotrophic environments have low amounts of available inorganic nutrients, low amounts of photosynthesis, low amounts of organic matter, and low numbers of animals, plants and micro-organisms.

Oligotrophic environments range from the surface of the open ocean, where nutrients are trapped in deep water, to certain lakes on erosion-resistant bedrock, to some soils. Soils may be oligotrophic because they are easily leached of nutrients, are anaerobic, are young soils, or are soils with the wrong nutrient ratio. For algae in fresh water, oligotrophy occurs when the available phosphate is below 5 $\mu g\ l^{-1}$ [48]. For heterotrophic micro-organisms, oligotrophy occurs when the carbon flux is low. Poindexter [35] suggests a flux of 1 mg C l^{-1} day^{-1} as the upper limit.

Oligotrophs are so effective at taking up low amounts of organic matter that some grow successfully in the distilled water in the laboratory [17]. This seemingly impossible feat occurs because the distilled water absorbs low amounts of organic compounds from the air. At the other end of the gradient are organisms called copiotrophs. These are heterotrophic organisms inhabiting rich environments that grow best with copious amounts of nutrients [35].

The study of oligotrophs is made difficult by their very low growth rate under natural conditions and by the aversion of many oligotrophs to high levels of nutrients. Many of the microbes that grow in oligotrophic environments will not grow on laboratory agar plates or even in the most dilute of laboratory media. They may grow in chemostat culture but are usually washed out at even the slowest practicable dilution.

As a result of these problems, two approaches to research have developed. One approach employs microbial species that will grow in the laboratory and extrapolates the results to what is happening in nature. The other approach employs natural populations and makes measurements of population numbers in natural soils and waters and of such things as growth of populations in short-term incubation experiments. Both approaches must be followed and combined wherever possible.

Oligotrophs in the laboratory

Research on laboratory cultures of bacteria leads to a number of conclusions about oligotrophs [35]. These organisms should be able to change their morphology in order to increase surface area and expose more of their nutrient binding sites to the dilute environment. The number of binding sites should be large per unit of surface, and the sites should have a low substrate specificity so that any type of organic molecule can be taken up. In addition, the sites should have a high affinity for the substrates so that they can be taken up at low concentrations. Once a nutrient is taken up by the cell, it must be used efficiently. Efficient use involves storage as a polymer such as a polysaccharide, polyphosphate or poly β-hydroxy alkanoate (PHAs, short-chain fatty acids). Nitrogen storage compounds are usually proteins and nucleic acids. All storage compounds are rapidly mobilized for cell maintenance when needed. A mechanism is also needed to cause the micro-organisms to store polymers even in periods of rapid growth and to direct the storage products to cell maintenance, rather than to cell growth, during periods of starvation.

The best metabolic strategy for oligotrophs is one of unbalanced growth, which takes advantage of short periods of nutrient abundance and does not waste organic compounds having less than optimal ratios of compounds or elements. Consequently, oligotrophs should have constitutive systems of catabolism to take immediate advantage of periods of relatively high nutrient intake. There should also be inducible catabolic systems to take advantage of some transient substrates.

Does the ideal oligotroph exist in nature? Soil arthrobacters, for example, are pleomorphic and change their surface-to-volume ratio during starvation. They utilize organic acids and are therefore competitive with copiotrophs during periods of high nutrient flux. Arthrobacters also accumulate large amounts of PHA as a reserve. During starvation they lower their endogenous metabolism and survive for long periods. Finally, they are resistant to desiccation and survive long periods without water.

Caulobacters provide another example of an idealized oligotroph (see Fig. 3.4.1). Under low nutrient conditions, they produce a slender stalk, a prostheca, that increases the surface area and allows more uptake of nutrients. Both PHA and polyphosphate are accumulated. These organisms are found only in oligotrophic aquatic habitats and are inhibited by high concentrations of organic nutrients [17].

Another way to examine oligotrophs is in chemostats. Matin and Veldkamp [32] used pond water to inoculate chemostats having dilution rates of 0.05 h^{-1} and 0.3 h^{-1}. A *Spirillum* sp. soon dominated the chemostat with the slow dilution rate and a *Pseudomonas* sp. the chemostat with the fast dilution rate. The limiting substrate was lactate and the organisms were isolated on 0.5% lactate medium. *Spirillum* sp. was obviously much more suited to oligotrophy than the *Pseudomonas* sp. (see Table 3.4.1).

Oligotrophs in nature

As most of the data on oligotrophs in nature are for communities, this is often helpful to an ecologist who is interested in the productivity or food web of an ecosystem

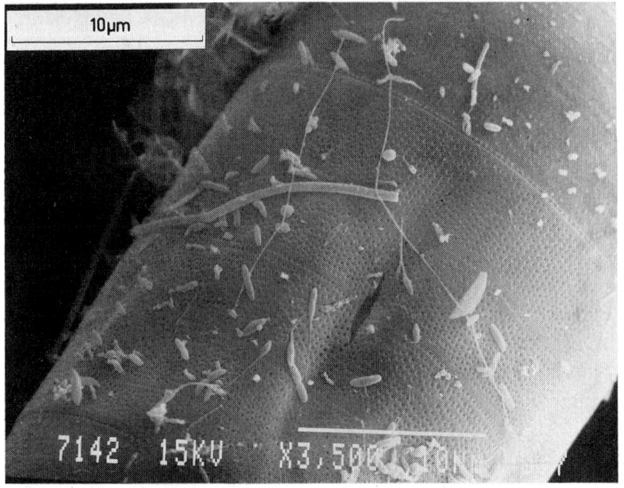

Fig. 3.4.1. *Caulobacter* (stalked bacteria) cells attached to the filamentous diatom *Melosira*. Photograph courtesy of J.S. Poindexter; taken at the Marine Biological Laboratory.

Table 3.4.1. Some characteristics of a *Spirillum* sp. and a *Pseudomonas* sp. selected for in a chemostat at various dilution rates*

Characteristic	*Spirillum*	*Pseudomonas*
Dilution rate	0.05 h^{-1}	0.30 h^{-1}
Retention time	20 h	3.3 h
Maximum generation time	2.9 h	1.6 h
Minimum generation time	100 h	12.5–20 h
Lactate uptake K_m	5.8 μM	20 μM
Lactate uptake V_{max}	29 nmol mg^{-1} min^{-1}	20 nmol mg^{-1} min^{-1}

* Data after Matin & Veldkamp [32].

but is not satisfactory for a microbiologist interested in how a particular species operates. Although the total numbers of microbes in these communities of soil or water may be easily counted with epifluorescence microscopy, only a few species have been identified either through their morphology (caulobacters) or through antibody enumeration. The total numbers of microbes are high in rich systems, more than 2×10^6 ml^{-1} in water and more than 1×10^9 g^{-1} in soil, and low in oligotrophic systems, as low as 0.040×10^6 in water and 0.001×10^9 in soils.

Oligotrophic bacteria are very small compared with laboratory populations; this small size gives the highest possible ratio of surface area to volume so that there are more uptake sites per unit of volume and so that the diffusion path from the surface to internal sites of utilization is decreased [46]. For example (see Fig. 3.4.2), most of the bacteria in the water column of the Irish Sea [44] are small mini-bacteria (0.04 μm^3) and the next most abundant group are cocci (0.2 μm^3). Other forms are less abundant but larger (rods 0.11 μm^3, large rods 0.45 μm^3) so that the mean volume per cell was 0.11 μm^3. These Irish Sea data also illustrate the problem of studying oligotrophic bacteria; even when the water sample is incubated *in situ* in cleaned dialysis bags with no added substrates, the bacteria rapidly increased in volume and began to divide. After 48 hours the average cell size had increased by 50% and the numbers by more than 100%. While this proves that all of the various groups of microbes are active, capable of growth, and able to change shape, this type of measurement indicates that some sort of activation is occurring and so the rate of growth calculated from the incubation is likely to be much higher than the natural rate. Much less activation and change occurred in the sample included at the cold temperatures and low light found at 60 m (see Fig. 3.4.2c and d).

Bacteria are capable of living at extremely low concentrations of dissolved organic carbon (DOC) — in fact, they themselves are responsible for the low concentrations. The clear sea water off California, for example [2], contains only 1.5 mg DOC l^{-1} (100 μM). If the natural population of bacteria is incubated with filter-sterilized sea water in a batch culture, they grow with a generation time of 9 hours and are able to use up about 10% of the DOC before growth stops (this amount is equivalent to about two generations of the original concentration of bacteria). If the same experiment is made but with continuous culture instead of with batch culture, mixed populations of bacteria grow at generation times of 6–39 hours, similar to generation times for natural populations [15]. In both types of culture, growth can only be detected by direct counting with the fluorescent microscope.

Only about 10% of the DOC has been characterized and this fraction is made up of simple compounds such as amino acids and sugars. In the same California waters, the total dissolved free amino acid concentration was 14–66 nM and individual amino acids were present at concentrations of a few nM [8]. Despite these low concentrations, bacteria are well adapted. For example, in the continuous culture mentioned above [15] the half saturation constant for uptake of leucine (K_t) was less than 3 nM, quite appropriate to the low levels found (0.5 nM). Another adaptation to these low levels is the ability to use several to many substrates simultaneously. This particular adaptation is reasonable given the low concentrations of any one substrate but it is difficult to demonstrate in natural waters. One type of evidence comes from our inability to grow micro-organisms from natural waters in continuous culture with the actual concentration of any given organic compound as the sole carbon source (about 20 μM was needed for growth). Another type of evidence comes from autoradiography of individual bacteria from water samples incubated with subnatural levels of tritiated organic compounds [19, 41]. A high percentage of the total population becomes labelled after incubation in the unaltered natural water plus a single labelled compound.

Laboratory studies of oligotrophic bacteria indicate that there is a competitive advantage for those able to store nutrients, usually as polymers such as polysaccharides or PHAs. This is presumed to be of advantage to bacteria in an environment where the amount of substrate fluctuates. So far, this prediction has not been proven and it may turn out that the changes of substrate

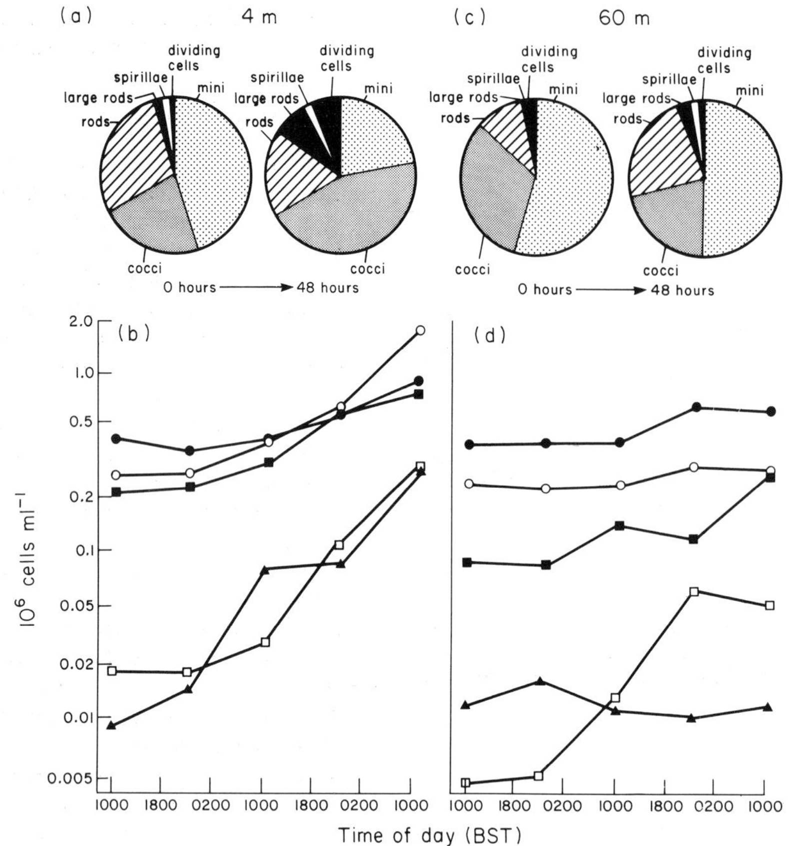

Fig. 3.4.2. Bacterial communities of the Irish Sea incubated *in situ* in 1 litre dialysis bags. Water samples collected and incubated at 4 m and 60 m; samples passed through a 3 μm pore size filter before incubation to remove grazers. (Redrawn with permission from Turley & Lochte [44].) (a) and (c) show the proportion of various morphological groups at the beginning and end of incubations. (b) and (d) show the numbers of mini-bacteria (●), cocci (○), rods (■), large rods (□) and dividing cells (▲).

from day to day are so small in oligotrophic environments that the bacteria cannot build up any storage products. There is another strategy available, inactivity, although the evidence for this is not yet complete. If the bacteria could drastically reduce their respiration rate when nutrients become too low, then this would seem an ideal adaptation. One type of evidence for this comes from studies over the entire day. In an oligotrophic lake in Canada, it has been found that thymidine incorporation into DNA, a measure of growth of bacteria, only occurs during daylight hours (J. Rudd, personal communication). Other evidence comes from autoradiography but data have been collected only from eutrophic conditions. For example, the rate of glucose uptake into microbes of a lake increased two- to fivefold over 12 hours while the bacterial numbers did not change [36]. Over a longer period, the number of bacteria incorporating a mixture of amino acids (active bacteria) changed dramatically from winter to summer in Lake Constance (see Fig. 3.4.3) while the total number of bacterial cells in the plankton (by fluorescence microscopy) changed much less. There was a close correlation ($r = 0.92$, P less than 0.05) between the numbers of active bacteria and the bacterial production [38] in this lake, which gives additional credence to the autoradiography data.

Almost all of the adaptations predicted for oligotrophic bacteria from laboratory measurements have been borne out in the field measurements. This is true even though

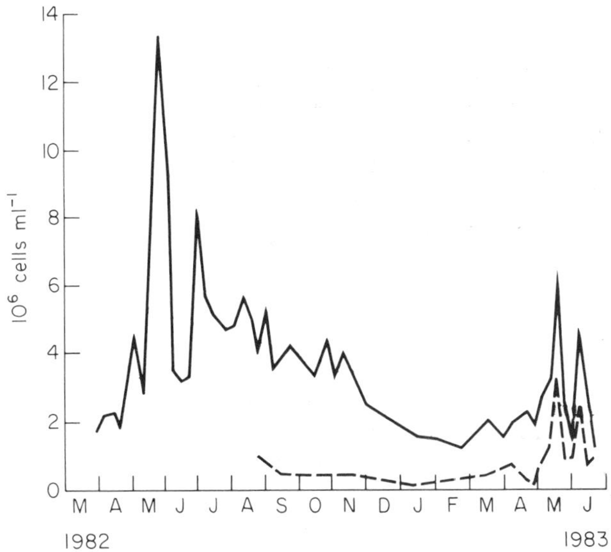

Fig. 3.4.3. Total numbers of bacteria (---) and numbers of bacteria actively incorporating labelled amino acids (by autoradiography) (——) in Lake Constance. The values are the mean for the top 3 m. Redrawn with permission from Simon [38].

the bacteria in nature usually exist under conditions many times more oligotrophic than those created in the laboratory.

3.4.3 ACIDOPHILES

Although most micro-organisms live in a narrow pH range around neutrality, some are able to live at a very low pH. These low pH organisms will not, in fact, grow near neutrality. They can be easily isolated from the water draining from coal deposits containing reduced iron and sulphur. Microbes in these acid mine drainages with a pH range of 1–4 oxidize the reduced iron and sulphur and produce ferric precipitates and sulphuric acid.

Many workers [6, 7, 30] have isolated acidophiles from hot springs in Yellowstone National Park, USA (*Sulfolobus acidocaldarius* and *Bacillus acidocaldarius*). Another interesting organism, *Thermoplasma acidophilum*, with no cell wall, was isolated from hot acid soils of coal refuse piles. What mechanisms have these microbes developed that allow them to live in such high proton concentrations, high enough to denature the essential structural and functional molecules of 'normal' organisms?

The genus *Thiobacillus*

The bacteria in this genus [31] are mostly chemolithotrophic (autotrophic), although some are mixotrophs (i.e. either autotrophic or heterotrophic). One important form, *T. ferrooxidans*, oxidizes reduced iron or reduced sulphur compounds for metabolic energy generation, a process coupled to CO_2 fixation by the Calvin cycle (see Chapter 2.5). At low pH, oxidation of reduced iron produces ferric oxide and hydroxide precipitates, accounting for the red and yellow colours of acid mine drainages. Oxidation of reduced sulphur compounds produces sulphuric acid, a good example of an organism creating its own extreme environment. If iron pyrite (FeS_2), a common compound in some coal deposits, is present, both the reduced iron and the reduced sulphur are oxidized to sulphuric acid and ferric precipitates. Another important form, *T. thiooxidans*, only oxidizes reduced sulphur compounds to sulphuric acid.

These bacteria can be isolated from acid mine drainage with a medium containing acidic inorganic salts and reduced iron (ferrous sulphate) or sulphur (colloidal sulphur) as the energy substrates. When ferric compounds precipitate and sulphuric acid is produced, the pH of the medium decreases to about pH 1.

How thiobacilli survive these extreme environments is not well understood. They must have some mechanism to exclude protons from their cytoplasm, yet the only enzyme found so far with an optimum at low pH is an iron oxidase bound to the cell membrane of *T. ferroxidans*. All of the cytoplasmic enzymes have pH optima of 5–8, which agrees with their intracellular pH of about 5.5. No special structures for proton exclusion have been found in thiobacilli, and their cell walls and membranes are structurally the same as other Gram-negative bacteria.

The outer cell envelope of thiobacilli includes a lipopolysaccharide (LPS), a unique feature that does not contain phosphorus or dideoxy sugars and is low in fatty acids [18, 47]. Although the organism also appears to contain a unique cyclic fatty acid, these features have yet to be directly related to acidophily.

Bacillus acidocaldarius

This bacterium is a typical sporeforming, aerobic bacillus that grows optimally at 60°C and pH 3. Although it has a typical Gram-positive cell wall, like other members of the genus *Bacillus*, it contains some unusual polar lipids in its membrane [30]. These *w*-cyclohexyl-C_{17} and -C_{19} fatty acids form 60–90% of the cellular fatty acids. Various polyprenols, squalene, pentacyclic triterpenes and menaquinones have been detected in the neutral lipid fraction. The glycolipids, 64% of total lipids, contain typical glycolipids and a unique pentacyclic triterpene-derived tetrol *N*-acylglucosaminoside (26%). It has been suggested but not proved [30] that the cyclohexyl fatty acids contribute to the thermal stability of the membrane, and the neutral lipids contribute to the acid stability (H^+ exclusion) of the membrane.

Sulfolobus

Sulfolobus acidocaldarius, a facultative chemolithotrophic sulphur-oxidizing bacterium, grows at pH 2–3 and 70–80°C. It is primarily responsible for acid production in sulphurous hot springs. The internal pH of *Sulfolobus*, around 6.3, indicates an ability to exclude protons.

Because these lobe-shaped organisms do not have the peptidoglycan cell wall typical of prokaryotes (see Fig. 3.4.4), they belong to the Archaebacteria [49, 50]. They exhibit proteinaceous cell walls containing a regular polyhedral array of proteins 13–15 nm in diameter in close association with the cell membrane.

The membrane lipids have an unusually high glycolipid content (68%) [30] and also contain long-chain C_{40} isoprenol glycerol diethers (see Fig. 3.4.5) instead of diester-linked fatty acids. These isoprenoid diethers, a main characteristic of Archaebacteria [49, 50], are found in other organisms isolated from extreme environments, such as *Thermoplasma* and the halophiles, and in the methanogenic bacteria.

Isoprenoid diethers can occur in a variety of cyclic configurations. Some of the *Sulfolobus* species contain various amounts of these lipids (see Table 3.4.2). These diphytanyl diether-linked lipids may contribute to membrane stability since ethers are less easily hydrolysed by acid, and the isoprenoids are more rigid (stable) than fatty acids because of their branching. Because these unique lipids are in close contact with the wall proteins, it may be that they remain in the proper liquid–crystal configuration at higher temperatures.

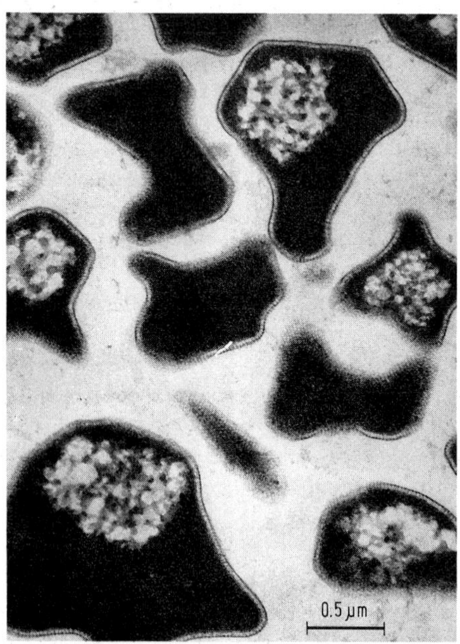

Fig. 3.4.4. Thin section electron micrograph of *Sulfolobus acidocaldarius* strain B6 showing its irregular lobate structure and glycogen storage granules. After Woese & Wolfe [50]. Photograph courtesy of W. Zillig.

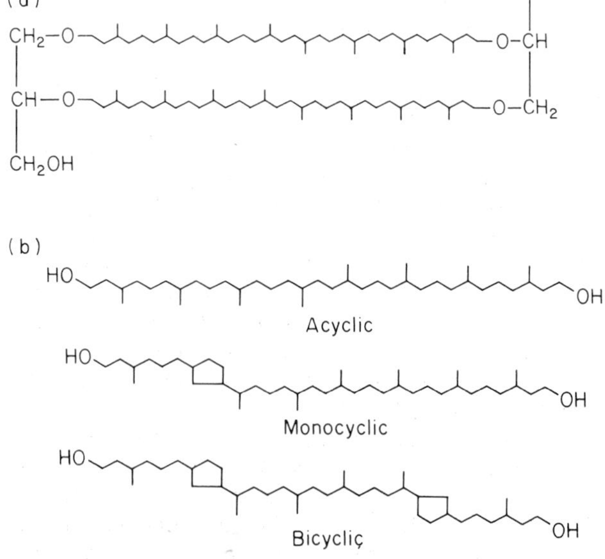

Fig. 3.4.5. (a) The structure of the diphytanyl diether lipid found in some thermoacidophilic bacterial membranes. (b) Structures of some diphytanyl hydrocarbons found in the lipids of some thermoacidophilic bacteria. After Brock [7].

Table 3.4.2. Composition of diphytanyl hydrocarbons in thermoacidophilic bacteria *

Organism	Component (% total hydrocarbons)		
	$C_{40}H_{82}$ (acyclic)	$C_{40}H_{80}$ (monocyclic)	$C_{40}H_{78}$ (bicyclic)
Thermoplasma acidophilum	65	32.5	2.5
Sulfolobus acidocaldarius 93–3	29.9	32.3	37.8
Sulfolobus acidocaldarius ex93–3	trace	14.4	85.6
Strain MT-3 (related to *Sulfolobus*)	3.6	18.2	71.6

* Adapted from Brock [7].

Thermoplasma acidophilum

This micro-organism is an obligate thermoacidophile originally isolated from a self-heating coal refuse pile at a pH of 2. It can live at temperatures from 45 to 65°C with an optimum around 60°C and an optimum pH of 2 [30]. It has no rigid cell wall and its internal pH is close to neutrality, again indicating a membrane-associated property that prevents protons from entering the cell.

Diphytanyl diether lipids form the major portion of the membranes of *T. acidophilum* (see Table 3.4.2), possibly accounting for some thermal as well as acidic stability. In the membrane proteins, the ratio of carboxyl (negatively charged) to amine (positively charged) groups of about 4:1 [30] suggests that the membrane is highly negative. When the net negative charge is altered by adding ethylene diamine groups, the membranes become stable over the complete pH range. Other work shows that the neutralizing cation is H^+; other mono- and divalent cations cannot substitute for H^+. *T. acidophilum* thus appears to exclude H^+ from its cytoplasm by stabilizing membranes with protons and rigid lipids.

There is no one universally evolved mechanism that accounts for obligate acidophily. The only common feature of acidophilic lipids is the presence of cyclic rings in the fatty acids and alkyl side chains. It seems reasonable to assume, however, that the presence of such lipids does not account totally for acidophily. One discussion of the bioenergetic considerations of living in a high proton environment [26] suggests that there are more subtle mechanisms at work.

3.4.4 ALKALINOPHILES

Soils contaminated with animal excreta or decaying proteins or containing alkaline minerals often provide extremely alkaline environments. Some alkaline lakes and springs have pH ranges of 8–11 [26]. Bacteria isolated from alkaline environments include *Bacillus, Nitrosomonas, Nitrobacter, Rhizobium, Agrobacterium* and *Flavobacterium* species. Some of these appear to grow best at high pH (9–11) and not at all well near neutrality. Eukaryotic organisms that can tolerate high pH include some fungal genera such as *Penicillium* and *Fusarium*, green algae of the genus *Chlorella*, cyanobacteria of the genera *Gleotheca, Microcystis, Arthrospira* and a *Plectonema* that can grow at pH 13, the highest pH recorded for growth of life forms [26].

The cytoplasmic pH of these organisms is near neutrality, suggesting that OH^- is excluded. While there is no information on the composition of the membranes and walls of these organisms, some of their extracellular enzymes exhibit high pH optima (between 8.5 and 11.5) for activity. It is likely that OH^- exclusion mechanisms do exist in these alkalinophiles or that there is some sort of inward H^+ pump to keep the pH of the cytoplasm near neutral [26]. Perhaps these organisms have a net positive charge on their cell surface, which would be neutralized by the excessive hydroxyl ions in the surrounding alkaline environment.

3.4.5 WATER STRESS ON MICRO-ORGANISMS

Availability of water for micro-organisms can be expressed in terms of water activity (a_w), the relative humidity (RH) expressed as a decimal fraction (see Sections 3.5.6 and 4.3.3) or as water potential (see Section 3.1.2). Some organisms that live in the desert and in high salt concentrations will serve as examples of matric and osmotic water stress on micro-organisms.

The effects of matric water on micro-organisms

Microbial survival in the extremely low moisture environment of deserts is dependent on matric waters from rain or dew in soil or rocks. In hot deserts, most of the matric water evaporates during the heat of the day so micro-organisms are dependent on water they absorb during the cooler nights or mornings.

There are also cold deserts such as the Dry Valleys (Ross Desert) of Antarctica where the air temperatures average about −10°C during the summer. There was some doubt as to whether micro-organisms even existed in the Antarctic soil as Uydess and Vishniac [45] found them while others [21] did not. Recently, Hirsch [17 and personal communication] has been successful in isolating a number of species, including some new species, on oligotrophic media. Many of these species had resting stages or formed spores, abilities helpful to their survival. It is now clear that the earlier findings [21] were dependent upon the isolation technique; oligotrophic micro-organisms colonized glass slides inserted into the soil but would not grow on copiotrophic media [45].

Another Antarctic desert site, the pore spaces of sandstones, is inhabited by a primitive but active microbial community [13, 14] (see Fig. 3.4.6). This microhabitat can be warmed by the sun to 20°C and the particles of quartz in the sandstone transmit enough light to allow photosynthesis to occur. The photosynthetic organisms are usually lichen algae, either green algae or cyanobacteria. The consumers of these primary producers are bacteria and fungi, including yeasts. These micro-organisms obtain water from the matric water of the sandstone which enters the pores after the melting of accumulated winter snow or of the occasional summer snowfall.

In the Antarctic, the right combination of a sunny and windless day and adequate matric water allows the sandstone to warm, the matric water to melt and the micro-organisms to metabolize. These conditions coincide only 2–10 days per year. Although the micro-organisms must be stressed by the frequent freezing and thawing, no special survival mechanisms have been found.

Lichen algae and fungi in these habitats need also to be resistant to desiccation [39]. Indeed, the most common Antarctic lichen alga from the sandstones is *Trebouxia* [13, 14], a species known to resist desiccation.

Photosynthesis in these Antarctic communities continues down to −6° to −8°C (see Fig. 3.4.7). It is possible that this occurs because the lichen algae produce polyols [39] which could act as a solute to hold the water in a liquid state.

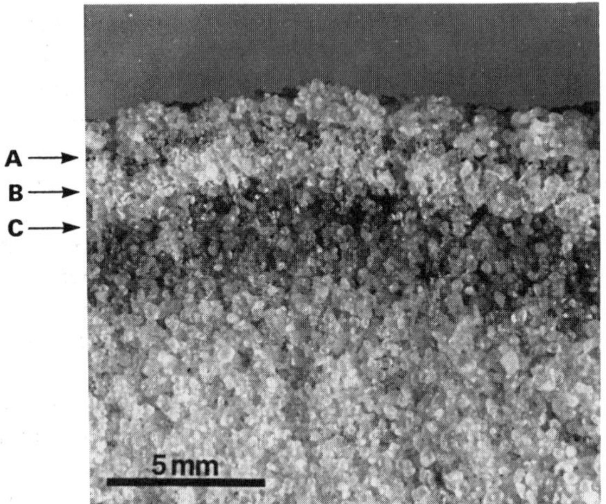

Fig. 3.4.6. A cross-sectional photograph of the cryptoendolithic microbiota found in the Beacon sandstone of the Antarctic dry valleys. The black zone (A) comprises the lichen, the white zone (B) contains filamentous fungi, and the green zone (C), which is not always found in these rocks, contains green algae. Decomposer bacteria and yeasts can be found in all zones. Photograph courtesy of E.I. Friedmann.

The effects of high osmotic pressure on micro-organisms

In many different contexts the high osmotic pressures of dissolved solutes affect microbes. For example, the high sugar content of jams and jellies lowers water activity (see Fig. 3.4.8) and prevents growth of all but a few species of fungi. In natural systems, the high salt environments of the Great Salt Lake and the Dead Sea contain obligate halophiles with specific adaptations.

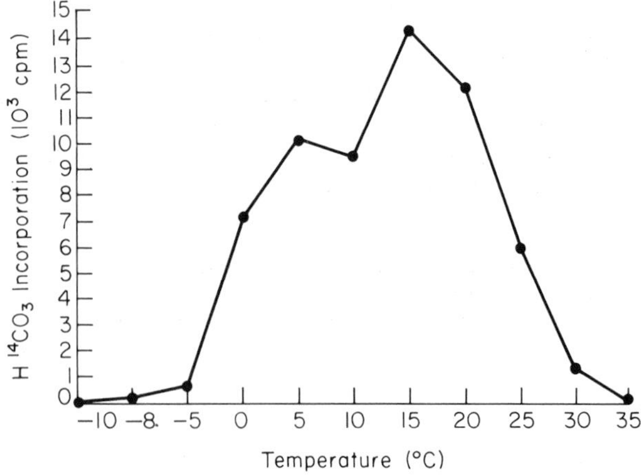

Fig. 3.4.7. The effect of temperature on the photosynthetic incorporation of $H^{14}CO_3$ into lipids by the cryptoendolithic microbiota from the Dry Valleys of Antarctica.

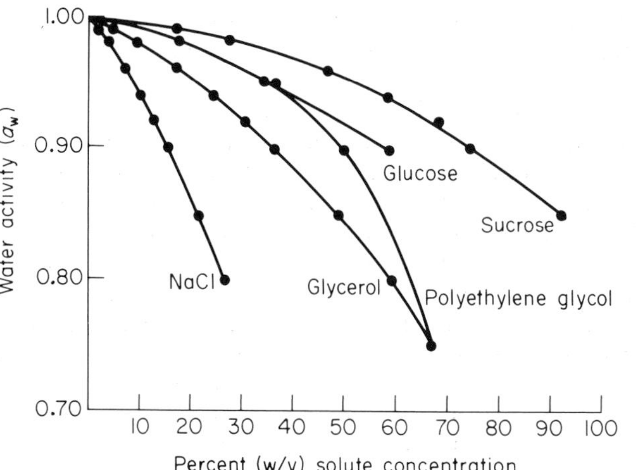

Fig. 3.4.8. The relationship between the concentrations of various solutes (g 100 ml^{-1}) and the water activities (a_w) of the solutions. After Kushner [29].

The osmotic pressure (π) exerted on a cell by the ionized solutes in an aqueous environment is: $\pi = \text{M}RT$, where M is the molarity of the solution, T is the absolute temperature and R is the universal gas constant. Even the relatively dilute solutions of sea water or physiological saline solution have a high π and a strong effect on cells (see Table 3.4.3).

Various bacteria, cyanobacteria, algae and fungi live in high salt environments (see Table 3.4.4), but the best studied are bacteria in the genera *Halobacterium* and *Halococcus* [28]. These are obligate halophilic bacteria that grow best at 2.5–5.0 M salt and do not grow in normal lab media. These bacteria are able to take up ions into the cell against a concentration gradient. For example, *Halobacterium cutirubrum* was shown to have an internal concentration of 0.8 M Na^+ and 5.32 M K^+ in an external medium of 0.8 M Na^+ and 0.05 M K^+ [28].

Table 3.4.3. Water activities (a_w) and osmotic pressures (π) of some common solutions

Solution	°C	a_w*	π (atm)
Water	—	1.00	0.0
Physiological saline	37	0.994	7.4
Sea water	25	0.98	25.2
Saturated NaCl	25	0.75	197.0
Saturated $CaCl_2{\cdot}6H_2O$	10	0.40	1233.4

* Adapted from Horowitz [20].

Table 3.4.4. Micro-organisms which can be considered as extreme halophiles *

Bacteria or Archaebacteria
Halobacterium
Halococcus
Ectothiorhodospira
Amoebacter
Cyanobacteria
Aphariocapsa
Algae
Dunaliella
Fungi
Cladosporium

* Adapted from Brock [7].

Another characteristic of halophilic bacteria is that, while most of their enzymes require high salt (e.g. those in the cell membranes), some of their enzymes are strongly inhibited by high salt concentrations [28]. The organisms appear to be able to isolate some enzymes from the damaging effects of high salt, probably by compartmentalization. For some enzymes, the high intracellular cation concentrations may be necessary for stabilization of charge. The cations neutralize the excess acidic amino acids found in many proteins and protein-containing components such as ribosomes (these may be 6–22 mol % excess over basic amino acids).

The cellular structures of the extreme halobacteria identify them as Archaebacteria (29, 49, 50). For example, the structure of the ribosomes is different from that of most other bacteria but like that of methanogens and some of the thermoacidophiles. Cell walls are also similar to those of Archaebacteria in that they do not contain muramic acid, diaminopimelic acid or *D*-amino acids. Instead, the walls contain glucose, galactose, glucosamine and 2-amino-2-deoxyguluronic acid, which can act like muramic acid. The lipid membranes of halophiles are also very much like those of Archaebacteria. About 63% of the total lipid is diphytanyl diether lipids (see Fig. 3.4.5), while 20% is glycolipid sulphate. The lipids are also negatively charged, thus requiring cations for stability.

The so-called purple membrane is unique to *H. halobium* and *H. cutirubrum*. In cells grown in the light with low aeration rates, this membrane can make up as much as 50% of the cell's wall–membrane complex. It contains a protein, bacteriorhodopsin, which has retinal (the visual pigment) as the prosthetic group. When these purple membranes are collected and used to make liposomes, they pump protons into the artificial vesicles in the presence of light and oxygen. If ATPase is added, ATP can be synthesized. This system is a useful model for the study of photophosphorylation [28, 29].

One genus of green algae, *Dunaliella*, grows in sea water and also in high salt concentrations [3]. It is able to maintain a very low intracellular salt concentration by producing glycerol to keep the osmotic pressure high. When a seawater-grown cell is placed in a high-salt environment, it immediately begins to produce glycerol. Some researchers have suggested that this quick response occurs when volume changes cause a tension on the membrane, and this triggers a membrane-bound enzyme to catalyse polysaccharides, such as starch, and produce glycerol via the glycolytic pathway. Two unique enzymes, dihydroxyacetone kinase and dihydroxyacetone reductase, help in the production of glycerol and the resynthesis of polysaccharides, respectively.

3.4.6 PSYCHROPHILES

Over 80% of the biosphere of earth is permanently cold [33]. The oceans cover 71% of the surface, and 90% of this, primarily the deep ocean, is colder than 5°C. The continent of Antarctica and lands near the Arctic Ocean are also permanently cold.

Micro-organisms live in these cold habitats, but most of these can also live at warmer temperatures. For this reason, Morita's [33] definition for a psychrophile, an organism that grows at temperatures from less than 0°C to 20°C with optimum growth at temperatures of 15°C or less, is quite unrestrictive. The cold habitats are populated by the same genera of bacteria found in most soils and waters as well as by algae, fungi, lichens and yeasts [5]. All the microbial processes are carried out in the cold habitats, albeit at slower rates than in warm habitats.

Microbial activity and growth is restricted at the low end of their temperature range by the freezing of water, but what stops growth at the high end of the range? In a general sense, it is likely that enzymes begin to denature above the temperature optimum. Many cellular mechanisms, such as protein synthesis and ribosome function, decline rapidly 1–2°C above the temperature optimum. Farrell and Campbell [11] showed that the temperature at which denaturation of a ribosome occurs is close to the temperature of maximum growth. When the temperature increased, there was a corresponding increase in the G + C content of the rRNA. Presumably, this increase gave thermal stability.

Another effect of temperature is on the fluidity of cell membranes. To maintain the fluidity necessary for nutrient transport, the fatty acids in the membrane become less saturated at low temperatures. At high temperatures the fatty acids become more saturated and sometimes more branched.

Just below the minimum temperature for growth the cell membrane becomes solid and transport stops. Enzymes also cease activity when ice forms but some species of algae and bacteria produce intracellular glycols that lower the freezing point of water and the minimum temperature for growth.

Micro-organisms living at low temperatures

The microbial community living near the surface of porous sandstones in the Antarctic includes lichens (*Buellia* and *Lecidea*), bacteria, yeasts and sometimes green algae (*Hemichlorus antarctica*) [13, 14] (see Fig. 3.4.6). In the laboratory, the community carried out photosynthesis between −8°C and +25–30°C. The optimum rate occurred at 15°C (see Fig. 3.4.7), thus making this community psychrophilic.

Diatoms inhabit another low temperature habitat, the sea ice of the Antarctic [34]. These algae live mostly at the bottom of the sea ice on the surface of large plates of ice that form a layer tens of centimetres thick. The annual temperature range is only from −2°C to 0°C. During the lighted months the algae are normal phototrophs. During the transition period from continuous light to continuous darkness, the algae store lipids and carbohydrates. During the dark months the algae use the stored materials and nearly cease growing.

3.4.7 THERMOPHILES

Natural thermal environments above 50°C occur in compost piles, coal refuse piles, volcanic dry-steam fumaroles, hot springs and sun-heated soil, rocks and litter. Man-made thermal environments include hot-water heaters, coffee machines, industrial cooling water towers and steam heating and condensate lines in buildings [42]. Many types of micro-organisms have adapted to these habitats — thermophiles are those that have optimal growth above 50°C. The hottest environments, above 90°C, are inhabited only by obligate thermophilic bacteria (see Table 3.4.5). In hot springs and in their outlet streams, a thermal gradient exists that allows bands of single species to develop. For this reason, much of the definitive work on thermophiles has been carried out by T.D. Brock and his associates in the thermal springs of

Table 3.4.5. The maximum temperature limits for representative groups of micro-organisms *

Group	Approx. max. temp. (°C)
Eukaryotes	
Protozoa	≤56
Algae	≤60
Fungi	≤62
Prokaryotes	
Cyanobacteria	≤73
Photosynthetic bacteria	≤73
Chemolithotrophic bacteria	≤90
Heterotrophic bacteria	≤90

* Adapted from Tansey & Brock [42].

Yellowstone Park, USA [6, 7, 42]. The thermophilic cyanobacteria have been studied by Castenholz [9, 10]. The molecular basis of thermophily has been discussed [1, 12, 16, 22].

Cell proteins of thermophiles are highly resistant to thermal denaturation. One possible reason is that stability arises from unique amino acid sequences. For example, aspartate, glutamate and arginine provide intersubunit stability in glyceraldehyde-3-phosphate dehydrogenase of thermophiles. Also, an analysis of a number of enzymes reveals a preference for alanine, threonine and arginine in thermophiles compared with mesophiles. It is also known that some enzymes need extrinsic factors, such as metal ions, coenzymes, substrates and charged macromolecules and membranes, for thermal stability [22].

As might be expected, other parts of the cellular machinery are also stable at high temperatures. In *Thermus thermophilus* the ribosomes and tRNA were found to be thermally stable [1]. In the tRNA this occurred because of increased numbers of G + C base pairs and their tight packing in one region of the molecule. As the temperature increased there was also increased thiolation (addition of sulphide bridges) of a ribothymidine base in the tRNA. In other organisms, it has been found that the higher the optimal temperature for growth, the higher the thermal denaturation temperature of the DNA [11].

The cell membranes also must remain fluid and functional at high temperatures. Changes in fatty acids may contribute. It has been found that cell membranes of *Thermus aquaticus* have more branched, more saturated and longer-chain fatty acids at higher temperatures than at low temperatures [7]. In the same experiment, a change in temperature of growth from 50°C to 75°C produced twofold increases in phospholipids and carotenoids and a fourfold increase in glycolipids. As described in the section on acidophiles, the unique diphytanyl lipids of *Sulfolobus* probably contribute to thermal as well as acid stability. In this organism, the glycolipids are also important and make up 68% of the lipid pool.

There is great debate about the maximum temperature for life. Since micro-organisms can grow in an autoclave at 140–150°C, there appears to be the possibility of life as long as water remains liquid. Organisms able to survive at high temperatures may live in the hot waters coming from vents in the ocean floor [23]. Because of the high pressure in the deep ocean, many hundreds of atmospheres, water remains liquid at temperatures as high as 350°C.

Baross and Deming [4] described bacteria isolated from these thermal vents and grown in the laboratory at high temperatures and high pressures. Electron micrographs showed pleomorphic micro-organisms with very thick cell walls. At 265 atm (26.8 Pa), numbers and protein content increased up to 300°C with doubling times of 8 hours at 150°C and 40 min at 250°C. However, their results have been called artefacts by Trent *et al.* [43], who found similar increases in numbers of particles and of protein in abiotic preparations when the growth media was subjected to 250°C and 265 atm.

3.4.8 BAROPHILES

At the average depth of the ocean, 3800 m, the pressure is 380 atm (38.4 Pa). The deepest depths are about 12,000 m [22, 23, 24]. Not only are there high pressures in the deep ocean, but there is no light, temperatures are less than 5°C, and nutrient concentrations are low. Micro-organisms there must be psychrophilic, oligotrophic and barophilic.

Barophilic organisms can grow at 600 or more atmospheres (60.6+ Pa). An obligate barophile grows better at 600 or more atmospheres than at 1–200 atm (0.1–20.2 Pa). Definitions are summarized in Fig. 3.4.9.

The first studies of barobiology used lab organisms, such as *E. coli*, and looked for physiological effects of increased pressures. Other studies used organisms from the surface ocean and incubated them at depth or brought deep water organisms up to the surface for tests. A much more realistic approach was pioneered by Jannasch *et al.* in 1976 [25]; they made a device to collect water samples from 6000 m and keep them under pressure while they were incubated with radiolabelled substrates. They found that the natural microbial population from 6000 m (600

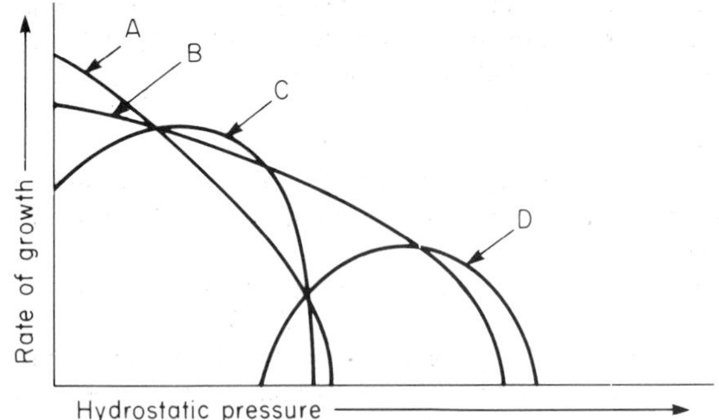

Fig. 3.4.9. The effect of hydrostatic pressure on microbial growth as it relates to the types of microbes isolated from deep-sea environments. A and B are barotolerant species; C is a barophile which has a growth response greater at high pressure than at 1 atmosphere; D is an obligate barophile which grows better at high pressure and not at all at 1 atmosphere. The shapes of the curves are arbitrary and for illustrative purposes. After Jannasch & Taylor [24].

atm or 60.6 Pa) metabolized glutamate and amino acids more slowly when incubated at 600 atm than at 1 atm and concluded that the organisms were barotolerant rather than barophilic.

A true barophile has been isolated from a decaying amphipod held at 580 atm (58.6 Pa) and 2–4°C for 5 months [51]. This spirillum will grow at 1 atm, where its generation time is 86 hours, but it has its optimum growth between 200 and 750 atm (20.2–75.7 Pa), where its generation time was 4–13 hours. Since true barophiles have also been isolated after the original water sample was decompressed [24], it appears that decompression is not necessarily fatal. One strain, isolated from 1035 m depth, even fits the description of an obligate barophile as it did not grow at 1, 173 or 346 atm (0.1, 17.5 or 35 Pa) but did grow at 690 atm (69.7 Pa) (generation time of 25 hours) and at 1035 atm (104.5 Pa) (generation time of 33 hours) [52].

The technical difficulties of growing barophiles in the laboratory under high pressures have been so great that little is known about the mechanisms that allow these micro-organisms to exist under such extreme conditions. The survival at hundreds of atmospheres of pressure of micro-organisms grown at 1 atm demonstrates, however, that adaptations to high pressure are likely to be minor compared with the adaptations of organisms existing at high temperatures or at high salt concentrations.

3.4.9 REFERENCES

[1] Amelunxen R.E. & Murdock A.L. (1978) Microbial life at high temperatures: mechanisms and molecular aspects. In *Microbial Life in Extreme Environments*, p.217 (ed. D.J. Kushner). Academic Press, New York.

[2] Ammerman J.W., Fuhrman J.A., Hagstrom A. & Azam F. (1984) Bacterioplankton growth in seawater: I. Growth kinetics and cellular characteristics in seawater cultures. *Mar. Ecol. Prog. Ser.* **18**, 31–9.

[3] Avron M. & Ben-Amotz, A. (1979) Metabolic adaptation of the alga *Dunaliella* to low water activity. In *Strategies of Microbial Life in Extreme Environments*, p.83 (ed. M. Shilo). Verlag-Chemie, Weinheim.

[4] Baross J.A. & Deming J.W. (1983) Growth of 'black smoker' bacteria at temperatures of at least 250°C. *Nature* **303**, 423–6.

[5] Baross J.A. & Morita R.Y. (1978) Microbial life at low temperatures: ecological aspects. In *Microbial Life in Extreme Environment*, p.9 (ed. D.J. Kushner). Academic Press, New York.

[6] Brock T.D. (1969) Microbial growth under extreme conditions. *Symp. Soc. Gen. Microbiol.* **19**, 15–41.

[7] Brock T.D. (1978) *Thermophilic Microorganisms and Life at High Temperature*. Springer-Verlag, New York.

[8] Carlucci A.F., Craven D.B. & Henrichs S.M. (1984) Diel production and microheterotrophic utilization of dissolved free amino acids in waters off southern California. *Appl. Environ. Microbiol.* **48**, 165–70.

[9] Castenholz R.W. (1969) Thermophilic blue-green algae and the thermal environment. *Bact. Revs.* **33**, 476–504.

[10] Castenholz R.W. (1973) The ecology of blue-green algae in hot springs. In *The Biology of Blue-Green Algae*, p.379 (ed. N.G. Carr & B.A. Whitton). Blackwell Scientific Publications, Oxford.

[11] Farrell J. & Campbell L.L. (1970) Thermophilic bacteria and bacteriophages. *Adv. Microbiol. Physiol.* **3**, 83–109.

[12] Friedman S.M. (1968) Protein-synthesizing machinery of thermophilic bacteria. *Bacteriol. Revs.* 32, 27–38.

[13] Friedmann E.I. (1982) Endolithic microorganisms in the Antarctic cold desert. *Science* **215**, 1045–53.

[14] Friedmann E.I. & Ocampo-Friedmann R. (1984) Endolithic microorganisms in extreme dry environments: analysis of a lithobiontic microbial habitat. In *Current Perspectives in Microbial Ecology*, p.177 (ed. M.J. Klug & C.A. Reddy). American Society for Microbiology, Washington, DC.

[15] Hagström A., Ammerman J.W., Henrichs S. & Azam F. (1984) Bacterioplankton growth in seawater: II. Organic matter utilization during steady-state growth in seawater cultures. *Mar. Ecol. Prog. Ser.* **18**, 41–8.

[16] Heinrich M.R. (ed.) (1976) *Extreme Environments: Mechanisms of Microbial Adaptations*. Academic Press, New York, NY.

[17] Hirsch P. (1979) Life under conditions of low nutrient concentrations. In *Strategies of Microbial Life in Extreme Environments*, p.357 (ed. M. Shilo) Verlag Chemie, Weinheim.

[18] Hirt W.E. & Vestal J.R. (1975) Physical and chemical studies of *Thiobacillus ferrooxidans* lipopolysaccharides. *J. Bacteriol.* **123**, 642–50.

[19] Hoppe H.-G. (1976) Determination and properties of actively metabolizing heterotrophic bacteria in the sea, investigated by means of microautoradiography. *Mar. Biol.* **36**, 291–302.

[20] Horowitz N.H. (1979) Biological water requirements. In *Strategies of Microbial Life in Extreme Environments,* p.15 (ed. M. Shilo). Verlag Chemie, Weinheim.

[21] Horowitz N.H., Cameron R.E. & Hubbard J.S. (1972) Microbiology of the dry valleys of Antarctica. *Science* **176**, 242–5.

[22] Jaenicke R. (1981) Enzymes under extremes of physical conditions. *Ann. Rev. Biophys. Bioengin.* **10**, 1–67.

[23] Jannasch H.W. (1984) Microbes in the oceanic environment. In *The Microbe 1984: Part II. Prokaryotes and Eukaryotes*, p.97 (ed. D.P. Kelly & N.G. Carr). Cambridge University Press, Cambridge.

[24] Jannasch H.W. & Taylor C.D. (1984) Deep-sea microbiology. *Ann. Rev. Microbiol.* **38**, 487–514.

[25] Jannasch H.W., Wirsen C.O. & Taylor C.D. (1976) Undecompressed microbial populations from the deep sea. *Appl. Environ. Microbiol.* **32**, 360–7.

[26] Krulwich T.A. & Guffanti A.A. (1983) Physiology of acidophilic and alkalophilic bacteria. *Adv. Microbiol. Physiol.* **24**, 173–214.

[27] Kushner D.J. (ed.) (1978) *Microbial Life in Extreme Environments*. Academic Press, New York.

[28] Kushner D.J. (1978) Life in high salt and solute concentrations: halophilic bacteria. In *Microbial Life in Extreme Environment*, p.318 (ed. D.J. Kushner). Academic Press, New York.

[29] Kushner D.J. (1985) The Halobacteriaceae. In *The Bacteria, Vol. VIII: Archaebacteria*, p.171 (ed. C.R. Woese & R.S. Wolfe). Academic Press, Orlando.

[30] Langworthy T.A. (1978) Microbial life in extreme pH values. In *Microbial Life in Extreme Environments*, p.279 (ed. D.J. Kushner). Academic Press, New York, NY.

[31] Lundgren D.G., Vestal J.R. & Tabita F.R. (1972) The microbiology of mine drainage pollution. In *Water Pollution Microbiology*, p.69 (ed. R. Mitchell). John Wiley, New York.

[32] Matin A. & Veldkamp H. (1978) Physiological basis of the selective advantage of a *Spirillum* sp. in a carbon-limited environment. *J. Gen. Microbiol.* **105**, 187–97.

[33] Morita R.Y. (1978) Psychrophilic bacteria. *Bacteriol. Revs.* **39**, 144–67.

[34] Palmisano A.C. & Sullivan C.W. (1982) Physiology of sea ice diatoms. 1) Response of three polar diatoms to a simulated summer–winter transition. *J. Phycol.* **18**, 489–98.

[35] Poindexter J.S. (1981) Oligotrophy: fast and famine existence. *Adv. Microbiol. Ecol.* **5**, 63–89.

[36] Riemann B. & Sondergaard M. (1984) Bacterial growth in relation to phytoplankton primary production and extracellular release of organic carbon. In *Heterotrophic Activity in the Sea*, p.233 (ed. J.E. Hobbie & P.J.leB. Williams). Plenum Press, New York.

[37] Shilo M. (1979) *Strategies of Microbial Life in Extreme Environments*. Verlag Chemie, Weinheim.

[38] Simon M. (1985) Specific uptake rates of amino acids by attached and free-living bacteria in a mesotrophic lake. *Appl. Environ. Microbiol.* **49**, 1254–9.

[39] Smith D.W. (1978) Water relations of microorganisms in nature. In *Microbial Life in Extreme Environments*, p.369 (ed. D.J. Kushner). Academic Press, New York.

[40] Smith D.W. (1982) Extreme natural environments. In *Experimental Microbial Ecology*, p.555 (ed. R.G. Burns & J.H. Slater). Blackwell Scientific Publications, Oxford.

[41] Tabor P.S. & Neihof R.A. (1982) Improved microautoradiographic method to determine individual microorganisms active in substrate uptake in natural waters. *Appl. Environ. Microbiol.* **44**, 945–53.

[42] Tansey M.R. & Brock T.D. (1978) Microbial life at high temperatures: ecological aspects. In *Microbial Life in Extreme Environments*, p.159 (ed. D.J. Kushner). Academic Press, New York.

[43] Trent J.D., Chastain R.A. & Yayanos A.A. (1984) Possible artefactual basis for apparent bacterial growth at 250°C. *Nature* **307**, 737–40.

[44] Turley C. & Lochte K. (1986) Diel changes in the specific growth rate and mean cell volume of natural bacterial communities in two different water masses in the Irish Sea. *Microbiol. Ecol.* **12**, 271–82.

[45] Uydess I.L. & Vishniac W.V. (1976) Electron microscopy of Antarctic soil bacteria. In *Extreme Environments: Mechanisms of Microbial Adaptation*, p.29 (ed. M.R. Heinrich). Academic Press, New York.

[46] Van Gemerden H. & Kuenen J.G. (1984) Strategies for growth and evolution of micro-organisms in oligotrophic habitats. In *Heterotrophic Activity in the Sea*, p.25 (ed. J.E. Hobbie & P.J.leB. Williams). Plenum Press, New York, NY.

[47] Vestal J.R., Lundgren D.G. & Milner K.C. (1973)

Toxic and immunological differences among lipopolysaccharides from *Thiobacillus ferrooxidans* grown autotrophically and heterotrophically. *Can. J. Microbiol.* **19**, 1335–9.

[48] Wetzel R.G. (1983) *Limnology*, 2nd edn. Saunders College Publishing, Philadelphia, PA.

[49] Woese C.R. (1981) Archaebacteria. *Sci. Amer.* **244**, 98–122.

[50] Woese C.R. & Wolfe R.S. (1985) *The Bacteria, Vol. VIII: Archaebacteria*. Academic Press, Orlando.

[51] Yayanos A.A., Dietz A.S. & VanBoxtel R. (1979) Isolation of a deep-sea barophilic bacterium and some of its growth characteristics. *Science* **205**, 808–10.

[52] Yayanos A.A., Dietz A.S. & VanBoxtel R. (1981) Obligately barophilic bacterium from the Mariana Trench. *Proc. Nat. Acad. Sci.* **78**, 5212–15.

3.5 Aerial dispersal and the development of microbial communities

3.5.1 THE ATMOSPHERE

Despite claims for the existence of an aerial plankton, the atmosphere is best considered as one of the principal media for the dispersal of micro-organisms. The gaseous components form the supporting medium; air movement provides the means of dispersal; precipitation aids deposition; radiation, temperature and humidity affect survival; most factors aid liberation of cells into the air; and the framework for our understanding of the behaviour of airborne particles is provided by the science of meteorology.

Structure of the atmosphere

The atmosphere changes discontinuously in physical properties and gaseous composition with height above the earth. It is formed of a series of concentric shells of which the innermost, the *troposphere*, comprises 80% of the mass of the atmosphere and is the most important for microbial dispersal. The troposphere varies in height from 17 km at the equator to 6–8 km at the poles and has a gaseous composition of 78% nitrogen, 21% oxygen and 0.03% carbon dioxide, with small amounts of other inert and pollutant gases and water vapour.

Structure of the troposphere

The troposphere can be subdivided into five zones, although their boundaries are seldom sharp. Fuller descriptions will be found in Gregory [42]. Change occurs most rapidly close to the ground.

The *laminar boundary layer* surrounds the surfaces of the earth and of all objects projecting from it including plant stems and leaves. It includes the still layer of air in contact with the surface and the adjacent layer that flows in streamlines without turbulence. Thickness of this layer varies with wind speed and surface roughness from less than 1 mm in high wind to 10 cm on a cloudy day to 10 m on a still, clear night. Air speed increases linearly with height above the surface; both air speed and temperature can fluctuate violently, with extreme heating by day and rapid cooling at night (see Section 3.5.6). Particles falling into this layer follow trajectories determined by air speed and gravity but are almost always protected from further dispersion. Additional boundary layers may form above vegetation as a whole, perhaps dependent on other boundary layers created by preceding surfaces upwind, giving *complex boundary layers* also referred to as *outer active surfaces* or *crown layers* [86].

A *local eddy layer* may be formed of stationary eddies

behind small roughnesses or in cup-shaped depressions [42].

The *turbulent boundary layer* is formed when eddies in the air flowing over objects projecting through the laminar boundary layer break away and travel downwind. These eddies add vertical and lateral components to the forward horizontal motion of the wind. The likelihood of turbulence can be predicted from the size of the object and the wind speed [42]. Turbulence mixes air much faster than molecular diffusion so that conditions in this layer are more equable with smaller diurnal temperature fluctuations than in the laminar boundary layer. Still, there are differences, and temperature, water vapour concentration and wind speed all change linearly with the logarithm of height. The thickness of this layer increases with wind speed both by thinning the laminar boundary layer and by pushing upwards into the transitional layer. It is thickest (150 m) on hot sunny days and thinnest on clear calm nights.

A *transitional layer* is formed above the turbulent boundary layer, extending to 500–1000 m altitude. In this layer, turbulence decreases with altitude and diurnal changes gradually disappear. The top of the transitional layer is the upper limit to which dust and micro-organisms are carried by turbulence.

In the *convective layer* temperature further declines with height to a minimum of −40 to −80°C with no diurnal variation. No frictional turbulence occurs but particles can reach this zone in convection currents (see Section 3.5.2).

3.5.2 PHYSICAL FEATURES OF THE AERIAL ENVIRONMENT

Air movement in the troposphere

Within the troposphere, air moves constantly in turbulent eddies which vary in size from about 1 cm across up to cyclones and anticyclones hundreds of kilometres across. The kinetic energy for all this movement comes from short-wave solar radiation. As it passes through the troposphere, little of this radiation is absorbed by the air but much is reflected from clouds. Of that which reaches the earth's surface, some is reflected but most is absorbed. Some of this energy is then re-radiated as longer wavelengths which can be absorbed by carbon dioxide and water vapour in the troposphere (18% of the original solar radiant energy).

A parcel of air lifted adiabatically (without adding or removing heat) is reduced in temperature by an average of 6°C for every km it rises. When the temperature gradient with altitude is less than this adiabatic lapse rate, air is stable. However, warming of air close to the ground causes a steeper gradient, the air becomes unstable and a convection current forms when air density becomes less than that of the overlying air that it replaces. Such currents may form vortices or bubbles; in summer one bubble per square kilometre may be released every 6–15 min and carry many micro-organisms. Each bubble can be only a few metres diameter at low altitude but they expand to 300 m across at 300 m and eventually to 2 km across; this gives rise to vertical velocities up to 25 m s^{-1}. Convection ceases at 3–15 km depending on temperature and water content of the air unless a layer of warmer air, a temperature inversion, inhibits upward movement at a lower altitude [42].

Overall, energy received by the earth is balanced by outward radiation. However, due to the angle of incidence of the sun's light more energy is absorbed in the tropics than at the poles. Energy therefore has to be transferred polewards from the tropics, mainly by wind movement. This transfer gives rise to basic wind systems resulting from convection, the earth's rotation and frictional drag but modified by differential heating of land, seasonal effects and mountains. Jet streams, strong narrow high altitude air movements, can also transport air rapidly around the earth and perhaps carry micro-organisms [47].

At night, wind speed tends to be less than by day, the laminar boundary layer thicker and the turbulent layer thinner.

Precipitation

The troposphere always contains enough water to provide a 2.5 cm deep layer over the earth; on average, this amount is evaporated and precipitated every 9 days. The form of precipitation and its intensity depends on concentration of water vapour, air temperature at different altitudes, wind, cloud height and other factors. Condensation starts on small particles, often less than 1 μm and hygroscopic (e.g. sea salt and particles from combustion). Droplets grow by collision until heavy enough to overcome rising air currents and reach the ground without evaporating. Drops may be 0.2–0.5 mm diameter in drizzle and usually 1–2 mm in rain but up to 5 mm from convective clouds.

Other features of the atmosphere of significance to micro-organisms

The water content of the air is usually referred to as

relative humidity, i.e. the ratio of its vapour pressure to that of saturated air at the same temperature. The actual vapour pressure of water in air may vary little over 24 hours but the saturation vapour pressure varies greatly with temperature, giving large changes in relative humidity. Relative humidity is a good indicator of the drying effect of air on plants and micro-organisms and of the availability of water for growth, sporulation and spore liberation. Atmospheric relative humidity may be 10–20% in deserts and 100%, even supersaturated, when fog, dew or frost forms. The lower limit for fungal growth is about 65% but bacteria require more moisture.

Long-wave radiation from the earth's surface leads to cooling, first of the ground then of the adjacent air. Often a temperature inversion is formed with stable conditions. Cold dense air below the inversion cannot mix with that above but can flow down slopes.

Ultraviolet radiation originates from the sun but much is reflected or absorbed before it reaches the earth's surface. However, micro-organisms carried into higher regions of the troposphere may be exposed to damaging doses.

Atmospheric temperatures range from up to about 40°C near the earth's surface to −80°C at the tropopause. Freezing temperatures may occur constantly above 3–5 km.

Radiation, desiccation and temperature all interact in affecting the survival of airborne micro-organisms. The manner of interaction is little understood, but desiccation and freezing may protect organisms from radiation damage.

3.5.3 DISPERSAL OF AIRBORNE PARTICLES

The dispersal of airborne micro-organisms involves: (a) liberation and take-off into the air; (b) dispersion in air currents; and (c) deposition on a surface at the end of the journey prior to germination and growth. A wide range of micro-organisms may be found in the air; many have developed particular mechanisms or adaptations which aid dispersal.

Liberation

Before it can become airborne, a particle has to overcome the adhesive forces attaching it to the surface, cross the laminar boundary layer and enter the turbulent boundary layer from where it can be carried to other parts of the troposphere. Viruses and bacteria are poorly adapted for liberation, while fungi have developed many mechanisms to enable their spores to become readily airborne. Some have tall sporophores which lift their spores well into or through the laminar boundary layer surrounding their substrate, e.g. mushrooms, toadstools and some myxomycetes. Others forcibly project spores into the air, while still others rely on passive mechanisms. The different methods of liberation and take-off have been reviewed by Ingold [59], Lidwell (in [47]) and Gregory [42]. Their characteristics are summarized below and illustrated in Fig. 3.5.1.

1 *Shedding under gravity*. Useful only where the spore is elevated on a stem, leaf or sporophore.

2 *Shedding in convection currents*. Spores of *Botrytis cinerea* and *Monilia sitophila* were carried to the top of 10–12 cm glass cylinders from cultures at the base by a temperature difference of at least 10°C.

3 *Deflation*. For various conidial fungi the minimum wind speed necessary to remove spores lies between 0.4 and 2.0 m s^{-1}; more are released as wind speed and turbulence increase and relative humidity decreases. However, the spores of rusts and smuts are released from lesions on the leaves of crops where wind speeds of even 0.5 m s^{-1} seldom occur [5]. This suggests that mechanical disturbance may be more important. By contrast, myxomycetes with raised fruiting bodies, e.g. *Dictydium* (see Fig. 3.5.1a), and lichens with cup-shaped podetia, e.g. *Cladonia*, may have spores or soredia removed by double eddy systems set up by wind of up to 2 m s^{-1} blowing across the top without any mechanical disturbance.

4 *Mechanical disturbance*. Mechanical disturbance is important to many bacteria and fungi without specialized liberation mechanisms. For instance, *Staphylococcus aureus* is probably chiefly spread on skin scales which become airborne as a result of showering, skin movement, friction with clothing and bed-making. Likewise, spores of actinomycetes and fungi abundant on mouldy fodder become airborne in vast numbers when the substrate is disturbed. In crops, shaking of leaves by wind may generate sufficient acceleration for the liberation of spores of plant pathogens such as *Erysiphe graminis* [5]. Wind may also cause adjacent leaves and stems to knock and release spores.

5 *Mist pick-up*. Minute water droplets in air currents may assist release of the spores of the plant pathogens *Pseudocercosporella herpotrichoides* and *Verticillium albo-atrum* (see Fig. 3.5.1b).

6 *Droplet liberation*. Apart from in rain, droplets can arise in many situations where a water film is broken. For example, talking, coughing and sneezing are all effective droplet producers which spread viruses and bacteria (see

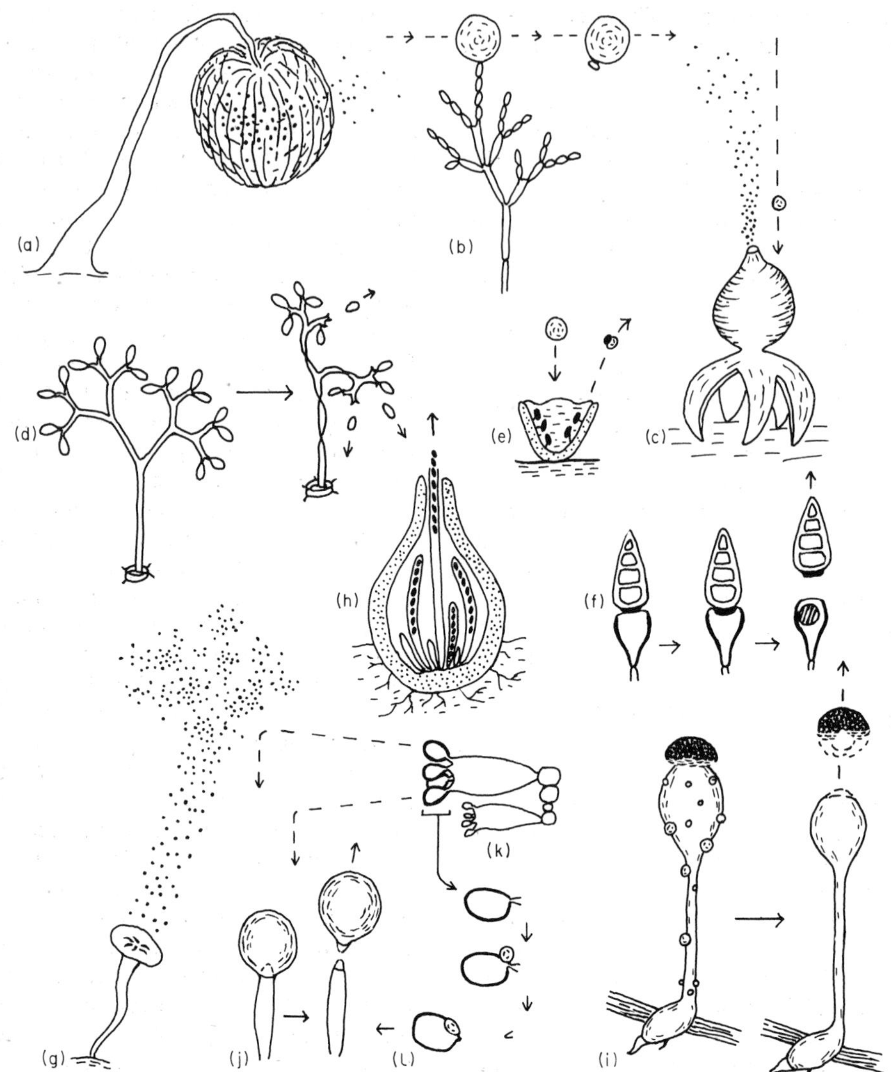

Fig. 3.5.1. Examples of spore liberation mechanisms. After Gregory [43], Gregory & Monteith [47] and Ingold [59]. (a) Deflation from raised fruiting body of *Dictydium* sp. Magnification ×20. Powdery spore mass enclosed within basket. (b) Mist pick-up of *Cladosporium* sp. spores. Magnification ×300. (c) Bellows mechanism of *Geastrum* sp. Magnification ×0.4. (d) Hygroscopic movements of conidiophore of *Peronospora tabacina*. Magnification ×130. (e) Splash cup of *Crucibulum vulgare*. Magnification ×1. (f) Water rupture in *Deightoniella torulosa*. Magnification ×250. (g) Squirt gun, discomycete type, in *Sclerotinia sclerotiorum*. Magnification ×0.7. (h) Squirt gun, pyrenomycete type, in *Sordaria fimicola*. Magnification × 70. (i) Squirting mechanism of *Pilobolus kleinii*. Magnification ×10. (j) Rounding of turgid cell in *Conidiobolus coronatus*. Magnification ×300. (k,l) Ballistospore discharge in *Agaricus campestris* (k, magnification ×370; l, magnification ×840).

also Section 3.5.4). In nature, droplets liberated by wind-blown waves or from waterfalls may carry bacteria and algae.

7 *Bellows mechanism.* The peridium covering the spore mass of Gasteromycetes like *Lycoperdon* (puff balls) and *Geastrum* (earth stars) is thin, flexible and unwettable. When struck by raindrops or drips from vegetation, it is depressed momentarily, which forces out a puff of air bearing millions of spores (see Fig. 3.5.1c).

8 *Rain tap and puff.* Rain falling at terminal velocity hitting a dry stem spreads radially over the surface at high speed, initially about 70 m s^{-1}. Within 2 ms it can spread over 2 cm. The rapidly moving water is preceded by a puff of air that disturbs the laminar boundary layer. The vibration as the raindrop strikes the stem, followed rapidly by the puff, transfers spores of fungi such as *Puccinia* into the turbulent layer [56].

9 *Hygroscopic movements.* Violent movements in response to rapid changes of relative humidity is characteristic of conidiophores of species of Phycomycetes, such as *Peronospora tabacina* and *Phytophthora infestans* (late blight of potatoes), and of elaters of Myxomycetes. These changes of humidity are most marked during the early morning as dew dries off plant foliage (see Fig. 3.5.1d). Recently, violent spore discharge has been observed in *Peronospora destructor* and an electrostatic mechanism proposed [72].

10 *Rain splash.* Some bacteria are embedded in slimes and many fungi imperfecti and some ascomycetes produce their spores in sticky, slimy masses (e.g. *Colletotrichum lindemuthianum*, cause of bean anthracnose). Other species of fungi, e.g. *Venturia inaequalis* (apple scab), have non-slimy spores which are still resistant to wind blowing but easily dispersed by rainsplash. Other species are easily spread by rain although they are also liberated without it, e.g. *Phytophthora infestans.*

The example of a raindrop falling on to a surface covered by a thin film of liquid is most relevant to dispersal of micro-organisms on plants [42]. The drop first pushes the film outwards until it becomes deflected upwards, moulded by surface tension into a crater. This breaks up into jets and rays of droplets at its periphery before it subsides. The maximum development of these jets occurs only 0.0035 s after impact and the original drop is dispersed over the surface of the crater. If the drop falls into deep liquid, the process is similar, but a pillar of liquid rises at the centre. The total number of droplets produced and the number carrying spores increases as the size and velocity of the incident drop increases and as the thickness of the film decreases. One splash may produce from 100 to 5000 droplets which total 25–200% of the volume of the incident drop. Droplets range from 5 to 2400 μm, with most < 100 μm. Droplets carrying spores of *Pseudocercosporella herpotrichoides* are mostly 200–400 μm but the greatest number of spores were carried in droplets > 1000 μm diameter. Droplet diameter is related linearly to the square root of the number of spores per droplet. Numbers of spore-carrying droplets and of spores deposited decrease with distance from the source. By contrast, droplets carrying conidia of *Fusarium solani* are mostly deposited 10–30 cm from the splash. Out of doors, wind may help to carry droplets further while the smallest may evaporate and any spores they contain be dispersed by wind [32, 42].

11 *Drip-splash.* Rain or mist on foliage may carry off spores in droplets larger than 5 mm. When a drop falls 31 cm, spores may splash over a 50–75 cm radius.

12 *Splash cup.* Cup shaped structures containing round peridioles are produced by the birds nest fungi (*Cyathus* spp. and *Crucibulum vulgare*) and podetia lined with soredia by the lichen *Cladonia pyxidata.* These utilize the energy of falling rain drops to project their contents about 1 m. However, they follow a definite trajectory and are hardly airborne (see Fig. 3.5.1e).

13 *Water rupture.* This is characteristically a method of spore liberation in pteridophytes and bryophytes, but it is also utilized by some fungi, e.g. *Deightoniella torulosa, Zygosporium oscheoides.* In *D. torulosa*, uneven thickening of the cell wall causes the thinner end wall to be drawn inwards. Rigidity of the side walls places the fluid contents under increasing tension until a break occurs, either because the cohesion between water molecules or their adhesion to the cell wall is overcome. A gas bubble is formed and rapidly enlarges, the distorted cell returns rapidly to its original shape, and the conidium is jerked off (see Fig. 3.5.1f). With *Z. oscheoides* the sporogenous cell is curved, with thickening on the convex side. Evaporation causes increasing curvature until a break occurs and the conidia are catapulted off.

14 *Air gun mechanism.* In *Sphagnum* moss, compression of air below the spore mass by contraction of the drying sporangium wall eventually breaks open the sporangium, ejecting a spore cloud. This mechanism is not known in fungi.

15 *Squirt gun mechanism.* Spore liberation by the bursting of a turgid cell is characteristic of ascomycete fungi [59]. The ascus containing the spores swells at maturity and the spores are shot from 0.2 to 40 cm, usually 1–2 cm, when it finally bursts at the top (see Fig. 3.5.1g–h). The distance travelled tends to increase with spore size.

16 *Squirting mechanism.* In the Phycomycete genera *Pilobolus* and *Basidiobolus*, in *Entomophthora muscae* and in the imperfect fungi *Nigrospora sphaerica* and *Pyricularia oryzae*, the spore mass or individual spores are discharged in a jet of liquid after the turgid cell on which they are borne splits (see Fig. 3.5.1i).

17 *Rounding of turgid cells.* The sudden rounding of flattened walls between two turgid cells causes the cells to spring apart, as in *Conidiobolus coronatus*, projecting the spores 0.5–4 cm (see Fig. 3.5.1j). The aecidiospores of rust fungi (Uredinales) are similarly discharged to a distance of several millimetres.

18 *Ballistospore discharge.* This is characteristic of Basidiomycetes and the Sporobolomycetaceae. The spores form asymmetrically on sterigmata of which are one to six on a cell (the basidium of the basidiomycetes). Immediately before release a drop of liquid forms at the hilum end of the spore. When the drop reaches a certain size, the spore and the drop are shot 0.01–0.02 cm (see Fig. 3.5.1k). No satisfactory mechanistic explanation of this type of spore discharge has yet been made. Although the distance of ballistospore discharge is small, many agarics and polypores have tall fruiting bodies. These allow the basidiospores, after discharge, to drop vertically until they emerge from the fruiting body and are carried away by air currents (see Fig. 3.5.11).

Dispersion

Dispersion of airborne spores can be considered at two levels: the fate of the individual spore and the behaviour of groups or clouds of spores. Both depend on physical characteristics of both spore and atmosphere. Important characteristics of the spore are its size, shape, degree of surface roughness, density and any electrostatic charges. Those of the environment include wind movement, turbulence, air viscosity, layering, convection, wind gradients near the ground and pattern of atmospheric circulation. Being heavier than air, spores tend to settle out of suspension under gravity but are also blown by wind and affected by electrostatic charges. Sedimentation is counteracted by upward motions of the air in turbulence and convection. Turbulence effectively dilutes clouds of spores but their behaviour may be predicted by the theories of eddy diffusion. The distance travelled by a spore or spores results from the interplay of these various factors, subject to the limitations of other meteorological phenomena.

SEDIMENTATION

The relative density of spores is usually about 1.1–1.2, so that they tend to sink in still air. They accelerate from rest until the surface drag caused by the viscosity of the air limits the rate of fall to a constant terminal velocity, often within one spore diameter [14].

The relation between size and terminal velocity of a smooth sphere in a viscous fluid is described by Stokes's Law. This relates terminal velocity to particle radius and density, acceleration due to gravity and the density and viscosity of air. A close approximation is given by $V_t = 0.0121\, r^2$ where V_t = terminal velocity of the sphere (cm s^{-1}) and r = radius (μm) in the range 1–50 μm. The V_t of fungal spores is close to that predicted by Stokes's Law (see Fig. 3.5.2), from 0.003 cm s^{-1} for *Coccidioides immitis* to 2.0–2.8 cm s^{-1} for *Helminthosporium sativum*. The V_t of spores of a particular species varies slightly, just as spore size varies within a characteristic range. Terminal velocity may also be modified by changes in

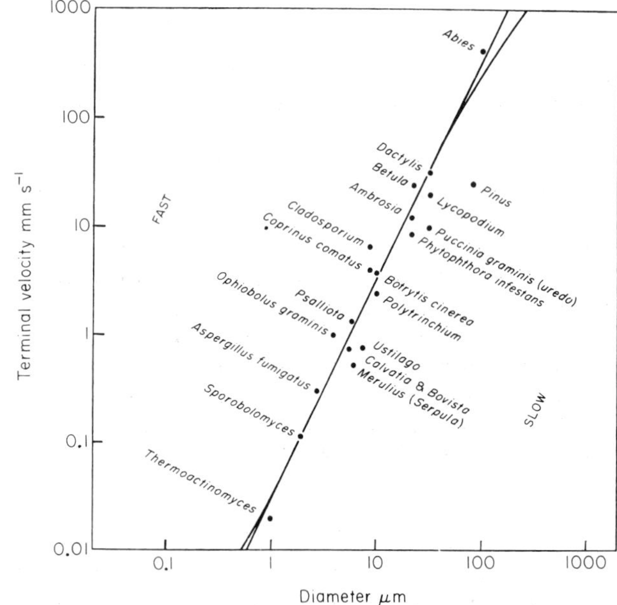

Fig. 3.5.2. Observed terminal velocities of fall of spores and pollen (mm s^{-1}) related to diameter (μm). The straight line represents expected terminal velocity of smooth spheres (density 1.00) from Stokes's Law. Davies's correction is shown at top right and Cunningham's correction at bottom left. Redrawn with permission from Gregory [42].

relative humidity (affecting the degree of hydration of the constituent colloids and changing spore size), by desiccation (the formation of gas bubbles changing the relative density), or surface roughening (increasing drag by creating eddies or even a layer of stationary air).

Non-spherical spores are subject to more drag than spheres of the same volume and relative density. Among suggestions for modifying Stokes's Law to accommodate non-spherical spores [39, 42] has been the use of a *dynamical shape factor* (α) relating drag, and thus actual settling speed (V_a), to that of a sphere of the same volume (V_s) and density, so that $V_a = V_s/\alpha$ [47]. For cylinders with length and breadth equal, $\alpha = 1.06$ while for those with length four times breadth, $\alpha = 1.32$.

Aggregation of particles also results in smaller terminal velocities than expected because drag is increased while trapped air may decrease the bulk density. Four spheres attached in line give $\alpha = 1.54-1.58$. Conversely, bacteria sometimes appear to have anomalously high terminal velocities because they are carried on rafts of skin flakes from animal epidermis.

WIND

In England the mean wind speed at 10 m is about 5 m s^{-1} while at Kew, London, 90% of hourly means exceed 1 m s^{-1} and 50% 3 m s^{-1}. Thus, the wind speed usually exceeds terminal velocity of spores by at least 100 times and often by more than 300 times. Below 10 m wind speeds are much less and at 2 m only 80% of those at 10 m. If there were only laminar flow, the distance travelled by spores would be determined by height of discharge or liberation, terminal velocity and wind speed. However, both turbulence and convection are usually present so that deposition patterns of spores of *Tilletia caries* and *Bovista plumbea* were similar although their diameters were 17 and 5.6 μm and their terminal velocities 1.4 and 0.24 cm s^{-1}, respectively [40, 99].

DIFFUSION OF SPORE CLOUDS

As a cloud of spores moves downwind from its source, it is spread both horizontally and vertically by eddies and diluted in the same way as a plume of smoke from a chimney. Its behaviour may be explained by the theories of eddy diffusion, a complex process that is still not fully understood. Eddies of the same order of magnitude as the spore cloud at that time are most effective in causing diffusion so that spores become more widely dispersed and their standard deviation larger. Diffusion is often greater horizontally than vertically near ground level because of the effects of temperature gradients, atmospheric stability and decreasing turbulence with height. The effects of variations in these factors can be seen in the different types of smoke plume; spore clouds will be affected similarly. For instance, large eddies cause looping of a smoke plume, as when parts of the plume rise in convection currents while other parts are carried down by compensating air movements, and move spore clouds bodily.

The most satisfactory theory to explain diffusion of spore clouds is that of Sutton (see [43]), which relates the standard deviation of spores in a cloud to distance from their source, a diffusion coefficient and a measure of turbulence. However, discrepancies occur over distances greater than 1 km. The diffusion coefficient, like turbulence, decreases with height, so that spore clouds are less widely dispersed as height increases. A different approach, Markov chain simulation, is most suitable to describe dispersal of plant pathogens within crop canopies [11].

For further discussion of eddy diffusion and other theories reference should be made to Gregory [39, 41, 42], Legg (in [11]), Pasquill [81] and Tyldesley (in [47]).

OTHER FACTORS

Spores falling in air have a small electrostatic charge but its significance in dispersion is uncertain [72]. Small particles also tend to move down a temperature gradient, to be repelled by hot and attracted by cold surfaces, and to move away from light.

Deposition

Deposition is the final stage in the airborne dispersal of micro-organisms. The micro-organism is returned to the boundary layer of plant or soil so that it can no longer be blown off by normal winds. Deposition may occur from dry air by sedimentation, boundary layer exchange, turbulent deposition, impaction or Brownian motion as well as in precipitation.

DEPOSITION FROM DRY AIR

Sedimentation of spores at their terminal velocity under gravity is only important in still air or at very low wind speeds. Out of doors, there is usually too much turbulence for it to be significant. Exceptions may occur in dense vegetation where wind speed is decreased or in stable conditions at night when the laminar boundary layer extends to a height of several metres.

Boundary layer exchange occurs when eddies carrying spores break from the turbulent into the laminar layer, remove spore-free air and leave spores behind. These can then sediment under gravity out of reach of other eddies. Turbulence is much more effective than diffusion in carrying spores to the laminar layer and this type of exchange allows even minute spores to sediment.

Turbulent air flowing over horizontal surfaces deposits spores faster than would be expected by sedimentation and at a rate that increases with wind speed. Of spores released at ground level, 90% are deposited by this means within 100 m. Since spores are deposited on both upper and lower sides of horizontal leaves, this cannot result from either impaction or sedimentation.

When wind encounters a twig or leaf, it is diverted around the obstacle. The inertia of airborne particles carries them briefly on towards the obstacle causing some to impact on the surface before they can be deflected to one side. Larger particles carry more inertia than small, and fast-moving particles more than slow-moving. Consequently, impaction is most efficient when large particles are blown at high speed towards small objects. The spores of many leaf pathogens, such as *Phytophthora infestans* and *Puccinia*, are large (30–100 μm) and can impact efficiently on stems and leaves while those of soil fungi, such as *Penicillium* spp., are small (5–10 μm) and must be deposited in other ways. *Ustilago* spores are also too small to impact efficiently on leaves but may impact well on the glumes and stigmas of the grass flowers they infect. Spores may also be trapped efficiently by fine hairs on plant surfaces.

Although impaction may be efficient on some obstacles, spores are not necessarily retained as many may bounce off unless the surface is wet or sticky. Once deposited, spores are not easily resuspended by further blowing.

Dense vegetation efficiently filters spores from an airstream both by impaction on stems and leaves and by slowing the wind so that sedimentation can occur. Some 70% of *Lycopodium* spores released near the top of a wheat crop canopy were deposited on plants within 10 m and 5% on the soil. Only 25% escaped above the crop within 1 m of the point of release and some of these were deposited within 10 m. When released at ground level, 95% of the spores were deposited in 10 m, 40% on the soil [87]. Smaller spores than *Lycopodium* impact with plant stems less efficiently and more readily blow through vegetation. Many fungi have spores about 10 μm diameter, which provides a reasonable compromise between dispersal and deposition in such situations.

Brownian motion, caused by the movement of gas molecules in the air, is important only with particles less than 0.1 μm diameter such as viruses. Deposition by Brownian motion is much less affected by the nature of the surface than deposition of larger particles by impaction.

An electrostatic charge, either positive or negative, occurs on basidiospores, ascospores and other dry spores. The initial charge, probably acquired during liberation, may later be masked by capture of atmospheric ions. The effect of the charge on many spores is likely to be negligible. For instance, *Ganoderma applanatum* spores in the normal fine weather electrical field of 1 volt cm^{-1} over a flat, negatively charged surface would gain only 0.05% of their terminal velocity. However, particles 0.5 μm in diameter could carry charges such that their deposition or suspension were controlled by the earth's field rather than by gravity.

DEPOSITION IN RAIN

Washing of air by rain rapidly terminates the dispersal process. Spores may impact on raindrops, be captured by cloud droplets or even form the nuclei of the drops. The efficiency with which spores will impact on raindrops is a function of the radius of the raindrops and the terminal velocities of both raindrop and particle. Raindrops have diameters up to 5 mm and terminal velocities from 2 to 9 m s^{-1}. The optimum size of spore for deposition in rain varies with the size of the drop. Collection efficiency is greatest for raindrops of about 2 mm diameter or slightly smaller [69, 98] with the optimum nearly the same for all spore sizes. However, efficiency of impaction decreases with decreasing spore size so that small spores, e.g. *Penicillium*, will be collected with a maximum efficiency of 15% by 2 mm drops and not at all by drops smaller than 1 mm. On the other hand, large spores, e.g. *Puccinia* and *Erysiphe*, will be collected by any raindrop with a maximum efficiency of 80% with 2.8 mm diameter drops. Even the spores of *Ustilago perennans*, which are difficult to wet, may be picked up by rain.

The concentration of particles remaining in the air decreases exponentially during washout. For 30 μm particles in air where rain is falling at 2 mm h^{-1}, it is calculated that 60, 36, 12 and 1% remain airborne after periods of 15, 30, 60 and 120 min, respectively, while

72% of 4 μm particles are still airborne after 120 min [39]. However, experimental observations suggest faster deposition [55]. If the drop evaporates before reaching the ground or if spores are only temporarily entrained in its wake, spores may only be moved from one level in the troposphere to another.

Rainfall is about ten times more efficient in depositing spores than sedimentation and impaction. However, on average rain falls for only about one-tenth of each day in lowland Britain, while in other regions it may fall only seasonally. Deposition in rain is most important when a spore cloud has diffused to its maximum height, while impaction of spores occurs mostly close to their source and sedimentation occurs only in calm conditions.

3.5.4 THE AERIAL ENVIRONMENT AND MAN

The aerial environment indoors

A building is effectively a walled-in portion of the environment, differing from that outdoors in patterns of air movement, humidity, temperature, buoyancy and perhaps also in gas composition.

AIR MOVEMENT

Air movement indoors results from ventilation, convection and buoyancy. An airstream entering a room is assumed immediately to mix thoroughly with the existing air, displacing a mixture of fresh and stale air. Thus the term 'one air change' commonly used to describe ventilation rate means the displacement of air equal in volume to the room but made up of equal parts of fresh and stale. It will thus not displace all airborne microbes.

Air flow is usually turbulent within rooms, and close to outlets is accelerated as it converges from all directions.

Ventilation results both from outdoor wind movement and from artificial means and applies air pressure at only a few points in contrast to outdoors. Wind blowing against a building creates a positive pressure and forces air through openings, such as cracks around doors and windows. On the leeward side, negative pressure sucks air out of the building in the same way. Changes in wind direction and force cause the air flow across a room to vary similarly. By contrast, artificial ventilation tends to stabilize the direction of flow [47].

Vertical motion to indoor air flow is provided by convection generated by heaters or by heating or cooling of walls. Walls warmer than room air generate updraughts and cooler walls downdraughts compensated by counter currents at the centre of the room. Since only small areas are heated, convection over heaters contrasts with that outdoors.

Movement of indoor air also results from the presence and movement of occupants. For instance, a person walking across a floor can propel dust particles into the air at up to 90 cm s^{-1} while heat from the body warms the adjacent air and causes convection [47]. The airflow reaches 0.5 m s^{-1} at head level and continues as a plume above the head for about 0.5 m.

Circulation of air within buildings is complex and can be greatly influenced by artificial ventilation or air-conditioning. However, convection currents alone can rapidly mix air through a building. In one experiment, spores of *Amorphotheca (Cladosporium) resinae*, released in a first floor room of a building, were trapped in the fourth floor hall within 5 min. Within 20 min they were found in rooms on the third and fourth floors with open doors and even in a second floor room with closed doors. Spores were still being trapped 2–4 hours later at most sites [19].

SOURCES OF AIRBORNE MICRO-ORGANISMS

Apart from human sources, airborne micro-organisms indoors may enter with air from outside, they may come from wood colonized by fungi such as *Serpula (Merulius) lacrymans* (dry rot) or they may originate from moulding on paintwork, wallpaper and food scraps as a result of condensation and dampness in underventilated homes. In farm buildings, stored fodder and grains can be large sources.

Droplets are produced by many of man's activities where a water film is broken and often in circumstances where they may carry bacteria or other micro-organisms. In homes and factories droplets may be produced by flushing a water closet or the use of air humidifiers; in the laboratory they are produced when pipetting, opening culture bottles and flaming inoculating loops, or from chemostats, homogenizers, centrifuges and mills [88].

BEHAVIOUR OF AIRBORNE MICRO-ORGANISMS

The concentration of airborne micro-organisms in an undisturbed room decreases with time. This die-away results from: (a) exchange with outdoor air; (b) deposition by various processes including sedimentation; and (c) death of the micro-organisms.

Because ventilation removes only a proportion of the microbial load, the number of airborne micro-organisms decreases exponentially in the ratio $1/e^n$ where e = the base of Napierian logarithms and n = number of air changes. Decreases of concentration by other means may also proceed logarithmically.

In the absence of ventilation, circulation of air by thermal convection enables a room to be regarded as a large stirred settling chamber [29] in which the deposition of particles is determined by their terminal velocity. Bourdillon *et al.* [8] integrated all processes contributing to the loss of airborne particles into a single expression, $n = n_0 e^{-Kt}$ in which n_0 and n are initial and final particle concentrations, t is time and K a constant expressing die-away of particles. The K can be subdivided into K_0, the death rate, K_R, the rate of removal by air exchange with outside, and K_s, the rate of sedimentation. Values of K for the die-away of bacteria in a bedroom with open windows can vary between the equivalent of 4.9 and 6.1 air changes h^{-1} [8].

Because ventilation only dilutes airborne microbes indoors and an outlet draws air from all directions it is difficult to contain harmful organisms and remove them from a room once they are dispersed. Use of a hood around the source and exhaust ventilation to an incinerator or filter are necessary to prevent their escape into the turbulent circulation of the room.

The human environment

SOURCES OF MICRO-ORGANISMS

An introduction to the superficial microflora of man may be found elsewhere [16, 94].

Bacteria commonly present on the skin become airborne when skin scales are released following bathing and as a result of friction with towelling, clothes, surgical dressings, bedding, etc. Bed-making can increase the concentration of airborne scales from the usual few thousand m^{-3} to near 400 000 m^{-3}. Although they average 8 μm in diameter they may persist in the air for up to 1 hour.

Micro-organisms are also liberated into the air by talking, coughing and sneezing, all efficient droplet-producing mechanisms as air is forced at high speed over moist surfaces. Droplet size depends on air velocity, fluid viscosity and losses of droplets by deposition in the respiratory passage. The sneeze is the most violent, ejecting about 10^6 droplets (10–100 μm diameter) from saliva at the front of the mouth. Wind speeds through the mouth may be as great as 10 m s^{-1}. The larger droplets do not really become airborne, while the smaller rapidly evaporate often in less then 1 s, leaving 'droplet nuclei' 2–10 μm diameter to disperse. Coughing twice or speaking loudly 4000 words will both produce about 10^4 droplets. Coughing atomizes pharyngeal secretion, which is much more viscous than saliva and so gives larger droplets. The risk of these droplets carrying bacteria or viruses increases as microbial concentration in the mouth increases [47].

THE RESPIRATORY SYSTEM AS A PARTICULATE SAMPLER

Each day the average human inhales some 10 m^3 of air along with all its airborne particles. About 10% of this volume comes from the convective flow of air over the body surface, which has been shown to contain 30–400% more micro-organisms coming from the skin or from ground level than the ambient air [51]. The fate of these particles has concerned doctors since Lister showed that exhaled air contained fewer bacteria than inhaled. That the respiratory system traps these particles is an accident of its function as a gas exchange system for the body. The way it does so, sorting particles into different size grades in different parts of the system, is a function of the anatomy of the respiratory tract and the terminal velocity of the inhaled particles [30].

The respiratory system consists of several clearly defined parts. The nasopharyngeal region includes the nose, with stout hairs guarding the openings against the largest particles, the complex folds of tissue, lined with mucus-producing and ciliated cells, behind and then the pharynx. From the pharynx, air enters the trachea which divides at the bottom into left and right main bronchi. The bronchi divide repeatedly, decreasing in diameter but increasing in total cross-sectional area so that airflow decreases from about 100 cm s^{-1} in the trachea to flows of the order of 10 cm s^{-1} in the tertiary bronchi. The bronchioles, following further subdivision, have diameters less than 1 mm and airflows of the order of 1 cm s^{-1}. To this point, the airways are lined with a moving stream of mucus which carries deposited particles to the throat. At the end of the system are terminal respiratory bronchioles and alveoli. The 10^7 alveoli are about 0.5 μm in diameter with a total surface area of about 30 m^2. In breathing, residual air that cannot be exhaled moves from alveoli to bronchioles and bronchi, mixes with tidal air and its microbial load, and carries it back to the alveoli.

The lung traps particles in a variety of ways, segregating those of different sizes (see Fig. 3.5.3). In the nasopharyngeal and tracheobronchial regions, impaction is most important, aided in the trachea by turbulence. For a given air flow, more particles are trapped in small airways than large because the likelihood of impaction is increased. Sedimentation becomes important for smaller particles in the deeper parts of the lung where the air flow has become slow. The smallest particles move by diffusion and can be deposited in the alveoli by Brownian motion. The nose retains particles larger than 10 μm while those smaller than 5 μm reach the alveoli. Few particles are deposited in the trachea and bronchi unless the mouth is used in breathing to bypass the nose. Optimum alveolar deposition occurs with particles 2–4 μm diameter. Particles about 0.5 μm diameter are deposited least efficiently because they sediment too slowly and are still too large for deposition by Brownian motion. Elongated particles, such as fibres or chains of spores perhaps rotating in the air flow, may be deposited by interception where one end of the particle touches the wall of an airway.

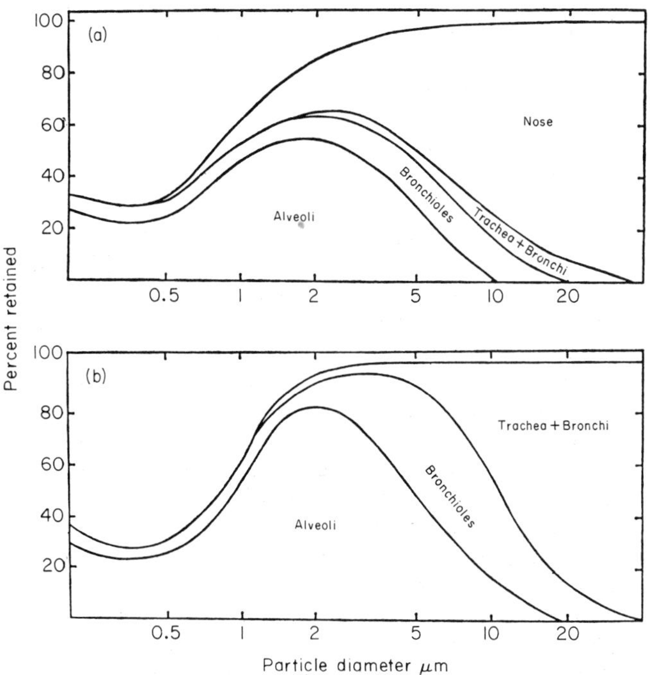

Fig. 3.5.3 Retention of inhaled particles in the respiratory tract, showing relation between particle size and site of deposition following (a) normal nose breathing and (b) mouth breathing. Redrawn with permission from Gregory [42].

PULMONARY DEPOSITION OF MICRO-ORGANISMS AND RESPIRATORY DISEASE

The segregation of airborne particles in the respiratory tract according to size has great relevance to the development of diseases acquired by inhalation. Both infections and allergies can be grouped into those affecting the upper respiratory tract and those affecting the alveolar region. This grouping can often be correlated with particle size. If this cannot be done, mechanisms for transporting the inoculum from site of deposition to site of infection or alternative methods of transmission must be sought.

Nasopharyngeal infections by viruses are associated with transport in coarse droplets liberated from the nose by sneezing. Only droplets larger than 20 μm may be trapped with a handkerchief, while very small droplets contain less inoculum and lose infectivity on drying, probably because the virus is damaged by desiccation. Infective droplets need to be deposited rapidly so that close personal contact is necessary for disease transmission [30, 47].

Most bacterial diseases also result from upper airways infection. As with viruses, some bacteria survive better in large droplets than in small, while others are transmitted on skin scales with a median diameter equivalent to unit density spheres of 12 μm diameter, half within the range 8–16 μm. They are thus ideal for deposition in the nose and throat and few penetrate to the lungs. Bacteria transported on skin scales include *Staphylococcus aureus* and perhaps *Streptococcus pyogenes. Staph. aureus* can flourish on tissue within the anterior orifice of the nose from where it can readily contaminate skin.

Lung infections by *Mycobacterium tuberculosis* and *Bacillus anthracis* appear to be associated with inhalation of single cells and spores, respectively, up to 3 μm long. These readily penetrate to the alveoli where even single cells can cause infection. *Myco. tuberculosis* seems exceptional among pathogenic bacteria in that it seems unable to establish itself in the upper respiratory tract [47].

Myxoviruses may similarly be exceptional among viruses in that they spread effectively, invade the lower respiratory tract and seem able to survive in small droplets.

Two types of hypersensitivity disease caused by airborne particles are generally recognized. Immediate allergic reactions (as in hay fever) affect the upper respiratory tract and cause rhinitis and asthma. Delayed reactions (as in farmer's lung) affect the lower respiratory tract and cause alveolitis and breathlessness. Particles causing immediate allergy include grass pollens up to

30 μm diameter and large fungal spores, e.g. *Alternaria* (many of which would mostly be deposited in the nose), *Cladosporium* and *Didymella* spp. (which can reach the bronchi). By contrast, those implicated in delayed hypersensitivity are mostly smaller than 5 μm, e.g. spores of actinomycetes (1 μm), *Aspergillus* and *Penicillium* spp. (3–5 μm). *Aspergillus fumigatus* may be associated with both types of reaction, perhaps because the spores often remain in groups or because the fungus is growing saprophytically on mucus in the airways.

It should not be assumed that the size of a particle at time of deposition is the same as when liberated or when being dispersed through the air. The drying of small droplets to form 'droplet nuclei' has been described earlier but fungal spores may also become dehydrated during dispersal, which affects their size and modifies the site of their deposition in the respiratory tract. Spores of *Lycoperdon giganteum* increase in size by about 5% when relative humidity is increased from 40% to 100% but their terminal velocity is increased by 50% because gas bubbles disappear [44].

Airborne micro-organisms and occupational respiratory disease

Occupational respiratory diseases are usually associated with the inhalation of such dusts as coal, silica or asbestos fibres. However, occupational diseases can also be caused by airborne micro-organisms. Perhaps the best known is inhalation anthrax or wool sorter's disease occurring in factories handling raw wool and hair, especially from goats and horses. *Bacillus anthracis* was found in the air of one factory, following an outbreak of the disease, with 25–30% of colony-forming units carried on particles less than 5 μm diameter. During an 8-hour shift a worker would have inhaled 600–2000 of these particles. Laboratory-acquired infection should also be recognized as a hazard to the pathologist, microbiologist and medical research worker handling pathogens and even fodders moulded with such fungi as *Aspergillus fumigatus*. The risks and ways to minimize them have been well described [88]. All micro-organisms are potentially hazardous and contaminated glassware should be sterilized before washing up.

Lately there has been increased recognition of the risks of allergy (hypersensitivity) to inhaled particles, particularly during the handling of agricultural crops and their products [62, 66, 67, 88]. Although such diseases were first described in the 18th century, the role of airborne micro-organisms has only recently been recognized.

Agricultural crops, both before harvest and during storage, constitute vast sources of spores. Disturbance during harvesting or after storage of hay or grain can disperse large numbers of these spores into the air so that they are inhaled by workers; these may develop either immediate or delayed allergies, depending on their immunological constitution [88]. Thus, in one survey, 20% of farm workers harvesting grain complained of being affected by grain dust and showed symptoms of both types of allergy [22]. In another survey [38], up to 8% of farmers had symptoms of farmer's lung, the classic form of allergic alveolitis, caused by thermophilic actinomycetes from hay which had been stored damp and had then heated spontaneously as a result of energy released by microbial respiration and perhaps through exothermic chemical reactions (further discussed in Section 3.5.6). Other examples are listed in Table 3.5.1.

Not all occupational allergies are associated with agricultural produce, however, or necessarily with viable micro-organisms. For instance, asthma may be caused by proteolytic enzymes from *Bacillus subtilis* and 'systemic influenza' by aerosols of tuberculin. Farmer's lung-like diseases have been caused by fungi contaminating the *Aspergillus niger* cultures used in citric acid fermentations and by free-living amoebae and perhaps bacteria, fungi and actinomycetes growing in humidifiers within air-conditioning systems.

The role of microbial toxins in occupational respiratory disease is still uncertain. Bacterial endotoxins have been implicated in byssinosis and those produced by *Erwinia herbicola* in respiratory disease of grain elevator workers. Mycotoxins have been suggested as the cause of acute respiratory symptoms in farm workers moving mouldy silage and exposed to massive doses of spores. However, their implication has not been confirmed. Mycotoxins were also suggested as the cause of pulmonary carcinoma in a worker handling peanuts (*Arachis hypogea*) infected with *Aspergillus flavus* and containing aflatoxin, a potent carcinogen.

3.5.5 METHODS IN AEROBIOLOGY

The study of airborne micro-organisms necessitates their separation from the air for microscopic classification and enumeration or for culturing, identification and counting. Many air-sampling instruments have been designed for industrial hygiene and gravimetric, chemical or radioactive determination of the collected dust, which may in some instruments be separated into different size grades.

Table 3.5.1. Occupational allergic diseases and their causes

Allergy	Organisms implicated	Source of dust
Immediate allergy		
Mushroom picker's lung	*Pleurotus ostreatus*	Mushrooms
Grain dust allergies	Various fungi	Cereal grain at harvest and after storage
Enzyme washing-powder allergy	*Bacillus subtilis*	Bacterial enzymes
Delayed allergy		
Farmer's lung	*Faenia rectivirgula*, *Thermoactinomyces vulgaris*	Mouldy hay and grain
Malt worker's lung	*Aspergillus clavatus*, *A. fumigatus*	Malting barley
Mushroom worker's lung	Probably actinomycetes	Mushroom compost
Bagassosis	*Thermoactinomyces sacchari*	Sugar cane bagasse
Suberosis	*Penicillium frequentans*	Mouldy cork
Maple bark pneumonitis	*Cryptostroma corticale*	Maple bark
Cheese washer's lung	*Penicillium casei*	Cheese
Farmer's lung-like syndrome	*Aspergillus, Penicillium* spp.	Contaminated citric acid fermentation
'Humidifier' lung	*Neisseria* and *Acanthomoeba* spp., *Thermoactinomyces* spp., *Cytophaga* spp.	Air conditioning system
Systemic 'influenza'	*Mycobacterium tuberculosis*	Tuberculin aerosol

Most have at best limited application in aerobiology. However, a range of instruments is now available for trapping airborne micro-organisms.

The different techniques utilize a variety of processes for trapping micro-organisms, of which sedimentation, filtration, impingement into liquids and impaction on solid surfaces are probably the most important. The range of instruments available has been reviewed elsewhere [32, 42, 73, 88]. However, some methods have serious limitations to their use. If the air spora, out of doors or in, is to be fully understood, it is important to choose the right method and to appreciate its limitations. For instance, *gravity slides* and *settle plates* have been widely used and are convenient and cheap. However, interpretation is difficult because catches cannot be related to volume of air sampled and rate of deposition varies with particle terminal velocity, wind speed and turbulence, with the edges of both devices giving rise to such aerodynamic effects as edge drift and shadowing and with turbulence causing more spores to be collected than expected from their terminal velocity. Correct sampling rate and alignment of suction traps is essential to ensure air enters the trap to prevent over- or underestimation of larger particles [30], although 'stagnation point' sampling may sometimes be useful [47].

'Bait' plants will often provide a more sensitive indicator of disease spread than any sampling device. These can be exposed in the crop for short periods where they effectively sample a large volume of air for both dry- and splash-dispersed spores, which can develop subsequently when the plants are incubated in conditions suitable for infection.

No one method of air sampling is ideal for all needs. A number of questions need to be answered before one or more are selected for an investigation. For instance:

1 *What are the reasons for the investigation*: to investigate the whole air spora, or a particular group of organisms; to determine health hazards or the presence of microbes in clean areas; or to relate numbers to the occurrence of allergies or spread of disease?

2 *What information is required*: the total number of airborne micro-organisms or the number of viable organisms; the number of cells or the number of particles; the size of airborne particles; changes of concentration with time; the presence and amount of airborne antigen?

3 *What are the characteristics of the organisms*: are they distinct microscopically or is culture necessary for identification; are selective isolation methods available; are the cells dispersed dry or in rain splash; do the cells resist desiccation; are concentrations likely to be large or small; what is the size range of particles?

4. *What is the smallest particle trapped efficiently by the proposed air sampler?*

Methods allowing microscopic assessment are the least selective and most suitable for studies of the total air spora. However, spores of few fungi can be identified to species, although some can be identified to genus. Also, bacteria may be carried on dust particles that obscure them on the slide. Isolation in culture enables specific identification and the use of media or incubation conditions specific for particular groups. However, obligate pathogens do not grow and some fungi either do not spore or produce spore forms in culture different from those that are airborne.

3.5.6 DEVELOPMENT OF MICROBIAL COMMUNITIES

Except, perhaps, in some extremely selective environments, micro-organisms rarely occur alone. Usually they are found in communities composed of both bacteria and fungi, are perhaps preyed upon by protozoa or mites and other arthropods, and may grow on a substrate that is itself alive. Micro-organisms may modify the environment by utilizing nutrients, by making complex nutrients more readily available, by changing pH or by producing metabolites toxic to other microbes or even to the host, or by causing death or hypersensitive responses. Aspects of the study of micro-organisms in their natural environments have been reviewed in Slater *et al.* [95].

The microflora of aerial plant parts

THE ENVIRONMENT

The grass or cereal plant provides an example of the complex of environments on leaf surfaces which differ in nutritional quality, in microclimate and in age from litter and tillers at the base to inflorescence at the top. Under farm management, only seed or cereal crops are allowed to develop to maturity; grass crops are usually grazed or cut for conservation at an earlier stage. In the wild, the stems remain until they collapse to form part of the litter horizon of the soil.

Leaf surfaces differ widely in surface characteristics that determine their suitability for trapping and retaining propagules and for microbial growth. Leaves differ in roughness, as a result of differences in wax deposits, venation and trichomes which affect the depth of the laminar boundary layer and microclimate around the leaf; they also differ in wettability and in nutritional status. The microclimate at the leaf surface may differ greatly from that of other leaves at different levels in the crop and from the atmosphere above the crop. Profiles of some environmental factors are shown in Fig. 3.5.4.

Leaf surface moisture, from rain, dew or guttation, may carry solutes from plant cells, micro-organisms and pollutants; these may either provide nutrients for the

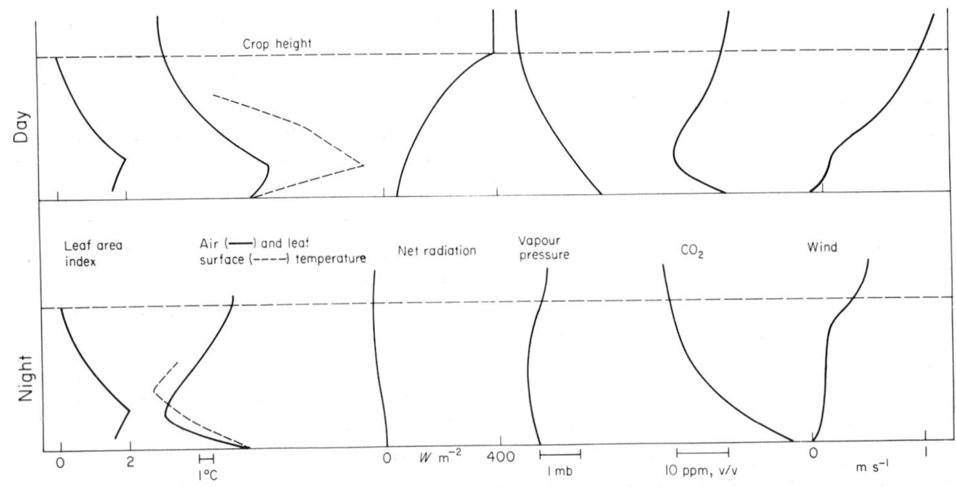

Fig. 3.5.4. Generalized profiles of leaf area index and climatological factors within and above a grass or cereal crop. After Caborn [15], Dickinson & Preece [26] and Preece & Dickinson [84].

superficial microflora or be inhibitory and alter the balance of the microflora. As senescence progresses, there is yellowing from the loss of chlorophyll, as well as loss of protein and an increase in degradative enzyme activity. The loss of nutrients from cells increases until membranes lose their integrity and the tissue becomes dry and brown with water availability controlled by rainfall and atmospheric humidity.

THE MICROFLORA OF LIVING LEAVES

Inoculum for the exposed surfaces of plants is continually provided by deposition of airborne and splash-dispersed micro-organisms. Some of these may be plant pathogens, but most are saprophytes that grow on leaves and litter. Their environment is now described as the phylloplane but our understanding of it is still fragmentary. Even for such a widespread and important family as the Gramineae, a synthesis has to be made from many observations separated in space and time and often referring to only part of the microflora [7, 26, 70, 84]. Methods of study have been reviewed by Apinis [1].

Bacteria are the first micro-organisms to colonize newly emerged leaves and inflorescences and remain numerically the most abundant group. Numbers are greatest close to the meristematic leaf bases while the tips may remain free of bacteria, perhaps reflecting differences in microclimate or the availability of nutrients. The bacterial flora of leaves is distinct from that of the soil. Yellow chromogens usually predominate, but populations of individual species can vary widely. Although often identified with *Erwinia herbicola*, many have been found to be *Xanthomonas campestris* or *Flavobacterium* with *Pseudomonas fluorescens* also abundant [4, 25, 37]. Lactic acid bacteria, important in silage, and actinomycetes seem rare in the phylloplane [100].

Yeasts are next to colonize the phylloplane, particularly the ballistospore, producing pink or mirror yeasts, *Sporobolomyces* spp. Yeast cells most often occur in the furrows between adjacent epidermal cells, particularly close to the leaf tips. However, on barley leaves late in their growth, these may be displaced from the centre of the lamina by *Tilletiopsis minor*. Similarly *Cryptococcus albidus* may replace *Cr. laurentii* on wheat flag leaves in late summer [33].

Last to colonize the phylloplane are filamentous fungi, both saprophytes and plant pathogens, although the latter usually spend only a short period solely within the phylloplane before penetrating the leaf surface and continuing development within the leaf. *Cladosporium* spp. are the earliest and most frequent fungal colonizers. The abundance of other fungi varies between different studies, perhaps because of differences in crop species, weather and location, but often *Alternaria, Verticillium, Acremonium* and *Epicoccum* spp. or, following lodging, *Fusarium* spp. are common [52].

Although the sequence of colonization remains the same, leaves formed in winter and early spring have few micro-organisms of any type until senescence. Later-formed leaves show slowly increasing populations until close to flowering and then a much faster rate of increase until senescence. Species composition may also change with season. In Wales, *Acremonium* spp. can be most abundant in spring, *Cladosporium* spp. in summer and *Phoma* spp. in autumn, although by trapping airborne spores only *Cladosporium* would be detected [63, 71]. In New Zealand, the seasons of *Cladosporium* and *Phoma* coincide [28].

Caution is always necessary when interpreting results of any single method of isolation unless the limitations are well known. For instance, direct plating of grain can indicate the presence of *Alternaria* spp. some 3–4 weeks before dilution plating. This probably indicates growth on or in the grain for a period before sporulation, as with the latent *Alternaria* infections reported on potato leaves [57].

THE MICROFLORA OF DEAD PLANT TISSUE

The initial colonization of dead tissue is a continuation of the succession observed in the phylloplane and has been studied in detail on flowering stems of the grasses *Dactylis glomerata* and *Agropyron repens* [58]. Five groups of organisms were recognized, fruiting successively on the stems (see Fig. 3.5.5).

Group I Phylloplane fungi and other species penetrating the leaf tissue and sporulating abundantly as the leaf dies. These species start at the base, spread upwards through the growing season and persist on the upper internodes. Many species also occur on ripening straw and grain.

Group II Sporulating on basal internodes from late summer.

Group III Fruit first on the bases of standing stems during the first summer after flowering and extending farther up collapsed stem.

Group IV Sporulation starts on lower internodes in early winter and spreads to upper internodes as stems collapse. These are more abundant during summer.

Group V Characteristic of upper nodes during the

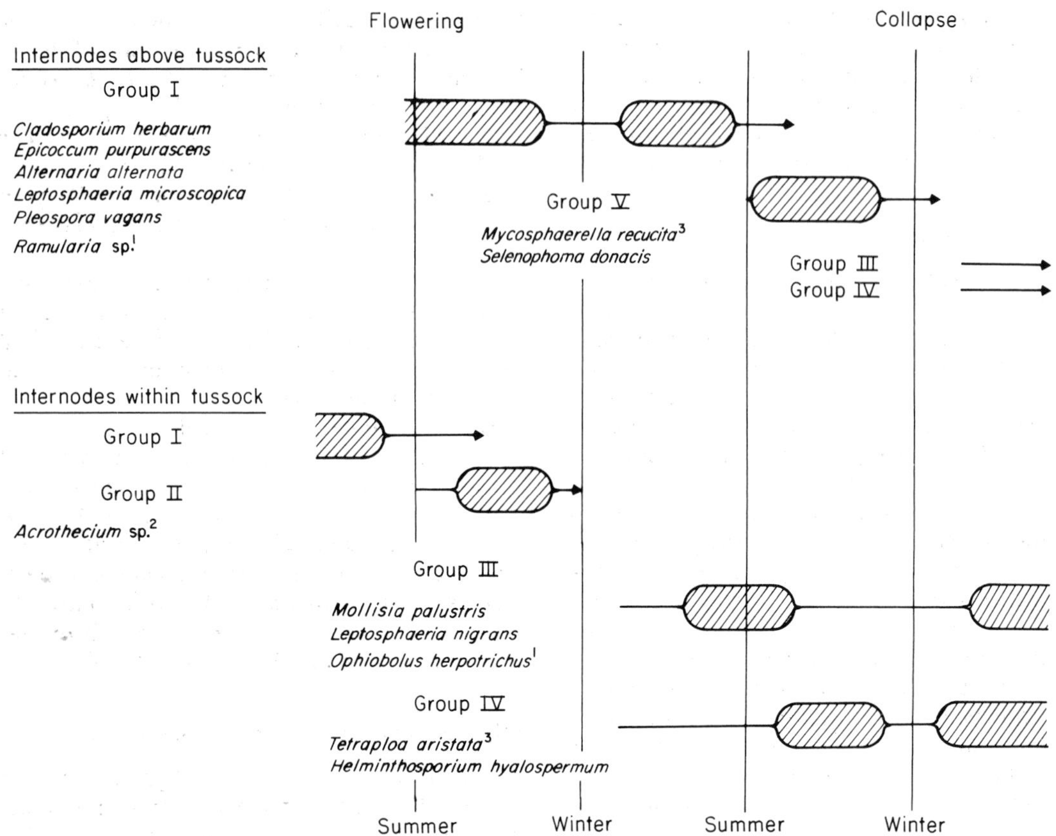

Fig. 3.5.5. Colonization of flowering stems of *Dactylis glomerata* and *Agropyron repens*. After Hudson [57]. 1, on *A. repens* only; 2, on *D. glomerata* only; 3, more abundant on *D. glomerata* than on *A. repens*. Broad lines indicate periods of greatest sporulation.

summer after flowering. Similar successions occur on tropical Gramineae [58, 89].

Order of sporulation does not necessarily equate with succession since species may differ in time taken to produce spores. *Aureobasidium pullulans*, for instance, is found on living leaves but does not fruit until the summer after flowering. Other species may be overlooked and fungi may also persist after spores have disappeared.

Colonization of leaf litter in New Zealand by *Pithomyces chartarum* is important because the toxic spores cause facial eczema of grazing sheep. It resembles Group I above in that it can only colonize tissue not previously utilized by other fungi. It is abundant for only a short period during late summer when temperatures are high, dead leaves are plentiful, rainfall is sufficient and it can overcome competition from *Cladosporium* and *Epicoccum* [12].

The microflora of cut grass

Grass cut for silage and hay may be left in the field to wilt for a few hours or a few days, perhaps with mechanical damage to the cuticle and stems to release sap and hasten drying. Bacteria rapidly multiply in the sap, especially lactic-acid bacteria of the genera *Streptococcus, Leuconostoc, Lactobacillus* and *Pediococcus*. The total number of bacteria may increase from about 10^6 to 10^8 g^{-1} fresh herbage between cutting and ensiling while lactic-acid bacteria increase from 10^2 to 10^5 g^{-1}.

To dry grass for hay often takes 3 days or more, depending on the weather. Rapid drying prevents further microbial growth and those micro-organisms originally present in the phylloplane decline in numbers. However, wet weather can lead to extensive colonization by *Cladosporium, Aureobasidium, Alternaria* and *Epicoccum* spp., representative of Group I colonizers of dead plant tissue.

The microflora of silage

Ensilage is a controlled fermentation of grass to preserve it for later feeding to farm animals. Objectives are to achieve rapid anaerobiosis and to encourage rapid multiplication of bacteria which ferment sugars to lactic acid and thus decrease pH to about 4. Volatile fatty acids, particularly acetic acid, may be produced from pentoses and citrate and also from sugars during poor fermentations.

Oxygen trapped in the herbage is rapidly utilized by aerobic respiration and the aerobic bacteria predominant at cutting rapidly disappear to be replaced by facultative anaerobes: *Escherichia, Klebsiella, Bacillus, Clostridium, Streptococcus, Leuconostoc,* and *Pediococcus* spp. All multiply rapidly at first but the balance of the microflora soon changes. First coliforms predominate, then *Leuconostoc* and *Streptococcus*, which produce lactic acid at pH 6.5–5, and finally, in good silage, *Lactobacillus* and *Pediococcus*, which are slow to produce acid above pH 5. Bacterial counts decline after the rapid growth phase, although much lactic acid is still being produced. The quantity of acid produced depends on the relative numbers of homofermentative and heterofermentative bacteria present. Although other groups decline as one becomes dominant, there is little evidence of strongly selective competition of antibiotic production. Cessation of growth is probably caused by lack of nutrients rather than amount of acid produced. Further description of the biochemical changes may be found in McDonald and Whittenbury [75] and McDonald [74].

If the silage is unstable through storing too wet or not attaining low enough pH, *Clostridium* spp. may start to grow. *Cl. butyricum* and other species ferment lactate to butyrate, raising the pH and allowing the proteolytic *Cl. bifermentans* and *Cl. sporogenes* to grow. These utilize amino acids, forming ammonia, amines and branched-chain fatty acids. *Clostridium* spp. are more tolerant of lactic acid at high water contents, but can be inhibited by wilting to 30% dry matter before ensiling. *Lactobacillus* spp. are still able to ferment a little down to 70% dry matter.

Clostridium and *Bacillus* spp. are restricted by low temperatures (22°C) but, as temperature is increased, *Bacillus* forms an increasing proportion of the microflora, becoming the only type at 45°C. *B. licheniformis* can sometimes predominate at 40°C and produce lactic and acetic acids, but less efficiently than *Lactobacillus*.

The response of *Lactobacillus* spp. to temperature depends on the predominant species. They sometimes grow as well as other bacteria at 30°C, sometimes best at 40°C.

Aeration of silage stimulates fresh microbial activity. Those silages containing 10^5 yeasts g^{-1} are unstable, deteriorate rapidly and give 10^{12} yeasts g^{-1} after only 2–3 days. Spontaneous heating may also occur with invasion of various fungi.

Ensiled high-moisture cereal grains tend to be less stable than grass silage and with aeration soon develop a fungal succession with increasing spontaneous heating. With little aeration yeasts, particularly *Hyphopichia burtonii* and *Hansenula anomala*, develop. As aeration and heating increase, first *Penicillium roquefortii* and then *Absidia corymbifera, Mucor pusillus, Aspergillus fumigatus, Humicola lanuginosa* and *Faenia rectivirgula* develop, the last with heating to 50–60°C.

Microbial communities in hay and grain

Microbial successions in hay and grain during storage are very similar. Although the succession in hay will be described in detail, reference will be made to grain as necessary.

Phylloplane micro-organisms rapidly decline in hay during storage, to be replaced by typical storage species. These are present at harvest in very small numbers but almost ubiquitously. The microflora that develops in storage depends on water activity modified by interactions with other factors. This process has certain similarities with composting.

Microbial growth in dry substrates is prevented by water activities (a_w) less than 0.65 (−59 MPa), equivalent in hay or grain to a water content of about 13% or in oil seeds of about 8%. Few fungi can grow at 0.7 a_w (−49 MPa) while most bacteria require 0.95 a_w (−7 MPa). Actinomycetes are more tolerant of dryness and grow in hay down to 0.90 a_w (−14 MPa).

Growth of micro-organisms in damp hay is accompanied by the output of heat from respiration; initially, perhaps, this is from plant cells but soon predominantly from microbial respiration. The maximum temperature attained by microbial heating is closely related to water availability in the hay. Together, temperature and water activity determine the pattern of colonization by micro-organisms (see Fig. 3.5.6). Heating is limited by drying and by utilization of readily available nutrients. Usually cooling follows, but under certain conditions exothermic chemical oxidation processes occur, raising the temperature above 200°C, and causing spontaneous ignition.

In dry hay with less than 0.93 a_w (−10 MPa = 25% water content), *Eurotium* spp. (anamorphs in the *Aspergillus glaucus* group) and *Wallemia sebi* predominate. These are typical xerophilic fungi with minima for growth of about 0.73 a_w (−43 MPa) and optima of about 0.93 a_w (−10 MPa) [75]. Their growth may be accompanied by spontaneous heating to a maximum of 30–35°C. As a_w and temperature increase they are successively replaced by other species of fungi and bacteria (see Fig. 3.5.7) until, above 0.99 a_w (−1.4 MPa ≡ 35% water content), heating exceeds 50°C and is sufficient for thermophilic micro-organisms to grow. These reach maximum numbers around 1 a_w (0 MPa ≡ 40% water content), with heating to 65°C, and characteristically include the thermotolerant fungus, *Aspergillus fumigatus*, the truly thermophilic fungus, *Humicola lanuginosa* and thermophilic actinomycetes, such as *Faenia rectivirgula*, which causes farmer's lung disease. Also *Bacillus licheniformis* is often numerous in the wettest hays [31].

Heating in damp hays is rapid and those with 40% water may reach maximum temperature within a few days (see Fig. 3.5.7) and then gradually cool to ambient over about 3 weeks. A rapid microbial succession occurs during the initial stages of moulding of such hays (see Fig. 3.5.8). Initially bacteria and yeasts multiply rapidly. *Micrococcus* spp., for instance, increase from 10^6 to 10^9 cells g^{-1} during their 1st day. Bacteria and yeasts are most numerous on the 2nd day and then decline as temperature increases above 40°C. Numbers of propagules decline to the 6th day although *Rhizomucor pusillus* reaches maximum sporulation on the 4th day and *Aspergillus fumigatus* on the 6th with a temperature of 55°C. These are supplemented by increasing numbers of *Paecilomyces variotii* and *Humicola lanuginosa* to at least the 5th day. Few actinomycetes can be detected in the first 3 days when the pH is still below 7.5 but numbers increase rapidly afterwards to give maxima of *Thermoactinomyces* spp. on the 5th day and of *Faenia rectivirgula* on the 7th. *Bacillus licheniformis* forms an increasing proportion of the bacteria isolated after the 5th day. The heavy sporulation of fungi and actinomycetes in these hays results in noticeable dustiness when disturbed. Counts of 10^9 propagules g^{-1} dry weight, perhaps 95% 'actinomycetes + bacteria', are not unusual. Good hay contains less than 5×10^6 propagules g^{-1}, mostly phylloplane or Group I litter ('field' fungi) (see Fig. 3.5.5). Antigenicity against sera from farmer's lung patients is first detectable after about 3 days, coinciding with sporulation of *F. rectivirgula* [65, 67].

Principal component analysis of data from stored grain has shown a similar initial decline of 'field fungi' followed by the development of storage species. However, *Peni-*

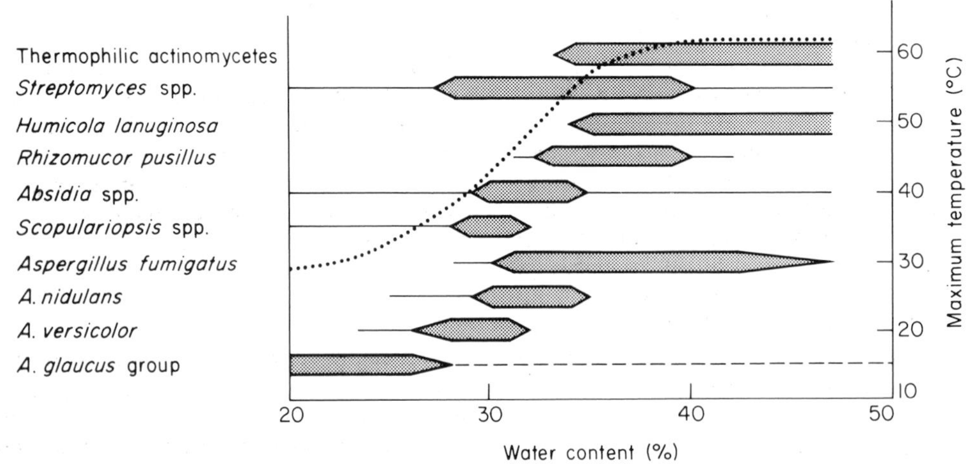

Fig. 3.5.6. Water content, heating and colonization of hay. Broad lines indicate water contents supporting heaviest sporulation; the dotted line indicates maximum temperature attained during spontaneous heating.

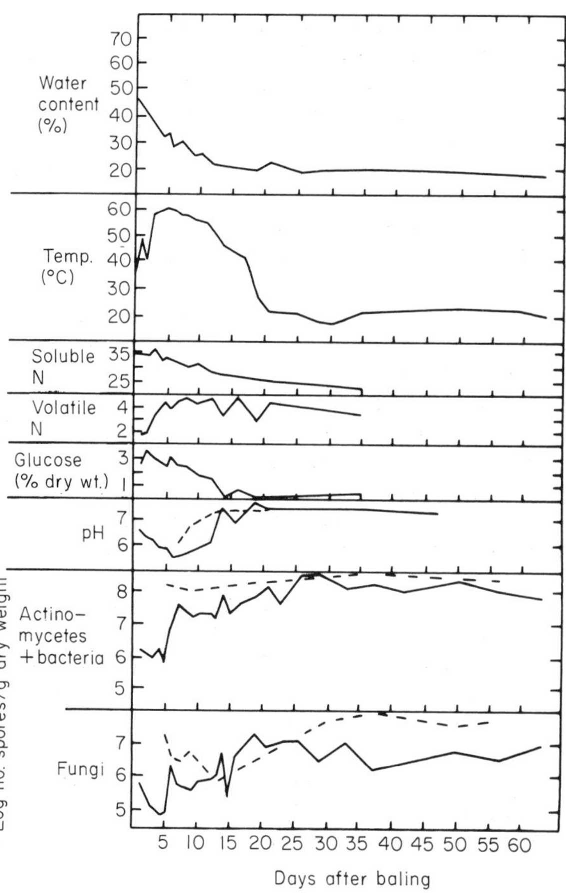

Fig. 3.5.7. Changes in hay baled at 46% water content. Soluble and volatile N are expressed as percentages of the total N content of the hay. Broken lines indicate changes in a faster moulding bale from the stack. After Gregory *et al.* [46].

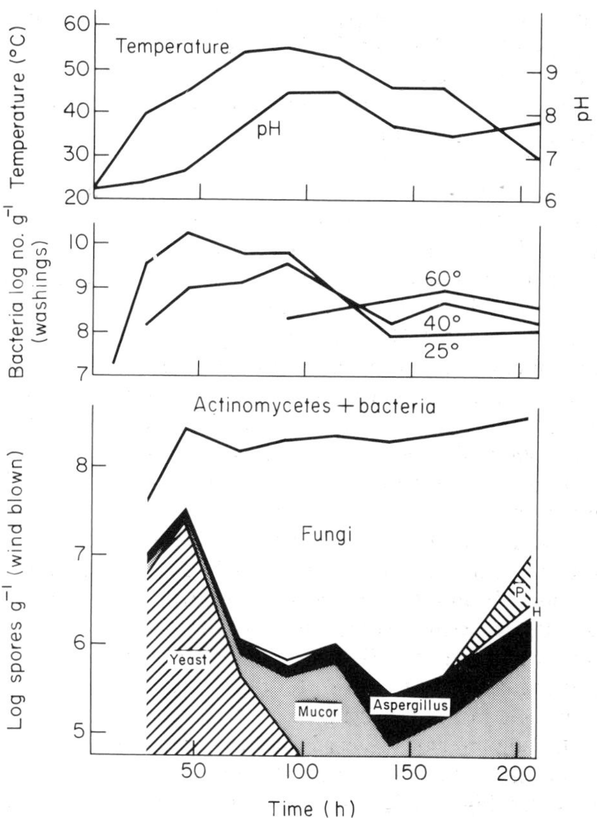

Fig. 3.5.8. Microflora succession in hay of 40% water content. After Festenstein *et al.* [31]. H, *Humicola lanuginosa*; P, *Paecilomyces variotii*.

cillium spp. and *Aspergillus flavus* may be more important in grain than in hay. Similar a_w/temperature relationships may be found in grain fungi as in hay although, in large grain bulks, microbial development may be modified by lack of oxygen or accumulation of carbon dioxide [53, 68, 76, 77, 78, 101]. Close interrelationships between populations of fungi and those of mites and insects may also occur. Mites may feed on fungi while their respiration releases water, favouring further fungal growth.

Factors affecting microbial successions

The pattern of colonization described above illustrates many of the factors affecting microbial successions on different substrates. The way in which successions develop is determined by the nutrient status of the substrate and changes in nutrient availability during colonization; changes in environmental conditions, especially temperature and water availability; changes in pH and gas composition; changes in host susceptibility; and interactions with other micro-organisms and arthropods.

NUTRIENT EFFECTS

Traditionally fungal colonization of plant material has been regarded as a succession from primary sugar fungi with simple nutrient requirements to cellulose decomposers to lignin decomposers. Secondary sugar fungi may occur in the last two phases, utilizing soluble carbon sources released by primary decomposers. In nature this succession is difficult to substantiate as it may reflect time taken to sporulate rather than order of colonization [48, 58]. Much will depend also on the initial nutrient status of the substrate.

In the phyllophane, leachates may be insufficient on the newly emerged leaf to support more than a few bacteria. However, these bacteria and subsequent yeasts may degrade surface waxes and cuticle and increase leaf permeability. At senescence, leaf degradation and loss of membrane integrity may lead to a flush of nutrients increasing fungal growth. Alternatively, production of phytoalexins (inhibitors of fungal growth) declines as the leaf ages, permitting the sequence from bacteria to yeasts and filamentous fungi.

Nutrients may also be provided in the phylloplane by pollen grains and spores. Fungus spore germination and germ-tube growth may be stimulated by pollen grains, perhaps causing the population increase observed at flowering [20, 34]. In contrast, leakage of amino acids and carbohydrates from *Botrytis cinerea* stimulates growth of inhibitory bacteria. Especially in the tropics, phylloplane bacteria may fix nitrogen or produce plant hormones such as indole-acetic acid and gibberellins. A further source of nutrients is animal excreta. For instance, urine favours colonization of litter by *Pithomyces chartarum* [59].

Once easily soluble carbohydrates are utilized in litter, cellulolytic enzymes possessed by many fungi become important but lignin-utilizing basidiomycetes are seldom evident.

Cutting grass for hay and silage also leads to a sudden flush of nutrients as sap is released by cutting and crushing. Bacteria are able to respond most rapidly to this. Only if drying is slow is a similar increase of fungi seen. However, once hay is in storage, yeasts first increase but are soon replaced by fungi and actinomycetes. There is evidence of part of a sugar fungus – cellulose decomposer – lignin decomposer succession. *Rhizomucor pusillus*, and perhaps yeasts, can be considered typical sugar fungi utilizing soluble carbohydrates. *Aspergillus fumigatus* can decompose cellulose, while *Humicola lanuginosa* is best described as a secondary sugar fungus and commensal of *A. fumigatus*, utilizing soluble carbohydrates released from cellulose and hemicellulose. Basidiomycetes in the lignin-decomposer role have not been found in hay but occur in wheat straw compost [17, 18, 50].

Overall nutrient changes in hay baled at 46% water content are shown in Fig. 3.5.7. Glucose was fully utilized within 15 days, perhaps limiting microbial activity and leading to decreased temperatures, although water was also becoming limiting. Soluble nitrogen, from proteolysis by plant enzymes during field drying, was lost more slowly than glucose by deamination forming volatile nitrogenous compounds which increased the pH. Fungal hyphae may themselves provide nutrient for actinomycetes that can decompose chitin [46, 65, 67].

ENVIRONMENTAL EFFECTS

The microclimate around the growing leaf or within the bulk of stored hay or grain exerts a considerable effect on the development of microbial communities. It is evident from Fig. 3.5.4. that conditions vary widely within a crop depending on position within the canopy and time of day or night. A leaf tip near the outer edge of the canopy can experience extremes of temperature, humidity and radiation, while at the base of a tussock conditions are more equable with lower wind speeds and higher humidity. Leaf surface temperatures near the top may vary from 15°C above ambient by day to 5°C below ambient at night, while at the base they may vary only between 1°C and 6°C above ambient. The base of the plant will be more subject to rainsplash from the soil and other vegetation than the top, ensuring inoculation with a variety of micro-organisms. As a consequence of all these factors, basal internodes decompose more rapidly and develop larger microbial populations than the upper parts of the plant.

In stored crops, the microclimate may be more equable but it is none the less subject to change as a result of microbial activity or through human intervention. The way in which water availability can determine the predominant micro-organisms in hay has been described above. However, water availability also determines the extent of microbial activity, which may be measured by the amount of respiration. Respiration releases energy as heat and also releases water from the substrate. Consequently, it may lead to spontaneous heating of the substrate if this heat cannot escape and it may increase water availability. The change in temperature in hay is shown in Figs. 3.5.5 and 3.5.7. Temperatures above 40°C which permit growth of thermophilic fungi and actinomycetes may be inhibiting to many mesophilic fungi. Few are able to grow over the temperature range of *Aspergillus fumigatus* [68].

Respiration also produces carbon dioxide which may accumulate in stored crops to such a level as to be inhibitory to some fungi. In moist grain silos, for instance, as much as 60% of the intergranular atmosphere may be carbon dioxide. Yeasts are most tolerant of high CO_2/low O_2 environments while *Penicillium roquefortii* is one of the most tolerant of the filamentous fungi, growing in the presence of 20% CO_2. High carbon dioxide concentrations inhibit spontaneous heating but heating may recommence with aeration [66, 78, 82].

EFFECTS OF pH CHANGE

Changes in pH resulting from microbial activity may be caused by fermentation of carbohydrates to acids, as in silage, or by the release of volatile nitrogen, as in hay. The pH of cut grass and good hay is 5–6 but in good silage this is decreased to about pH 4 by *Lactobacillus* and other bacteria. At higher pH levels *Clostridium* spp. can grow, first producing butyric acid and later causing proteolysis. In damp hay, the pH is increased to 7–8 which is favourable to actinomycetes. Without fungal growth, actinomycetes will only grow following ammonia treatment. Change in pH may also alter the a_w limits at which storage fungi can grow [76].

CHANGES IN HOST SUSCEPTIBILITY

Changes in the environment may affect not only the microbial population on a plant but also the plant itself, which may affect its susceptibility to fungal colonization. Thus the stress of high temperatures and drought on maize and groundnuts may lead to increased infection by *Aspergillus flavus* and increased contamination with aflatoxin before harvest [54].

INTERACTIONS WITH ARTHROPODS

As in other environments, arthropods may coexist with micro-organisms in the grain ecosystem with a variety of effects. Some beetles may feed on the germ of the grain, rendering it more susceptible to fungal invasion, while other insects and mites feed on the fungi colonizing the grain and thus restrict their development. Mites are favoured by spontaneous heating of damp grain, which also favours fungal development but *Aspergillus restrictus* and *Wallemia sebi* may be pathogenic to some mites. However, some insects can invade grain too dry for fungi but then produce heating and moisture by their respiration which may be sufficient to permit fungal growth subsequently, especially if condensation occurs in the cooler upper layers of the stored grain. Conversely, mycotoxins produced by some fungi may affect insect growth and development.

COMMUNITY DEVELOPMENT

Environments are usually characterized by a diversity of micro-organisms. Some may exert a dominant influence on the community (*dominants*) while others (*associates*) are dependent on the dominants for their survival. In addition, there may be *incidental* micro-organisms which are indifferent to both dominants and associates. Organisms may also be classified according to their activity in the community as primary utilizers, capable of assimilating major carbon and energy sources in the substrate, and secondary organisms which cannot do this but depend on other substrates, perhaps metabolic or lytic products of the primary utilizers.

Communities can be classified according to the biological mechanisms involved into those whose structure is dependent on:

1 Provision of specific nutrients by different members of the community.
2 Alleviation of inhibition by the removal of inhibitory metabolites.
3 Interactions altering individual population growth parameters leading to more competitive and/or more efficient community performance.
4 A combined metabolic activity that is not expressed by any individual population alone.
5 Co-metabolism.
6 Hydrogen transfer reactions.
7 The interaction of several primary species.

Further discussion may be found in Bull and Slater [13].

Colonization of a new substrate depends on the ability of an organism to compete with the existing microflora and its ability to withstand subsequent invasions. A range of microbial interactions have been recognized and have been described in Chapter 2.1. Interactions may not remain constant under all conditions but may be modified by substrate, water availability and temperature [77].

At the start of colonization there are likely to be excess nutrients in the ecosystem, leading to a diversity of species. A succession develops in response to environmental change or as a consequence of invasion by new species. The success of a new species invading a colonized substrate will depend on its own competitive ability, the amount of inoculum and the competitive abilities of the established microflora. The fewer the substrates and the greater the competition, the more inoculum is likely to be needed for successful invasion. Competition is difficult to assess in natural conditions as it is uncertain to what extent antimicrobial substances are produced or whether the whole surface is equally susceptible to colonization. Only 0.1–2% of the living leaf surface is colonized by micro-organisms so that competition for space would not seem to be limiting. Nevertheless, principal component analysis of the variables affecting populations on ripening wheat suggests there is competition and antagonism. One group, *Alternaria, Cochliobolus* and *Acremonium* spp., seem to compete spatially for the same niche, as do members of a second group, *Cladosporium* and *Epicoccum* spp.; antagonism occurs between the two groups. *Alternaria* and *Fusarium* may also be antagonistic [77, 101]. A different interaction has been noted in the case of *Drechslera dictyoides* on ryegrass and *Septoria nodorum* on wheat, which are inhibited by phylloplane bacteria and fungi, respectively [3, 27, 93]. Both bacteria and fungi inhibit spore germination and retard germ-tube growth while germ tubes are lysed by bacteria. By contrast, *Aureobasidium pullulans* can reverse spore stimulation by pollen [70]. Another type of inhibition, that of *S. nodorum* by *Aureobasidium pullulans, Sporobolomyces roseus* and *Cryptococcus laurentii*, results from nutrient competition [36]. Infection of wheat leaves by *S. nodorum* or *Cochliobolus sativus* was stimulated when aphid honey dew was added to inoculum. Simultaneous inoculation with the three yeast-like fungi and *Cladosporium herbarum* could decrease this stimulation if populations of the saprophytes were sufficiently large and eliminate it if inoculated 4–5 days before the pathogens [35].

The phylloplane microflora may be increased by plant pathogens where pustules rupture the epidermis (e.g. *Puccinia* spp.) or where premature senescence and death result. Where growth of the pathogens continues in the necrotic tissue, the subsequent succession may also be modified. By contrast, colonies of *Erysiphe* spp. on leaf surfaces do not cause stimulation.

Microbial communities are much less stable than communities of higher plants. When change occurs in the environment, micro-organisms are much more likely to react by changing their community structure.

MICROBIAL DIVERSITY

Diversity describes the assemblage of species in a community and is a measure of the entropy of the community or the degree of randomness or disorder. An index of diversity measures the degree of uncertainty that an individual picked at random belongs to a particular species in that community. Various indices have been proposed but the most widely used is the Shannon index, which measures both species richness and equitability components of community diversity.

Measurement of diversity gives insight into the ecological functioning of a community, describing the heterogeneity of information contained within that community. This can be measured in terms of species present, physiological characteristics or genetic variation. Diversity has a large influence on the stability of a community. Normally a stable community develops through a series of stages, a succession, to a certain level of diversity in which functional niches in the habitat are filled by member populations of the community. If there is great divergence from this level of diversity in either direction, the community will be subject to continuous or dramatic change. Changes in the environment cause changes in community structure which may lead to changes in diversity.

Diversity in microbial populations has chiefly been studied in aquatic environments with reference to algae and protozoa. Identification by morphological criteria aids the study of diversity and as a consequence few studies have been done with bacteria although the development of numerical taxonomic methods could assist. Studies of fungi also appear to be lacking. However, actinomycetes in soil maintain stable populations over long periods unless there is a major environmental change, such as a rain storm. Diversity was less following rain. Populations were modified in plant rhizospheres but the effect varied with species.

Further information will be found in Atlas [2] and in Wicklow and Carroll [101].

NICHES

An individual habitat is seldom uniform. Different parts may differ in their microclimate, nutrient availability and susceptibility to microbial colonization. As a consequence micro-organisms are seldom uniformly distributed within the habitat but occupy particular niches. A *niche* may be defined as a multidimensional space with axes representing all the environmental variables affecting a species' survival and reproduction. The *fundamental niche* is determined by the organism's genotype and is the total niche in which it can survive and reproduce. However, the actual niche occupied by the organism is usually limited by competition with other organisms and predation to the *realized niche*. The fundamental niche may also be expanded through symbiosis, e.g. lichens. Laboratory studies of organisms in pure culture relate to the fundamental niche whereas those of mixed cultures and natural habitats refer to the realized niche.

A niche may be studied in relation to each individual environmental parameter, but it must not be forgotten that each interacts with others. For example, in hay and grain, temperature interacts with water activity and also with insect and mite infestation and with nutritional and morphological characteristics of the substrate. Thus, *Penicillium* species characteristically colonize grain at low temperatures and high water activities and often preferentially colonize the germ part of the grain. Interactions between micro-organisms are likely to be most intense where the fundamental niches of two or more species overlap. If the interactions are weak and each can utilize the substrate under the existing environmental conditions equally well, a number of species may coexist. However, if interactions are strong, the system may support few species and one may become dominant. Often species dominate closer to the limits for their growth than to the optimum conditions. For instance, *Eurotium* spp. predominate in hay and grain where the a_w is less than 0.93, the optimum for their growth.

Further discussion of niches in relation to fungal communities may be found in Wicklow and Carroll [101].

Effects of microbial colonization on plants and animals

PLANTS

The deleterious effects of stem and leaf pathogens on growth and yield are well known. However, plant growth may also be affected deleteriously by phylloplane micro-organisms, perhaps because they produce metabolites that hasten senescence or because some are minor pathogens. Broad-spectrum fungicides that decrease this microflora can extend the life of leaves and increase grain yield [52, 92].

ANIMALS AND MAN

Grazing animals ingest many micro-organisms, some of which can produce metabolites that affect the rumen microflora causing a condition called ill-thrift [10]. Some leaf-spotting fungi may cause oestrogenic substances to be produced by leaves of infected legumes, and fungi such as *Fusarium* spp. could perhaps contribute to the oestrogenicity sometimes found in *Lolium perenne* (ryegrass). Other fungal toxins cause more severe disease, e.g. facial eczema of sheep.

Many fungi on grasses and cereals are allergenic, particularly *Cladosporium, Sporobolomyces* and *Alternaria* spp. *Didymella exitialis* seems almost ubiquitous on ripening barley crops and its ascospores may be an important cause of late-summer asthma in England [49]. Dust from combine harvesters, mostly fungal spores, can also cause respiratory symptoms [22].

Storage moulds can cause disease in man and animals by infection, by provoking allergy and by poisoning with toxic metabolites. *A. fumigatus* is well known as a cause of infections like pulmonary aspergillosis, particularly of young poultry, and mycotic abortion of cattle. Other causes of mycotic abortion include *Absidia corymbifera, Rhizomucor pusillus* and *Mortierella wolfii*, sometimes found in rotting silage. Thermophilic actinomycetes, especially *Faenia rectivirgula*, can cause farmer's lung, a form of allergic alveolitis, while many of the fungi in hay and grain can cause immediate allergy. Toxins which affect the kidneys may be produced by *Penicillium viridicatum*, and toxins of *Fusarium* spp. have oestrogenic effects and cause feed refusal, emesis and possibly haemorrhagic disease. Although all toxins may cause acute disease, perhaps of greater importance are the subclinical effects caused by intermittent small doses that could affect rumen microflora and immunity to other diseases. Ill effects of fungi can also result from depletion of nutrients or impaired digestibility. Carbohydrate levels, protein levels and digestibility, and micro-nutrient and vitamin levels may all be affected [64].

Development of the outdoor air spora

CHARACTERISTICS

Fungal spores are almost always present in the air but their numbers and types vary with time of day, weather, season, vegetation and geographical location. The variety of airborne spores and abundance of many types was not realized before the introduction of the automatic volumetric spore trap because of the limitations of settle plates and slides (see Section 3.5.5). Many of these spores are illustrated by Gregory [42].

On a fine summer day in rural England, *Cladosporium* spores predominate in the air with peak concentrations of 10^5 spores m^{-3} [45, 55]. Also present may be spores of *Alternaria, Epicoccum, Erysiphe, Puccinia* and *Ustilago* spp. At night, these are replaced by spores of *Sporobolomyces* (up to 10^6 m^{-3}), *Tilletiopsis*, basidiomycetes and ascomycetes. Rain can have a marked effect on the air spora, depending on its intensity and persistence. At first, *Cladosporium, Puccinia* and similar dry-spored fungi increase sharply as a result of 'tap and puff' (see Section 3.5.3). As rain continues, these are washed out of the air and replaced by abundant ascospores. Splash-dispersed spores may also be present although these are not easily trapped and many hardly become airborne since the drops follow definite trajectories.

Fungal spores are not the only microbes in outdoor air. Bacteria and actinomycetes have been detected by culture methods, such as the Andersen sampler (see Section 3.5.5), while algal cells, lichen soredia, spores of liverworts, mosses and ferns, pollen grains of angiosperms and gymnosperms and even amoeboid cysts can be seen on trap slides.

DIURNAL PERIODICITY

As might be expected from the characteristic day and night populations of airborne spores and the different methods of spore liberation, the time when different types are released varies. Five patterns have been described (see Fig. 3.5.9):

Daytime release

1 Post-dawn: liberation mechanisms dependent on decreasing or rapidly changing relative humidity, e.g. *Phytophthora infestans, Deightoniella toruloides*.

2 Middle day: associated with the warmest period when wind speed and turbulence are greatest, e.g. *Cladosporium, Alternaria, Ustilago, Erysiphe* spp.

3 Double peak: possibly resulting from the removal of spores by convection near midday faster than they are produced, e.g. *Helminthosporium, Curvularia, Tetraploa, Cladosporium* spp. and bacteria.

Night-time release

4 Post-dusk: maximum in late evening, reasons uncertain, e.g. *Ustilaginoidea virens, Tilletia* spp.

5 Night: water required for release, e.g. *Sporobolomyces* spp., ascospores and basidiospores.

Different studies show slight variations in pattern or timing of maxima of particular species, e.g. *Cladosporium* spp., which perhaps reflect local weather conditions or distance from spore sources.

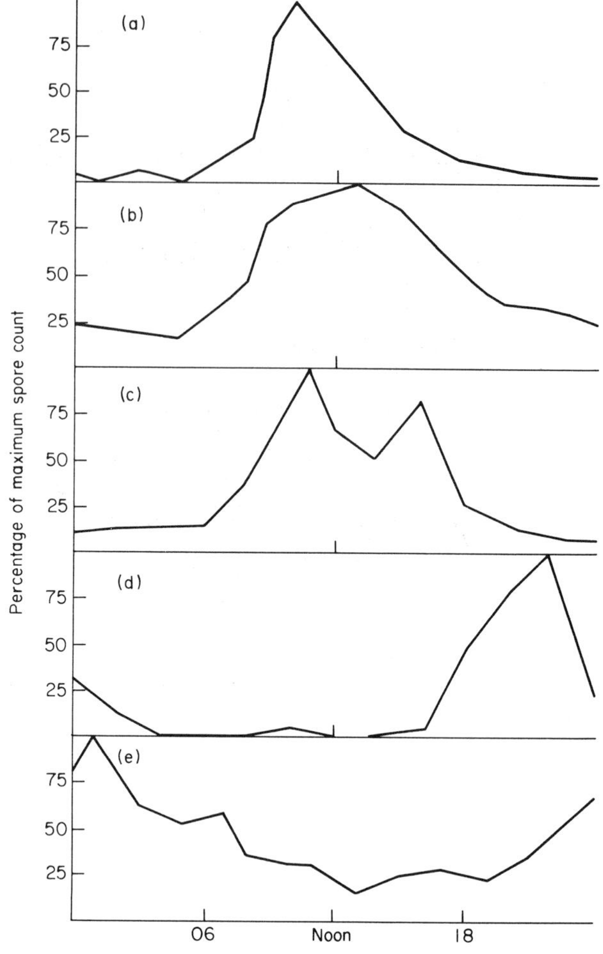

Fig. 3.5.9. Mean diurnal periodicities of airborne fungus spores illustrating different patterns of release. After Hirst [54], Shenoi & Ramalingam [90] and Sreeramulu & Ramalingam [97]. (a) Early-morning pattern: *Phytophthora infestans*; (b) midday pattern: *Cladosporium* spp.; (c) double-peak pattern: *Tetraploa* spp.; (d) post-dusk pattern: tetraspore type; and (e) night pattern: fusiform ascospores.

SEASONAL VARIATION

Usually, airborne spores are fewest in temperate regions during winter and spring and most abundant in summer when *Cladosporium* and *Sporobolomyces* are most numerous. In tropical areas, spores are often most abundant in air in the wet season with *Cladosporium* and basidiospores most abundant [97]. Many other micro-organisms, in particular the plant pathogens, occur in distinct seasons. For instance, in England *Erysiphe* spores occur in early summer when the disease is most abundant on cereals; most *Ustilago* are found during the flowering period of their grass hosts and *Phytophthora infestans* occurs in August or September when late-blight epidemics occur on potatoes (see Fig. 3.5.10a). *Alternaria* spp. are most numerous at harvest time when a second peak of *Cladosporium* spp. may occur. Ascospores of *Didymella exitialis* also occur close to harvest, disappearing abruptly afterwards, while many basidiospores have autumn maxima. In India, occurrence of *Nigrospora, Pyricularia* and other spore types is correlated with the growth stage of the rice crop, although they are usually most abundant with the wet season crop (see Fig. 3.5.10b) [97].

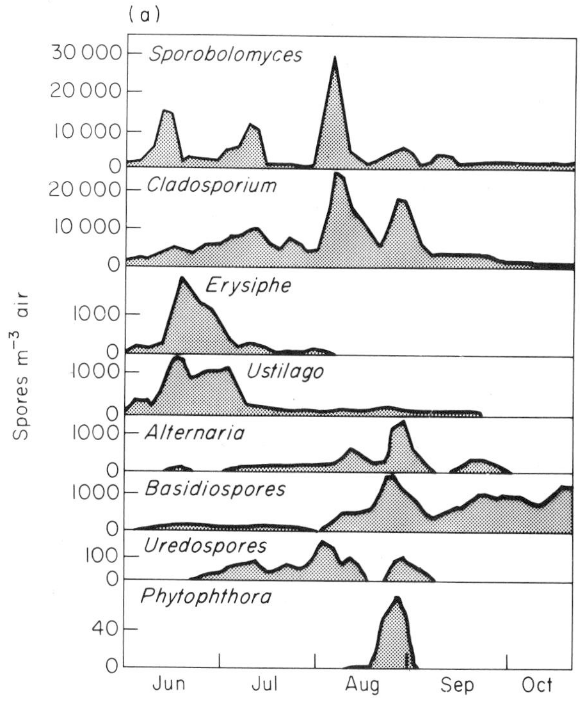

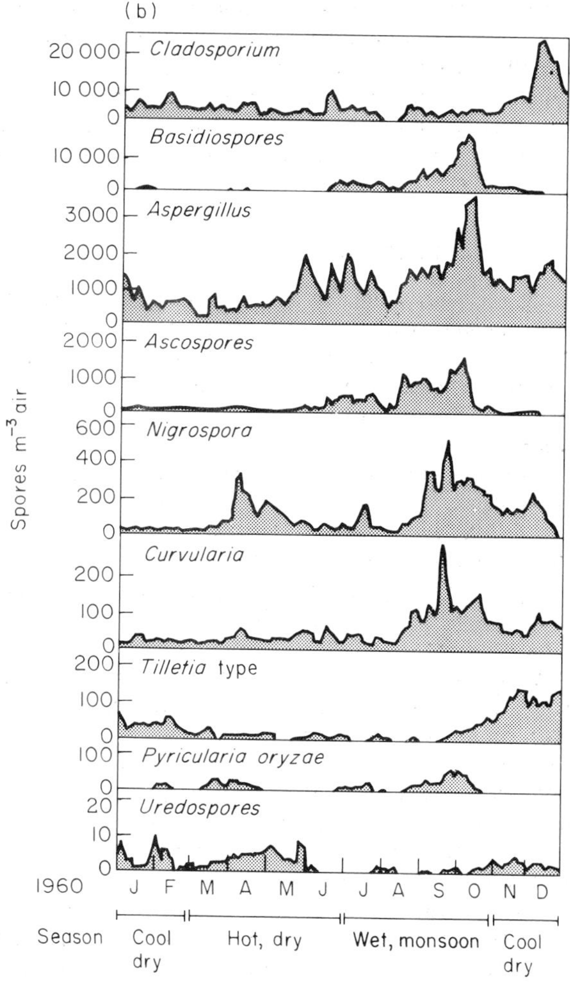

Fig. 3.5.10. Seasonal periodicities of some spores in (a) England; after Gregory & Hirst [45]; (b) India; after Sreeramulu & Ramalingam [97].

GEOGRAPHICAL LOCATION

The predominant spore types are remarkably widespread in their distribution. *Cladosporium* is the most abundant spore type over the whole year in temperate and most tropical regions, although perhaps exceeded by other spore types in some seasons of the year and by *Alternaria* spp. in warm, dry regions. Local spore sources may, however, dominate the air spora nearby.

Pastures and cereal crops provide vast sources of *Cladosporium, Sporobolomyces, Alternaria, Erysiphe, Puccinia, Ustilago* spp. and other spores, but near woods basidiospores of *Ganoderma applanatum* may be as numerous as *Cladosporium*. In tropical regions, the air spora is supplemented by many fungi uncommon or absent in temperate areas, such as *Nigrospora, Curvularia, Pyricularia oryzae, Deightonella torulosa* and *Spegazzinia* spp., while in arctic regions ascomycetes and basidiomycetes predominate [21].

SPREAD OF AIRBORNE MICRO-ORGANISMS

The spread of fungal spores has important implications for the epidemiology and transmission of plant diseases and has been studied widely. Most studies have been limited to short-distance spread because of the difficulties of finding and recognizing small numbers of spores at a long distance from their source.

Once spores are airborne the spore cloud diffuses through the atmosphere while losing spores by deposition on to vegetation and to the ground by impaction and sedimentation. Most large fungal spores are deposited within 100 m of their source, but perhaps 5–10% are carried away from the ground to travel long distances unless washed out of the air by rain. Small organisms are carried further and higher than large, so that although bacteria and fungi may each contribute 50% of the catch at 700 m altitude, bacteria form 90% above 1600 m.

The short distance spread of fungus spores can be demonstrated graphically by the spread of disease from a focus, as in potato late blight. A primary stem infection gives rise to new infections on neighbouring plants, most usually occurring on those closest to the source plant. This develops into a characteristic dispersal gradient with numbers of lesions decreasing rapidly at first, then more slowly with increasing distance from the source; this pattern corresponds closely to the expected deposition of spores under normal turbulence conditions. For instance, in a potato crop before secondary spread of disease there may be about 300 lesions (100 leaves)$^{-1}$ at 30 m from the source, 100 at 60 m, 30 at 90 m and 10 at 120 m. Power law and exponential models have been proposed to fit such data. One such model is $\log y = \log a + b \log x$ where y is the amount of disease, x the distance from the source, a is a constant and b the regression coefficient which is usually negative. With secondary spread, the gradient becomes flatter and if spores come from a distant source there will be no gradient ($b = 0$). Gradients are usually steeper from a point source than from a larger source and from ground level sources than from above ground sources [41, 42].

Within a crop, impaction also helps to remove spores from the air, causing steeper gradients, but spores do not always behave predictably. Although data for *Lycopodium* spores fitted a model for predicting spore deposition within cereal crops, *Erysiphe* spores were deposited much more rapidly than expected, probably because spores stuck together in clumps and fell more rapidly [87].

Spores of plant pathogens sometimes give information on long-distance dispersal when they are trapped before disease is present in local crops. For instance, spores of *Puccinia graminis* (stem rust) originating in Europe were trapped in south Wales before the disease occurred on wheat there. Spores were most numerous at 900 m altitude, indicating long travel over rust-free regions where spores were deposited from the base of the cloud without replenishment [47].

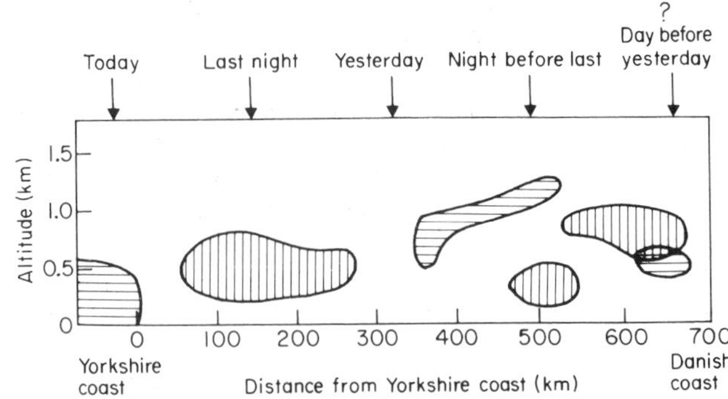

Fig. 3.5.11. Positions of peak concentrations of *Cladosporium* spp. spores (horizontal hatching) and damp air spore types (vertical hatching) at various altitudes over the North Sea downwind of the English coast, 16 July 1964. Interpreted as windborne remnants of alternating day and night spore clouds liberated from the English land surface. After Gregory [43] and Gregory & Monteith [47].

Other evidence of long-distance dispersal comes from the study of spore clouds over the North Sea which originated in Britain. On average, westerly winds blow across the British Isles in half a day. Trapping spores at different altitudes in a morning flight from Yorkshire to Denmark showed that numbers of *Cladosporium* spores (daytime release) rapidly decreased with distance. After 60 km from land they were replaced by *Sporobolomyces* and ascospores characteristic of the night spora. *Cladosporium* was again common after 250 and 600 km and *Sporobolomyces* after 500 km. The densest areas of these clouds are shown in Fig. 3.5.11. With distance from the English coast, the spore clouds became progressively more dilute [47].

The indoor air spora

As out of doors, the numbers and types of micro-organisms in indoor air depend on whether there are local sources. In the absence of a source, numbers are usually smaller than outside but the types are similar. Indoor sources may include the coughs, sneezes and epidermal scales of the human inhabitants, the air humidification equipment or the fungi growing in condensation or on damp timber. For instance, active fruiting bodies of the dry rot fungus

Serpula lacrymans can give up to 3.6×10^5 spores m^{-3} nearby and up to 10^4 spores m^{-3} elsewhere in a house.

In farm buildings, stored crops can constitute massive spore sources. Moving grain or handling and feeding mouldy hay can release dense clouds of spores into the air. Their numbers and types will depend on the degree of deterioration of the produce but concentrations up to 3×10^9 propagules m^{-3} have been found with up to 98% actinomycete spores and bacteria capable of deep penetration into the lung. Large spore concentrations can also occur at docks, in grain, cotton, jute and saw mills and in cork factories where mouldy material is handled [62]. Spore concentrations exceeding 10^6 m^{-3} are common in these occupational situations, where forms of allergic alveolitis occur.

Epidemiology and forecasting

Epidemiology is the study of the development of disease in populations and how this responds to environmental change and to preventive measures. It is the sum of factors controlling the presence or absence of a disease or pathogen. These are the phytopathological and medical aspects of ecology and their study can aid our understanding of the ecology of saprotrophic microbes. Epidemiology concerns the interaction between the pathogen, its host and the environment and can be applied equally to infectious and to non-infectious diseases, whether or not the cause is known. In medicine the transmission, development and control of diseases have often been known before the cause has been identified, although in plant pathology the reverse is more common and the reaction of the pathogen to environment has been emphasized. The whole pathosystem should be considered, giving as much emphasis to the host and its associated microflora as to the pathogen [85].

Aerobiology is concerned with the dispersal of pathogens and so can make valuable contributions to understanding the epidemiology of human and animal as well as plant diseases. Infectious agents may be detected and quantified in aerosols and their persistence in air and relationship with disease determined [30]. Six types of disease with aerogenic transmission were recognized by Sinnecker [91], depending on the resistance of the infective agent in droplets, droplet nuclei or dust; their ability to infect from secondary aerosols formed when dust was resuspended; the host range; and whether there were alternative means of transmission. Even non-infectious diseases can occur in epidemic proportions as a result of airborne dispersal of the infective agent from local sources. Thus, many people in some desert areas of the USA have antibodies against *Coccidioides immitis*, cause of coccidioidomycosis, which grows in soil and is carried in airborne dust. Also, disturbance of faecal accumulations in which *Histoplasma capsulatum* has grown can cause large outbreaks of histoplasmosis down-wind of blackbird roosts in the USA.

In plant pathology, epidemic spread of disease is well known in such diseases as potato late blight (*Phytophthora infestans*) and rice *Helminthosporium* leaf blight (*Drechslera oryzae*). Sometimes these have been so severe as to cause famine, e.g. Ireland, 1845; West Bengal, 1943. Aerobiological methods have been used widely to determine numbers of spores of plant pathogens in the air, when these are released, weather conditions favouring spore release and infection and spore dispersal gradients over short and long distances. Results may then be correlated with changes in microclimate induced by crop growth and occurrence of lesions on plants in the crop, or on 'bait' plants' rate of foliage destruction and resistant varieties. Different aspects of plant disease epidemiology are discussed in a number of texts [23, 79, 80, 83, 87, 102].

Disease forecasting is an extension of plant disease epidemiology and it has been developed to predict when control measures can be used most economically. For an epidemic to develop there must be a population of susceptible hosts, an inoculum of the pathogen and a suitable environment for infection. So forecasting criteria develop from epidemiological studies and for different diseases may include stage of growth of the crop, levels of inoculum of the pathogen and weather criteria, particularly temperature and humidity.

For potato late blight, the criteria for predicting an outbreak, defined as 0.1% foliage infected, are: relative humidity exceeding 75% for 48 hours or 90% for 11 hours and a temperature of at least 10°C for 48 hours and, except in southwest England and south and west Wales, a starting date of the last week of June, before which forecasts are ignored. This date coincides with when potato plants meet between the rows, considerably modifying the microclimate within the crop so that it is more favourable for sporulation, spread and infection of *Phytophthora infestans* [6, 96].

At present, forecasting criteria only indicate that spread of infections can already have occurred. Thus, sprays applied during the incubation period may prevent new lesions occurring but the initial infections, already protected within the leaf, continue to develop. To prevent initial infections, an ability to predict periods of high humidity or rain further ahead than is presently possible is needed [24].

Forecasting of animal disease is rarely possible. An exception is foot-and-mouth disease virus, where spread can be predicted. Long-distance dispersal depends on: a high output of virus, usually associated with disease in pigs; low dispersion of virus — when surface air is stable and winds light so that there is little atmospheric turbulence; high survival of virus — when relative humidity exceeds 60%; and large numbers of susceptible livestock, especially cattle, exposed to the virus for many hours. Over short distances, spread is predicted by estimating eddy diffusion (see Section 3.5.3) during the period when virus is being emitted in order to calculate the dosage inhaled by exposed cattle [11].

Models in aerobiology and epidemiology

Mathematical models are developed by breaking down processes into separate small systems and defining these and their interrelationships in mathematical terms. In aerobiological systems, they may be used to predict the appearance of a disease or its spread and amount, to evaluate the major factors influencing this spread, or to estimate losses. To achieve a useful and reliable model, many factors have to be taken into account demanding detailed study of the pathogen, of the effects of environmental factors on spore production and their liberation, dispersal, deposition and germination, of the host–pathogen relationship, of the availability of susceptible tissue and of effects of crop growth on microclimate. The model becomes more complex if the fungus produces both anamorph and teleomorph stages on the host with different requirements for sporulation and dispersal. Models may differ greatly in their nature and complexity.

Conceptual models aim to show the sequence of events in the disease cycle, significant factors that affect the system (time scale, crop growth, environmental or seasonal variable) and interrelationships with other systems. They can be used to decide what type of model would meet the objectives of the study and to plan the project leading to its construction.

Empirical models were used in early studies to relate disease development to weather. They were based on detailed observations of interactions between pathogen and environmental factors. Early empirical models were used for potato late blight (*Phytophthora infestans*) forecasting, first relating outbreak to environmental conditions and stage of crop growth [6] and later to synoptic weather patterns in the North Atlantic [9].

Simulation models attempt to simulate each stage of the disease process using information derived from experimental observations while *analytical models* attempt to describe an epidemic by mathematical equations. Both types of model help to give a better understanding of the processes and variables involved in the progress of an epidemic, indicating, where it deviates from field results, that other factors need to be taken into account and suggesting areas for further experimentation.

Synoptic models use statistical methods based on the relative frequency of different factors. These may be derived from experimental studies of the effects of these factors on different stages of the disease cycle and the establishment of significant correlations. Synoptic models require less time and resources to construct than simulation models but may not give such good results. However, statistical approaches are sometimes necessary, even in simulation models where the interaction of host, pathogen and environment is difficult to define, perhaps because of a long period of development.

Regression models may be used in the study of parts of the disease cycle, such as the effects of environmental factors on sporulation, or in the construction of spore dispersal or disease gradients. Simple linear regressions may be useful in relating two factors but more complex polynomial relationships or multiple regressions may be necessary where more than two factors are related. Transformations may also enable the conversion of complex relationships into linear ones to allow easier comparison, e.g. the log–log transformation used to compare spore dispersal and disease gradients [42].

Extensive discussion of the use of models in plant disease epidemiology and aerobiology may be found in Kranz [61], Edmonds [30] and Fitt and McCartney [32].

3.5.7 REFERENCES

[1] Apinis A.E. (1977) Techniques available for microbiological investigation of grassland vegetation and produce. In *Biodeterioration Investigation Techniques*, p. 201 (ed. A.H. Walters). Applied Science Publishers, London.

[2] Atlas R.M. (1984) Diversity of microbial commuinities. *Adv. Microb. Ecol.* **7**, 1–47.

[3] Austin B., Dickinson C.H. & Goodfellow M. (1977) Antagonistic interactions of phylloplane bacteria with *Drechslera dictyoides* (Drechsler) Shoemaker. *Can. J. Microbiol.* **23**, 710–15.

[4] Austin B., Goodfellow M. & Dickinson C.H. (1978) *Adv. Microb. Ecol.* **7**, 1–47.

[3] Austin B., Dickinson C.H. & Goodfellow M. (1977) Antagonistic interactions of phylloplane bacteria with *Drechslera dictyoides* (Drechsler) Shoemaker. *Can. J. Microbiol.* **23**, 710–15.

[4] Austin B., Goodfellow M. & Dickinson C.H. (1978) Numerical taxonomy of phylloplane bacteria isolated from *Lolium perenne. J. Gen. Microbiol.* **104**, 139–55.

[5] Bainbridge A. & Legg B.J. (1976) Release of barley-mildew conidia from shaken leaves. *Trans. Br. Mycol. Soc.* **66**, 495–8.

[6] Beaumont A. (1947) Dependence on the weather of the date of outbreak of potato blight epidemics. *Trans. Br. Mycol. Soc.* **31**, 45–53.

[7] Blakeman J.P. (1981) *Microbial Ecology of the Phylloplane*. Academic Press, London.

[8] Bourdillon R.B., Lidwell O.M. & Lovelock J.E. (1948) *Studies in Air Hygiene*. Med. Res. Council Special Rept. Series No. 262. HMSO, London.

[9] Bourke P.M.A. (1957) *The Use of Synoptic Weather Maps in Potato Blight Epidemiology*. Tech. Note 23, Irish Meteorol. Service, Dublin.

[10] Brewer D., Calder F.W., MacIntyre T.M. & Taylor A. (1971) Ovine ill-thrift in Nova Scotia. I. The possible regulation of rumen flora in sheep by the fungal flora of permanent pasture. *J. Agric. Sci. Camb.* **76**, 465–77.

[11] Brooksby J.B. (ed.) (1983) *The Aerial Transmission of Disease*. Royal Society, London. Also *Phil. Trans. R. Soc. Lond. B* **302**, 439–604.

[12] Brooks P.J. (1963) Ecology of the fungus *Pithomyces chartarum* (Berk & Curt.) M.B. Ellis in pasture in relation to facial eczema disease of sheep. *N.Z. J. Agric. Res.* **6**, 147–228.

[13] Bull A.T. & Slater J.H. (1982) *Microbial Interactions and Communities*, Vol. I. Academic Press, London.

[14] Buller A.H.R. (1909) *Researches on Fungi*, Vol. 1. Longmans, Green, Toronto.

[15] Caborn J.M. (1973) Microclimates. *Endeavour,* **32**, 30–3.

[16] Campbell R. (1977) *Microbial Ecology*. Blackwell Scientific Publications, Oxford.

[17] Chang Y. (1967) The fungi of wheat straw compost. II. Biochemical and physiological studies. *Trans. Br. Mycol. Soc.* **50**, 667–77.

[18] Chang Y. & Hudson H.J. (1967) The fungi of wheat straw compost. I. Ecological studies. *Trans. Br. Mycol. Soc.* **50**, 649–66.

[19] Christensen C.M. (1950) Intramural dissemination of spores of *Hormodendrum resinae. J. Allergy* **21**, 409–13.

[20] Chu-Chou M. & Preece T. (1968) The effect of pollen grains on infections caused by *Botrytis cinerea* Fr. *Ann. Appl. Biol.* **62**, 11–22.

[21] Cole G.T. & Kendrick B. (1981) *Biology of Conidial Fungi*. Academic Press, New York.

[22] Darke C.S., Knowelden J., Lacey J. & Ward A.M. (1976) Respiratory disease of workers harvesting grain. *Thorax* **31**, 294–302.

[23] Day P.R. (ed.) (1977) The genetic basis of epidemics in agriculture. *Ann. N.Y. Acad. Sci.* **287**, 1–400.

[24] De Weille G.A. (1965) The epidemiology of plant disease as considered within the scope of agrometeorology. *Agr. Meteorol.* **2**, 1–15.

[25] Dickinson C.H., Austin B. & Goodfellow M. (1975) Quantitative and qualitative studies of phylloplane bacteria from *Lolium perenne. J. Gen. Microbiol.* **91**, 157–66.

[26] Dickinson C.H. & Preece T. (ed.) (1976) *Microbiology of Aerial Plant Surfaces*. Academic Press, London.

[27] Dickinson C.H. & Skidmore A.M. (1976) Interactions between germinating spores of *Septoria nodorum* and phylloplane fungi. *Trans. Br. Mycol. Soc.* **66**, 45–56.

[28] Di Menna M.E. & Parle J.N. (1970) Moulds on leaves of perennial rye grass and white clover. *N.Z. J. Agric. Res.* **13**, 51–68.

[29] Dimmick R.L. & Akers A,B. (ed.) (1969) *An Introduction to Experimental Aerobiology*. Wiley (Interscience), New York.

[30] Edmonds R.L. (ed.) (1979) *Aerobiology: the Ecological Systems Approach*. US/IBP Synthesis Series 10. Dowden, Hitchinson & Ross, Stroudsburg, Pennsylvania.

[31] Festenstein G.N., Lacey J., Skinner F.A., Jenkins P.A. & Pepys J. (1965) Self-heating of hay and grain in Dewar flasks and the development of farmer's lung antigens. *J. Gen. Microbiol.* **41**, 389–407.

[32] Fitt B.D.L. & McCartney H.A. (1986) Spore dispersal in relation to epidemic models. In *Plant Disease Epidemiology: Population Dynamics and Management*, p.311 (ed. K.J. Leonard & W.E. Fry). Macmillan, New York.

[33] Flannigan B. & Campbell J. (1977) Pre-harvest mould and yeast floras on the flag leaves, bracts and caryopsis of wheat. *Trans. Br. Mycol. Soc.* **69**, 485–94.

[34] Fokkema N.J. (1971) The effect of pollen in the phyllosphere of rye on colonization by saprophytic fungi and on infection by *Helminthosporium sativum* and other leaf pathogens. *Neth. J. Pl. Path.* **77**, Suppl. 1, 60.

[35] Fokkema N.J., Riphagen I., Poot R.J. & de Jong C. (1983) Aphid honeydew, a potential stimulant of *Cochliobolus sativus* and *Septoria nodorum* and the competitive role of saprophytic microflora. *Trans. Br.*

Mycol. Soc. **81**, 355–63.

[36] Fokkema N.J. & van der Meulen (1976) Antagonism of yeast-like phyllosphere fungi against *Septoria nodorum* on wheat leaves. *Neth. J. Pl. Path.* **82**, 13–16.

[37] Goodfellow M., Austin B. & Dickinson C.H. (1976) Numerical taxonomy of some yellow-pigmented bacteria isolated from plants. *J. Gen. Microbiol.* **21**, 219–33.

[38] Grant I.W.B., Blyth W., Wardrop V.E., Gordon R.M., Pearson J.C.G. & Mair A. (1972) Prevalence of farmer's lung in Scotland: a pilot survey. *Brit. Med. J.* **1**, 530–4.

[39] Gregory P.H. (1945) The dispersion of airborne spores. *Trans. Br. Mycol. Soc.* **21**, 26–72.

[40] Gregory P.H. (1966) Dispersal. In *The Fungi* Vol. 2, p.709 (ed. G.C. Ainsworth & A.S. Sussman). Academic Press, London.

[41] Gregory P.H. (1968) Interpreting plant disease dispersal gradients. *A. Rev. Phytopath.* **6**, 189–212.

[42] Gregory P.H. (1973) *Microbiology of the Atmosphere*, 2nd edn. Leonard Hill, Aylesbury.

[43] Gregory P.H. (1976) *Outdoor Aerobiology*. Oxford Biology Readers No. 62. Oxford University Press, London.

[44] Gregory P.H. & Henden D.R. (1976) Terminal velocity of basidiospores of the giant puffball (*Lycoperdon giganteum*). *Trans. Br. Mycol. Soc.* **67**, 399–407.

[45] Gregory P.H. & Hirst J.M. (1957) The summer air spora at Rothamsted in 1952. *J. Gen. Microbiol.* **17** 135–52.

[46] Gregory P.H., Lacey M.E., Festenstein G.N. & Skinner F.A. (1963) Microbial and biochemical changes during moulding of hay. *J. Gen. Microbiol.* **30**, 75–88.

[47] Gregory P.H. & Monteith J.L. (1967) *Airborne Microbes*. Cambridge University Press, London.

[48] Harley J.L. (1971) Fungi in ecosystems. *J. Ecol.* **59**, 653–68.

[49] Harries M.G., Lacey J., Tee R.D., Cayley G.R. & Newman Taylor A.J. (1985) *Didymella exitialis* and late summer asthma. *Lancet.* **1**, 1063–6.

[50] Hedger J.N. & Hudson H.J. (1974) Nutritional studies of *Thermomyces lanuginosus* from wheat straw compost. *Trans. Br. Mycol. Soc.* **62**, 129–43.

[51] Hers J.F.P. & Winkler K.C. (eds.) (1973) *Airborne Transmission and Airborne Infection*. Oosthoek Publishing Company, Utrecht.

[52] Hill R.A. & Lacey J. (1983a) The microflora of ripening barley grain and the effects of late fungicide application. *Ann. Appl. Biol.* **102**, 455–65.

[53] Hill R.A. & Lacey J. (1983b) Factors determining the microflora of stored barley. *Ann. Appl. Biol.* **102**, 467–83.

[54] Hirst J.M. (1953) Changes in atmospheric spore content: diurnal periodicity and the effects of weather. *Trans. Br. Mycol. Soc.* **36**, 375–93.

[55] Hirst J.M. & Stedman O.J. (1963) Dry liberation of fungus spores by raindrops. *J. Gen. Microbiol.* **33**, 335–44.

[56] Holloman D.W. (1967) Observations on the phylloplane flora of potatoes. *Eur. Potato. J.* **10**, 53–61.

[57] Hudson H.J. (1965) The ecology of fungi on plant remains above the soil. *New Phytol.* **67**, 837–74.

[58] Ingold C.T. (1966) Spore release. In *The Fungi*, Vol. 2, p. 679 (ed. G.C. Ainsworth & A.S. Sussman). Academic Press, London.

[59] Ingold C.T. (1971) *Fungus Spores: Their Liberation and Dispersal*. Clarendon Press, Oxford.

[60] Keogh R.G. (1973) *Pithomyces chartarum* spore distribution and sheep grazing patterns in relation to urine-patch and inter-excreta sites within rye grass-dominant pastures. *N.Z. J. Agric. Res.* **16**, 353–5.

[61] Kranz J. (ed.) (1974) *Epidemics and Plant Diseases — Mathematical Analysis and Modelling*. Springer Verlag, New York.

[62] Lacey J. (1975a) Occupational factors in allergy. In *Allergy '74*, p.303 (ed. M.A. Ganderton & A.W. Frankland). Pitman Medical, London.

[63] Lacey J. (1975b) Airborne spores from pastures. *Trans. Br. Mycol. Soc.* **64**, 265–81.

[64] Lacey J. (1975c) Risks from using mouldy fodders. *Trans. Br. Mycol. Soc.* **65**, 171–84.

[65] Lacey J. (1978a) The ecology of actinomycetes in fodders and related substrates. In *Nocardia and Streptomyces*, p.161 (ed. M. Mordarski, W. Kurylowicz & J. Jeljaszewicz). Gustav Fischer Verlag, Stuttgart.

[66] Lacey J. (1978b) The microflora of grain dusts. In *Occupational Pulmonary Disease: Focus on Grain Dust and Health*, p.417 (ed. J.A. Dosman & D.J. Cotton). Academic Press, New York.

[67] Lacey J. (1981) Airborne actinomycete spores as respiratory allergens. In *Actinomycetes* (ed. K.P. Schaal & G. Pulverer). Gustav Fischer Verlag, Stuttgart. *Zentbl. Bakt. Mikrobiol. Hyg.* **1**, Suppl. 11, 245–50.

[68] Lacey J., Hill S.T. & Edwards M.A. (1980) Microorganisms in stored grains; their enumeration and significance. *Trop. Stored Prod. Inf.* **39**, 19–33.

[69] Langmuir J. (1948) The production of rain by a chain reaction in cumulus clouds at temperatures above freezing. *J. Meteorol.* **5**, 175–92.

[70] Last F.T. & Warren R.C. (1972) Non-parasitic microbes colonizing green leaves: their form and functions. *Endeavour* **31**, 143–50.

[71] Latch G.C.M. & MacKenzie E.H.C. (1977) Fungal flora of rye grass swards in Wales. *Trans. Br. Mycol. Soc.* **68**, 181–4.

[72] Leach C.M. (1982) Active spore discharge in *Peronospora destructor. Phytopathology* **72**, 881–5.

[73] Lioy P.J. & Lioy M.J. (1983) *Air Sampling Instruments for Evaluation of Atmospheric Contaminants*, 6th edn. American Conference of Governmental Industrial Hygienists, Cincinnati, Ohio.

[74] McDonald P. (1976) Trends in silage making. In *Microbiology in Agriculture, Fisheries and Food*, p. 109 (ed. F.A. Skinner & J.G. Carr). Academic Press, London.

[75] McDonald P. & Whittenbury R. (1973) The ensilage process. In *Chemistry and Biochemistry of Herbage*, Vol. 3, p. 33 (ed. G.W. Butler & R.W. Bailey). Academic Press, London.

[76] Magan N. & Lacey J. (1984a) The effect of temperature and pH on the water relations of field and storage fungi. *Trans. Br. Mycol. Soc.* **82**, 71–81.

[77] Magan N. & Lacey J. (1984b) The effect of water activity, temperature and substrate on interactions between field and storage fungi. *Trans. Br. Mycol. Soc.* **82**, 83–93.

[78] Magan N. & Lacey J. (1984c) The effects of gas composition and water activity on growth of field and storage fungi and their interactions. *Trans. Br. Mycol. Soc.* **82**, 305–14.

[79] Meredith D.S. (1973) Significance of spore release and dispersal mechanisms in plant disease epidemiology. *Ann. Rev. Phytopath.* **11**, 313–42.

[80] Palti J. & Kranz J. (eds.) (1980) *Comparative Epidemiology: a Tool for Better Disease Management.* Centre for Agricultural Publishing and Documentation, Wageningen.

[81] Pasquill F. (1974) *Atmospheric Diffusion* 2nd edn. Ellis Horwood Ltd., Chichester.

[82] Pelhate J. (1980) Oxygen depletion as a method in grain storage: microbiological basis. In *Controlled Atmosphere Storage of Grain*, p.133 (ed. J. Shejbal). Elsevier, Amsterdam.

[83] Pennypacker S.P. & Madden L.V. (eds.) (1980) Epidemiology. *Protection Ecology* **2**, 157–284.

[84] Preece T. & Dickinson C.H. (eds.) (1971) *Ecology of Leaf Surface Micro-organisms*. Academic Press, London.

[85] Robinson R.A. (1976) *Plant Pathosystems*. Springer Verlag, Berlin.

[86] Rosenberg N.J. (1974) *Microclimate: the Biological Environment*. John Wiley, New York.

[87] Scott P.R. & Bainbridge A. (eds.) (1978) *Plant Disease Epidemiology*. Blackwell Scientific Publications, Oxford.

[88] Shapton D.A. & Board R.G. (eds.) (1972) *Safety in Microbiology*. Academic Press, London.

[89] Sharma P.D. (1973) Succession of fungi on decaying *Setaria glauca* Beauv. *Ann. Bot.* **37**, 203–9.

[90] Shenoi M.M. & Ramalingam A. (1975) Circadian periodicities of some spore components of air at Mysore. *Arogya J. Health Sci.* **1**, 154–6.

[91] Sinnecker H. (1976) *General Epidemiology* (translated by N. Walker). John Wiley, London.

[92] Skidmore A.M. & Dickinson C.H. (1973) Effect of phylloplane fungi on the senescence of excised barley leaves. *Trans. Br. Mycol. Soc.* **60**, 107–16.

[93] Skidmore A.M. & Dickinson C.H. (1976) Colony interactions and hyphal interference between *Septoria nodorum* and phylloplane fungi. *Trans. Br. Mycol. Soc.* **66**, 55–64.

[94] Skinner F.A. & Carr J.G. (eds.) (1974) *The Normal Microbial Flora of Man.* Academic Press, London.

[95] Slater J.H., Whittenbury R. & Wimpenny J.W.T. (eds.) (1953) *Microbes in Their Natural Environments*. Society for General Microbiology Symposium 34. Cambridge University Press, Cambridge.

[96] Smith L.P. (1956) Potato blight forecasting by 90 per cent humidity criteria. *Plant Path.* **5**, 83–7.

[97] Sreeramulu T. & Ramalingam A. (1966) A two-year study of the air-spora of a paddy field near Visakhapatnam. *Indian J. Agric. Sci.* **36**, 111–32.

[98] Starr J.R. & Mason B.J. (1966) The capture of airborne particles by water drops and simulated snow crystals. *Q. J. R. Met. Soc.* **92**, 490–9.

[99] Stepanov K. (1935) Dissemination of infective diseases of plants by air currents. *Bull. Pl. Prot. Leningr.* Ser. 2. Also *Phytopathology N.* **8**, 1–68.

[100] Stirling A.C. & Whittenbury R. (1963) Sources of the lactic acid bacteria occurring in silage. *J. Appl. Bact.* **26**, 86–90.

[101] Wicklow D.T. & Carroll G.C. (1981) *The Fungal Community: Its Organization and Role in the Ecosystem*. Marcel Dekker, New York.

[102] Zadoks J.C. & Schein R.D. (1979) *Epidemiology and Plant Disease Management*. Oxford University Press, Oxford.

Part 4
Economic Microbial Ecology

A fool and his words are soon parted; a man of genius and his money.

William Shenstone 1714–1763
On Reserve

4.1 Symbiotic nitrogen fixation and nitrogen cycling in terrestrial environments

4.1.1 TAXONOMY

Organisms capable of N_2 fixation are found exclusively in the Prokaryotae, scattered throughout the 19 groups of bacteria. They are widely distributed among the eubacteria and cyanobacteria and include some of the streptomyces. The taxonomy of these organisms is continually being updated. However, the N_2-fixing organisms are difficult to place in the various taxa as they may have characters which span many different categories.

The symbiotic N_2-fixing micro-organisms are divided into several groups which include the leaf associations and root nodules of the *Alnus, Cycas, Parasponia* and leguminous type. This chapter will deal with the leguminous root nodule symbiosis, since many of these interactions are economically important. Species of *Rhizobium* involved in legume symbioses (see Table 4.1.1.) recently underwent a taxonomic change [79].

Taxonomic consideration of the N_2-fixing organisms involves a number of characteristics, which include nucleic acid homology, hybridization, melting point and base composition as well as protein patterns by electrophoresis. Other characteristics include chromosomal and extrachromosomal elements and potential for genetic exchange as well as serological characteristics utilizing cell surface antigens and purified proteins.

4.1.2 GLOBAL ESTIMATES OF N_2 FIXATION

Estimates of inputs for nitrogen fixation and for losses through leaching and denitrification are difficult to obtain on a global scale. Some estimates of N_2 fixation in various habitats and for the globe are given in Tables 4.1.2 and 4.1.3. Total biological N_2 fixation estimates of 149–175 g N $\times 10^{12}$ from the two sources agree reasonably well. The National Research Council [105] estimates of annual worldwide biological N_2 fixation on agricultural lands are identical to those shown in Table 4.1.3.

4.1.3 MEASUREMENT OF N_2 FIXATION

Accurate and simple estimates of N_2 fixation are difficult to achieve simultaneously under all circumstances. Some

Table 4.1.1. Leguminous type root nodules [79, 129]

Family	Genus	Species	Host
Rhizobiaceae	*Rhizobium*	*fredii*	*Glycine* (soybean)
		leguminosarum	
		biovar *phaseoli*	*Phaseolus* (bean)
		biovar *trifolii*	*Trifolium* (clover)
			Vicia (broad bean)
		biovar *viceae*	*Lathyrus* (pea)
			Lens (lentils)
			Pisum (pea)
			Vicia (broad bean)
		loti	*Lotus* (trefoil)
			Lupinus (lupin)
			Cicer (chickpea)
		meliloti	*Melilotus* (sweet clover)
			Medicago (lucerne)
			Trigonella (fenugreek)
	Bradyrhizobium	*japonicum*	*Glycine* (soybean)
		sp. (host genus)	*Acacia*
			Arachus (peanut)
			Cajanus (pigeon pea)
			Centrosema (butterfly pea)
			Cicer (chickpea)
			Leucaenea
			Lotus (trefoil)
			Lupinus (lupin)
			Macroptilium
			Mimosa
			Ornithopus (seradella)
			Sesbania
			Vigna (cowpea)

Table 4.1.2. Rates of symbiotic N_2 fixation in various habitats

	Quantity fixed (kg N ha^{-1} y^{-1})	Method	Reference
Legumes			
Medicago sativa L. (N. America)	138–224	^{15}N ID	[65]
Trifolium repens L. (N. America)	85–100	^{15}N ID	[25]
Glycine max (L.) Merr. (N. America)	21–155	^{15}N Diff.	[57]
Phaseolus vulgaris L. (N. America)	40–125	^{15}N ID	[119]
Phaseolus vulgaris L. (Brazil)	24–65	^{15}N ID	[125]
Cicer arietinum L. (Syria)	14–120	^{15}N ID	[118]
Pisum sativum L. (N. America)	166–189	^{15}N ID	[118]
Non-legumes			
Alnus rubra	140–209		[168]
Hippophae rhamnoides L. (sand dune)	2–58	^{15}N	[138]
Casuarina (Senegal)	52	N_T	[133]
Azolla (rice paddy, Indonesia)	83–125	C_2H_2	[13]
Dryas drummondii (glacial area, Alaska)	12	N_T	[133]

ID = isotope dilution; N_T = nitrogen total; C_2H_2 = acetylene reduction.

Table 4.1.3. Estimates of N_2 fixed on a global scale

Type of fixation	Nitrogen fixed (10^{12} g N yr^{-1}*) Hardy [59]	CAST [28]
Non-biological		
Industrial — fertilizers and industrial use	50	57
Combustion	20	20
Lightning	10	10
Total non-biological	80	87
Biological		
Natural processes on agricultural land	89	89
Natural processes under forest and unused land	50	60
Sea	36	—
Total biological	175	149
Total nitrogen fixation	255	237

* 10^{12} g $year^{-1}$ is equivalent to 10^6 metric tons.

methods provide a short-term rate estimate (acetylene reduction and ureide measurement) while others (difference and isotope dilution) provide a seasonal estimate. Rate estimates must be integrated over the entire growing season.

Direct methods

Perhaps the oldest method of estimating N_2 fixation is by difference in total plant N between a N_2-fixing crop (N_F) and a non-N_2-fixing control (N_C). The N_C is usually either a non-nodulating legume or a non-legume. The basic assumption of this method is that the control crop must assimilate the same quantity of soil N as the fixing crop. This is often not the case because of differences in rooting patterns, especially between legumes and non-legumes. If a more detailed N budget to account for N losses is desired, reasonable estimates can be obtained through long-term studies using actual values for fertilizer input, soil N, crop N removal, and estimates of N_2 fixation [136]. Estimates of total N fertilizer loss by this approach vary from 30 to 50% and are within other loss estimates [3, 60]. Methods using isotopes of N ($^{15}N_2$, $^{13}N_2$) are useful in the laboratory for estimation of N_2 fixation and to verify the $N_2:C_2H_2$ ratio [15]. However, field estimates of N_2 fixation have almost exclusively utilized the ^{15}N isotope in which ^{15}N-labelled fertilizer or plant residue is incorporated into the soil of plots containing a legume and non-fixing crop. It is assumed that the plants do not discriminate between ^{14}N and ^{15}N in soil and take up both forms in proportion to the amounts available [51]. Considerable discussion has centred around the appropriate control and the necessity of adding N to the non-fixing control [150, 162]. In addition, the difference in the rate of seasonal N uptake between the N_2-fixing plants and control can result in errors [163]. In some instances, however, reasonably good estimates of N_2 fixation have been obtained using cereals or grasses as controls [137].

Estimates of N_2 fixation by natural ^{15}N abundance are not yet suitable for routine analysis because factors involved in isotope discrimination by plants are not yet fully understood and because a high level of analytical precision is needed [154].

Indirect methods

The acetylene reduction (AR) assay is one of the most widely used methods to estimate N_2 fixation because it is convenient, simple and sensitive. Detailed procedures [62] and limitations [15, 147] have been described. The assay provides an indirect estimate of N_2 fixation by measuring the electron flow through nitrogenase during the reduction of acetylene to ethylene. Factors that contribute to inaccurate estimates of N_2 fixation by AR include an inaccurate estimate for moles C_2H_2 reduced to moles N_2 fixed. Furthermore, in the presence of C_2H_2, all the electrons through nitrogenase are used to produce C_2H_4, whereas some H_2 may be produced during N_2 reduction [131]. Consequently, values obtained from AR may overestimate N_2 fixation. This discrepancy may be less for systems that possess a Hup^+ recycling trait (see Section 4.1.6) [40]. Further errors in estimating N_2 fixation by AR are due to diurnal variations in light, moisture and temperature [135]. It has also been found that in rhizosphere and soil assays of low nitrogenase activity C_2H_2 inhibits the oxidation of indigenous C_2H_4 [37]. Under such conditions, C_2H_2 reduction usually overestimates N_2 fixation even though a value for the indigenous C_2H_4 accumulated is measured prior to the exposure to C_2H_2 [161]. Many scientists agree that the AR technique has limitations for quantitative estimates in the field, but can be used with caution as a qualitative measure [92]. However, recent studies [103] have shown a reduced nitrogenase activity in nodules using AR in detopped and disturbed roots. Consequently, they suggest that it is best to check the AR activity using intact roots in a flow-through system prior to assaying detopped plants in a disturbed system.

The association between ureides in the xylem sap or plant tissue has been used to estimate symbiotic N_2 fixation [70, 108]. The amount of ureide N relative to N in other compounds is proportional to the amount of N_2 fixed [94] and varies with plant species and cultivar, bacterial species and strain, and environmental conditions [108]. Since ureides are the dominant form of xylem N for soybean (*Glycine max* [L.] Merr.), pigeon pea (*Cajanas cajan* [L.] Millsp.), common bean (*Phaseolus vulgaris* L.) and *Vigna* spp., xylem sap analysis can be used to estimate N_2 fixation [95]. However, exudation rates can fluctuate and in field situations collection may be difficult. Even for the known ureide producers, ureide levels may not be proportional to N_2 fixed because the relationship is sensitive to environmental fluctuations [108].

4.1.4 ENUMERATION OF SYMBIOTIC N_2-FIXING MICRO-ORGANISMS

Serology

Serological techniques to identify N_2-fixing organisms rely on the antigenic characteristics of the bacterial cell wall and other cellular constituents. Techniques include agglutination, immunodiffusion, enzyme-linked immunosorbent assay (ELISA) and immunofluorescence (IF) [44, 153]. The sensitivity of the test depends primarily on the methodology used and is limited by the quality of the derived antibody. Identification is based on serotype rather than specific strain characteristics and therefore cross reaction is common between strains.

Monoclonal antibodies (MCA) may lead to greater specificity of serological techniques and may aid in identification of N_2-fixing strains [166]. Each MCA is specific for a single antigenic determinant which allows strains to be identified based on minute differences. There are several advantages of MCA over polyclonal antiserum. One is that only a small quantity of antigen is required to produce a continual supply of pure antibody. Another is that serological relationships previously not recognized by polyclonal antisera may be uncovered by MCA. The development of MCA for a given strain, however, may take well over a year, which is longer than the three to six week period required for the development of polyclonal antibodies.

The enzyme-linked immunosorbent assay (ELISA) technique utilizes the antigen/antibody reaction coupled to a colorimetric enzyme assay [14]. It has been shown to be reliable and specific [99]. A large number of samples using only a small volume of bacterial or nodular suspension can be screened within a short time period using ELISA.

Immunofluorescence (IF), the conjugation of fluorescein and antibody, has been used to identify and enumerate a large number of different bacteria [23]. One limit to this technique is nonspecific staining of background and soil particles. This can be lessened by pretreatment with a gelatin–rhodamine conjugate and other techniques [23]. Various flocculants and dispersing agents as well as density centrifugation can be used to separate the bacteria from soil particles more efficiently [82, 18]. Enumeration of bacteria from soil by IF is generally limited to greater than 10^4 cells g^{-1} soil.

Antibiotic resistance

Intrinsic resistance to antibiotics (IAR) or other inhibitors has been used to identify and differentiate strains [80]. It is based on the ability of strains to grow in the presence of low concentrations of various inhibitors. Relatively good agreement has been obtained between serology and IAR [83, 145] which suggests that IAR can be used as a prescreening procedure for selecting antigens in serological studies. No alteration of bacterial function occurs using IAR methods.

The development of spontaneous mutants resistant to antibiotics has allowed identification and enumeration of strains. Since changes in the populations of introduced strains can be assessed on selective media, these mutants can be used in ecological studies [27, 78]. One criterion is that the resistance characteristic must be stable. Another is that symbiotic and ecological characteristics of the resistant population must be similar to that of the parent. Antibiotic resistant strains must be characterized prior to use, since the attributes of the population may have been altered with selection [146]. Often double or triple antibiotic resistant strains or antibiotic resistance in combination with other identification techniques are used to enhance the specificity of identification.

Antibiotic resistant strains can also be obtained by insertion of transposons into the genetic material of N_2-fixing bacteria [16]. These mutant strains, which carry resistance to various antibiotics can be isolated and plated on medium containing various levels of antibiotics and are useful in ecological studies once their other attributes are identified. They can also be used in genetic studies and can be enumerated by DNA probe. The insertion of a transposable element usually leads to a stable antibiotic resistant characteristic, although other characteristics of the parent may be altered.

DNA probe

Detection of specific gene sequences in environmental samples by DNA probe is possible with DNA–DNA colony hybridization [10, 128]. This approach has also been used to identify strains of *Rhizobium* in nodules [74]. The DNA probe can be used as a genotypic marker for identifying specific gene sequences and is useful in ecological studies for evaluating genetic exchange in the environment. When used in combination with Tn5 or other transposable elements which confer antibiotic resistance, the DNA probe can serve as a genotypic and phenotypic marker for enumerating soil organisms [50].

Most probable number

One of the earliest methods used for the isolation and enumeration of N_2-fixing symbionts is the plant infectivity–most probable number (MPN) technique. Isolation or enumeration of the symbiont by this technique is based on the ability of the organism to form nodules on a given host from a serial dilution of soil or growth medium. Values obtained from the plant infectivity method depend on the strain, host, type of enumeration and growth system used [26].

4.1.5 ECOLOGY OF SYMBIOTIC N_2-FIXING MICRO-ORGANISMS

Biotic and abiotic factors

Factors important in the survival and function of symbiotic N_2-fixing micro-organisms in soil differ with the organism and environment. In general, acidic soil conditions result in lower population densities than neutral or alkaline conditions [121]. For this reason liming of acid soils usually results in increasing population densities and greater nodulation of the host [33, 93]. Changes in interstrain competition [55] and nodule occupancy [45] were observed with changes in pH. High temperatures reduced both survival [33] and nodulation [157]. Soil type [98] and water content [98, 110] influenced survival of rhizobia.

The population increase often observed when rhizobia are inoculated into sterile soil is attributed to the lack of competition or antagonism by other organisms [152]. Actinomycetes, bacteria and fungi have been identified as antagonists of some rhizobia and can influence survival of rhizobia either by competition or by toxin production [81, 116]. Increases in protozoan populations [34], and the presence of phage [116] are correlated with declines in rhizobial populations.

The rhizosphere environment

In soil devoid of roots, rhizobia grow poorly unless a carbon source is added. The rhizosphere is an extremely favourable habitat for rhizobial growth and usually results in a substantial increase in population relative to bulk soil. The extent of this stimulation is related to environment, soil type, plant variety, plant age and rhizobial strain [120]. Stimulation of rhizobia occurs in the non-leguminous rhizosphere of oat (*Avena sativa* L.), maize (*Zea mays* L.) and wheat (*Triticum aestivum* L.) [104, 120]. On one occasion, the early rhizobial population in the non-legume rhizosphere was greater than that in the host rhizosphere [111]. Population densities of *Bradyrhizobium japonicum* and *Rhizobium leguminosarum* biovar *viceae* were greater in the rhizosphere of pea (*Pisum sativum* L.) than in the rhizosphere of soybean during early plant growth [21]. Pena-Cabriales and Alexander [111] found *B. japonicum* to be stimulated more than *R. leguminosarum* biovar *phaseoli* in the rhizosphere of red clover (*Trifolium pratense* L.) and kidney bean. The above studies suggest that there is little justification for the idea that preferential stimulation of host-specific rhizobia is necessary for successful symbiosis.

Competition

Firmly entrenched in the ecological studies of rhizobia is the concept of competition. In microbial ecology, competition occurs when two or more populations are striving for the same resource with an adverse effect on one or both populations. However, in rhizobial ecology, competition usually refers to the interaction between similar strains of rhizobia with the ultimate success being the proportion of nodules occupied by one strain. Factors which influence competition are soil type, host, strain and environmental factors.

Competition is of utmost concern in rhizobial ecology and agriculture because the end result of inoculation should be to supplant the indigenous population and establish the more desirable strain in the nodule. One way to do this is to inoculate at extremely high levels and saturate the system with the desired strain [156]. However, this strategy is not practical for legume inoculation. Attempts to develop strains which 'outcompete' the indigenous strain have met with limited success, but only where the indigenous population is absent, or less than 10^3 cells g^{-1} soil [100, 101]. More often, however, indigenous populations outcompete and do not allow the inoculated strain to establish or occupy the nodule to any great extent.

In field studies involving three serogroups of *B. japonicum*, serogroup 123 occupied more soybean nodules than each of the others even though the rhizosphere population of 123 at planting did not differ from the other two [104]. All three serogroups were stimulated by the rhizosphere, while serogroup 123 was stimulated the least.

Time of exposure to a host may be critical in determining the strain which forms and occupies the nodule. In studies with the Afghanistan pea, co-inoculation of two strains of *R. leguminosarum* biovar *viceae*, strain TOM (nodulating) and strain PF_2 (non-nodulating), resulted in no nodules being formed. However, if the strain TOM was inoculated as many as three days before the non-nodulating strain, nodules were formed [159]. In other studies, nodule occupancy of *B. japonicum* could be altered by prior exposure of soybean roots to one strain before inoculation of a second strain. Strain 110 dominated the nodules when co-inoculated with strain 138. However, if strain 138 was inoculated as little as 24 hours before strain 110, strain 138 dominated the nodules [84]. Population dynamics during early plant growth may play a part in determining competition and the outcome of nodule occupancy.

Molecular basis for competition

Our ultimate understanding of symbiosis is likely to be based on molecular mechanisms. Motility and chemotaxis of the bacterium [107] as well as the attachment and recognition of the rhizobia and host may be involved. Ames and Bergman [4] compared the nodulating ability of motile and non-motile strains of *R. meliloti* and concluded that the motile strain had a definite competitive advantage and occupied more nodules than the non-motile strain.

Adhesion may be important in the recognition process between the host and bacterium. Strains of *R. meliloti* which firmly attached to the root surface were more competitive than those strains which only loosely adhered [76]. Microfibrils [72] and pili [151] of rhizobia have been suggested as being possible adhesins. An 80% reduction in nodulation by *B. japonicum* was observed when cells were treated with antiserum against pili. These adhesins may function by allowing the exchange of recognition molecules between the host and bacterium or by allowing modification of cell surfaces by extracellular degrading enzymes. The molecular basis of cellular recognition is not fully understood, but is clearly an event which needs further study.

One hypothesis in the recognition phenomena is that lectins are involved in the recognition process [12, 36, 56]. There may be some sort of communication between the legume lectin and polysaccharides on the compatible rhizobial cell surface. Although numerous data address lectin binding, there is no conclusive evidence which supports or denies the role of lectins in recognition [22, 35, 56, 123]. Other problems which weaken the lectin hypothesis are the nonspecific attachment of rhizobia to the host [115], rhizobial species which bind lectin from plants they do not nodulate [165], and soybean lines which produce no lectin but still nodulate [117]. Lectin binding ability is transient, related to root age and age of bacterial culture [20, 76]. While recognition is definitely a critical step in the infection process, the molecular mechanisms are not fully understood.

4.1.6 CARBON AND NITROGEN DYNAMICS IN N_2 FIXATION

Energy costs

Symbiotic N_2 fixation costs the plant photosynthate to support N_2 fixation and NH_4^+ assimilation at the site of N_2 fixation. This cost has been estimated to be from 15% to 30% of the total assimilation capacity of the plant [130] so there is concern that these processes may reduce plant yield. Therefore, numerous studies have compared the relative costs of N_2 fixation and reduction of nitrate.

In both the industrial and the biological processes, a catalyst and energy are required. The biological catalyst is nitrogenase, which is capable of reducing N_2, C_2H_2 and other organic compounds [132]. The overall reaction for purified nitrogenase preparations is [49, 167]:

$$N_2 + 8H^+ + 8e^- + 16ATP \rightarrow 2NH_3 + 16ADP + 16P_i + H_2$$

One mol H_2 is formed for each mol N_2 reduced. A minimum of four molecules of ATP is consumed per pair of electrons transferred for the reduction of N_2 to NH_3 or for the formation of H_2 in the absence of N_2 [167]. Reduced ferredoxin serves as the reductant in the above reaction.

Efficiency of coupling free energy from the oxidation of glucose through glycolysis and the TCA cycle to the synthesis of ATP through the electron transport chain is only about 40% [97, 130]. The theoretical cost of N_2 reduction has been estimated to be from 0.5 to 0.75 mol glucose mol^{-1} N_2 reduced [97, 130]. These glucose equivalent values do not include the additional costs of nodule growth and maintenance, of carbon skeletons, and of ATP and reductant required for NH_4^+ assimilation. Estimates of the *in vitro* cost of N_2 fixation have been compared to assimilation of NO_3^- (see Table 4.1.4). The cost of incorporation of NH_4^+ is common to both N_2 fixation and NO_3^- assimilation, but is included in the nodule measurements. Although different legumes translocate different N compounds, a value of 3.3 mol ATP or 0.09 mol glucose mol^{-1} NH_4^+ has been suggested by Atkins *et al.* [6]. The total cost of N_2 fixation in terms of glucose equivalents has been estimated [97] to be 1.26 mol glucose mol^{-1} N_2, which is slightly below those values shown in Table 4.1.4.

The above estimates reflect the cost of N fixation *in vitro* and should not be extrapolated to the actual costs within the intact nodule. Many attempts have been made to compare the relative energy costs of N_2 fixation and NO_3^- assimilation. One approach compares the growth rates of nodulated plants grown in N-free medium to those grown with combined N. Another approach is based on respiration measurements where it is assumed that the reduction and assimilation of N_2 are the primary functions of nodules and that the respiratory activities of these tissues reflect the costs of N_2 fixation. Estimates of energy costs based on respiratory CO_2 efflux indicate that costs calculated by including both roots and nodules may be overestimated, whereas nodule respiration based on detached nodules could underestimate the costs (see Table 4.1.5). The values presented agree reasonably well, con-

Table 4.1.4. Estimates of the *in vitro* cost (glucose equivalents)* of N_2 fixation, NO_3^- utilization and NH_4^+ assimilation (after Schubert & Ryle [132])

	Source of N	
	N_2	NO_3^-
Glucose equivalents used for:		
Reductant generation	0.25	0.66
ATP production	0.50	—
Transport	—	?
NH_4^+ assimilation		
ATP	0.13	0.13
Reductant	0.08	0.08
Carbon skeletons	0.5–0.66†	0.5–0.66†
Total	1.5–1.6	1.4–1.5

* Assumes a P : 0 ratio of 2 and that ammonia is assimilated and exported as asparagine.
† The smaller value corresponds to estimates in which oxaloacetate is produced via the carboxylation of phosphoenolpyruvate (Christeller *et al.* [29]).

Table 4.1.5. Estimates of respiratory CO_2 efflux related to symbiotic nitrogen fixation (after Mahon [97])

Method and species	Mean cost (mol CO_2 mol^{-1} NH_4^+)	Reference
Root + nodule respiration per N accumulation		
Pisum sativum L.	6.9	[102]
Trifolium repens L., *Glycine max* (L.) Merr., *Vigna unguiculata* (L.) Walp.	7.7	[126]
Lupinus albus L.	11.9	[109]
Nodule respiration per N accumulation		
Glycine max (L.) Merr.	8.6	[24]
Pisum sativum L.	1.7	[102]
Vigna unguiculata (L.) Walp., *Lupinus albus* L.	3.0	[89]
Trifolium repens, Glycine max (L.) Merr., *Vigna unguiculata* (L.) Walp. (estimated)	4.1	[126]
Regression of root and nodule respiration on C_2H_2 reduction		
Pisum sativum L., *Phaseolus vulgaris* L., *Glycine max* (L.) Merr., *Vicia faba* L.	7.5	[96]
Glycine max (L.) Merr.	5.8	[71]

sidering the wide variation in legumes studied and methods used.

The energy requirements for N_2 fixation and for utilization of combined nitrogen have been compared by Schubert [130]. In general, costs of N_2 fixation are either equal to or greater than costs of nitrate reduction although there is no clear-cut conclusion. Further consideration must be given to the reduction of nitrate in the chloroplasts during photosynthesis [90].

Evolution and recycling of H_2

The ATP-dependent evolution of H_2 catalysed by nitrogenase, which was reported during fixation in soybean [73], can account for about 25% of the energy consumed in the overall nitrogenase reaction. Later studies [39, 40] showed that some species of *Rhizobium* do not evolve H_2 and possess the capacity to recycle H_2 through hydrogenase, a membrane-bound enzyme. Figure 4.1.1 shows the relationship between H_2 evolution and nitrogenase in nodules and H_2 oxidation in nodules having a H_2 recycling capacity. The trait, known as Hup^+, is present in a minority of strains [131] and is found more commonly in the *Bradyrhizobium* sp. (*Vigna*) and *B. japonicum* [49]. Strains of *Rhizobium* possessing this trait have also been shown to possess autotrophic growth using H_2 and CO_2 as sources of energy and carbon, respectively [143]. Three potential advantages of the Hup^+ system have been suggested [41]: (a) O_2 is utilized and therefore the O_2-sensitive nitrogenase can be protected; (b) it may prevent the accumulation of H_2 in the nodules; and (c) metabolism of the H_2 evolved will recover a part of the energy expended by ATP to produce H_2.

Studies that compare Hup^+ and Hup^- strains of *B. japonicum* [2, 58] showed higher N_2 fixation by Hup^+ than Hup^- strains, whereas others [106, 144] showed no difference in Hup^+ and Hup^- strains of *R. leguminosarum* biovar *viceae*. However, since these Hup^+ and Hup^- strains may not be isogenic with respect to H_2 recycling capacity, other traits may influence their respective N_2-fixing capacities.

Transfer of the Hup^+ gene to strains of *Rhizobium* already efficient in symbiotic N_2 fixation would seem a worthwhile research objective. There was a significant 11% increase in N_2 fixation by a Hup^+ revertant of a Hup^- mutant derived originally from *B. japonicum* Hup^+ strain USDA DES122 [48]. Research [38] with *R. leguminosarum* biovar *viceae* showed that the recombinant plasmid pIJ1008 (= pVW5JI, pRL6JI), which confers a Hup^+ phenotype, increased N_2 fixation from 31 to 128% in N_2-fixing Hup^- recipient strains. However, Tn5-mob mutagenesis of the Hup^- plasmid pRL6JI produced Hup^- phenotypes which did not differ significantly in N_2 fixation from the Hup^+ strains [32]. Cunningham and colleagues [32] concluded that the improvement of symbiotic properties associated with the transfer of pIJ1008 to Hup^-

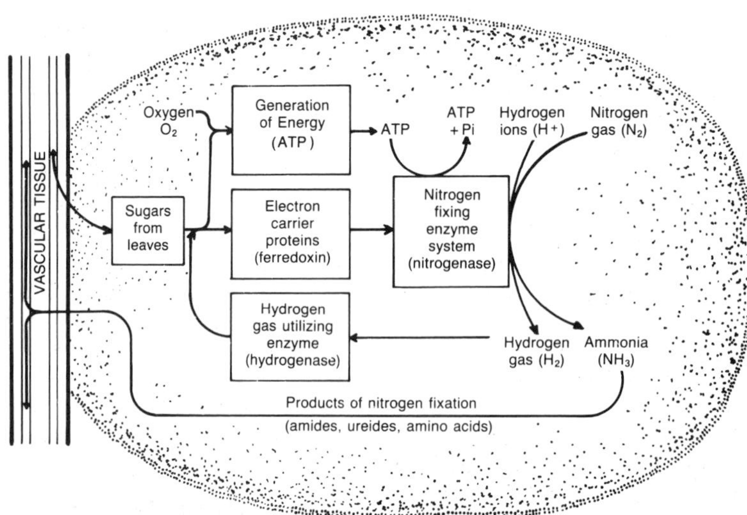

Fig. 4.1.1. The relationship between the system for N_2 fixation and H_2 recycling in a legume nodule. Nodules formed by strains lacking H_2 recycling evolve H_2 into the atmosphere.

R. leguminosarum biovar *viceae* strain in the study by DeJong and colleagues [38] was due to some trait other than the Hup^+ phenotype. Further studies with isogenic traits need to be continued. It is not likely that a comparison of random Hup^+ and Hup^- strains will contribute much to the understanding of the significance of the Hup^+ trait in N_2 fixation.

Table 4.1.6. Efficiency of recovery of soil and applied N for several crops after 50 years of management (after Smith [136])

Fertility management practice	Percent recovery*
Inorganic fertilizer	46–58
Manure (13 tons ha^{-1} yr^{-1})	52–87
Crop rotations (3-,4- and 6-year)	84–103 (57–76)†
6-year rotation + manure and inorganic fertilizer	66 (57)
Continuous timothy	101

* (N removed in crops + soil N after 50 years) ÷ (soil N at start + N added) × 100.
† Values in parentheses represent N recovery using an estimate of 112 kg N ha^{-1} yr^{-1} fixed by clover in rotation.

4.1.7 NITROGEN CYCLING IN LEGUME CROPPING SYSTEMS

Nitrogen budgets

Considerable effort has often gone into estimating N_2 fixation of legumes in agricultural systems without considering the overall N economy of the system [69]. Nitrogen that is fixed is subject to the same fate as fertilizer or soil nitrogen. In most studies, a complete balance-sheet of gains, transfers and losses is not usually possible since all gains and losses cannot be measured with equal ease and precision. A balance-sheet can be estimated as suggested [53, 69].

$$\Delta N = F_L + F_F + M + DR + A - C - L - V - E$$

where: ΔN = change in the soil N over time; F_L = rate of symbiotic N_2 fixation; F_F = rate of non-symbiotic N_2 fixation; M = N added in fertilizer, seed or manures; DR = N added in dust and rainfall; A = N added via ammonia (NH_3) fixation by plants; C = harvested N (grain and vegetation); L = N lost through leaching; V = gaseous losses of N through denitrification, burning, ammonia volatilization; and E = N lost through erosion.

Mathematical manipulation and omission of terms F_F, M, DR, A, V and E reveals that $F_L = \Delta N + L + C$. Under relatively long-term studies in the Australian pasture system, reasonably good estimates of F_L can be made where leaching losses and crop removal tends to be small [53]. In situations where leaching and denitrification are substantial and where fertilizers have been added to increase the standing vegetation, such an approach is not useful to estimate F_L. In long-term rotation studies, an estimate can be made of the relative recovery of fertilizer and legume N as shown in Table 4.1.6. Loss of N from inorganic N fertilizer ranged from 40% to 50%, which compares favourably with estimates by Allison [3] and Hardy [60]. Greater recovery of added N was obtained with longer crop rotations using clover and proportionately less row crops. From the long-term Broadbalk experiments at Rothamsted, accretion of 23–29 kg of soil N ha^{-1} was attributed largely to N_2 fixation by blue-green algae (18–24 kg N ha^{-1}) [77, 164]. In plots that were allowed to go to wilderness over a period of 81 years, a mean accumulation of 34 kg N ha^{-1} $year^{-1}$ was attributed to non-symbiotic N_2 fixation in the rhizosphere [164]. These values are realized, however, under conditions of little disturbance and without supplemental N additions. An excellent review on soil N budgets is available [91].

While legumes are often lauded for their N contribution to crops following in rotation, N budgets reveal that seed legumes, especially, may result in a net negative balance of soil N. Forage legumes may provide for a more favourable N balance [64, 65]. An estimate of N_2 fixation can be determined if a value is known for N removed in the forage. Data from long-term lysimeter studies also provide some estimates [91]. The readers are directed to an excellent review on the subject [88], which lists estimates of N_2 fixation for the forage legumes ranging from 100 to over 200 kg N ha^{-1} yr^{-1}.

Although few studies are available which provide an N budget for forage legumes, cutting management determines the enrichment of soil by fixed N_2 (see Table 4.1.7). If two harvests were taken, a net loss of 5 kg N ha^{-1} occurred, whereas a net input of 53 and 102 kg N ha^{-1} was realized for one and no harvests, respectively. These studies point out the importance of harvest management on the accretion of N and suggest that the beneficial response of crops following legumes may be due only in part to actual N carry-over from the legume.

Nitrogen transfer between crop communities

Several reviews are available on the transfer of N between legumes and other crops [67, 158]. When healthy legumes

Table 4.1.7. Nitrogen budget for seeding year alfalfa showing the allocation of symbiotically-fixed and soil-derived N among plant parts, and the input of N to the soil with two management options*

	Seeding year harvests		
	First: 12 July	Second: 30 Aug	Third: 20 Oct
Herbage yield (tonnes dry matter ha^{-1})	3.5	3.05	1.16
Total N yield (herbage and crown and roots) ($kg\ N\ ha^{-1}$)	118	127	59
Total N_2 fixed ($kg\ N\ ha^{-1}$)	57	102	34
Herbage	52	74	22
Roots and crown	5	28	12
N from soil	61	25	25
Herbage	54	18	16
Roots and crown	7	7	9
Management options			
Plough 20 October			
N input $harvest^{-1}$ ($kg\ ha^{-1}$)	−49	+10	+34
cumulative N input ($kg\ ha^{-1}$)	−49	−39	−5
Plough 30 August			
N input $harvest^{-1}$ ($kg\ ha^{-1}$)	−49	+102	—
cumulative N input ($kg\ ha^{-1}$)	−49	+53	—

* Adapted from Heichel *et al.* [65], as shown in Heichel & Barnes [63].

Table 4.1.8. Yield of maize grain in response to interplanting with legumes and to four rates of application of N fertilizer*

	Maize grain yield (tonnes ha^{-1})			
Fertilizer applied ($kg\ N\ ha^{-1}$)	Control	Cowpea interplanted	Mungbean interplanted	Calopo interplanted
0	1.79	1.85	3.08	1.85
45	3.08	2.75	3.07	3.05
90	3.42	2.75	2.75	3.07
135	2.58	2.92	1.96	2.92

* Data of Agboola & Fayemi [1], as shown in Heichel & Barnes [63].

and non-legumes are grown together, there is usually little exchange of N, although there is competition between these species for uptake of mineral soil N. Legumes grown in grass mixtures are often less competitive in uptake of soil N and may increase their symbiotic N fixation capacities as a result. Some interspecies transfer of N is believed to occur through mineralization of root exudates or from sloughed or dead nodules during the growing season [63].

Differences in N transfer between species may be due to senescence of nodules, roots and stolons after harvest [63] although little turnover apparently occurs after alfalfa [31, 148]. Evidence of N transfer from legume to maize (*Zea mays* L.) has been suggested (see Table 4.1.8). The results suggest that, with maize interplanted with mungbean (*Phaseolus aureus* L.), the mungbean contributed an equivalent of 45 $kg\ N\ ha^{-1}$ to the maize. Little response of maize was elicited from either cowpea (*Vigna sinensis* L. Endl. ex. Hassk.) or calopo (*Calopogonium mucunoides* Desv.). The above studies do not accurately assess whether N is actually being transferred from the legume to the non-legume. Rather it is inferred that the response is due to N from the legumes. Clearly, these studies should be verified with ^{15}N.

Studies in Nigeria by Eaglesham [46] show an increase in N content and concentration in maize intercropped with a legume compared to monoculture maize grown at low levels of fertilization (see Table 4.1.9). From ^{15}N enrichment values, it is apparent that intercropped maize had access to a source of N not available to monoculture

Table 4.1.9. N accumulation by sole cropped and intercropped maize (after Eaglesham [46])

	Treatment (kg N ha^{-1})	N content (mg $plant^{-1}$)	% N	% Atom excess ^{15}N
Sole crop	0	469	0.70	—
Intercrop	0	915*	0.92†	—
Sole crop	25	426	0.58	0.7727
Intercrop	25	782*	0.83‡	0.3714*
Sole crop	100	810	0.67	0.8867
Intercrop	100	989 NS	0.70 NS	0.6225 NS

Statistical comparisons are within N treatments and columns. NS = not significantly different; * = significantly different at $P = 0.05$; † = significantly different at $P = 0.01$; ‡ = significantly different at $P = 0.0001$.

maize. These results are not typical for all legumes and especially for soybean, which has a characteristically high harvest N index. In another study, no direct evidence of N transfer was obtained from soybean to an associated non-legume [155].

The maize/bean intercropping association is practised widely throughout Latin America [127]. For example, 75% of the beans produced in Latin America are grown principally with maize [68]. Reports of N_2 fixation in bean range from near zero to 65 kg N ha^{-1} in tropical regions to an excess of 100 kg N ha^{-1} in temperate regions [119]. Few data are available on the interspecies transfer of N from bean to maize [124]. Since N deficiency is common throughout the maize/bean growing region, some mineral N would clearly benefit the maize but is likely to reduce the symbiotic N_2 fixation of the bean in association. Additional factors of competition for light, nutrients and water in the association are in need of further study.

Nitrogen budgets in seed legumes

Contrary to popular opinion, many seed legumes export more N in the seed at harvest than what is added by N_2 fixation. Consequently, there is a loss instead of accretion. This is particularly true for soybean, which can export 130 kg N ha^{-1} [46, 63]. Positive budgets have been reported for soybean [19] but only under a very low level of soil N. Accretion of 54 kg N ha^{-1} has been reported for cowpea by Eaglesham [46]. Budgets for cool-season pulses grown in the Pacific northwest, USA, have varied from a negative 48 to a positive 54 kg N ha^{-1} [137]. Relatively high N accretion values of about 50 kg N ha^{-1} for faba bean (*Vicia faba* L.) in this study reflect the relatively low harvest N index for this crop. Breeding programmes which increase the harvest N index will obviously decrease the accretion.

Nitrogen transfer to subsequent crops

Legume N can be mineralized to amino compounds and to ammonia which is then nitrified. A sizable fraction of the legume residue, however, will remain in the soil organic N pool. In Australia, from 10 to 17% of the N from medic (*Medicago littoralis* Rhode) was available to a subsequent wheat crop while 72–78% was found in the soil organic N pool [86]. Later studies [85] showed slightly higher utilization (20–28%) of medic N by a first wheat crop. A second wheat crop utilized <5% of the original medic N, a value close to release from stabilized residues.

Yield response following legumes

Perhaps the information most significant to the farmer is the potential yield benefit to non-legumes following legumes. While yield increases are often attributed to residual N from legumes, N budgets for legumes and especially for seed legumes are often negative as previously shown. Therefore the yield response of cereals and grasses following legumes can be attributed not only to N but also to other influences which can be termed rotational.

NITROGEN EFFECTS

These effects are attributed to direct N response to the decomposition of legume residues or to residual soil N not taken up by the legume (sparing effect).

ROTATIONAL EFFECTS

These responses are attributed to effects other than N such as beneficial changes in soil properties [8], reduction in diseases and pests [30], reduction in toxic substances in crop residues [9] and release of growth-promoting sub-

Table 4.1.10. Maize grain yield as influenced by previous crop and rate of fertilizer N addition at the Southern Experiment Station, Waseca, Minn. (average 1975–80)*

	Maize grain yield (tonnes ha^{-1})			
Nitrogen rate (kg ha^{-1})	Maize previously	Soybean previously	Wheat previously	Wheat–alfalfa previously
0	4.45	6.89	6.85	7.25
45	5.80	8.14	8.00	8.00
90	6.46	8.70	8.78	8.80
135	6.85	8.97	8.78	9.05
180	7.36	9.37	8.98	9.24
225	7.53	9.37	9.20	9.11

* Data of Langer & Randall [87], as cited by Heichel & Barnes [63].

stances [122]. The benefits of rotating crops are clearly evident from a 6-year study in Minnesota (see Table 4.1.10). Yield of maize at all N levels following wheat was equal to maize yield following soybean, which suggests that benefits were largely from rotation and not from direct effects of N. Cook [30] suggests that crops grown without adequate rotation can appear nutrient-deficient because of poor root health leading to less efficiency in the utilization of soil nutrients. Yield of maize grown in rotation with soybean in midwest USA is commonly 10–15% higher than maize grown continuously.

4.1.8 PRODUCTION POTENTIAL OF LEGUMES

While production potentials of legumes and cereals are often compared, legume yield improvement lags far behind that of cereals. Soybean production in the USA has increased only 14–21% in the past 25 years, while cereal yields have increased 50–75% during this period [140]. On a worldwide basis, from a third to half of the increase in cereal yields during the past 30 years can be attributed to the use of N fertilizer [47, 59]. The relatively greater cereal yields in the more developed countries contrast with much lower yields and lower fertilizer additions for the developing countries. To meet the food needs of the expanding world population, it has been estimated that cereal production will need to double from the current 1300 to 2600 million metric tons [61]. Yet unpredictable factors such as availability and cost of fossil fuel-derived fertilizers, political stability and economics may prevent these goals from being attained. For example, the cost of fertilizer N increased greatly in recent years. It was 31% of the total non-solar energy input in maize production in 1970 as compared to 6% in 1945 [114].

Dietary needs of much of the world's population are met by a balance of cereals and grain legumes. While the yields of grain legumes are low compared to that of cereals, legumes are much higher in protein and require greater amounts of energy [134]. The study of biological nitrogen fixation (BNF) has long been devoted to improving crop productivity worldwide. Notable increases in production have been observed after introduction of rhizobia, especially in regions where the host rhizobia were not native. However, in many instances, BNF may not be the limiting factor [5]. Rather, soil acidity, drought, pests, diseases and poor soil fertility limit production.

In many developing countries, farmers regard legumes as a secondary crop. For example, chickpea (*Cicer arietinum* L.) and lentil (*Lens culinaris* Medik.) are grown in marginal and arid areas in the Near and Far East. In addition to the general agronomic neglect of legumes in fertilizer and irrigation, the legume research effort has been small compared with research on maize and other cereals. In 1976, there were more than 400 maize breeders in the USA compared with only 25 for soybean [112]. In addition, legumes have a narrow genetic base as compared with cereals [139], which limits 'hybrid vigour' in legume breeding. These factors all contribute to the general lag in seed legume production as compared with cereals.

4.1.9 GENETIC IMPROVEMENT OF SYMBIOTIC N_2 FIXATION

Conventional selection and breeding

A successful symbiosis requires input from both host and symbiont. Therefore, genetic variability in both can be exploited as a means of improving the partnership [113].

Wide variation in N_2 fixation and host specificity has been shown for various strains of *Rhizobium* [19, 101]. While significant progress has been made in screening naturally-occurring *Rhizobium* isolates for N_2 fixation, the inability of otherwise effective strains to compete for nodule occupancy against an indigenous population still remains a major obstacle in achieving maximum N_2 fixation.

Exploitation of the host is considered a viable option for improving symbiotic N_2 fixation [141] in both forage and seed legume breeding programmes. A first requisite for improving N_2 fixation through conventional plant breeding is genetic variability in the current germplasm. Research on alfalfa [11, 66] has shown heritable differences in C_2H_2 reduction that were useful for separate phenotypic recurrent selections. After two cycles, selections for low and high levels of C_2H_2 reduction, nodule mass and plant dry matter were positively correlated to total Kjeldahl N of alfalfa in the greenhouse. However, field studies showed that selected plants showed no increase in either N_2 fixation or yield. Similar selections [113, 142] were made for forage yield and N content in a preliminary field trial using a single strain of *Rhizobium* in the presence or absence of NH_4NO_3. Only plants high in both N concentration and dry weight were selected, since these parameters are often negatively correlated. The significance of this approach as a viable selection method is still to be tested in the field, although preliminary results are promising.

Genetic studies to improve N_2 fixation in seed legumes have been conducted for the common bean [7] using the inbred backcross line method. Lines from crosses of low N_2-fixing commercial varieties and high N_2-fixing parents were selected for increased N_2 fixation by C_2H_2 reduction and seed yield. The selected lines fixed more N_2 (^{15}N isotope dilution) and had a higher percentage of N derived from the atmosphere and seed yield than their parents. Lines differed considerably with regard to the proportion of nodules in the primary and secondary root area.

Role of molecular biology

Molecular genetics has been used to identify the *nif* and other genes important in N_2 fixation such as genes for root hair curling [42], the Hup^+ trait [32], and the leghaemoglobin genes. Nodule formation involves the expression of about 25 genes whose expression has not been observed in other tissues [17]. These specific plant genes encode for nodulins [149]. Most nodulin genes are generally expressed before and during symbiotic N_2 fixation while a few are expressed earlier [17]. Useful tools for studying the expression of nodulin genes are nodule-specific cDNA clones [52] selected from a cDNA library constructed from poly A RNA that was isolated from nodules. In fast growing *Rhizobium* species, *nod* genes required for nodulation are located on the Sym plasmid. Curing of this plasmid and introduction of cloned DNA containing the *nod*-region cluster resulted in the formation of nodules by strains of *Rhizobium* [43].

One other approach is to develop NO_3^--insensitive plant mutants that nodulate and fix N_2 in the presence of soil nitrate. Through mutagenesis with ethyl methane-sulphonate, a supernodulating mutant of a pea variety 'Rondo' was produced that developed substantially more nodule mass and nodule number in the presence or absence of 15 mM KNO_3 [75]. The phenotype was stable in the M_3 and M_4 generations and was controlled by the shoot. Through chemical mutagens, a supernodulation mutant *nts* 1007 of soybean, variety 'Bragg' was produced that was well nodulated at either zero or 5 mM NO_3^- [54]. However, there was no increase in either total shoot or root dry weight. The supernodulating trait may act as an additional drain on the plant for photosynthate.

The use of NO_3^--insensitive N_2-fixing genotypes, which presumably leaves more soil NO_3^- for subsequent crops, may be realistic under low rainfall conditions where NO_3^- storage is more likely. Under high rainfall conditions, residual NO_3^- may be leached and denitrified, which may not provide much gain for subsequent crops. This strategy would also not be feasible in environments where ground water supplies are already contaminated with excess NO_3^-. When bean, cowpea and other legumes are intercropped with cereals and grasses, 'extra' soil NO_3^- would benefit the non-legumes.

4.1.10 FUTURE PROJECTIONS

Biological N_2 fixation can be studied in many ways, from the more basic aspects of the biochemistry of nitrogenase to the ecology of *Rhizobium*. The ultimate goal is to better understand the process of biological N_2 fixation and to better determine its role in microbial ecology and agroecosystems. Of paramount importance is the significance of BNF in food production. Projections indicate that food production will need to double in the next half century to meet the increased competition for limited energy, water and land resources, the rising costs of labour, the demands for improved diets and population growth [160]. These projections come at a time of grain and oil surpluses and falling prices. Several countries — China, Indonesia, India — which were dependent on food imports are now net exporters of some food commodities.

In developing countries, improvement of BNF is sometimes limited by the relatively small number of farmers that use inoculation either because inoculants are not available, or they are of poor quality, or because of the problems which may result from inadequate infrastructure for its distribution. The relatively low success rate of inoculant strains in the nodules limits the potential benefits from inoculation for many legumes grown worldwide.

Future research to improve BNF will need the principles and tools of molecular biology to better understand the nature of host specificity and competition for nodule sites. Both the host and the endophyte need to be studied. The rhizosphere needs to be exploited as the centre stage where a number of actors can perform in concert. Selection, enhancement and genetic engineering of organisms that can reduce root diseases and increase productivity and BNF through co-inoculation with specially selected or designed strains of *Rhizobium* should be encouraged. This process needs to involve the selection of mycorrhizae and other organisms that can improve plant nutrition, especially in soils low in phosphorus and other nutrients. Research on BNF and cropping systems which addresses the sequence and interplanting of crops rather than the individual commodities needs to be encouraged. Agroforestry systems with N_2-fixing perennial species need to be identified and established to reduce the loss of top soil and the desertification process and to serve as a source of fuel and food, especially in parts of the tropics.

4.1.11 REFERENCES

[1] Agboola A.A. & Fayemi A.A.A. (1972) Fixation and excretion of nitrogen by tropical legumes. *Agron. J.* **64**, 409–12.

[2] Albrecht S.L., Maier R.J., Hanus F.J., Russell S.A., Emerich D.W. & Evans H.J. (1979) Hydrogenase in *Rhizobium japonicum* increases nitrogen fixation by nodulated soybeans. *Science* **203**, 1255–7.

[3] Allison F.E. (1966) The fate of nitrogen added to soils. *Adv. Agron.* **18**, 219–58.

[4] Ames P. & Bergman K. (1981) Competitive advantages provided by bacterial motility in the formation of nodules by *Rhizobium meliloti*. *J. Bact.* **148**, 728–9.

[5] App A. & Eaglesham A. (1982) Biological nitrogen fixation — problems and potential. In *Biological Nitrogen Fixation Technology for Tropical Agriculture*, p.1 (ed. P.H. Graham & S.C. Harris). Papers presented at a workshop held at CIAT, March 9–13, 1981. Centro Intrernacional de Agricultura Tropical, Cali, Colombia.

[6] Atkins C.A., Herridge D.F. & Pate J.S. (1979) The economy of carbon and nitrogen in nitrogen-fixing annual legumes. In *Isotopes in Biological Fixation*, p.211. International Atomic Energy Agency, Vienna.

[7] Attewell J. & Bliss F.A. (1985) Host plant characteristics of common bean lines selected using indirect measures of N_2 fixation. In *Nitrogen Fixation Research Progress*, p.3 (ed. H.J. Evans, P.J. Bottomley & W.E. Newton). Nijhoff, Dordrecht.

[8] Baldock J.O., Higgs R.L., Paulsen W.H., Jackobs J.A. & Schrader W.D. (1981) Legume and mineral nitrogen effects on crop yields in several crop sequences in the upper Mississippi Valley. *Agron. J.* **73**, 885–90.

[9] Barber S.A. (1972) Relation of weather to the influence of hay crops on subsequent corn yields on a Chalmers silt loam. *Agron. J.* **64**, 8–10.

[10] Barkay T., Fouts D. & Olson B.H. (1985) Preparation of a DNA probe for detection of mercury resistance genes in Gram-negative bacterial communities. *Appl. Environ. Microbiol.* **49**, 686–92.

[11] Barnes D.K., Heichel G.H., Vance C.P., Viands D.R. & Hardarson G. (1981) Successes and problems encountered while breeding for enhanced N_2 fixation in alfalfa. In *Genetic Engineering Conservation of Fixed Nitrogen*, p.233 (ed. J.M. Lyons, R.C. Vallentine, D.A. Phillips, D.W. Rains and P.C. Hoffaker). Plenum Press, New York.

[12] Bauer W.D. (1981) Infection of legumes by *Rhizobia*. *Ann. Rev. Plant Physiol.* **32**, 427–49.

[13] Becking J. H. (1974) Frankiaceae. In *Bergey's Manual of Determinative Bacteriology*, p.701 (ed. R.E. Buchanan & N.E. Gibbons). Williams & Wilkins, Baltimore.

[14] Berger, J.A., May S.N., Berger L.R. & Bohlool B.B. (1979) Colorimetric enzyme-linked immunosorbent assay for identification of strains of *Rhizobium* in culture and in the nodules of lentils. *Appl. Environ. Microbiol.* **37**, 642–6.

[15] Bergerson F.J. (1980) Measurement of nitrogen fixation by direct means. In *Methods for Evaluating Biological Nitrogen Fixation*, p.65 (ed. F.J. Bergersen). John Wiley, New York.

[16] Beringer J.E., Beynon J.L., Buchanan-Wollaston A.V. & Johnston A.W.B. (1978) Transfer of the drug-resistance transposon Tn5 to *Rhizobium*. *Nature (London)* **276**, 633–4.

[17] Beringer J.E., Bisseling T.A. & Larue T.A. (1988) Improving symbiotic nitrogen fixation through the genetic manipulation of *Rhizobium* and legume host plants. In *World Crops: Cool Season Food Legumes*

(ed. R.J. Summerfield). Nijhoff, Dordrecht. (In press).

[18] Bezdicek D.F. & Donaldson M.D. (1981) Flocculation of *Rhizobium* from soil colloids for enumeration by immunofluorescence. In *Microbial Adhesion to Surfaces*, p.297 (ed. R.C.W. Berkeley, J.M. Lynch, J.Melling, P.R. Rutter & B. Vincent). Ellis Harwood, London.

[19] Bezdicek D.F., Evans D.W., Abebe B. & Witters R.W. (1978) Evaluation of peat and granular inoculum for soybean yield and N fixation under irrigation. *Agron. J.* **70**, 865–8.

[20] Bhuvaneswari T.V., Mills K.K., Christ D.K., Evans W.R. & Bauer W.D. (1983) Effects of culture age on symbiotic infectivity of *Rhizobium japonicum*. *J. Bacteriol.* **153**, 443–51.

[21] Bohlool B.B., Kosslak R.M. & Woolfenden R. (1984) The ecology of *Rhizobium* in the rhizosphere: survival, growth and competition. In *Advances in Nitrogen Fixation Research*, p.287 (ed. C. Veeger & W.E. Newton). North Holland, Amsterdam.

[22] Bohlool B.B. & Schmidt E.L. (1974) Lectins: a possible basis for specificity in the *Rhizobium*–legume root nodule symbiosis. *Science* **185**, 269–71.

[23] Bohlool B.B. & Schmidt E.L. (1980) The immunofluorescence approach in microbial ecology. *Adv. Microb. Ecol.* **4**, 203–42.

[24] Bond G. (1941) Symbiosis of leguminous plants and nodule bacteria. I. Observations on respiration and on the extent of utilization of host carbohydrates by the nodule bacteria. *Ann. Bot.* **5**, 313.

[25] Broadbent F.E., Nekashime T. & Chang G.C. (1982) Estimation of nitrogen fixation by isotope dilution in field and greenhouse experiments. *Agron. J.* **74**, 625–8.

[26] Brockwell J. (1963) Accuracy of a plant infection technique for counting populations of *Rhizobium trifolii*. *Appl. Microbiol.* **11**, 377–83.

[27] Brockwell J., Schwinghammer E.A. & Gault R.R. (1977) Ecological studies of root-nodule bacteria — V. A critical examination of the stability of antigenic and streptomycin-resistance markers for identification of strains of *Rhizobium trifolii*. *Soil Biol. Biochem.* **9**, 19–24.

[28] CAST (Council for Agricultural Science and Technology) (1976) *Effect of Increased Nitrogen Fixation on Stratographic Ozone*. Report No. 53, Iowa State University, Ames, Iowa.

[29] Christeller J.T., Laing W.A. & Sutton W.D. (1977) Carbon dioxide fixation by lupin root nodules. I. Characterization, association with phosphoenolpyruvate carboxylase and correlation with nitrogen fixation during nodule development. *Plant Physiol.* **60**, 47–50.

[30] Cook R.J. (1984) Root health: importance and relationship to farming practices. In *Organic Farming: Current Technology and its Role in a Sustainable Agriculture*, p.111 (ed. D.F. Bezdicek, J.F. Power, D.R. Keeney & M.J. Wright). Amer. Soc. Agron. Special Publication No. 46. Madison, WI.

[31] Cralle H.T. & Heichel G.H. (1981) Nitrogen fixation and vegetation regrowth of alfalfa and birdsfoot trefoil after successive harvests of fibral debudding. *Plant Physiol.* **67**, 898–905.

[32] Cunningham S.D., Kapulnik Y., Brewin N.J. & Phillips D.K. (1985) Uptake hydrogenase activity determined by plasmid pRL6J1 in *Rhizobium leguminosarum* does not increase symbiotic nitrogen fixation. *Appl. Environ. Microbiol.* **50**, 791–4.

[33] Danso S.K.A. & Alexander M. (1974) The survival of *Rhizobium* in soil. *Soil Sci. Soc. Am. J.* **38**, 86–9.

[34] Danso S.K.A., Keya S.O. & Alexander M. (1975) Protozoa and the decline of *Rhizobium* populations added to soil. *Can. J. Microbiol.* **21**, 884–95.

[35] Dazzo F.B. & Hubbell D.H. (1976) Concanavalin A: lack of correlation between binding to *Rhizobium* and specificity in the *Rhizobium*–legume association. *Plant Soil* **43**, 713–17.

[36] Dazzo F.B. & Truchet G.L. (1983) Interaction of lectins and their saccharide receptors in the *Rhizobium*–legume symbiosis. *J. Membrane Biol.* **75**, 1–16.

[37] De Bont J.A.M. (1976) Bacterial degradation of ethylene and the acetylene reduction test. *Can. J. Microbiol.* **22**, 1060–2.

[38] DeJong T.M., Brewin N.J., Johnston A.W.B. & Phillips D.A. (1982) Improvement of symbiotic properties in *Rhizobium leguminosarum* by plasmid transfer. *J. Gen. Microbiol.* **128**, 1829–38.

[39] Dixon R.O.D. (1967) Hydrogen uptake and exchange by pea root nodules. *Ann. Bot.* **31**, 179–88.

[40] Dixon R.O.D. (1968) Hydrogenase in pea root nodule bacteroids. *Arch. Mikrobiol.* **62**, 272–83.

[41] Dixon R.O.D. (1972) Hydrogenase in legume root nodule bacteroids: occurrence and properties. *Arch. Mikrobiol.* **85**, 193–201.

[42] Djordjovic M.A., Schofield P.R., Ridge R.W., Morrison N.A., Bassam J., Plazinski J., Watson J.M. & Rolfe B.J. (1985) *Rhizobium* nodulation genes involved in root hair curling (Hac) are functionally conserved. *Plant Mol. Biol.* **4**, 147–60.

[43] Downie J.A., Ma Q.S., Knight C.D., Hombrecher G. & Johnston A.W.B. (1983) Cloning of the symbiotic region of *Rhizobium leguminosarum*: the nodulation genes are between the nitrogenase genes and a *nifA*-like gene. *Embo Journal* **2**, 947–52.

[44] Dudman W.F. (1977) Serological methods and their application to dinitrogen-fixing organisms. In *A Treat-*

ise on Dinitrogen Fixation. Section IV: Agronomy and Ecology, p.487 (ed. R. Hardy & A. Gibson). John Wiley, New York.

[45] Dughri M.H. & Bottomley P.J. (1984) Soil acidity and the composition of an indigenous population of *Rhizobium trifolii* in nodules of different cultivars of *Trifolium subterraneum* L. *Soil Biol. Biochem.* **16**, 405–11.

[46] Eaglesham A.R.J. (1982) Assessing the nitrogen contribution of cowpea (*Vigna unguiculata*) in monoculture and intercropped. In *Biological Nitrogen Fixation Technology for Tropical Agriculture*, p.641 (ed. P.H. Graham & S.C. Harris). Papers presented at a workshop held at CIAT, March 9–13. Centro Internacional de Agricultura Tropical, Cali, Colombia.

[47] Engibous J.C. (1975) Possible effects of fertilizer shortages on food grain production. In *Impact of Fertilizer Shortage: Focus on Asia*, p.193. Asian Productivity Organization, Tokyo.

[48] Evans H.J., Hanus F.J., Russell S.A., Harker A.R., Lambert G.R. & Dalton D.A. (1985) Biochemical characterization, evaluation and genetics of H_2 cycling in *Rhizobium.* In *Nitrogen Fixation and CO_2 Metabolism*, p.3 (ed. P.W. Ludden & J.E. Burris). Elsevier, New York.

[49] Evans H.J., Russell S.A., Hanus F.J. & Ruiz-Argueso T. (1987) The importance of hydrogen cycling in nitrogen fixation by pulses. In *World Crops: Cool Season Food Legumes*, (ed. R.J. Summerfield). Nijhoff, Dordrecht (In press).

[50] Frederickson J.K., Bezdicek D.F. & Li S.-W. (1986) Utility of transposable elements as markers for detection and enumeration of bacterial candidates for genetic engineering. In *Terrestrial Environments: Joint ASM–CSM Symposium on Environmental Insult and Recovery of Stressed Systems*, p.19. Toronto, Canada.

[51] Fried M. & Dean L.A. (1952) A concept concerning the measurement of available soil nutrients. *Soil Sci.* **73**, 263–71.

[52] Govers F., Moerman M., Nap J.P., van Kamman A. & Bisseling T. (1985) Expression of nodule specific sequences during pea root nodule development. In *Nitrogen Fixation Research Progress*, p.62 (ed. H.J. Evans, P.J. Bottomley & W.E. Newton). Nijhoff, Dordrecht.

[53] Greenland D.J. (1977) Contribution of microorganisms to the nitrogen status of tropical soils. In *Biological Nitrogen Fixation in Farming Systems of the Tropics*, p.13 (ed. A. Ayanaba and P.J. Dart). Wiley, New York.

[54] Gresshoff P.M., Day D.A. & Delves A.C. (1985) Plant host genetics of nodulation and symbiotic nitrogen fixation in pea and soybean. In *Nitrogen Fixation Research Progress*, p.19 (ed. H.J. Evans, P.J. Bottomley & W.E. Newton). Nijhoff, Dordrecht.

[55] Gross D.C., Vidaver A.K. & Klucas R.V. (1979) Plasmids, biological properties and efficacy of nitrogen fixation in *Rhizobium japonicum* strains indigenous to alkaline soils. *J. Gen. Microbiol.* **114**, 257–66.

[56] Halverson L.J. & Stacey G. (1986) Signal exchange in plant–microbe interactions. *Microbiol. Rev.* **50**, 193–225.

[57] Ham G.E. & Caldwell A.C. (1978) Fertilizer placement effects on soybean seed yield, N_2 fixation and ^{32}P uptake. *Agron. J.* **70**, 779–83.

[58] Hanus F.J., Albrecht S.L., Zablotowicz R.M., Emerch D.W., Russell S.A. & Evans H.J. (1981) Yield and N content of soybean seed as influenced by *Rhizobium japonicum* inoculants possessing the hydrogenase characteristics. *Agron. J.* **73**, 368–72.

[59] Hardy R.W.F. (1975) Fertilizer research with emphasis on nitrogen fixation. In *Proceedings of 24th Annual Meeting of Agriculture Research Institute*, Nat. Acad. of Sci., Washington, DC.

[60] Hardy R.W.F. (1976) Potential impact of current abiological and biological research on the problem of providing fixed nitrogen. In *Proceedings of the 1st International Symposium on Nitrogen Fixation*, Vol. 2, p.693 (ed. W.E. Newton & C.J. Nyman). Washington State University Press, Pullman, WA.

[61] Hardy R.W.F. (1977) Increasing crop productivity: agronomic and economic considerations on the role of biological nitrogen fixation. In *Report of the Public Meeting in Genetic Engineering for Nitrogen Fixation*, p.77. Nat. Acad. Sci., Washington, DC.

[62] Hardy R.W.F., Burns R.C. & Holsten R.D. (1973) Applications of the acetylene-ethylene assay for measurement of nitrogen fixation. *Soil Biol. Biochem.* **5**, 47–81.

[63] Heichel G.H. & Barnes D.R. (1984) Opportunities for meeting crop nitrogen needs from symbiotic nitrogen fixation. In *Organic Farming: Current Technology and its Role in a Sustainable Agriculture*, p.49 (ed. D.F. Bezdicek, J.F. Power, D.R. Keeney & M.J. Wright). Am. Soc. Agron., Special Publication No. 46, Madison, WI.

[64] Heichel G.H., Barnes D.K. & Vance C.P. (1981a) Nitrogen fixation of alfalfa in the seedling year. *Crop Sci.* **21**, 330–5.

[65] Heichel G.H., Barnes D.K. & Vance C.P. (1981b) Nitrogen fixation by forage legumes and benefits to the cropping systems. In *Proceedings 6th Annual Symposium*, Minn. Forage and Grassland Council, St Paul, MN.

[66] Heichel G.H., Vance C.P. & Barnes D.K. (1981c). Evaluating elite alfalfa lines for N_2-fixation under field

conditions. In *Genetic Engineering of Symbiotic Nitrogen Fixation and Conservation of Fixed Nitrogen*, p.217 (ed. J.M. Lyons, R.C. Vallentine, D.P. Phillips, D.W. Rains & R.C. Hoffaker). Plenum Press, New York.

[67] Henzell E.F. & Vallis I. (1977) Transfer of nitrogen between legumes and other crops. In *Biological Nitrogen Fixation in Farming Systems in the Tropics*, p.73 (ed. A. Ayanaba & P.J. Dart). John Wiley, New York.

[68] Hernandez-Bravo G. (1973) Potentials and problems of production of dry beans in the lowland tropics. In *Potentials of Field Beans and Other Food Legumes in Latin America*, p.144. CIAT, Cali, Colombia.

[69] Herridge D.F. (1982) A whole-system approach to quantifying biological nitrogen fixation by legumes and associated gains and losses of nitrogen in agricultural systems, In *Biological Nitrogen Fixation Technology for Tropical Agriculture*, p.593 (ed. P.H. Graham & S.C. Harris). Papers presented at a workshop held at CIAT, March 9–13, 1981. Centro Internacional de Agricultura Tropical, Cali, Colombia.

[70] Herridge D.F., Atkins G.A., Pate J.S. & Rainbird R.M. (1978) Allantoin, and allantoic acid in the nitrogen economy of the cowpeas (*Vigna unguiculata* (L.) Walp). *Plant Physiol.* **62**, 495–8.

[71] Heytler P.G. & Hardy R.W.F. (1979) Energy requirement for N-fixation by rhizobial nodules in soybeans. *Plant Physiol. Supplement* **63**, 84.

[72] Higashi S. & Abe M. (1980) Scanning electron microscopy of *Rhizobium trifolii* infection sites on root hairs of white clover. *Appl. Environ. Microbiol.* **40**, 1094–9.

[73] Hoch G.E., Little H.M. & Burris R.H. (1957) Hydrogen evolution from soybean root nodules. *Nature* **179**, 430–1.

[74] Hodgson A.L.M. & Roberts W.P. (1983) DNA colony hybridization to identify *Rhizobium* strains. *J. Gen. Microbiol.* **129**, 207–12.

[75] Jacobsen E. & Feenstra W.J. (1984) A new pea mutant with efficient nodulation in the presence of nitrate. *Plant Science Letters* **33**, 337–44.

[76] Jansen van Rensburg H. & Strijdom B.W. (1982) Root surface association in relation to nodulation of *Medicago sativa*. *Appl. Environ. Microbiol.* **44**, 93–7.

[77] Jenkinson D.S. (1977) The nitrogen economy of the Broadbalk experiments. I. Nitrogen balance in the experiments. In *Rothamsted Exp. Sta. Report for 1976*, Part 2, p.103.

[78] Jones D.G. & Bromfield E.S.P. (1978) A study of the competitive ability of streptomycin and spectinomycin mutants of *Rhizobium trifolii* using various marker techniques. *Ann. Appl. Biol.* **88**, 445–87.

[79] Jordan D.C. (1984) Rhizobiaceae. In *Bergey's Manual of Systematic Bacteriology*, Vol. 1. p.234 (ed. N.R. Krieg & J.G. Holt). Williams & Wilkins, Baltimore.

[80] Josey D.P., Beyon J.L., Johnston A.W.B. & Beringer J.E. (1979) Strain identifiation in *Rhizobium* using intrinsic antibiotic resistance. *J. Appl. Bacteriol.* **46**, 343–50.

[81] Keya S.O. & Alexander M. (1975) Regulation of parasitism by host density: the *Bdellovibrio–Rhizobium* interrelationship. *Soil Biol. Biochem.* **7**, 231–7.

[82] Kingsley M.T. & Bohlool B.B. (1981) Release of *Rhizobium* spp. from tropical soils and recovery for immunofluorescence enumeration. *Appl. Environ. Microbiol.* **42**, 241–8.

[83] Kingsley M.T. & Bohlool B.B. (1983) Characterization of *Rhizobium* sp. (*Cicer arietinum* L.) by immunofluorescence, immunodiffusion, and intrinsic antibiotic-resistance. *Can. J. Microbiol.* **29**, 518–26.

[84] Kosslak R.M., Bohlool B.B., Dowdle S. & Sedowsky M.J. (1983) Competition of *Rhizobium japonicum* strains in early stages of soybean nodulation. *Appl. Environ. Microbiol.* **46**, 870–3.

[85] Ladd J.N., Amato M., Jackson R.B. & Butler J.H.A. (1983) Utilization by wheat crops of nitrogen from legume residues decomposing in soils in the field. *Soil Biol. Biochem.* **15**, 231–8.

[86] Ladd J.N., Oades J.M. & Amato M. (1981) Distribution and recovery of nitrogen from legume residues decomposing in soils sown to wheat in the field. *Soil Biol. Biochem.* **13**, 251–6.

[87] Langer D.K. & Randall G.W. (1981) Corn production as influenced by previous crop and N rate. *Agronomy Abstracts*, p.182. Amer. Soc. Agron., Madison, WI.

[88] Larue T.A. & Patterson T.G. (1981) How much nitrogen do legumes fix? *Ad. Agron.* **34**, 15–38.

[89] Layzell D.B., Rainbird R.M., Atkins C.A. & Pate J.S. (1979) Economy of photosynthate use in nitrogen-fixing legume nodules. *Plant Physiol.* **64**, 888–91.

[90] Lea P.J. & Miflin B.J. (1974) Alternative route for nitrogen assimilation in higher plants. *Nature* **251**, 614–16.

[91] Legg J.O. & Meisinger J.J. (1982) Soil N budgets. In *Nitrogen in Agricultural Soils*, p.503 (ed. F.J. Stevenson). Am. Soc. Agron., Madison, WI.

[92] Lethbridge G., Davidson M.S. & Sparling G.P. (1982) Critical evaluation of the acetylene reduction test for estimating the activity of nitrogen-fixing bacteria associated with the roots of wheat and barley. *Soil Biol. Biochem.* **14**, 27–35.

[93] Lowendorf H.S., & Alexander M. (1983) Selecting *Rhizobium meliloti* for inoculation of alfalfa planted in acid soils. *Soil Sci. Soc. Am. J.* **47**, 935–8.

[94] McClure P.R., Israel D.W. & Volk R.J. (1980) Evaluation of the relative ureide content of xylem sap

as an indicator of N_2 fixation in soybeans. *Plant Physiol.* **66**, 720–5.
[95] McNeil D.C. (1982) Quantification of symbiotic nitrogen fixation using ureides: a review. In *Biological Nitrogen Fixation Technology for Tropical Agriculture*, p.609 (ed. P.H. Graham & S.C. Harris). Papers presented at a workshop held at CIAT, March 9–13, 1981. Centro Internacional de Agricultura Tropical, Cali, Colombia.
[96] Mahon J.D. (1979) Environmental and genotypic effects on the respiration associated with symbiotic nitrogen fixation in peas. *Plant Physiol.* **63**, 892–7.
[97] Mahon J.D. (1983) Energy relationships. In *Nitrogen Fixation Volume 3: Legumes*, p.299 (ed. W.J. Broughton). Clarendon Press, Oxford.
[98] Marshall K.C. (1964) Survival of root-nodule bacteria in dry soils exposed to high temperatures. *Aust. J. Agric. Res.* **15**, 273–81.
[99] Martensson A.M., Gustafsson J. & Ljunggren H.D. (1984) A modified, highly sensitive enzyme-linked immunosorbent assay (ELISA) for *Rhizobium meliloti* strain identification. *J. Gen. Microbiol.* **130**, 247–53.
[100] Materon L.A. & Hagedorn C. (1984) Competitiveness of *Rhizobium trifolii* strains associated with red clover (*Trifolium pratense* L.) in Mississippi soils. *Appl. Environ. Microbiol.* **44**, 1096–101.
[101] May S.N. & Bohlool B.B. (1983) Competition among *Rhizobium leguminosarum* strains for nodulation of lentils (*Lens esculenta*). *Appl. Environ. Microbiol.* **45**, 960–5.
[102] Minchin F.R. & Pate J.S. (1973) The carbon balance of a legume and the functional economy of its root nodules. *J. Exp. Bot.* **24**, 259–71.
[103] Minchin F.R., Sheehy J.E. & Witty J.F. (1985) Factors limiting N_2 fixation by the legume–*Rhizobium* symbiosis. In *Nitrogen Fixation Research Progress, Proceedings of the 6th International Symposium on Nitrogen Fixation*, p.285 (ed. H.H. Evans, P.J. Bottomley & W.E. Newton). Nijhoff, Dordrecht.
[104] Moawad H.A., Ellis W.R. & Schmidt E.L. (1984) Rhizosphere response as a factor in competition among three serogroups of indigenous *Rhizobium japonicum* for nodulation of field grown soybeans. *Appl. Environ. Microbiol.* **47**, 607–12.
[105] National Research Council (NRC) (1978) *Nitrates: an Environmental Assessment.* Nat. Acad. Sci., Washington, DC.
[106] Nelson L.M. (1983) Hydrogen cycling by *Rhizobium leguminosarum* isolates and growth and nitrogen contents of pea plants (*Pisum sativum* L.). *Appl. Environ. Microbiol.* **45**, 856–61.
[107] Parke D., Rivelli M. & Ornston L.N. (1985) Chemotaxis to aromatic and hydromatic acids: comparison of *Bradyrhizobium japonicum* and *Rhizobium trifolii. J. Bact.* **163**, 417–22.
[108] Pate J.S., Atkins C.A., White S.T., Rainbird R.M. & Woo K.C. (1980) Nitrogen nutrition and xylem transport of nitrogen in ureide-producing grain legumes. *Plant Physiol.* **65**, 961–5.
[109] Pate J.S., Layzell D.B. & Atkins C.A. (1979) Economy of carbon and nitrogen in a nodulated and non-nodulated (NO_3-grown) legume. *Plant Physiol.* **64**, 1083–8.
[110] Pena-Cabriales J.J. & Alexander M. (1979) Survival of *Rhizobium* in soils undergoing drying. *Soil Sci. Soc. Am. J.* **43**, 962–6.
[111] Pena-Cabriales J.J. & Alexander M. (1983) Growth of *Rhizobium* in unamended soil. *Soil Sci. Soc. Amer. J.* **47**, 81–4.
[112] Pendleton J.W. (1976) *The Key to Maximum Soybean Production.* Extension Bulletin Asian and Pacific Council Food and Fertilizer Technology Centre no. 82. Taipei, Taiwan.
[113] Phillips D.A. & Teuber L.R. (1985) Genetic improvement of symbiotic nitrogen fixation in legumes. In *Nitrogen Fixation Research Progress*, p.11 (ed. H.J. Evans, P.J. Bottomley & W.E. Newton). Nijhoff, Dordrecht.
[114] Pimental D. (1976) Food, nitrogen and energy. In *Proceedings of the First International Symposium on Nitrogen Fixation*, Vol. 2, p.656 (ed. W.E. Newton & C.J. Nyman). Washington State University Press, Pullman, WA.
[115] Pueppke S.G. (1984) Adsorption of slow- and fast-growing rhizobia to soybean and cowpea roots. *Plant Physiol.* **75**, 924–8.
[116] Pugashetti B.K., Angle J.S. & Wagner G.H. (1982) Soil microorganisms antagonistic towards *Rhizobium japonicum. Soil Biol. Biochem.* **14**, 45–9.
[117] Pull S.P., Pueppke S.G., Hymowitz T. & Orf J.H. (1978) Soybean lines lacking the 122,000 dalton seed lectin. *Science* **200**, 1277–9.
[118] Rennie R.J. (1985) Nitrogen fixation in agriculture in temperate regions. In *Nitrogen Fixation Research Progress*, p.659 (ed. H.J. Evans, P.J. Bottomley & W.E. Newton). Nijhoff, Dordrecht.
[119] Rennie R.J. & Kemp G.A. (1983) N_2-fixation in field beans quantified by ^{15}N isotope dilution. II. Effect of cultivars of beans. *Agron. J.* **75**, 645–9.
[120] Reyes V.G. & Schmidt E.L. (1979) Population densities of *Rhizobium japonicum* strain 123 estimated directly in soil and rhizospheres. *Appl. Environ. Microbiol.* **37**, 854–8.
[121] Rice W.A., Penney D.C. & Nyborg M. (1977) Effects

of soil acidity on rhizobia numbers, nodulation and nitrogen fixation by alfalfa and red clover. *Can. J. Soil Sci.* **57**, 197–203.

[122] Ries S.K., Wert V., Sweeley C.C. & Leavitt R.A. (1977) Triacontanol: a new naturally occurring growth regulator. *Sci.* **195**, 1339–41.

[123) Robert F.M. & Schmidt E.L. (1985) A comparison of lectin-binding activity in two strains of *Rhizobium japonicum. FEMS Microbiology Letters* **27**, 281–5.

[124] Ruschel A.P., Salati E. & Vose P.B. (1979) Nitrogen enrichment of soil and plant by *Rhizobium phaseoli–Phaseolus vulgaris* symbiosis. *Plant Soil* **51**, 425–9.

[125] Ruschel A.P., Vose P.B., Matsui E., Victoria R.L. & Saito S.M.T. (1981) Field evaluation of N_2-fixation and nitrogen utilization by *Phaseolus* bean varieties determined by ^{15}N isotope dilution. *Plant Soil* **65**, 397–407.

[126] Ryle G.J.A., Powell C.E. & Gordon A.J. (1979) The respiratory costs of nitrogen fixation in soybean, cowpea, and white clover. I. Nitrogen fixation and the respiration of the nodulated root. *J. Exp. Bot.* **30**, 135–44.

[127] Saito S.M.T. (1982) The nitrogen relationships of maize/bean associations. In *Biological Nitrogen Fixation Technology for Tropical Agriculture*, p.631 (ed. P.H. Graham & S.C. Harris). Papers presented at a workshop held at CIAT, March 9–13, 1981. Centro Internacional de Agricultura Tropical, Cali, Colombia.

[128] Sayler G.S., Shields M.S., Tedford E.T., Breen A., Hooper S.W., Sirotkin K.M. & Davis J.W. (1985) Application of colony hybridization to the detection of catabolic genotypes in environmental samples. *Appl. Environ. Microbiol.* **49**, 1295–303.

[129] Scholla M.H. & Elkan G.H. (1984) *Rhizobium fredii* sp. nov., a fast-growing species that effectively nodulates soybeans. *Int. J. Syst. Bacteriol.* **34**, 484–6.

[130] Schubert K.R. (1982) The energetics of biological nitrogen fixation. In *Conference on the Energetics of Biological Fixation of Dinitrogen*, April 27–30, 1980. American Society of Plant Physiologists, Michigan State University.

[131] Schubert K.R. and Evans H.J. (1976) Hydrogen evolution: a major factor affecting the efficiency of nitrogen fixation in nodulated symbionts. *Proceedings of the National Academy of Sciences* **73**, 1207–11.

[132] Schubert K.R. & Ryle G.J.A. (1980) The energy requirements for nitrogen fixation in nodulated legumes. In *Advances in Legume Science*, p.85 (ed. R.J. Summerfield & A.H. Bunting). Royal Botanic Gardens, Kew.

[133] Silvester W.B. (1977) Dinitrogen fixation by plant associations excluding legumes. In *A Treatise on Dinitrogen Fixation Section IV: Agronomy and Ecology*, p.141 (ed. R.W.F. Hardy & A.H. Gibson). Wiley–Interscience, New York.

[134] Sinclair J.R. & Dewit C.K.J. (1975) Photosynthate and nitrogen requirements for seed production by various crops. *Science* **189**, 565–7.

[135] Sloger C., Bezdicek D.F., Milberg R. & Boonkerd N. (1975) Seasonal and diurnal variations in N_2 (C_2H_2) fixing activity in field systems. In *Nitrogen Fixation by Free-Living Microorganisms*, p.271 (ed. W.D.P. Stewart). IBP, Cambridge University Press, Cambridge.

[136] Smith G.E. (1942) *Sanborn Field: Fifty Years of Field Experiments with Crop Rotations, Manure and Fertilizers.* Missouri Agr. Exp. Sta. Bull. no. 458. Columbia, Missouri.

[137] Smith S.C., Bezdicek D.F., Cheng H.H. & Turco R.F. (1987). Seasonal N_2 fixation by cool-season pulses based on several ^{15}N methods. *Plant and Soil* **97**, 3–13.

[138] Stewart W.D.P. & Pearson M.C. (1967) Nodulation and nitrogen-fixation by *Hippophae rhammnoides* L. in the field. *Plant and Soil* **26**, 348–59.

[139] Summerfield R.J. (1979) The contribution of physiology to breeding for increased yields in grain legume crops. In *Opportunities for Increasing Crop Yields*, p.51 (ed. R.E. Hurd, P.V. Biscoe & C. Dennis). Assoc. Appl. Biol., Putnam Advanced Publishing Program, Boston.

[140] Summerfield R.J., Minchin F.R. & Roberts E.H. (1978) Relation of yield potential in soybeans (*Glycine max* (L.) Merr.) and cowpea (*Vigna unguiculata* (L.) Walp). In *Proceedings BCPC/BPGRG Symposium — Opportunities for Chemical Plant Growth Production*, p.215. British Crop Protection Council, Croydon.

[141] Tan G.Y. (1981) Genetic variation for acetylene reduction rate and other characters in alfalfa. *Crop Sci.* **21**, 485–8.

[142] Teuber L.R., Levin R.P., Sweeney C. & Phillips D.A. (1984) Selection for N concentration and forage yield in alfalfa. *Crop Sci.* **24**, 553–8.

[143] Tilak K.V.B.R., Schneider K. & Schlegel H.G. (1984) Autotrophic growth of strains of *Rhizobium* and properties of isolated hydrogenase. *Curr. Microbiol.* **10**, 49–52.

[144] Truelsen T.A. & Wyndaele R. (1984) Recycling efficiency in hydrogenase uptake positive strains of *Rhizobium leguminosarum. Physiol. Plant* **62**, 45–50.

[145] Turco R.F. & Bezdicek D.F. (1987) Diversity within two serogroups of *Rhizobium leguminosarum* native to soils in the Palouse of eastern Washington. *Ann. Appl. Biol.* **111**, 103–14.

[146] Turco R.F., Bezdicek D.F. & Moorman T.B. (1986)

Effectiveness and competitiveness of spontaneous antibiotic resistant mutants of *R. leguminosarum* and *R. japonicum*. *Soil Biol. Biochem.* **18**, 259–62.

[147] Turner G.L. & Gibson A.H. (1980) Measurement of nitrogen fixation by indirect means. In *Methods for Evaluating Biological Nitrogen Fixation*, p.111 (ed. F.J. Bergersen). John Wiley, New York.

[148] Vance C.P., Heichel G.H., Barnes D.K., Bryan J.W. & Johnston L.E. (1979) Nitrogen fixation, nodule development, and vegetative regrowth of alfalfa (*Medicago sativa* L.) following harvest. *Plant Physiol.* **64**, 1–8.

[149] Van Kammen A. (1984) Plant genes involved in nodulation. In *Advances in Nitrogen Fixation Research*, p.587 (ed. C. Veeger & W.E. Newton). Nijhoff, Dordrecht.

[150] Vasilas B.L. & Ham G.E. (1984) Nitrogen fixation in soybeans: an evaluation of measurement techniques. *Agron. J.* **76**, 759–64.

[151] Vesper S.J. & Bauer W.D. (1986) Role of pili (fimbriae) in attachment of *Bradyrhizobium japonicum* to soybean roots. *Appl. Environ. Microbiol.* **52**, 134–41.

[152] Vidor C. & Miller R.H. (1980) Relative saprophytic competence of *Rhizobium japonicum* strains in soils as determined by the quantitative fluorescent antibody technique (FA). *Soil Biol. Biochem.* **12**, 483–7.

[153] Vincent J.B. (1970) *A Manual for the Practical Study of Root-nodule Bacteria*. IBP Handbook no. 15. Blackwell Scientific Publications, Oxford.

[154] Vose P.B., Ruschel A.P., Victoria R.L., Saito S.M.T. & Matsui E. (1982) ^{15}N research as a tool in biological nitrogen fixation research. In *Biological Nitrogen Fixation Technology for Tropical Agriculture*, p.575 (ed. P.H. Graham & S.L. Harris). Papers presented at a workshop held at CIAT, March 9–13, 1981. Centro Internacional de Agricultura Tropical, Cali, Colombia.

[155] Wahua T.A.T. & Miller D.A. (1978) Effects of intercropping on soybean N_2-fixation and plant composition on associated sorghum and soybeans. *Agron. J.* **70**, 292–5.

[156] Weaver R.W. & Frederick L.F. (1974) Effect of inoculum rate on competitive nodulation of *Glycine max* (L.) Merr. II. Field studies. *Agron. J.* **66**, 233–6.

[157] Weber D.F. & Miller V.L. (1972) Effect of soil temperature on *Rhizobium japonicum* serogroup distribution in soybean nodules. *Agron. J.* **64**, 796–8.

[158] Whitney A.S. (1977) Contribution of forage legumes to the nitrogen economy of mixed swards. A review of the relevant Hawaiian research. In *Biological Nitrogen Fixation in Farming Systems in the Tropics*, p.89 (ed. A. Ayanaba & P.J. Dart). John Wiley, New York.

[159] Winarno R. & Lie T.A. (1979) Competition between *Rhizobium* strains in nodule formation: interaction between nodulating and non-nodulating strains. *Plant Soil* **51**, 135–42.

[160] Wittwer S.H. (1985) Crop productivity — research imperatives: a decade of change. In *Crop Productivity — Research Imperatives Revisited*, p.1 (ed. M. Gibbs & C. Carlson). International conference held at Boyne Highlands Inn, October 13–18 and Arlie House, December 11–18, 1985.

[161] Witty J.F. (1979) Acetylene reduction assay can overestimate nitrogen-fixation in soil. *Soil Biol. Biochem.* **11**, 209–10.

[162] Witty J.F. (1983) Estimating N_2-fixation in the field using ^{15}N-labelled fertilizer: some problems and solutions. *Soil Biol. Biochem.* **15**, 631–9.

[163] Witty J.F. (1984) Acetylene-induced changes in the oxygen diffusion resistance and nitrogenase activity of legume root nodules. *Ann. Bot.* **53**, 13–20.

[164] Witty J.F., Day J.M. & Dart P.J. (1977) The nitrogen economy of the broadbalk experiments. II. Biological nitrogen fixation. In *Rothamsted Exp. Sta. Report for 1976,* Part 2, p.111.

[165] Wong P.O. (1980) Interactions between rhizobia and lectins of lentil, pea, broad bean, and jack bean. *Plant Physiol.* **65**, 1049–52.

[166] Wright S.F., Foster J.G. & Bennett O.L. (1986) Production and use of monoclonal antibodies for identification of strains of *Rhizobium trifolii*. *Appl. Environ. Microbiol.* **52**, 119–23.

[167] Yates M.A. (1980) Biochemistry of nitrogen fixation. In *The Biochemistry of Plants*, p.1 (ed. P.K. Stumpf & E.E. Conn). Academic Press, New York.

[168] Zavitkovski I. & Newton M. (1968) Effect of organic matter and combined nitrogen on nodulation and nitrogen fixation in red alder. In *Biology of Alder*, p.209 (ed. J.M. Trappe, J.F. Franklin, R.F. Tarrant & G.M. Hansen). Pacific Northwest Forest and Range Experiment Station, Forest Service, USDA, Portland.

4.2 Biological control

4.2.1 INTRODUCTION

Some of the environmental problems created by recalcitrant chemical molecules are outlined in Chapter 4.5. While non-degradable residues of some chemical pesticides can have a major impact on the environment, even those pesticides which are readily metabolized can disrupt the ecosystem. Pesticides (taken here to include fungicides and bacteriocides) with broad-spectrum activity not only control the target pest or disease but may also affect the natural enemies of pests and diseases. Thus, insecticidal control of the bollworm, *Heliothis armigera*, on cotton in Sudan has led to outbreaks of the whitefly, *Bemisia tabaci*, as a consequence of the elimination of natural control agents [53]. The resurgence of the apple sucker (*Psylla mali*) in Canada was probably attributable to the use of organic mercuric fungicides to control orchard fungal diseases that also killed the insect-parasitic fungus, *Zoophthora radicans* [16].

Populations of pests and plant pathogens can develop resistance to pesticides, particularly in intensive agriculture where the frequent use of chemicals imposes enormous selection pressure. The development of resistance and the resurgence of pests and diseases induces the response of higher doses of chemical, more frequent application and the need for new generations of pesticides.

The production of new compounds is not beyond the skills of the synthetic organic chemist but costs are rising as the chances of finding new active molecules decrease and registration requirements become increasingly stringent. As a result, agrochemical companies are less and less interested in costly product development for minor use chemicals.

These are some of the reasons which have stimulated agrochemical and biotechnological companies to increase their interest and investment in biological control, i.e. the use of pathogens, predators and parasites as pest and disease control agents. Microbial pathogens, including bacteria, fungi, viruses and protozoa, provide a major potential source of biological control agents although at present they account for only 1–2% of the total pesticide market [62]. Some of the relative merits of microbial and chemical control methods are outlined in Table 4.2.1.

The following account presents several examples of microbial control in practice, including strategies for their use and the mode of action and efficacy of selected micro-organisms and viruses. Microbial control of insect and mite pests has been exploited more rapidly than other areas. The subject is well described in two recent texts [14, 69]. A sourcebook on disease control is that by Cook and Baker [23]. It is beyond the scope of the present chapter to consider the use of plant pathogens to control weeds, an option for reduced herbicide usage, and the reader is referred to the text by Charudattan and Walker [21]. It should be noted that commercial microbial herbicides have been developed to control the milkweed vine, *Morrenia odora*, in citrus groves (*Phytophthora palmivora*) and northern jointvetch, *Aeschynomene virginica*, in rice and soybean (*Colletotrichum gloeosporioides*).

4.2.2 PESTS

Strategies for pest control

The ways in which microbial agents can be used to control pests fall broadly into three categories:

1 Regular application: the pathogen does not maintain itself in the host population or environment.

Table 4.2.1. Comparison of chemical and microbial pesticides (from Lisansky [73] with permission)

	Chemical	Microbial
Product development		
New product discovery	Screen 15 000 compounds, discover afterwards what targets they control	Target selected on market need; microbial control agents often easy to find
Cost of R & D	£12 000 000	£400 000
Market size required for profit	£30 000 000 yr^{-1} to recoup investment; limited to major crops	Markets less than £600 000 yr^{-1} can be profitable due to low development cost
Patentability of product	Well established	Recent precedents are encouraging
Product use		
Kill	Often 100%	Usually 90–95%
Speed	Usually rapid	Can be slow
Spectrum	Generally broad	Generally narrow
Resistance	Often develops	None yet shown, but microbes also adaptable
Product safety		
Toxicological testing	Lengthy and expensive — £3 000 000	£40 000
Environmental hazard	Various well known examples in the last 25 years, e.g. DDT; 2,4,5,T	None yet shown
Residues	Interval before harvest usually required	Crop may be harvested immediately

2 Limited release: there is some degree of colonization of the pathogen in the target host population.
3 Manipulation of enzootic pathogens [74].

REGULAR APPLICATION

In many crops, economic damage thresholds are low and the high pest densities that are required to establish epizootics of a pathogen in the pest population cannot be tolerated. Cultivation and crop harvesting dilute or remove the pathogen from the ecosystem, permitting little or no carry-over from one year to another. Certain useful pathogens may also have little capacity to spread or persist in the environment. In these circumstances, there is little option but to apply the pathogen at frequent intervals. Microbial control agents used in this way should be highly virulent and easy to produce in large quantities. Good examples include *Bacillus thuringiensis* and some viruses infecting pests of short-duration crops.

LIMITED RELEASE

The strategy in which limited releases of a pathogen are made is most likely to succeed in stable ecosystems, such as forests and grasslands, where high economic thresholds permit greater tolerance of pest damage. There are some examples of classical biological control (e.g. virus control of pine sawfly in Canada) where a microbial pathogen has continued since its introduction to be a satisfactory natural regulator of its host [11]. Such natural regulators appear rare. Although epizootics of naturally occurring pathogens may arise, the reduction in the host population often occurs only after severe crop damage has been inflicted. More frequent releases of pathogens are then required. The success of the limited release strategy will depend on effective generation-to-generation transmission of the pathogen, good persistence of the pathogen in the environment and a capacity to spread. Many microbial pathogens form a spore or other type of resistant resting stage that enables them to survive in the absence of a host.

MANIPULATION OF ENZOOTIC PATHOGENS

In certain circumstances, the microbial control of a pest can best be achieved by encouraging a naturally-occurring pathogen. For example, the virus control of the pasture pests *Wiseana* spp. in New Zealand can be achieved by manipulating pasture cultivation techniques (described below). In another example, repeated use of a site for cereal crops eventually inhibits the growth of cereal cyst nematodes. This inhibition is probably attributable to a build-up of natural populations of nematode-parasitic fungi [60].

The potential for long-term control of pests provided

Table 4.2.2. Comparative properties of the principal microbial agents used to control insect and mite pests

	Bacteria (*B. thuringiensis*)	Nematodes (insect-parasitic rhabditids)	Fungi (Deuteromycetes)	Viruses (baculoviruses)	Protozoa (Microsporidia)
Host range	Principally Lepidoptera and Diptera. Some strain specificity	Very broad. >1000 insect species	Very broad, but some strain specificity	Principally Lepidoptera and Hymenoptera. Viruses may be genus- or species-specific	Often broad within insect family
Mode of entry	Oral	All body openings; occasionally cuticular	Usually cuticular	Oral	Oral
Speed of kill	30 min–1 day	1–2 days	4–7 days	3–10 days	Often chronic disease, >4 days
Stability	Spores killed by UV. Crystals persist longer. Good persistence in soil	Unstable to UV and desiccation. Reasonable persistence in soil	Unstable to UV. Good spore survival in soil	Unstable to UV. Prolonged persistence of occlusion bodies in soil	Unstable to UV
Capacity to spread	Does not spread	Localized spread within soil. Seeks out host	Spores spread by air currents, rain and host movement	Spread by vertical transmission and/or passive vectors (e.g. birds)	Often transmitted vertically through the egg
Humidity requirements	Humidity not limiting	Free water required for nematode spread	High humidity essential for spore germination	Humidity not limiting	Humidity not limiting
Method of production	*In vitro*	*In vitro*	*In vitro*	*In vivo*	*In vivo*
Ideal ecosystem	Field and forest: vector control (e.g. mosquitoes, blackflies)	Soils, composts	Soil, protected crops and other humid environments	Forests, pastures and other stable ecosystems	Forests, pastures
Strategy of use	Regular application	Regular application and limited release	Regular application and limited release	Regular application, limited release. Enzootic disease manipulation	Limited release

by stable ecosystems also creates opportunities for the use of pathogens that are less easy to produce and less pathogenic than those used in strategies involving regular treatment applications. Pathogens of low virulence can generally survive at lower host density and have a lower threshold density for the development of epizootics than virulent pathogens [1].

Regardless of the strategy used, effective microbial control requires an understanding of the mode of action, pathogenicity and population biology of the pest–pathogen interaction. Some of the properties and potentials of the different groups of pathogens used for the control of invertebrates are summarized in Table 4.2.2 and described in more detail below.

Bacteria as pest control agents

Strains of *Bacillus* spp. were the first microbial pathogens to be commercialized and are now the most widely used of any of the pathogen groups (see Table 4.2.3).

Table 4.2.3. Bacterial candidates for pest control

Species	Active ingredient	Target pests	Method of production	Availability of commercial product*
Bacillus thuringiensis var. *kurstaki*; var. *aizawai*	Crystal δ-endotoxin (and spores) β-exotoxin	Caterpillars (Lepidoptera) Houseflies	*In vitro* *In vitro*	Worldwide Finland, Eastern Europe
Bacillus thuringiensis var. *israelensis*	Crystal δ-endotoxin (and spores)	Mosquitoes and blackflies	*In vitro*	Worldwide
Bacillus thuringiensis var. *tenebrionis*; var. *san diego*	Crystal δ-endotoxin (and spores)	Beetles	*In vitro*	
Bacillus sphaericus (strains 1593, 2297 and 2362)	Spores (some with crystal toxin)	Mosquitoes	*In vitro*	
Bacillus popilliae	Spores	Japanese beetle (*Popillia japonica*)	*In vivo*	USA
Pasteuria (= *Bacillus*) *penetrans*	Spores	Nematodes (e.g. *Meloidogyne* spp.)	*In vivo*	

* Indicates countries where these pathogens have been sold and marketed as commercial products.

BACILLUS THURINGIENSIS

Bacillus thuringiensis (*B.t.*) was first discovered infecting moth larvae in a flour mill in Germany in 1911. Since then, many *B.t.* strains with different pathogenicities have been isolated from a wide range of lepidopterans. Commercial registration in the USA, UK and many other countries now permits the use of *B.t* at unlimited dosage on food crops, up to and including the day of harvest [17].

Mode of action

B.t. is an aerobic, toxin-producing bacterium that, at sporulation, produces both a spore and a large proteinaceous crystal which is commonly bipyramidal in shape (see Fig. 4.2.1). This crystal, the δ-endotoxin, is composed largely of polypeptides with a relative molecular mass (M_r) in the range 120 000–140 000 (dependent on strain). These molecules are inert protoxins. However, when a susceptible larva is fed a mixture of spores and crystals, the crystals dissolve in the alkaline gut juice of the insect and are degraded by proteases, releasing toxic polypeptides (generally within the range M_r = 58 000–80 000). Further degradation of the protein (e.g. below M_r = 30 000) leads to loss of toxin activity [76]. Recent studies with one *B.t.* strain [29, 66] have suggested that the toxin of Lepidoptera-active strains interacts with glycoproteins in the plasma membrane of gut cells, destroying regulation of ion exchange. The consequences in susceptible larvae are that the gut epithelium lyses, the muscles of gut and mouthparts are paralysed, feeding stops and death occurs 30 min–3 days after ingestion [16]. In most susceptible species of Lepidoptera, ingestion of the toxin crystal alone seems sufficient to ensure death.

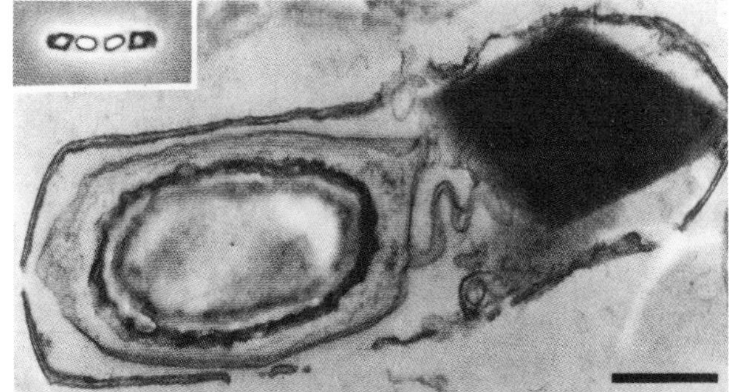

Fig. 4.2.1. Thin section of endospore and proteinaceous δ-endotoxin of *Bacillus thuringiensis* var. *kurstaki*. Inset shows light micrograph of two cells each with spore and crystal. Scale bar = 500 nm. Reproduced with permission from Johnson *et al.* [56].

However, in some insects (e.g. the wax moth, *Galleria mellonella*) toxicity is enhanced by the presence of *B.t.* spores in the ingested material. Most commercial products contain a mixture of spores and crystals.

It has been known for some time that different lepidopteran pests differ in their response to distinct strains of *B.t.* [28]. Few naturally occurring strains are highly pathogenic for all the species in the complex of lepidopteran pests that occurs on any one crop (see Table 4.2.4). Reasons for this include different properties of the toxic polypeptides in the crystals of different strains. In addition, proteolysis of the crystal protein in the larval gut, essential for activation of the protoxin, may differ between host species. Insects may also differ in their susceptibilities to the same toxin. Finally, the food ingested by the insect together with the crystals may contain components (e.g. lectins, tannins) which can neutralize the toxins to some degree.

Some strains (such as the common commercially used HD-1 strain) have a relatively broad-spectrum activity against a number of important pests, while other strains are more effective than HD-1 against some species but inactive against others (see Table 4.2.4). Even though the biochemical basis of these differences is not yet fully understood, useful improvements in biological activity have been obtained by screening naturally occurring strains [28]. Further improvements in potency have also been made by the production of new strains by mutation [19]. Even greater improvements are likely from knowledge now being obtained on the genetic basis of strain potency (see below).

Table 4.2.4. Relative activities* of four distinct *Bacillus thuringiensis* strains against four lepidopteran pests

Insect pest	*B. thuringiensis* strain			
	a (HD1)	b	c	d
Heliothis virescens	42	2	100	52
Heliothis armigera	90	6	100	49
Spodoptera littoralis	7.5	100	1	24
Mamestra brassicae	15	100	2	32

* Activities are expressed as relative potencies, i.e. as percentage activities of the highest potency (100) with each species [91].

Practical use

Although *B.t.* is known to cause natural epizootics in artificially dry environments (e.g. granaries) where spore-laden cadavers survive a long time [18], it is otherwise not a highly infectious natural pathogen. The principal reason for this is that it fails to spread.

This general failure to spread means that *B.t.* must be used as a microbial insecticide and applied frequently. Because it has no contact action and must be ingested to be effective, it should be applied to sites and at temperatures where larvae are actively feeding. Like other microbial control agents, the bacterium is inactivated by the ultraviolet (UV) component of sunlight, but the crystal toxin persists for longer periods on foliage than spores. The HD-1 strain of the bacterium is the most widely used microbial pathogen of insects, being applied against lepidopterous pests of horticulture, agriculture and forestry [28, 76, 86].

Although several *B.t.* strains exhibited limited pathogenicity for insects other than Lepidoptera, it was only in 1977 that Goldberg and Margalit [40] reported the isolation of a new variety (*B.t. israelensis*) with improved activity against mosquitoes and simuliid blackflies. Urged on by the need to develop alternative methods to control pesticide-resistant blackfly vectors of river blindness (onchocerciasis) in West Africa, this *B.t.* strain was commercialized within six years of its discovery and has made a major contribution to blackfly and mosquito control [20]. *B.t. israelensis* also produces a δ-endotoxin crystal which contains three major proteins (M_r 230 000, 130 000 and 28 000) with probable insecticidal activity [111]. There is some dispute over the insecticidal role of the smallest protein but it is certainly cellulolytic and appears to exert its effect by interacting with phospholipids in the gut cell plasma membrane [108]. Recently, varieties (*B.t. tenebrionis* and *san diego*) active against beetles have been discovered [49, 67, 71].

Genetics of δ-endotoxin production

Recent studies on the genetic basis of crystal protein production in *B.t.* have provided methods for the future development of improved *B.t.* strains. Suspicions that the gene(s) responsible for the production of the protein crystal was located on a plasmid arose from the observation that *B.t.* strains in culture frequently lost the ability to produce crystals when exposed to temperatures above their normal growth range. Direct evidence of plasmid involvement came from the work of Schnepf and Whiteley [103], who cloned plasmid DNA from the HD-1 strain into *Escherichia coli*, obtained expression of δ-endotoxin activity and identified in the transformed bacteria a polypeptide antigenically related to the crystal protein and with a similar size (M_r 130 000). The production of toxin-gene-specific probes has shown that crystal protein genes are present in many *B.t.* strains on plasmids of various sizes [68] and, in some strains, possibly on the bacterial chromosome [64].

These results suggested that the transfer of plasmids between *B.t.* strains could provide recombinants with altered potencies. By means of a conjugation-like technique [41], plasmids can be exchanged between different parental strains to obtain recombinants (transcipients), some of which possess the combined toxic properties of both parents [91]. Such techniques make possible the tailoring of *B.t.* strains for improved toxicity and efficacy against a broader range of lepidopteran pests.

The cloning and expression of *B.t.* toxin genes in other bacteria [103, 113] have recently led to the construction of a transformed *Pseudomonas fluorescens* rhizosphere bacterium which expresses a *B.t.* δ-endotoxin and may have potential for the control of root-feeding lepidoptera [4]. It has also proved possible to transfer and express the toxin gene in tobacco plants to produce plants that are largely immune to attack by the tobacco hornworm, *Manduca sexta*. This strategy is likely to be extended to a range of other crop species [110, 118].

β-*exotoxin*

Other types of toxin molecules have been isolated from some *B.t.* strains [33]. The most interesting of these is the β-exotoxin. In contrast to the δ-endotoxin, it is thermostable and highly toxic for a wide range of insect species and orders. Its broader spectrum of activity was explained when its structure was confirmed as an adenine nucleotide analogue of ATP [105]. Concern that the toxin may have some toxicity for vertebrates (e.g. chickens) has prevented its widespread development as a pesticide. In the USA, commercial *B.t.* products must be free of β-exotoxin [33]. However, it is used in the USSR and in Finland [17], where it has proved effective for housefly control in intensive animal-rearing enclosures.

OTHER BACTERIA

Bacillus sphaericus

This aerobic, spore-forming bacterium is found commonly throughout the world in soil and aquatic environments. Some strains of *B. sphaericus* are pathogens for larval stages of mosquitoes through a toxin associated with proteinaceous inclusions and spores, as well as the cell wall and cytoplasm of vegetative cells immediately prior to sporulation. There is some evidence, as with *B.t.*, that the *B. sphaericus* toxin is produced as a protein protoxin and digested to a toxic form in the midgut. The primary toxin is a protein with an M_r of 43 000–55 000; similar proteins exist in all the most toxic strains (1593, 2297 and 2362). After ingestion of sporulated cells by mosquito larvae, the gut epithelium distends and the gut is paralysed, followed by epithelial cell lysis and larval death [114]. These strains have higher activity against certain mosquitoes (particularly *Anopheles* and *Culex*) than *B.t. israelensis*, though they will not affect blackflies. One potential advantage is that some strains appear able to grow saprophytically in heavily polluted water (e.g. sewage disposal units), providing opportunities for the maintenance or increase of inoculum in the absence of the insect host [15, 26].

Bacillus popilliae

In contrast to *B.t.*, which kills the host by toxin action and has little capacity to spread, *Bacillus popilliae* (isolated from the Japanese beetle, *Popillia japonica*), kills the insect host by infection instead of by toxins, spreads naturally and persists even in low density host populations. Spores applied to pastures in the USA during the 1930s promoted disease in the beetle population which was still present 25–30 years later. Thus, the bacterium provides an example of a classical biocontrol agent. There has, however, been recent pest resurgence in some treated areas, which may be attributable to the development of resistance in the insect population or an attenuation of the bacterium [63].

The progress of the disease is slow after larvae ingest the bacterial spores but eventually large cadavers result which are packed with spores that have good survival in the soil. The bacterium is an obligate parasite and it has not proved possible to produce spores adequately *in vitro*, though two small companies in the USA produce spores in living larvae.

Pasteuria (= Bacillus) penetrans

Pasteuria penetrans is an obligate parasite of some plant-parasitic nematodes that, like *B. popilliae*, can be cultured only *in vivo* at present. However, the capacity of the bacterial spore to survive for several years in soil could make it a useful pathogen for treating localized nematode outbreaks in small quantities of soil [60]. Unlike the bacterial pathogens discussed above, the endospores of *P. penetrans* need not be ingested to be effective. They adhere to the cuticle of juvenile nematodes (e.g. root knot nematodes, *Meloidogyne* spp.) and penetrate the cuticle by the production of a germ tube. Infected nematodes eventually become filled with persistent spores and egg production in female nematodes ceases [60, 102].

Nematodes as vectors of bacterial pathogens

Nickle and Welch [87] mention nine major nematode groups which have insects as intermediate or definitive hosts. A description of the majority of these is outside the scope of this chapter, but successful biocontrol has been achieved with members of three groups. The neotylenchid nematode, *Deladenus siricidicola*, has been used in Australia with great success for controlling the imported woodwasp, *Sirex noctilis*, on *Pinus radiata* [6]. The mermithid *Romanomermis culicivorax* was produced in the USA during the 1970s for mosquito control, though efficacy and production problems have restricted its use [95]. The third group includes rhabditid nematodes in the genera *Heterorhabditis* and *Steinernema* (= *Neoaplectana*) which are symbiotically associated with insect-pathogenic bacteria in the genus *Xenorhabdus* [117]. These large Gram-negative, rod-shaped, facultative anaerobic bacteria are held in a pouch in the intestine of the nematode (see Fig. 4.2.2) and in nature are associated only with these nematodes.

Free-living rhabditid nematode larvae in the third of four larval stages are relatively resistant to desiccation and can survive in damp soil without an insect host for several months. When they encounter — or are attracted to — a susceptible host, the nematodes enter the host through the mouth or anus and pass into the haemocoele by penetrating the gut wall. *Heterorhabditis* spp. may enter directly through the cuticle. Once inside the host, the symbiotic bacteria multiply and are released into the insect haemocoele. Further bacterial multiplication occurs and the insect host is killed by septicaemia within 24–48 hours of invasion. *Xenorhabdus* spp. produce antibiotics which prevent invasion of the host by other microorganisms and delay the breakdown of the host cuticle. This allows the nematode to feed and develop on the

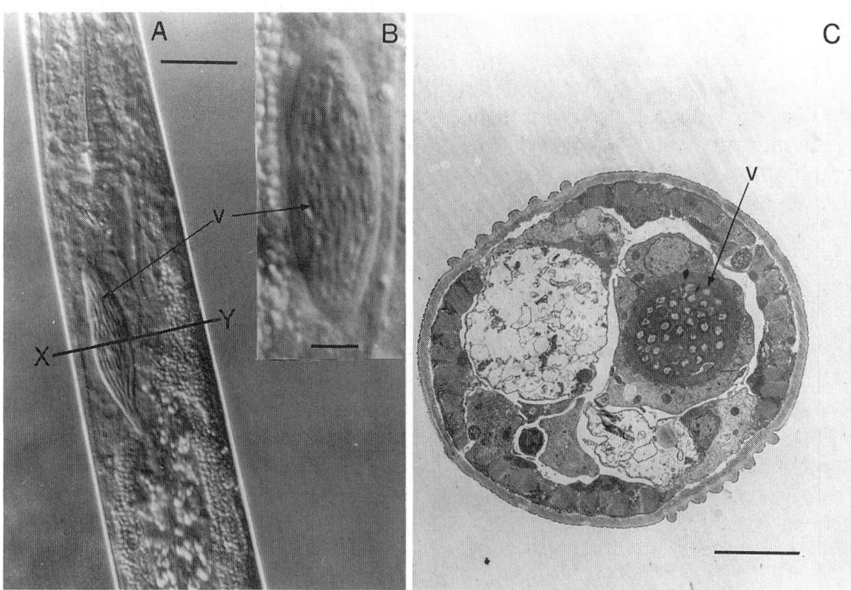

Fig. 4.2.2. Structure of the intestinal vesicle in the insect-parasitic nematode, *Steinernema bibionis*, showing the organization of the vesicle and the bacteria (*Xenorhabdus nematophilus*) contained within it. (A) Living specimen, photographed using differential interference contrast optics, showing the intestinal vesicle (v) and the X–Y axis through which the section (shown in C) was cut. Scale bar = 20 μm. (B) Intestinal vesicle at higher magnification. Scale bar = 5 μm. (C) Whole transverse section viewed using the transmission electron microscope, showing the intestinal vesicle (v) packed with bacteria. Scale bar = 4 μm. Reproduced with permission from Bird & Akhurst [10].

bacteria and decomposing host tissues and to complete its complex life cycle [117]. About 10 days after invasion hundreds or thousands of infective third-stage larval nematodes may be released from the cadaver.

This fascinating and complex relationship between rhabditid nematode and bacterium has shown considerable promise for biocontrol. Insect-parasitic rhabditid nematodes are already sold for pest control in the USA and Holland and a worldwide network of field trials has been conducted with nematodes produced in Australia. The ideal target insects are pests which live in soil and other moist and protected environments. The temperature range for activity of most strains is between 10°C and 32°C, and low humidity and high solar radiation are limiting factors in their use. Recent successes have been reported for the control of cutworms (*Agrotis* spp.) in lettuce [75], black vine weevil (*Otiorhynchus sulcatus*) on a range of container-grown plants [8] and dipterous pests of mushrooms [96]. One nematode species, *Steinernema feltiae* (= *Neoaplectana carpocapsae*), has been reported to kill more than 1000 species of insects [87]. At present, nematodes are applied in the same way as microbial pesticides, taking advantage, where possible, of their capacity to search for insect hosts, spread and persist.

Both the bacteria and the nematodes can be grown *in vitro* using diets of homogenized animal tissues [7] or nutrient broth and soybean flour [117]. Provided that rearing systems can reliably produce large numbers of infective nematodes, these parasites show considerable potential for insect biocontrol. In addition, the insect-pathogenic properties of the bacterium alone warrant further investigation.

Fungi as pest control agents

Several strains of pathogenic fungi have been commercialized for the control of insect and mite pests as well as of plant-parasitic nematodes (see Table 4.2.5). While more than 150 fungi have been reported attacking nematodes or their eggs (see Chapter 3.3), in excess of 400 species have been isolated from insects and mites [43]. A much smaller number have been given serious consideration as microbial control agents.

FUNGI PATHOGENIC FOR INSECTS AND MITES

Fungi are virtually unique amongst pathogens of insects and mites as most do not have to be ingested but are able to penetrate directly through the arthropod cuticle. They therefore have potential for the control of sap-feeding arthropods which would not ingest pathogens on the plant surface. The main groups of fungal pathogens are considered below.

Deuteromycetes

Deuteromycete fungi have received most attention as biocontrol agents because they are amongst the easiest to produce *in vitro* and several have a broad host range (see Table 4.2.5). For example, *Metarhizium anisopliae* has more than 200 known hosts amongst Coleoptera, Lepidoptera, Orthoptera and Hemiptera, while *Beauveria* spp. have been identified from about 500 host species, principally Lepidoptera and Coleoptera [43]. However, individual strains of the same species may not all have the same pathogenicity or host range.

Mode of action. Although a few strains enter the host through the gut and respiratory tract, the majority invade both terrestrial and aquatic insects through the cuticle. The infective propagule is the conidiospore. The pathogen host range may be influenced by specificity in the ability of spores to adhere to the cuticle, to germinate, and to penetrate into the haemocoele by mechanical or enzymatic means. In the haemocoele the fungus may be met by humoral and cellular defence mechanisms which can encapsulate the invading fungus [43]. Once past these defences, hyphal growth continues, often with the production of a yeast-like phase (blastospores) which allows the fungus to spread throughout the insect. When the insect is dead (following hyphal growth, toxin production or both), sporulation occurs with the development of conidia on the surface of the insect [32]. *Verticillium lecanii* is unusual in that it will sporulate freely on live insects (see Fig. 4.2.3), providing greater opportunities for spread of the disease in the insect population.

Relative humidity (RH) is the most critical environmental factor influencing successful fungal parasitism. Amongst Deuteromycetes infecting terrestrial insects, the lower limit for spore germination is probably about 93% RH and some may require a film of water to germinate. Natural epizootics usually coincide with periods of high humidity and rainfall. This, together with temperature optima for growth in the range of 20–30°C, makes deuteromycete strains most appropriate for use against pests in tropical and sub-tropical regions or in other environments where RH, in particular, is high.

Practical use. The first fungal species to be successfully commercialized, *Verticillium lecanii*, was developed in

Table 4.2.5. Principal fungal candidates for arthropod and nematode pest control

Species	Main target pests	Growth *in vitro*	Commercial scale production
Insect pathogens			
Deuteromycetes			
Aschersonia aleyrodis	Whitefly	+	—
Beauveria bassiana	Colorado beetle	+	USSR
Beauveria brongniartii	Cockchafer	+	—
Culicinomyces clavosporus	Mosquito larvae	+	—
Hirsutella thompsonii	Rust mites	+	USA
Metarhizium anisopliae	Beetles, bugs	+	Brazil
Nomuraea rileyi	Caterpillars	+	—
Tolypocladium cylindrosporum	Mosquito larvae	+	—
Verticillium lecanii	Aphids, whitefly	+	UK
Zygomycetes			
Conidiobolus, Entomophthora, Erynia, Zoophthora spp. (etc.) (many species of Entomophthorales)	Aphids, mites, planthoppers, caterpillars, crickets	±	—
Oomycetes			
Lagenidium giganteum	Mosquito larvae	+	—
Chytridiomycetes			
Coelomomyces spp.	Mosquito larvae	–	—
Nematode pathogens			
Deuteromycetes			
Arthrobotrys spp.	*Meloidogyne* spp.; *Ditylenchus* sp.	+	France
Dactylella oviparasitica	*Meloidogyne* adults and eggs	+	—
Paecilomyces incognita	*Meloidogyne* spp.	+	—
Verticillium chlamydosporum	Cyst nematode ♀ and eggs	+	—
Chytridiomycetes			
Catenaria auxiliaris	Cyst nematode ♀	–	—
Nematophthora gynophila	Cyst nematode ♀	–	—

the UK in the early 1980s to control aphids and scale insects (e.g. whitefly) on glasshouse crops [42]. In the protected environment of a glasshouse it proved possible to generate the required temperature and humidity levels for effective spore germination and growth. Single spray treatments at relatively low pest density proved sufficient to control the aphid, *Myzus persicae*, on chrysanthemums for the duration of the crop (3 months) with effective spread of spores ensuring the prolonged control [42]. As conidia are difficult to produce in quantity by liquid fermentation, the commercial product is based on dried blastospores, supplemented with a substrate which allows the fungus to grow and sporulate to a limited extent on foliage in the absence of an insect host. This can increase the spore inoculum by as much as 40-fold [43]. *V. lecanii* is an extremely widespread insect pathogen, strains of which will infect nematodes [44] and even act as antagonists of some rust diseases [107].

Other important deuteromycete insect pathogens include *Metarhizium anisopliae*, which has been widely used as a microbial pesticide for the control of spittle bugs (e.g. *Mahanarva posticata*) on sugar cane in Brazil [37]. *Beauveria bassiana* has been widely used against Colorado potato beetle in the USSR, and *Hirsutella thompsonii* was produced as a commercial trial-product for a short period to control citrus rust mites in Florida [79]. All these species are released as conidiospores, grown on semi-solid media based on cereal grain or bran.

The extensive spread and recycling of introduced fungal pathogens over several seasons at the site of introduction is rare. The best-documented example is provided by the control of the pasture pest, the cockchafer *Melolontha melolontha*, in France and Switzerland by *Beauveria brongniartii*. The perennial, undisturbed nature of the pasture ecosystem provides opportunities for colonization and durable residual efficacy [37].

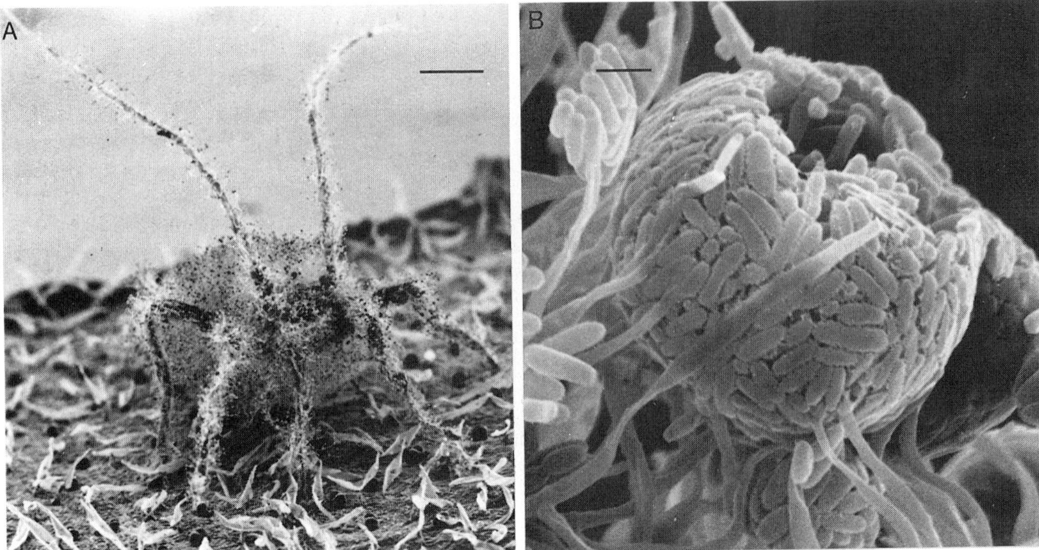

Fig. 4.2.3. (A) Scanning electron micrograph of an aphid at an advanced stage of infection with *Verticillium lecanii*. Scale bar = 200 μm. (B) Higher magnification of conidiospores of *V. lecanii* produced during sporulation on the surface of an infected aphid. Scale bar = 5 μm. Photographs courtesy of Peter Atkey.

Entomophthorales

Of the other classes of fungal insect pathogens, the Entomophthorales (Zygomycetes) contain by far the largest number of known strains [115]. Under favourable conditions (e.g. high relative humidities) they produce conidiospores which are forcibly discharged from the surface of the dead host. Under less favourable conditions, some species produce thick-walled resting spores, often within the host cadaver, which can survive for long periods. Members of the Entomophthorales appear more effective in temperate climates as the temperature optima for growth are generally lower than for Deuteromycetes [43]. As for Deuteromycetes, high humidities are also required for germination of conidia and resting spores. Many strains are difficult to culture; their conidia are often short-lived and reliable mass production of resting spores has proved difficult. For these reasons, field applications have been few and there are no commercial products [115].

Oomycete and chytridiomycete fungi

These groups contain few insect pathogens. One, *Lagenidium giganteum*, is a mosquito pathogen capable of saprophytic growth in aquatic environments, and producing motile zoospores that can infect host larvae [34]. Its full potential as a control agent is still being evaluated.

Coelomomyces spp. are obligate parasites, principally isolated from mosquito larvae. They have a complex life cycle which involves a crustacean (copepod or ostracod) intermediate host. Infection in mosquito larvae leads to the production of large numbers of thick-walled sporangia (see Fig. 4.2.4). After meiosis, sporangia release haploid meiospores which infect the crustacean intermediate host, producing a haploid thallus and uniflagellate gametes. Gametes of opposite mating type fuse and the resulting diploid zoospores infect the mosquito larvae. Although *Coelomomyces* spp. produce extensive natural epizootics, the life cycle complicates *in vitro* production of the fungus and hence limits its potential as a manipulable microbial pathogen.

NEMATODE PATHOGENIC FUNGI

A great variety of relationships exist between fungi and soil-dwelling plant-parasitic nematodes. Many of the fungi concerned are obligate parasites; they are difficult to isolate and assessment of their effects in the complex soil habitat is not easy [60, 83]. The fungi of interest for

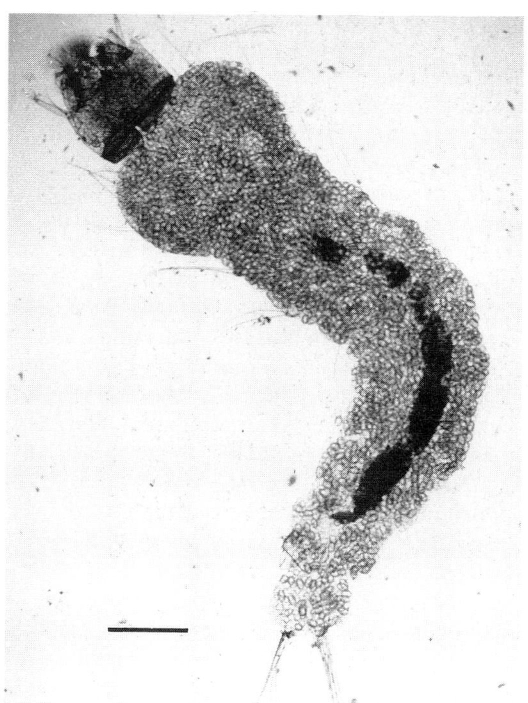

Fig. 4.2.4. A larva of the mosquito *Anopheles gambiae* filled with thick-walled sporangia of *Coelomomyces indicus*. Scale bar = 1 mm. Preparation courtesy of Dr Wellington Otieno.

nematode biocontrol fall broadly into two groups: 'predatory' or nematode-trapping fungi and endoparasitic fungi.

Nematode-trapping fungi

These predatory fungi trap nematodes by adhesive devices (e.g. hyphal nets), non-adhesive snares and hyphal loops. In many species a toxin is produced that immobilizes the nematode almost immediately, followed by penetration of hyphae through the cuticle. Some species produce antibiotics which prevent the development of competing micro-organisms [83].

Trapping fungi are generally rather non-specific, i.e. once traps are produced, a range of nematode species can be captured with equal efficiency [60]. The fungi can grow saprophytically and, to be used effectively for biocontrol, it is necessary to establish an actively growing mycelium before the nematode-susceptible crop is planted. This is not as easy as it sounds. Most trapping fungi do not colonize soils rapidly and are often poor competitors with other soil microflora [83]. An added carbohydrate source is usually necessary to help overcome the mycostatic influences of soil micro-organisms on spore germination and hyphal growth [60].

Although some predatory fungi produce chlamydospores which may allow them to survive adverse conditions, most do not and it is unclear how they survive for prolonged periods [61]. In fact, most strains persist in field soils for only a few weeks after application, although *Arthrobotrys* spp. have been known to survive for up to five years.

Such problems have not stopped the small-scale commercial production in France of two *Arthrobotrys* spp. (see Table 4.2.5) for the control of *Ditylenchus myceliophagus* in mushrooms and *Meloidogyne* spp. on tomatoes. In the latter case, applications of fungi growing on cereal grain are made to the soil surface 30 days before planting the crop. However, the inconsistent and relatively small positive effects obtained [61] suggest that the products have limited use.

Endoparasitic fungi of nematodes

The most promising endoparasitic fungi include both obligate and facultative parasites of nematodes that attack eggs or egg-producing adults (see Table 4.2.5). These strains produce conidia or resting spores which are surrounded by adhesive material which allows the spores to attach to the nematode cuticle [55].

The Chytridiomycetes, *Nematophthora gynophila* and *Catenaria auxiliaris*, have motile zoospores with positive tropisms for female cyst nematodes [83]. Such fungi are widely distributed in northern Europe where soil moisture is not a limiting factor. Under adverse conditions, zoospores of *N. gynophila* can encyst and later give rise to another motile stage, increasing the chances of infection [59]. The thick-walled resting spores of this species can survive at least 5 years in the absence of cyst nematode females [60]. Fungal pathogens of nematode eggs, including *Verticillium chlamydosporium*, are facultative parasites that can survive in soil by parasitizing the eggs of other invertebrates as well as by chlamydospore production [60].

Once they have infected the host, endoparasitic fungi take longer to kill the host than do trapping fungi. Thus, *N. gynophila* takes up to 7 days at 13°C to kill female cyst nematodes [59]. In general, many fungal endoparasites are scarcely more specific than the trapping fungi, although *N. gynophila* will parasitize females of *Heterodera* spp., but not *Globodera rostochiensis* or most root-knot nematodes (*Meloidogyne* spp.) [60].

The most specific of the endoparasitic fungi, e.g. *N. gynophila* and *C. auxiliaris*, have not yet been produced *in vitro*. For this reason, field trials have largely been restricted to the use of facultative parasites for the control of *Meloidogyne* spp. eggs (see Table 4.2.5). *Paecilomyces lilacinus* has been used in trials in Peru in attempts to control the potato root-knot nematode *Meloidogyne incognita* [102]. Inoculation of the soil at planting with conidia produced *in vitro* on cereal grain appeared to provide some measure of control, with fungal survival of at least one year [61].

Several endoparasites exert natural control of nematodes in crop monocultures so it may be possible to manipulate the environment to encourage fungal growth. *Dactylella oviparasitica* is a natural pathogen of *Meloidogyne* spp. in peach orchards, and the combined activities of *V. chlamydosporium* and *N. gynophila* are largely held responsible for natural reductions in cereal cyst nematode (*Heterodera avenae*) in land intensively cultivated for cereal production [61].

At the present time, no introduced fungal strains have shown a sustained capacity to control nematodes. Before this sustained control can be achieved, we must attain a satisfactory understanding of the complex soil environment and of the conditions for fungal colonization.

Pest control by viruses

In this chapter, we have largely restricted our examples of microbial agents to their use for the control of invertebrate pests and plant pathogens. However, one of the best-known examples of biocontrol with viruses is in the control of a vertebrate pest, the European rabbit.

MYXOMATOSIS

The natural reservoir of myxoma virus (the causal agent of myxomatosis) is the tropical forest rabbit of South America, *Sylvilagus brasilienis*, in which the virus is relatively avirulent and produces localized benign fibromas. However, when the virus was transmitted to European rabbits (*Oryctolagus cuniculus*) it was extremely virulent [35]. The virus is relatively host-specific and affects only some rabbit species; there are only very rare cases of infection in European hares. The disease is spread mechanically by insect vectors, principally mosquitoes (see Section 2.2.2).

Following its introduction into Australia in 1859, the European rabbit spread rapidly until it became a major agricultural pest. In an attempt to control it, strains of myxoma virus were released in Australia in 1950 and in Europe in 1952 [70]. The virus liberated was highly virulent and more than 99% of infected rabbits died. The virus spread rapidly, particularly during the summer when the insect vectors were active. Although it was anticipated that the disease would die out during the winter, through reduced insect activity and lower numbers of susceptible hosts, the virus became established and survived the winter months in many areas. This survival was due, at least in part, to a selection for viruses of lower virulence. Attenuated myxoma viruses appeared within a year of its introduction to Australia and, within 3–4 years, became the dominant strains [35]. Infection with these strains enables rabbits to survive in an infectious state for weeks rather than days, hence improving the opportunity for transmission. This reduction in virulence has been accompanied by an increase in genetic resistance of the rabbits [70, 85, 100].

Myxomatosis provides an interesting example of the limited release strategy for microbial pathogens; there was extensive spread from the original release sites aided by mobility of both the host and the vectors. It also illustrates that pest resistance to disease can arise, just as pests may become resistant to chemical pesticides. Further, microbial agents themselves are capable of natural selection and genetic change and so substantial outbreaks of virulent myxoma virus still occur [100]. None the less, it seems likely that myxomatosis could eventually become a relatively benign disease in the European rabbit as it is in its natural host in South America [85].

VIRUS CONTROL OF INSECT PESTS

Well over 1200 virus isolates have been shown to cause disease in more than 800 species of insects and mites [84]. While these viruses have been grouped into several categories on the basis of their morphological and biochemical properties [92], only baculoviruses have been given extensive consideration as microbial control agents [90].

Properties of baculoviruses

The hosts of baculoviruses are restricted to a small number of insect orders (principally Lepidoptera and Hymenoptera) and a few crustaceans and mites. Many isolates infect only a few closely related insect species and are often highly virulent. Virus particles are comprised of rod-shaped nucleocapsids (containing DNA) surrounded by a unit membrane. In some members of the group, the virus particles are packaged singly (granulosis viruses or GVs) or in large numbers (nuclear polyhedrosis viruses or NPVs) within large proteinaceous crystals produced late in infection (see Fig. 4.2.5). The main component of these crystalline occlusion bodies (OBs) is a matrix protein comprised of a single polypeptide species with an M_r of about 30 000. The OBs provide the means by which virus infectivity is preserved outside the host (see Section 2.2.5) and, in the practical use of NPVs and GVs for pest control, it is the occlusion bodies that have been applied as microbial pesticides. One baculovirus group does not produce OBs; these viruses are less stable outside the host and alternative methods have been used to introduce such viruses into the insect pest population (see Fig. 4.2.6).

Infection takes place after susceptible insect larvae eat food contaminated with virus. With GV and NPV infections, the matrix protein of the occlusion body dissolves in the insect gut and releases virus particles which infect and multiply in gut epithelial cells. In Lepidoptera the infection quickly spreads to other tissues, while in Hymenoptera (e.g. sawflies) the infection is confined to the gut [89].

Some of the practical limitations to the use of baculoviruses as pesticides are imposed by their mode of infection. These limitations include the need for pest larvae to eat the virus before infection can occur. There is no contact or systemic action and virus application methods must provide good coverage and distribute the virus to parts of the plant where the target pests are feeding. In this respect the application problems are similar to those experienced with the use of *B. thuringiensis*. However, unlike *B. thuringiensis*, baculovirus infections rarely cause

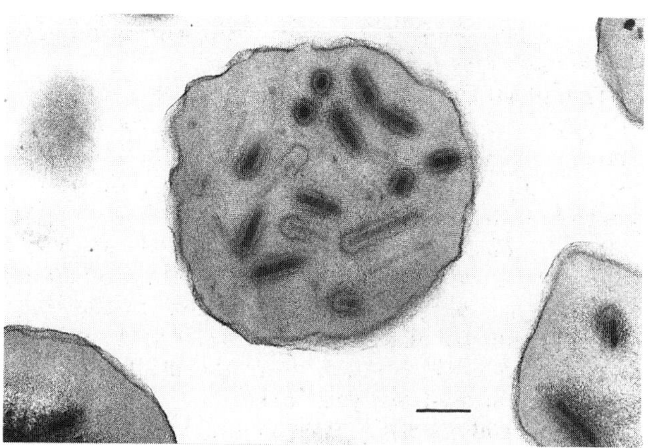

Fig. 4.2.5. Thin section of an occlusion body of a nuclear polyhedrosis virus from the cotton bollworm, *Heliothis zea* (scale bar = 200 nm). Numerous enveloped virus particles are enclosed within a crystalline matrix protein. Photograph courtesy of Margaret K. Arnold.

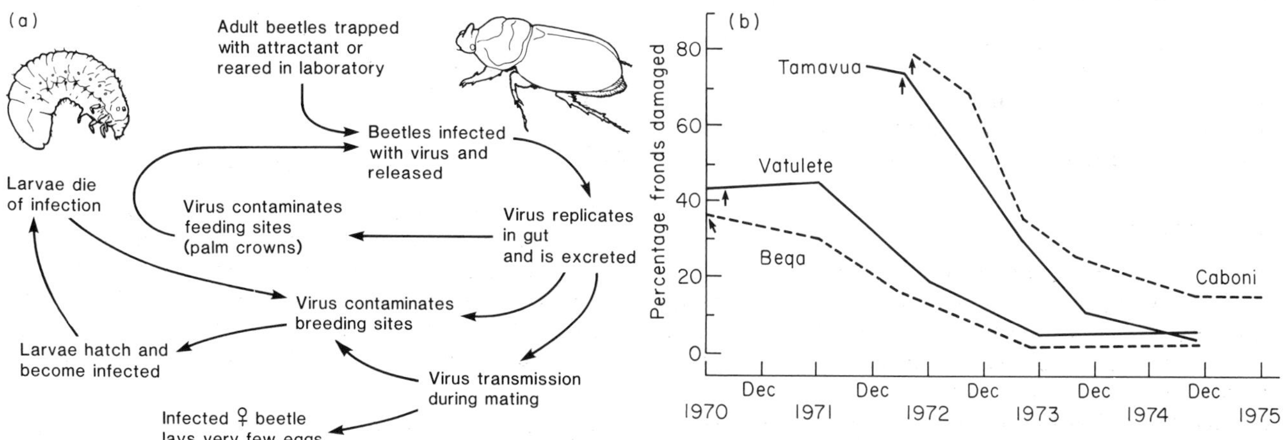

Fig. 4.2.6. The control of coconut rhinoceros beetle, *Oryctes rhinoceros*, by a baculovirus. (a) Diagrammatic representation of the transmission of *Oryctes* virus following the release of infected adult beetles. Reproduced with permission from Payne [90]. (b) The effect of virus control on damage to coconut palms in several Fijian islands. Arrows indicate time of virus introduction. Palm damage by *Oryctes* falls very significantly 12–24 months after virus introduction as the virus spreads through the beetle population. Reproduced with permission from Bedford [9].

rapid cessation of feeding and pest larvae may cause considerable crop damage before the insects die. Death of an insect as a direct consequence of virus infection is unlikely to occur less than 3–4 days after infection even with the most virulent of currently available virus strains. One consequence of this is that most baculoviruses which have been successfully developed to an industrial scale of production infect pests of forests, where economic thresholds are high and a degree of cosmetic damage to the crop can be tolerated. Crop damage can be restricted by applying virus to control young larvae, which are consistently more susceptible to virus infection than older ones (see Fig. 4.2.7). Such use of virus requires an adequate pest-monitoring programme and critical timing of virus application. Although these are considerable practical constraints on the use of viruses, it should be remembered that, unlike *B. thuringiensis*, NPV and GV infections result in the production of huge quantities of often highly infectious virus OBs. Diseased larvae can contain up to 10^{10} NPV OBs or well in excess of 10^{11} of the smaller GV OBs. Under suitable conditions of host density, these can promote rapid spread and persistence of infection within the host population.

Many baculoviruses are highly host-specific and may seem ideal candidates for selective pest control because of the reduced ecological hazard resulting from their use. However, pest control problems can rarely be solved by the application of one highly selective agent, and there is considerable interest in some baculoviruses with less selective host ranges. A virus from the alfalfa looper, *Autographa californica*, has attracted most interest with a recorded host range of 43 species of pest Lepidoptera in 11 families [90].

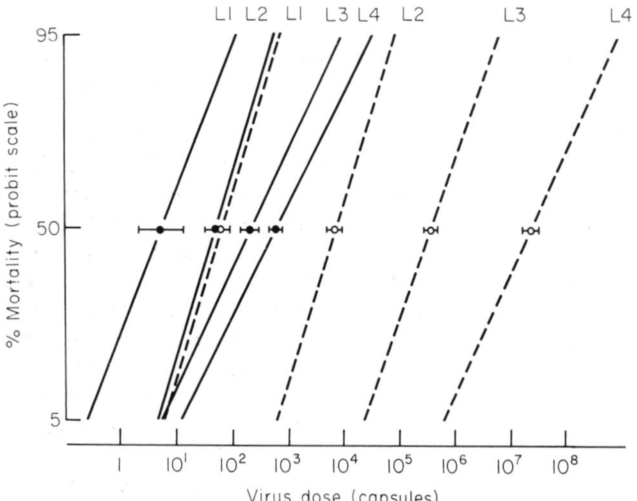

Fig. 4.2.7. Susceptibilities of first (L1), second (L2), third (L3) and fourth (L4) -instar larvae of *Pieris brassicae* (– – –) and *Pieris rapae* (——) to a granulosis virus. Older larvae are less susceptible to virus infection and *P. rapae* larvae are more susceptible to the virus than *P. brassicae* larvae. 95% confidence limits are shown at the median lethal doses (LD_{50}s). Reproduced with permission from Payne *et al.* [93].

PRACTICAL USE OF BACULOVIRUSES

When viruses and most other microbial pesticides are applied to a crop by conventional application methods, there are a number of environmental factors which influence the survival of infective virus. The most important of these for baculoviruses are the effects of UV radiation from sunlight and adverse plant-surface effects such as alkaline dew on cotton leaves [30]. Other environmental factors, such as rain and normal ambient temperatures, do not influence the stability of the occluded baculoviruses (NPVs and GVs). Without formulations to protect the virus against UV, spray applications have to be made more frequently or the virus concentration in the spray has to be increased. Despite these difficulties, baculoviruses have been employed quite successfully for biocontrol, both as short-term selective pesticides and for longer-term pest suppression (see Table 4.2.6).

Short-term pest control by baculoviruses has concentrated on the regular application of the most highly pathogenic viruses against the youngest and most susceptible larval stages of the pest. An NPV of cotton bollworms (*Heliothis* spp.) was produced commercially for several years for use on cotton [54]. At present, a GV is being produced to control the apple pest, codling moth (*Cydia pomonella*). This GV provides a good example of the relative advantages and disadvantages of using a virus as an insecticide. Amongst the advantages are that virus applications aimed at first instar larvae match the control achievable by chemical pesticides [38]. Population reductions of up to 100% were reported following four sprays of 10^{11} occlusion bodies per tree [52]. In addition, use of the highly specific virus does not harm natural parasites and predators in an orchard. As a consequence, outbreaks of the fruit tree red spider mite and woolly aphid (which normally *increase* after chemical pesticide treatment) do not occur [39]. However, because the virus takes several days to kill the pest, some superficial cosmetic fruit damage continues. Also, other caterpillar pest species in the orchard are not susceptible to the virus and may resurge once the grower uses virus rather than chemical pesticides to control the codling moth [39]. The

Table 4.2.6. Principal baculovirus candidates for insect pest control

Virus	Target hosts	Ecosystem	Commercial-scale production*
Nuclear polyhedrosis viruses			
Anticarsia gemmatalis NPV (soybean looper)	*A. gemmatalis*	Soybean	Brazil
Autographa californica NPV (alfalfa looper)	*Orgyia pseudotsugata* *Trichoplusia ni*	Forests Cabbage	USA USA
Gilpinia hercyniae NPV (spruce sawfly)	*G. hercyniae*	Forests	—
Heliothis spp. NPV (cotton bollworm)	*Heliothis* spp.	Cotton, sorghum, maize	USA
Lymantria dispar NPV (gypsy moth)	*L. dispar*	Forests	USA, USSR
Mamestra brassicae NPV (cabbage moth)	*Mamestra, Heliothis, Diparopsis* spp.	Cabbage, cotton, other agricultural crops	France, USSR
Neodiprion sertifer NPV (pine sawfly)	*N. sertifer*	Forests	Canada, Finland, UK, USA, USSR
Neodiprion lecontei	*N. lecontei*	Forests	Canada
Orgyia pseudotsugata NPV (Douglas fir tussock moth)	*O. pseudotsugata*	Forests	USA, Canada
Panolis flammea NPV (pine beauty moth)	*P. flammea*	Forests	—
Spodoptera littoralis NPV (Cotton leafworm)	*S. littoralis*	Cotton	France
Granulosis viruses			
Cydia pomonella GV (codling moth)	*C. pomonella*	Orchards	USA, UK, W. Germany, France
Non-occluded baculoviruses			
Oryctes rhinoceros virus (coconut rhinoceros beetle)	*O. rhinoceros*	Coconut plantations	—

* Indicates countries where these viruses have been produced on a large scale, though not all are commercially available at present.

full advantage of the virus can be gained only by its integration with other selective agents (other microbial pathogens or selective chemicals) able to control the other pests. Unfortunately, the virus does not establish itself within the insect population in the orchard [52], probably because little virus is produced in the young infected larvae and the pathogen will not recirculate readily within the orchard ecosystem.

The problem of cosmetic damage is less important in forests and plantation crops, where insect viruses have considerable potential as control agents. Many of the viruses produced on an industrial scale have been for the control of forest pests (see Table 4.2.6). On rare occasions, the introduced virus may become established and regulate the pest over an extended period. One of the best examples of classical biological control with a virus comes from the accidental introduction of an NPV into populations of the European spruce sawfly, *Gilpinia hercyniae*, in Canada [11]. Since its initial diagnosis in 1936, it has spread rapidly through the sawfly population and now combines with parasites to keep the pest population below the economic threshold. Likewise, the baculovirus of the coconut rhinoceros beetle (*Oryctes rhinoceros*) has been successfully deployed by releasing virus-diseased adults, which readily transmit infection (see Fig. 4.2.6). As this virus is difficult to produce in large quantities and does not form occlusion bodies, its survival and transmission in the environment is ensured by a close association with its host. Unlike most baculoviruses, which infect larvae only, the *Oryctes* virus infects both larvae and adults and can persist and spread very effectively [9, 90].

Finally, prolonged pest control by some viruses can be ensured by manipulation of an enzootic virus (one that occurs naturally at a low level in the population). Kalmakoff and Crawford [57], studying the control of the lepidopteran New Zealand pasture pests, *Wiseana* spp., concluded that a successful balance between virus (an NPV) and the host could be obtained by maintaining the numbers of larvae in the soil at 15–25 m^{-2} (still below the economic threshold for damage). A long-term interaction between insects and virus was assured, provided that the soil was not disturbed by cultivation methods which led to virus dilution and inactivation. In fact, if virus was applied against young larvae in the pasture, the infection level was lowered, presumably because the overall amount of virus produced in the ecosystem was reduced. Successful biocontrol of these pests can thus be maintained by the manipulation of cultivation techniques rather than by the application of exogenous virus. Such natural regulation of pests by viruses is rare and it is more commonly necessary for viruses to be introduced regularly to produce limited epizootics and short-term control.

Protozoa

Most orders of insects are hosts to protozoan pathogens, some of which undoubtedly play a significant role in the natural suppression of insect populations. However, because insect-pathogenic Protozoa generally produce chronic rather than acute disease, they do not create such dramatic results as some viruses, bacteria and fungi [116]. Interest has largely been confined to a limited number of Microsporidia, in particular, *Nosema locustae* for grasshopper control [48], *Vairimorpha* (= *Nosema*) *necatrix* a broader-spectrum disease agent infectious for many Lepidoptera [81], *Nosema fumiferanae* (for spruce budworm control) and, more recently, *Octosporea muscadomesticae* for control of the Australian sheep blowfly, *Lucilia cuprina* [24].

Infection in susceptible hosts is initiated by the ingestion of protozoan spores. Microsporidia are obligate parasites that require living cells for their development, and the gut or fat body provide the main foci of infection. The disease is often not highly pathogenic but significantly reduces the rate of development of the insect and its fecundity [116]. Natural spread and survival of the pathogens may be assured by transmission within the host egg (e.g. *N. fumiferanae*) and by the presence of a number of alternative hosts (e.g. *V. necatrix*).

Nosema locustae has received most attention for biocontrol purposes and is now produced commercially on a small scale in the USA. More than 60 species of grasshoppers and crickets are susceptible to the disease. Although in nature it is generally uncommon (usually <1% individuals infected), when spores were applied in a wheat-bran bait infection levels of up to 40% were obtained [48]. In the season following application, infections were also common and the disease spread substantially, suggesting that it may have some potential for long-term control. However, it is generally felt that protozoan pathogens are unlikely to be major candidates for biocontrol and they should be considered within an overall integrated control programme where they may predispose the host to attack by other factors [116].

4.2.3 DISEASES

Strategies for disease control

The development of microbial control of diseases has lagged behind insect biocontrol, both at the practical and scientific levels. The only two successful commercial exploitations are the control of crown gall and *Heterobasidion* root rot of pine, which are discussed further below. This is somewhat disappointing because the principles of microbe/microbe interactions are well established (see Chapters 2.1 and 2.4). Antibiosis and production of cell-wall degrading enzymes are examples of biocontrol strategies which are receiving attention as modes of action at present, and both of these can be exclusive of mycoparasitism and physical restriction as the antagonist action. The indirect mechanisms of cross-protection or hypovirulence, where inoculation of the plant with a nonpathogen can provide protection, are also known.

The rhizosphere [77, 78, 112] may be a better target than the phylloplane for disease biocontrol because the water relations are likely to be more favourable. Also there is a greater incentive for development by the agrochemical industry because relatively few plant-protection chemicals are soil-acting. Rather than catalogue all the types of microbe/pathogen interactions which can occur, most of which have only been studied *in vitro*, a few examples showing commercial promise are discussed. A more extensive account of the range of antagonisms which have been tested in the field is available [23] but it does not include viruses as antagonists, a topic still at an early stage.

Bacteria as antagonists

CROWN GALL

Crown gall is caused by the soil bacterium *Agrobacterium radiobacter* var. *tumefaciens*, which enters fruit trees and roses through wounds and induces unregulated cell division, leading to massive gall formation (see Fig. 4.2.8) [58]. Some strains were found to be non-pathogenic and in some experiments with tomato no galls developed when the ratio of pathogen to non-pathogen was one or less. The non-pathogenic strain 84 was selected for further study as a potential biocontrol agent. In experiments with peach, inoculation of seeds with this strain gave 78% control of crown gall, 95% control was achieved with root inoculation and 99% from combined seed and root inoculation. The treatment has been used worldwide, often achieving 100% control with natural disease levels, and is the only bacterial disease biocontrol agent available commercially.

The biocontrol activity is by the production of an antibiotic, agrocin 84, which is a substituted fraudulent adenine nucleotide:

Agrocin 84

R is α- or β-1-O-D-glucofuranoside

M^+ is monovalent cation

The evidence for the antibiotic involvement is that non-producers are ineffective and only the pathogens sensitive to the antibiotic in the laboratory can be controlled. Production of agrocin 84 is coded for on a non-conjugative plasmid but strain 84 also has another conjugative plasmid that codes for the catabolism of nopaline, an amino acid derivative found only in crown-gall tissue. Conjugation allows the transfer of the plasmid to a non-pathogenic recipient bacterium. The conjugative plasmid also mobilizes the agrocin 84 plasmid. Thus the recipient can produce agrocin 84 and control the disease.

The genes controlling pathogenicity in *Agrobacterium* are carried on a large plasmid, the Ti plasmid. Genes controlling sensitivity to the antibiotic are located on the same plasmid and the effectiveness of control is probably caused by this close linkage. However, this also presents potential for breakdown in the system. For example, in

Fig. 4.2.8. Crown gall caused by *Agrobacterium tumefaciens* on the rootstock of cherry. Photograph courtesy of Dr C.M.E. Garrett, AFRC Institute of Horticultural Research.

Greece there is evidence that agrocin 84 production and resistance has been transferred to pathogenic recipients. This transfer only occurs in the presence of the nopaline contained in the crown-gall tissues; in Australia, where control has been very effective, there is little or no gall tissue present and little or no plasmid transfer. However, a mutant strain 84 with a defective plasmid transfer or mobilization system should be sought to avoid the appearance of agrocin-producing strains of the pathogen. This strategy might also aid in other situations where the inoculum has failed to give protection, such as on grape-vine and cane fruits.

TAKE-ALL OF WHEAT

Take-all (*Gaeumannomyces graminis* var. *tritici*) is recognized as a major pathogen of wheat, barley and other grasses. When the plant is infested, darkened root systems are formed, but the farmer usually sees the disease when plants are killed and 'white heads' are formed against a background of green plants. There has been extensive study of the disease for a century [2] and it has been a major target of those interested in biological control procedures [23].

In normal cropping the pathogen present in crop residues becomes displaced by saprophytes. The disease can be controlled by crop rotation using a non-susceptible crop (e.g. potatoes, beets, maize, alfalfa, beans) after a wheat or barley crop. Various agronomic practices such as thorough tillage can reduce disease incidence but, even in monoculture, the disease is naturally reduced after a few years, a phenomenon known as 'take-all decline'. It has generally been believed that antagonistic micro-organisms build up under these conditions. From these 'decline soils' pseudomonads which are antagonistic to take-all *in vitro* and also when applied to cereal seed in field trials have been obtained. Some population dynamics studies using antibiotically marked bacterial strains have demonstrated that these antagonistic bacteria can build up to large populations in the soil following inoculation. At this stage little is known about the mode of action of the bacteria. Pathologists appear to have been more interested in disease control *per se* rather than the microbe/microbe interaction.

A further mechanism of take-all control has been achieved in England [27] by using cross-protection. In this approach the control is achieved by an avirulent fungus (*Phialophora graminicola*) which is closely related to the pathogen. The cross-protecting fungus builds up under grass and this can be followed by a wheat crop. However, the general applicability of this is unclear and it will certainly depend on the natural levels of the cross-protecting fungus present in the soil.

PLANT GROWTH-PROMOTING RHIZOBACTERIA (PGPR)

A mode of action of PGPR has been proposed and many trials have been carried out on the biocontrol action of those organisms against recognized pathogens and deleterious organisms ('minor' pathogens) on potatoes, radishes, sugar beets, melons, lettuces and lima beans [104].

In essence, the mode of action proposed is that the bacteria produce a siderophore, pseudobactin, which chelates iron and thereby makes it unavailable both to the native microflora of the rhizosphere and to pathogens [65]. The main evidence for this hypothesis is the observation that addition of iron to the system eliminated the beneficial effect and that bacterial mutants which did not produce the siderophore were not active in biocontrol. It was demonstrated that pseudobactin also has antibiotic properties but the relative importance of iron deprivation and antibiosis to the pathogen have never been clearly established.

It has now been established that the iron deprivation could limit the effectiveness of *Trichoderma hamatum*, which is a natural antagonist of fungal pathogens [51], and also that the pseudobactin limits the uptake of ferric iron (labelled with ^{59}Fe) into the plant [5]. These opposing effects of siderophore production cast doubt on the total ecological significance and demonstrate clearly that the total ecosystem should be considered in the design and evaluation of biocontrol agents. With this in mind, it is perhaps not surprising that, whereas inoculation increased potato growth by 500% in the glasshouse and yield increased up to 17% in the field, the effects have generally been variable.

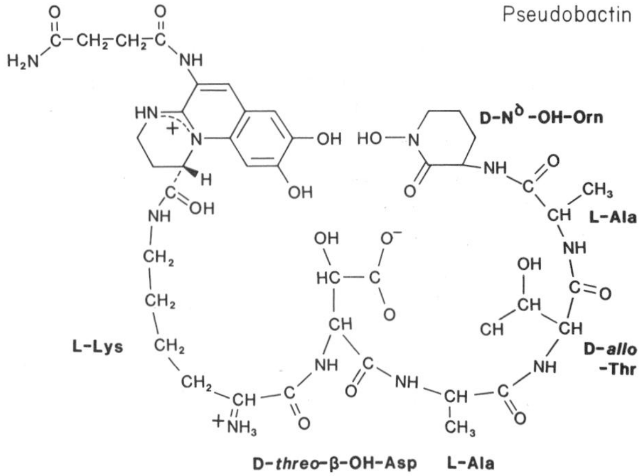

Pseudobactin

L-Lys D-N^{δ}-OH-Orn L-Ala D-*allo*-Thr D-*threo*-β-OH-Asp L-Ala

Another rhizosphere pseudomonad with specific beneficial properties is *Pseudomonas fluorescens* strain Pf-5 [50]. This produces an antibiotic, pyoluteorin, which is active against the *Pythium ultimum*-induced damping-off of cotton seedlings. This is only effective as a seed treatment because the antibiotic was absorbed and inactivated by the soil colloids when added directly to soil.

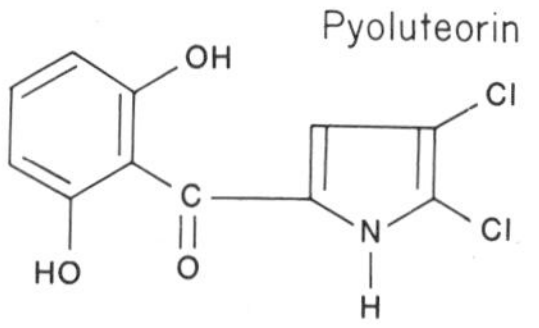

Pyoluteorin

MUSHROOM BLOTCH

One of the most serious and widespread diseases of the cultivated mushroom, *Agaricus bisporus*, is bacterial blotch caused by *Pseudomonas tolaasii*. It causes a browning of the cap, usually as large spots, which reduces the shelf-life of the mushroom and makes it unsaleable. Partial control of the disease can be achieved chemically with sodium hypochlorite or formalin. Biological control would be a good possibility with such a crop because there is a high degree of control of environment in mushroom-growing and therefore it could be expected that the performance of the antagonist would be more reliable. Studies in Australia [31] and in the United Kingdom [36] have shown that bacterial antagonists of the pathogen can be isolated from mushroom caps and from in and around mushroom houses. These have been shown to be more effective in controlling the pathogen *in vitro* and in cropping trials than the chemical agents. Whereas the antagonists do produce antibiotic-inhibition zones on agar, the mode of action is unclear at this stage.

EPIPHYTIC ICE-NUCLEATION BACTERIA

Some herbaceous annuals, flowers of fruit trees and fruit of many species of plants are killed by ice formation within their tissues. Injury to these frost-sensitive forms may be avoided by allowing the water in and on the plant to cool well below the freezing point, the so-called supercooling property of water.

Under natural conditions, ice formation on plant surfaces is inefficient because of a lack of ice nuclei which trigger or catalyse the reaction. If there are no ice nuclei, water can supercool to approximately −40°C outside the plant or to −5°C or −6°C inside the plant. When water is cooled below these temperatures or when nuclei are present, ice begins to form in or on frost-sensitive plants, spreads rapidly both intercellularly and intracellularly, and mechanically disrupts cell membranes. When special epiphytic bacteria (*Pseudomonas syringae, P. fluorescens* and *Erwinia herbicola*) are present on plant surfaces, they catalyse the reaction at temperatures as warm as −1°C [72]. These forms are called ice-nucleation bacteria; although they may make up less than 10% of all epiphytes, their numbers are often sufficient to cause damage.

Bactericides are one option for treatment but a more selective treatment is to spray the plant with bacteria which will inhibit the growth of the ice-nucleation bacteria. These antagonistic forms, called non-ice-nucleation-active bacteria, thus augment the natural antagonism of the bacterial flora. Bacterial mutants which are particularly effective in this respect have been generated but have not yet been tested in the field (see Fig. 4.2.9). The biggest delay is due to the environmentalist lobby in the USA which is against the introduction of genetically engineered organisms into the environment, but testing has now started.

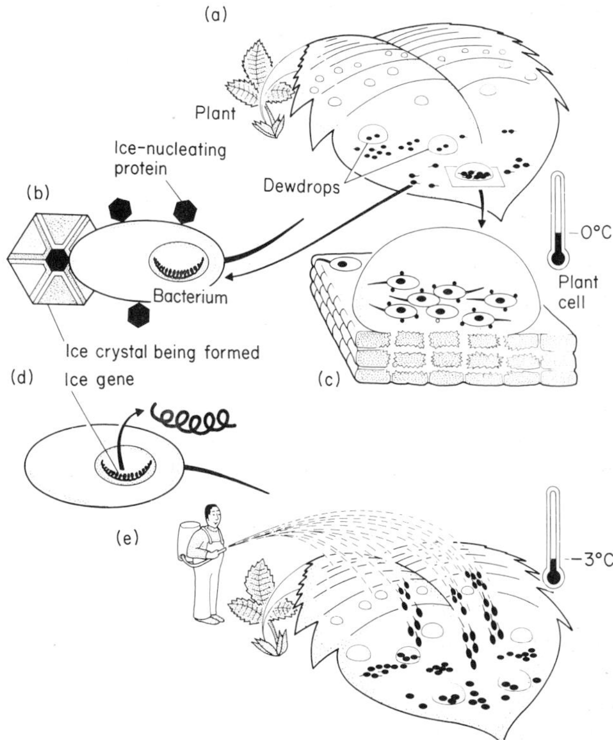

Fig. 4.2.9. The use of genetically manipulated *Pseudomonas syringae* to overcome frost damage on leaves. After a *Time* diagram by Joe Lertola. (a) Ice-forming bacteria on leaf surface. (b) Below freezing, protein on bacterial membrane acts as nucleus for ice crystals. (c) As morning dew becomes ice, fluid in plant cells freezes and crushes them. (d) Ice-gene is removed from bacterial DNA using recombinant DNA. (e) When manipulated bacteria are sprayed on plant, temperatures down to −3°C can be tolerated without damage.

Fungi as antagonists

HETEROBASIDION ROOT ROT OF PINE

The first commercial exploitation of a biocontrol agent against plant diseases was in 1963 when *Peniophora gigantea* was used to control *Heterobasidion* (*Fomes*) *annosum*, a root rot of pine [98]. The pathogenic fungus can break down cellulose and lignin in infected roots, causing extensive decay that may lead to wind-throw, decreased growth, heart rot and death. The fungus usually infects freshly cut stumps in moist conditions. Asexual oidia of the antagonistic fungus are distributed in the forest as dehydrated tablets or in fluid suspensions; it grows well on wood but does have a limited ability to cause a white rot. The antagonism is apparently by pre-emption (establishment before *H. annosum*) and by hyphal interference. This relatively straightforward action of the antagonist is still being exploited commercially.

TRICHODERMA AS A GENERAL ANTAGONIST

One of the most extensively studied areas of biological disease control has been the use of *Trichoderma viride, T. harzianum* and *T. hamatum*. The subject area, along with the use of the related genus *Gliocladium*, has been reviewed [88]. A commercial formulation of two strains of *T. viride* with nutrients and metabolites (Binab T) is available and this is claimed to control silver leaf disease (*Chondrostereum purpureum*), Dutch elm disease (*Ceratocystis ulmi*), tree stem and root rot (*Heterobasidion annosum*), internal rot of utility poles (*Lentinus lepideus*) and honey fungus (*Armillaria mellea*). However, most of the attention in recent years has focused on the use of *Trichoderma* spp. against root and damping-off diseases of crop plants. Patents have been granted on selected strains, particularly those with fungicide resistance.

The hyphal coiling phenomenon of *Trichoderma* spp. (see Fig. 4.2.10) has been recognized for a long time. The coiling *per se* may do little damage but this can lead to penetration by the fungus of the pathogen cell walls.

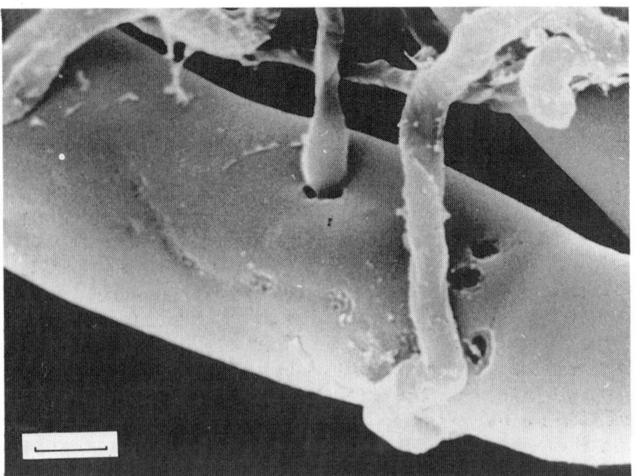

Fig. 4.2.10. Hyphal coiling and penetration of *Rhizoctonia solani* by *Trichoderma harzianum*. Scale bar = 2 μm. Photograph courtesy of C. Ridout, University of Hull and AFRC Institute of Horticultural Research.

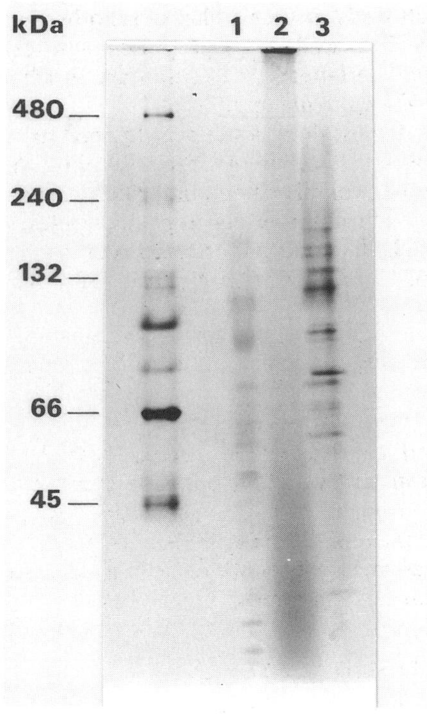

Fig. 4.2.11. Separation of extracellular protein induced in *Trichoderma harzianum*. Lane 1, fungus grown on glucose; lane 2, fungus grown on *Rhizoctonia solani* cell wall concentrate; lane 3, fungus grown on cell walls of *R. solani*. Marker proteins alongside. Reproduced with permission from Ridout *et al.* [97].

The penetration is readily observed on nutrient-poor media but on nutrient-rich media it is less evident, demonstrating the necrotrophic action of *Trichoderma. In vivo*, *Trichoderma* acts as a necrotroph (deriving organic compounds from killed cells of its symbiotic host) and there is no evidence of biotrophic nutrition (derivation of organic compounds from living cells of its symbiotic host).

The penetration of the cell walls of the pathogen almost certainly involves the production of cell wall-degrading enzymes. When *T. harzianum* or *T. viride* is grown on cell walls of *Rhizoctonia solani* in preference to glucose as sole carbon source, the electrophoretic separation of extracellular proteins indicates that several extra protein bands are induced (see Fig. 4.2.11) [97]. The enzymes responsible for the cell-wall degradation are thought to be β1−3 glucanase and chitinase but there is not a clear correlation between the activity of these enzymes and the biocontrol activity. With so many other proteins induced by the cell walls, it seems likely that other enzymes will be involved as well, perhaps proteinases, cellulases or xylanases. The latter two enzymes are produced at high levels in *Trichoderma* spp.

The antagonist does not appear to need to break down the pathogen cell wall in order to control it. The antagonist produces a volatile compound which is active against a wide range of soil-borne fungal pathogens [22]. The compound smells like coconut and has been characterized in a strain of *T. harzianum* as 6-pentyl-2H-pyran-2-one.

Provided consistency of response can be obtained, *Trichoderma* spp. probably offer the greatest prospect as broad-spectrum biocontrol agents [101]. At present the major obstacle is the inconsistency of response but Fig. 4.2.12 illustrates a trial where a beneficial response to inoculation was observed.

SCLEROTIAL DISEASES

Several soil-borne plant pathogens produce sclerotia, which are masses of fungal mycelia, commonly black and of various sizes depending on species. For example, *Rhizoctonia* spp. produce sclerotia and have been a target for *Trichoderma* as described in the preceding section. Also, *Sclerotinia* spp. and *Sclerotium* spp. have been targets for other mycoparasites, such as *Coniothyrium minitans* [109], in biological control procedures. *C. minitans* is not usually considered a soil inhabitant as it survives for only about 18–24 months but it has been used to control *Sclerotinia sclerotiorum* on sunflower and beans, *S. trifoliorum* on clover and as a seed treatment against *Sclerotium cepivorum* on onion. It also attacks sclerotia of *Sclerotinia minor, Botrytis cinerea, B. fabae, B. narcissicola* and *Claviceps purpurea*. The effect on the sclerotia can be dramatic; in some cases there is a destruction of over 90% and large yield-increases result.

The hyphae of *C. minitans* penetrate and decay sclerotia, reducing the amount of inoculum of the pathogen. As in the case of *Trichoderma* spp., β1–3 glucanase and chitinase are produced by the antagonist and this may be responsible for the lysis of the mycelia of the pathogen.

The antagonist can be grown on potato dextrose agar to produce a suitable spore preparation. On a large scale it is more practical to use a solid-substrate fermentation using a substrate such as autoclaved milled rice or wheat

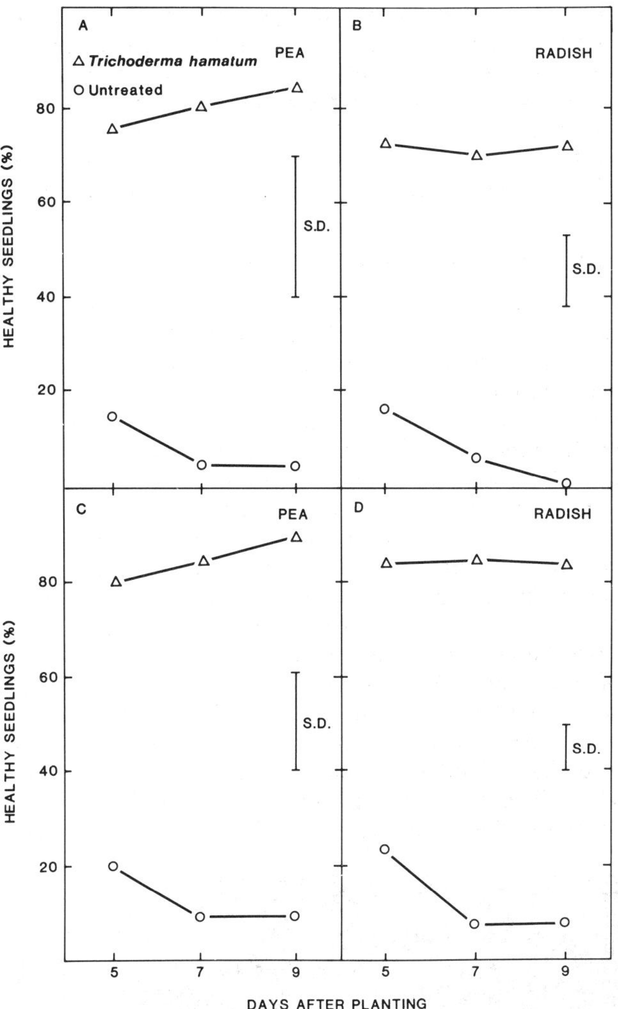

Fig. 4.2.12. Percentage of healthy pea (A and C) or radish (B and D) seedlings at various intervals after planting. Seedlings grew either from seeds treated with *Trichoderma hamatum* (A and B) or from non-treated seeds in soil in which seeds of the same species had been planted with treated seeds in the previous planting cycle (C and D). Peas were planted in soil containing *Pythium* spp. indigenous to Fort Collins clay soil, while radishes were planted in soil to which *Rhizoctonia solani* had been added. The bars labelled S.D. represent the overall standard deviation for the experiment shown. Redrawn with permission from Harman *et al.* [45].

grain in plastic bags at 20°C. However, this takes approximately 6 weeks and there is clearly scope to develop a more rapid and sophisticated procedure if the agent becomes a commercial proposition.

The biocontrol activity of *C. minitans* has been compared with *Gliocladium roseum*, which is a parasite of non-sclerotial hyphae [78]. For spring and autumn crops on a site where disease incidence was high, soil applications of *C. minitans* before planting gave very good disease control, comparable with that using fungicide (vinclozoline) sprays fortnightly. In contrast *G. roseum* was less satisfactory on a site where disease incidence was lower; the effectiveness depends on disease level, soil microbial populations affecting the antagonist and pathogen, humidity and temperature. Also, there are indications that the two antagonists act against different stages of the disease. When *G. roseum* spore suspensions were sprayed on lettuce plants grown in trays shortly after inoculation of the pathogen, they decreased lesion development more effectively than *C. minitans* but not when they were applied later. It appears that there is a complex interaction of antagonist effectiveness at different stages of the pathogen life cycle. *C. minitans* seems to be particularly effective in decreasing the number of sclerotia which could serve as a source of pathogen inoculum for successive crops.

Sporodesmium sclerotivorum is a biotrophic mycoparasite (in contrast to *C. minitans*, which is necrotrophic) which has also been successfully used to control sclerotial diseases, particularly white rot of onion (*Sclerotium cepivorum*) and lettuce drop (*Sclerotinia minor*) [3]. A patent has been published on a method of producing the inoculum using living sclerotia as a substrate.

EPIPHYTIC YEASTS ON THE PHYLLOPLANE

Whereas the usefulness of bacteria and filamentous fungi is well proved for biocontrol on leaf surfaces, the exploitation of yeasts is still in the exploration stage [12]. The question is whether common leaf saprophytes can antagonize biotrophic pathogens, spores of which are largely independent of exogenous nutrients for germination. In one case, populations of *Sporobolomyces* and *Cryptococcus* on wheat flag leaves in the field increased from 100 to 10 000 cells cm^{-2} within a few days by spraying yeasts together with nutrients on to just unfolded leaves, whereas over 2 weeks was needed to get similar cell densities of naturally occurring yeasts on untreated leaves. The early inoculations of the yeasts suppressed *Cochliobolus sativus* and *Septoria nodorum* co-inoculated as pathogens but later inoculation showed no advantage as presumably the natural antagonistic population of yeasts had developed. Natural colonization of yeasts of fruits and flowers may be less than that on leaves so there may be greater scope for inoculation.

Little is known about the mode of antagonistic action of yeasts. An antibiotic material has been separated from *Sporobolomyces ruberrimus* but there appears scope for more study of yeast metabolites in relation to biocontrol. Yeasts may well prove suitable biocontrol agents for commercial production for they are relatively easy to culture and tolerant of environmental extremes such as anaerobic conditions.

4.2.4 CONCLUSIONS

In this chapter it is not possible to do more than highlight some of the examples of microbial control agents that can be used for pest and disease control. The main advantage of such biological control agents lies in their selectivity and hence the reduced ecological hazard arising from their use. It has not been possible here to dwell on safety aspects such as the effects of these microbial agents on non-target organisms; these are considered in detail in several excellent reviews [14, 46, 99]. It should be stressed, however, that appropriate safety tests must be an integral part of any microbial control programme. Current evidence suggests that the responsible use of the pathogens described above does not adversely affect non-target vertebrates, plants and beneficial invertebrates such as predators, parasites and pollinators.

The use of microbial agents at present forms a very small part of the total pesticide market. We cannot ignore the fact that many of the microbial pesticides available today fail to meet the expectations of an agrochemical industry used to the production of synthetic chemicals and of growers attuned to the rapid and broad-spectrum pest and disease-knockdown achieved by many chemical pesticides. Further research is needed with most microbial pathogens to improve efficacy, production methods and ability to be stored for long periods. Some pathogens which can be produced only *in vivo* are unlikely to have widespread use unless they can be applied in small amounts to exert long-term biocontrol. Improved environmental survival of introduced pathogens will need further studies on protection from UV light in the aerial environment and the avoidance of mycostasis in the soil. The recent detection of host strains which have developed resistance to viruses or *B. thuringiensis* can also not be ignored [13, 80]. However, these observations should be viewed in the context that the resistance described so far represents a shift in the mean response level of host populations in laboratory or other artificial environments rather than the appearance of non-susceptibility in a natural host population.

Major future improvements in efficacy should be obtained by taking advantage of the naturally occurring wide differences in virulence between closely related fungal, bacterial, viral and protozoan strains and nematode species. The strain selection programme for increased virulence in *B. thuringiensis*, which has yielded major improvements in potency, is a good example [28]. Further genetic variation can be induced by mutation and genetic manipulation, and improvements in *B. thuringiensis* efficacy have already been achieved in this way [91]. In the same way, improvements in baculovirus pathogenicity seem likely to arise from studies on genetic recombination [25] and the insertion of new genes into precise positions in baculovirus genomes [82, 94, 106]. As with both bacteria and viruses, important traits for biological control are found in different deuteromycete isolates; these may allow strain manipulation in fungi [47]. As in many areas of genetic engineering, we need a better understanding of those genetic characteristics that are important in the determination of pathogenicity and host range.

Pest biocontrol has advanced more rapidly than disease control, probably because there have been fewer target organisms studied and there has generally been more extensive consideration of modes of action. If disease biocontrol is to advance, it is essential that the 'spray and pray' approach is rejected and that mode-of-action studies be coupled to field evaluations. The control of crown gall is an excellent example of success in this area.

Finally, individual microbial control agents will not solve all pest and disease problems. They need to be used within integrated pest and disease management programmes, where their compatibility with chemical pesticides and other management practices must be carefully evaluated.

4.2.5 REFERENCES

[1] Anderson R.M. & May R.M. (1981) The population dynamics of microparasites and their invertebrate hosts. *Phil. Trans. R. Soc. B* **291**, 451–524.

[2] Asher M.J.C. & Shipton P.J. (eds.) (1981) *Biology and Control of Take-All.* Academic Press, London.

[3] Ayers W.A. & Adams P.B. (1981) Mycoparasitism of sclerotia of *Sclerotinia* and *Sclerotium* species by *Sporodesmium sclerotivorum. Can. J. Microbiol.* **25**, 17–23.

[4] Beardsley T. (1984) Monsanto goes ahead with trials. *Nature (Lond.)* **312**, 686.

[5] Becker J.O., Hedges R.W. & Messens E. (1985) Inhibitory effect of pseudobactin on the uptake of iron by higher plants. *Appl. Environ. Microbiol.* **49**, 1090–5.

[6] Bedding R.A. (1984a) Nematode parasites of Hymenoptera. In *Plant and Insect Nematodes*, p.755 (ed. W.R. Nickle). Marcel Dekker, New York.

[7] Bedding R.A. (1984b) Large scale production, storage and transport of the insect-parasitic nematodes *Neoaplectana* spp. and *Heterorhabditis* spp. *Ann. Appl. Biol.* **104**, 117–20.

[8] Bedding R.A. & Miller L.A. (1981) Use of a nematode, *Heterorhabditis heliothidis*, to control black vine weevil, *Otiorhynchus sulcatus*, in potted plants. *Ann. Appl. Biol.* **99**, 211–16.

[9] Bedford G.O. (1981) Control of the rhinoceros beetle by baculovirus. In *Microbial Control of Pests and Plant Diseases 1970–1980*, p.409 (ed. H.D. Burges). Academic Press, London.

[10] Bird A.F. & Akhurst R.J. (1983) The nature of the intestinal vesicle in nematodes of the family Steinernematidae. *Int. J. Parasitol.* **13**, 599–606.

[11] Bird F.T. & Elgee D.E. (1957) A virus disease and introduced parasite as factors controlling the European spruce sawfly, *Diprion hercyniae* (Htg.) in central New Brunswick. *Can. Entomol.* **89**, 371–8.

[12] Blakeman J.P. & Fokkema N.J. (1982) Potential for biological control of plant diseases on the phylloplane. *Ann. Rev. Phytopathol.* **20**, 167–92.

[13] Briese D.T. & Podgwaite J.D. (1985) Development of viral resistance in insect populations. In *Viral Insecticides for Biological Control*, p.361 (ed. K. Maramorosch & K.E. Sherman). Academic Press, New York.

[14] Burges H.D. (ed.) (1981). *Microbial Control of Pests and Plant Diseases 1970–1980.* Academic Press, London.

[15] Burges H.D. (1982) Control of insects by bacteria. *Parasitology* **84**, 79–117.

[16] Burges H.D. (1986a) Production and use of pathogens to control insect pests. *J. Appl. Bacteriol., Symposium Suppl.*, 127–37.

[17] Burges H.D. (1986b) Impact of *Bacillus thuringiensis* on pest control with emphasis on genetic manipulation. *MIRCEN J.* **2**, 101–20.

[18] Burges H.D. & Hurst J.A. (1977). Ecology of *Bacillus thuringiensis* in storage moths. *J. Invertebrate Path.* **30**, 131–9.

[19] Burges H.D., Jarrett, P. & Li R. (1984) Bacterial insecticides: *Bacillus thuringiensis. Annual Report of the Glasshouse Crops Research Institute 1983*, 91–4.

[20] Burges H.D. & Pillai J.S. (1986) Microbial bioinsecticides. In *Microbial Technology in the Developing World*, p.121 (ed. E. DaSilva, Y. Domergues, E.J.

Nyns & C. Ratledge). Oxford University Press, Oxford.

[21] Charudattan R. & Walker H.L. (eds.) (1982) *Biological Control of Weeds with Plant Pathogens*. John Wiley, New York.

[22] Claydon N., Allan M., Hanson J.R. & Avent A.G. (1987) Antifungal alkyl pyrones of *Trichoderma harzianum*. *Trans. Brit. Mycol. Soc.* **88**, 503–13.

[23] Cook R.J. & Baker K.F. (1983) *The Nature and Practice of Biological Control of Plant Pathogens*. American Phytopathological Society, St. Paul.

[24] Cooper D.J., Pinnock D.E. & Bateman S.M. (1983) Susceptibility of *Lucilia cuprina* (Wiedemann) (Diptera: Calliphoridae) to *Octosporea muscaedomesticae* Flu. *J. Aust. Ent. Soc.* **22**, 292.

[25] Croizier G. & Quiot J.M. (1981) Obtention and analysis of two genetic recombinants of baculoviruses of Lepidoptera, *Autographa californica* Speyer and *Galleria mellonella* L. *Ann. Virol.* **132**, 3–18.

[26] Davidson E.W. (1982) Bacteria for the control of arthropod vectors of human and animal disease. In *Microbial and Viral Pesticides*, p.289 (ed. E. Kurstak). Marcel Dekker, New York.

[27] Deacon J.W. (1976) Biological control of the take-all fungus, *Gaeumannomyces graminis*, by *Phialophora radicicola* and similar fungi. *Soil Biol. Biochem.* **8**, 275–83.

[28] Dulmage H.D. & Co-operators (1981) Insecticidal activity of isolates of *Bacillus thuringiensis* and their potential for pest control. In *Microbial Control of Pests and Plant Diseases 1970–1980*, p.193 (ed. H.D. Burges). Academic Press, London.

[29] Ellar D.J., Thomas W.E., Knowles B.H., Ward S., Todd J., Drobniewski F., Lewis J., Sawyer T., Last D. & Nichols C. (1985) Biochemistry, genetics and mode of action of *Bacillus thuringiensis* delta-endotoxins. In *Molecular Biology and Microbial Differentiation*, p.230 (ed. J.A. Hoch & P. Setlow). Proceedings of International Spore Conference, No. 9. American Society for Microbiology, Washington, DC.

[30] Evans H.F. & Harrap K.A. (1982) Persistence of insect viruses. In *Virus Persistence*, p.57 (ed. B.W.J. Mahy, A.C. Minson & G.K. Darby). Symposia of the Society for General Microbiology, Vol. 33. Cambridge University Press, Cambridge.

[31] Fahy P.C., Nair N.G. & Bradley J.K. (1981) Epidemiology and biological control of bacterial blotch caused by *Pseudomonas tolaasii*. *Mushroom Science* **11**, 343–52.

[32] Faull J.L. (1986) Fungi and their role in crop protection. In *Biotechnology and Crop Improvement and Protection*, p.141 (ed. P.R. Day). Monograph No. 34, British Crop Protection Council.

[33] Faust R.M. & Bulla L.A. Jr. (1982) Bacteria and their toxins as insecticides. In *Microbial and Viral Pesticides*, p.75 (ed. E. Kurstak). Marcel Dekker, New York.

[34] Federici B.A. (1981) Mosquito control by the fungi *Culicinomyces, Lagenidium* and *Coelomomyces*. In *Microbial Control of Pests and Plant Diseases 1970–1980*, p.555 (ed. H.D. Burges). Academic Press, London.

[35] Fenner F., McAusalan B.R., Mims C.A., Sambrook J. & White D.O. (1974) *The Biology of Animal Viruses*. Academic Press, New York.

[36] Fermor T.R. & Lynch J.M. Unpublished.

[37] Ferron P. (1981) Pest control by the fungi *Beauveria* and *Metarhizium*. In *Microbial Control of Pests and Plant Diseases 1970–1980*, p.465 (ed. H.D. Burges). Academic Press, London.

[38] Glen D.M. & Payne C.C. (1984) Production and field evaluation of codling moth granulosis virus for control of *Cydia pomonella* in the United Kingdom. *Ann. Appl. Biol.* **104**, 87–98.

[39] Glen D.M., Wiltshire C.W., Milsom, N.F. & Brain P. (1984) Codling moth granulosis virus: effects of its use on some other orchard arthropods. *Ann. Appl. Biol.* **104**, 99–106.

[40] Goldberg L.J. & Margalit J. (1977) A bacterial spore demonstrating rapid larvicidal activity against *Anopheles sergentii, Uranothaenia unguiculata, Culex univittatus, Aedes aegypti* and *Culex pipiens*. *Mosq. News* **40**, 67–70.

[41] Gonzalez J.M. Jr., Brown B.J. & Carlton B.C. (1982) Transfer of *Bacillus thuringiensis* plasmids coding for δ-endotoxin among strains of *B. thuringiensis* and *B. cereus*. *Proc. Natl. Acad. Sci. USA* **79**, 6951–5.

[42] Hall R.A. (1981) The fungus *Verticillium lecanii* as a microbial insecticide against aphids and scales. In *Microbial Control of Pests and Plant Diseases 1970–1980*, p.483 (ed. H.D. Burges). Academic Press, London.

[43] Hall R.A. & Papierok B. (1982) Fungi as biological control agents of arthropods of agricultural and medical importance. *Parasitology* **84**, 205–40.

[44] Hänssler G. & Hermans M. (1981) *Verticillium lecanii* as a parasite on cysts of *Heterodera schachtii*. *Z. PflKrankh. PflSchutz*. **88**, 678–81.

[45] Harman G.E., Chet I. & Baker R. (1980) *Trichoderma hamatum* effects on seed and seedling disease induced in radish and pea by *Pythium* spp. or *Rhizoctonia solani*. *Phytopathology* **70**, 1167–72.

[46] Harrap K.A. (1982) Assessment of the human and ecological hazards of microbial insecticides. *Parasitology* **84**, 269–96.

[47] Heale J.B. (1982) Genetic studies on fungi attacking

insects. In *Invertebrate Pathology and Microbial Control*, p.25 (ed. C.C. Payne & H.D. Burges). Proceedings 3rd International Colloquium on Invertebrate Pathology, Brighton.

[48] Henry J.E. & Oma E.A. (1981) Pest control by *Nosema locustae*, a pathogen of grasshoppers and crickets. In *Microbial Control of Pests and Plant Diseases 1970–1980*, p.573 (ed. H.D. Burges). Academic Press, London.

[49] Herrnstadt C., Soares G.G., Wilcox E.R. & Edwards D.L. (1986) A new strain of *Bacillus thuringiensis* with activity against coleopteran insects. *Biotechnology* **4**, 305–8.

[50] Howell C.R. & Stipanovic R.D. (1980) Suppression of *Pythium ultimum*-induced damping-off of cotton seedlings by *Pseudomonas fluorescens* and its antibiotic, pyoluteorin. *Phytopathology* **70**, 712–15.

[51] Hubbard J.P., Harman G.E. & Hadar Y. (1983) Effect of soilborne *Pseudomonas* spp. on the biological control agent, *Trichoderma hamatum*, on pea seeds. *Phytopathology* **73**, 655–9.

[52] Huber J. & Dickler E. (1977) Codling moth granulosis virus: its efficiency in the field in comparison with organophosphorus insecticides. *J. Econ. Entomol.* **70**, 557–61.

[53] Hussey N.W. (1980) Crop protection: a challenge in applied biology. *Ann. Appl. Biol.* **96**, 261–74.

[54] Ignoffo C.M. & Couch T.L. (1981) The nucleopolyhedrosis virus of *Heliothis* species as a microbial insecticide. In *Microbial Control of Pests and Plant Diseases 1970–1980*, p.330 (ed. H.D. Burges). Academic Press, London.

[55] Jansson H-B., Van Hofsten A. & Von Mecklenburg C. (1984) Life cycle of the endoparasitic fungus *Meria coniospora*: a light and electron microscopic study. *Antonie van Leeuwenhoek J. Microbiol.* **50**, 321–7.

[56] Johnson D.E., Niezgodski D.M. & Twaddle G.M. (1980) Parasporal crystals produced by oligosporogeneous mutants of *Bacillus thuringiensis* ($spo^{-}Cr^{+}$). *Can. J. Microbiol.* **26**, 486–91.

[57] Kalmakoff J. & Crawford A.M. (1982) Enzootic virus control of *Wiseana* spp. in the pasture environment. In *Microbial and Viral Pesticides*, p.435 (ed. E. Kurstak). Marcel Dekker, New York.

[58] Kerr A. (1980) Biological control of crown gall through production of agrocin 84. *Plant Disease* **25**, 17–23.

[59] Kerry B.R. (1980) Biocontrol: fungal parasites of female cyst nematodes. *J. Nematol.* **12**, 253–9.

[60] Kerry B.R. (1984) Nematophagous fungi and the regulation of nematode populations in soil. *Helminth. Abstr. Ser. B* **53**, 1–14.

[61] Kerry B.R. (1987) Biological control. In *Principles and Practice of Nematode Control in Crops*, p.233 (ed. R.H. Brown & B.R. Kerry). Academic Press, Sydney.

[62] Klausner A. (1984) Microbial insect control: using bugs to kill bugs. *Biotechnology* **2**, 408–19.

[63] Klein M.G. (1981) Advances in the use of *Bacillus popilliae* for pest control. In *Microbial Control of Pests and Plant Diseases 1970–1980*, p.183 (ed. H.D. Burges). Academic Press, London.

[64] Klier A., Fargette F., Ribier J. & Rapoport G. (1982) Cloning and expression of the crystal protein genes from *Bacillus thuringiensis* strain *berliner* 1715. *EMBO J.* **1**, 791–9.

[65] Kloepper J.W., Leong J., Teintze M. & Schroth M.N. (1980) Enhanced plant growth by siderophores produced by plant-growth-promoting rhizobacteria. *Nature (Lond.)* **286**, 885–6.

[66] Knowles B.H., Thomas W.E. & Ellar D.J. (1984) Lectin-like binding of *Bacillus thuringiensis* var. *kurstaki* lepidopteran-specific toxin is an initial step in insecticidal action. *FEBS Letters* **168**, 197–202.

[67] Krieg A., Huger A., Langenbruch G.A. & Schnetter W. (1983) *Bacillus thuringiensis* var. *tenebrionis*: ein neuer, genüber larven von Coleopteren wirksomer pathotyp. *Z. Angew. Entomol.* **96**, 500–8.

[68] Kronstad J.W., Schnepf H.E. & Whiteley H.R. (1983) Diversity of locations for *Bacillus thuringiensis* crystal protein genes. *J. Bacteriol.* **154**, 419–28.

[69] Kurstak E. (ed.) (1982) *Microbial and Viral Pesticides*. Marcel Dekker, New York.

[70] Kurstak E. & Tijssen P. (1982) Microbial and viral pesticides: modes of action, safety, and future prospects. In *Microbial and Viral Pesticides*, p.3 (ed. E. Kurstak). Marcel Dekker, New York.

[71] Langenbruch G.A., Krieg A., Huger A.M. & Schnetter W. (1985). Erste feldversuche zur Bekämpfung der larven des Kartoffelkäfers (*Leptinotarsa decemlineata*) mit *Bacillus thuringiensis* var. *tenebrionis*. *Med. Fac. Landbouww. Rijksuniv. Gent* **50**, 441–9.

[72] Lindow S.E. (1983) Methods of preventing frost injury caused by epiphytic ice-nucleation-active bacteria. *Plant Disease* **67**, 327–33.

[73] Lisansky S.G. (1984) Biological alternatives to chemical pesticides. *World Biotechnology Report* **1**, 455–66. Online Publications, Pinner.

[74] Longworth J.F. & Kalmakoff J. (1982) An ecological approach to the use of insect pathogens for pest control. In *Microbial and Viral Pesticides*, p.425 (ed. E. Kurstak). Marcel Dekker, New York.

[75] Lössbroeck T.G. & Theunissen J. (1985) The entomogenous nematode *Neoaplectana bibionis* as a biological control agent of *Agrotis segetum* in lettuce. *Entomol. Exp. Appl.* **39**, 261–4.

[76] Lüthy P., Cordier J-L. & Fischer H-M. (1982) *Bacillus thuringiensis* as a bacterial insecticide: basic considera-

tions and application. In *Microbial and Viral Pesticides*, p.35 (ed. E. Kurstak). Marcel Dekker, New York.

[77] Lynch J.M. (1987) Biological control within microbial communities of the rhizosphere. In *Ecology of Microbial Communities*, p.58 (ed. M. Fletcher, T.R.G. Gray & J.G. Jones). Cambridge University Press, Cambridge.

[78] Lynch J.M. & Ebben M.H. (1986) The use of microorganisms to control plant disease. *J. Appl. Bact., Symposium Suppl.* **61**, 1155–265.

[79] McCoy C.W. (1981) Pest control by the fungus *Hirsutella thompsonii*. In *Microbial Control of Pests and Plant Diseases 1970–1980*, p.499 (ed. H.D. Burges). Academic Press, London.

[80] McGaughey W.H. (1985) Insect resistance to the biological insecticide *Bacillus thuringiensis*. *Science* **229**, 193–5.

[81] Maddox J.V., Brooks W.M. & Fuxa J.R. (1981) *Vairimorpha necatrix*, a pathogen of agricultural pests: potential for pest control. In *Microbial Control of Pests and Plant Diseases 1970–1980*, p.587 (ed. H.D. Burges). Academic Press, London.

[82] Maeda S., Kawai T., Obinata M., Fujiwara H., Horiuchi T., Saeki Y., Sato Y. & Furusawa M. (1985) Production of human α-interferon in silkworm using a baculovirus vector. *Nature (Lond.)* **315**, 592–4.

[83] Mankau R. (1980) Biocontrol: fungi as nematode control agents. *J. Nematol.* **12**, 244–52.

[84] Martignoni M.E. & Iwai P.J. (1981) A catalogue of viral diseases of insects, mites and ticks. In *Microbial Control of Pests and Plant Diseases 1970–1980*, p.897 (ed. H.D. Burges). Academic Press, London.

[85] Mills S. (1986) Rabbits breed a growing controversy. *New Scientist* **109**, 50–4.

[86] Morris O.N. (1982) Bacteria as pesticides: forest applications. In *Microbial and Viral Pesticides*, p.239 (ed. E. Kurstak). Marcel Dekker, New York.

[87] Nickle W.R. & Welch H.E. (1984) History, development and importance of insect nematology. In *Plant and Insect Nematodes*, p.627 (ed. W.R. Nickle). Marcel Dekker, New York.

[88] Papavizas, G.C. (1985) *Trichoderma* and *Gliocladium*: biology, ecology and potential for biocontrol. *Ann. Rev. Phytopathol.* **23**, 23–54.

[89] Payne C.C. (1982) Insect viruses as control agents. *Parasitology* **84**, 35–77.

[90] Payne C.C. (1986) Insect pathogenic viruses as pest control agents. *Fortschr. Zool.* **32**, 183–200.

[91] Payne C.C. & Jarrett P. (1984) Microbial pesticides: selection and genetic improvement. *Proc. 1984 British Crop Protection Conference — Pests and Diseases*, 231–8.

[92] Payne C.C. & Kelly D.C. (1981) Identification of insect and mite viruses. In *Microbial Control of Pests and Plant Diseases 1970–1980*, p.61 (ed. H.D. Burges). Academic Press, London.

[93] Payne C.C., Tatchell G.M. & Williams C.F. (1981) The comparative susceptibilities of *Pieris brassicae* and *P. rapae* to a granulosis virus from *P. brassicae*. *J. Invertebr. Path.* **38**, 273–80.

[94] Pennock G.D., Shoemaker C. & Miller L.K. (1984) Strong and regulated expression of *Escherichia coli* β-galactosidase in insect cells with a baculovirus vector. *Mol. Cell Biol.* **4**, 399–406.

[95] Petersen J.J. (1984) Nematode parasites of mosquitoes. In *Plant and Insect Nematodes*, p.797 (ed. W.R. Nickle). Marcel Dekker, New York.

[96] Richardson P.N. & Hughes J.G. (1984) Prospects for the biological control of mushroom pests. *Annual Report of the Glasshouse Crops Research Institute 1983*, 96–100.

[97] Ridout C.J., Coley-Smith J. & Lynch J.M. (1986) Enzyme activity and electrophoretic profile of extracellular protein induced in *Trichoderma* sp. by cell walls of *Rhizoctonia solani*. *J. Gen. Microbiol.*, **132**, 2345–52.

[98] Rishbeth J. (1979) Modern aspects of biological control of *Fomes* and *Armillaria*. *Eur. J. For. Pathol.* **9**, 331–40.

[99] Rogoff M.H. (1982) Regulatory safety data requirements for registration of microbial pesticides. In *Microbial and Viral Pesticides*, p.645 (ed. E. Kurstak). Marcel Dekker, New York.

[100] Ross J. & Tittensor A.M. (1986) Influence of myxomatosis in regulating rabbit numbers. *Mammal Rev.* **16**, 163–8.

[101] Ruppel, E.G., Baker, R., Harman, G.E., Hubbard, J.P., Hecker, R.J. & Chet, I. (1983) Field tests of *Trichoderma harzianum* Rifai aggr. as a biocontrol agent of seedling diseases in several crops and *Rhizoctonia* root rot of sugar beet. *Crop Protection* **2**, 399–408.

[102] Sayre R.M. (1986) Pathogens for biological control of nematodes. *Crop Protection* **5**, 268–76.

[103] Schnepf H.E. & Whiteley H.R. (1981) Cloning and expression of the *Bacillus thuringiensis* crystal protein gene in *Escherichia coli*. *Proc. Natl. Acad. Sci. USA* **78**, 2893–7.

[104] Schroth M.N. & Hancock J.G. (1981) Selected topics in biological control. *Ann. Rev. Microbiol.* **35**, 453–76.

[105] Sebesta K., Farkas J., Horska K. & Vankova J. (1981) Thuringiensin, the beta-exotoxin of *Bacillus thuringiensis*. In *Microbial Control of Pests and Plant Diseases 1970–1980*, p.249 (ed. H.D. Burges). Academic Press, London.

[106] Smith G.E., Summers M.D. & Fraser M.J. (1983)

Production of human beta interferon in insect cells infected with a baculovirus expression vector. *Mol. Cell Biol.* **3**, 2156–65.

[107] Spencer D.M. (1980) Parasitism of carnation rust (*Uromyces dianthi*) by *Verticillium lecanii. Trans. Br. Mycol. Soc.* **74**, 191–4.

[108] Thomas W.E. & Ellar D.J. (1983) Mechanism of action of *Bacillus thuringiensis* var. *israelensis* insecticidal δ-endotoxin. *FEBS Letters* **154**, 326–68.

[109] Tribe H.T. (1957) On the parasitism of *Sclerotinia trifoliorum* by *Coniothyrium minitans. Trans. Brit. Mycol. Soc.* **40**, 489–99.

[110] Vaeck M., Reynaerts A., Höfte H., Jansens S., De Beuckeleer M., Dean C., Zabeau M., Van Montagu M. & Leemans J. (1987) Transgenic plants protected from insect attack. *Nature (Lond.)* **328**, 33–7.

[111] Visser B., Van Workum M., Dullemans A. & Waalwijk C. (1986) The mosquitocidal activity of *Bacillus thuringiensis* var. *israelensis* is associated with Mr 230 000 and 130 000 crystal proteins. *FEMS Microbiol. Lett.* **30**, 211–14.

[112] Whipps J.M. & Lynch J.M. (1986) The influence of the rhizosphere on crop productivity. *Adv. Microb. Ecol.* **9**, 187–244.

[113] Whiteley H.R., Schnepf H.E., Kronstad J.W. & Wong H.C. (1984) Structural and regulatory analysis of a cloned *Bacillus thuringiensis* crystal protein gene. In *Genetics and Biotechnology of Bacilli*, p.375 (ed. A.T. Ganesan & J.A. Hoch). Academic Press, London.

[114] WHO (1985) *Informal Consultation on the Development of Bacillus sphaericus as a Microbial Larvicide.* TDR/BCV/SPHAERICUS/85.3. World Health Organization, Geneva.

[115] Wilding N. (1981) Pest control by Entomophthorales. In *Microbial Control of Pests and Plant Diseases 1970–1980*, p.539 (ed. H.D. Burges). Academic Press, London.

[116] Wilson G.G. (1982) Protozoans for insect control. In *Microbial and Viral Pesticides*, p.587 (ed. E. Kurstak). Marcel Dekker, New York.

[117] Wouts W.M. (1984) Nematode parasites of Lepidopterans. In *Plant and Insect Nematodes*, p.655 (ed. W.R. Nickle). Marcel Dekker, New York.

[118] Yanchinski S. (1985) Plant engineered to kill insects. *New Scientist* **105**, 25.

4.3 Microbial spoilage of foods

Food, which is either a biological system or the product of one, is still subject to biochemical reactions, some of which may lead to changes in its desirability as a foodstuff. When sufficient changes have accumulated to make the product undesirable, the food is said to be spoiled. Thus spoilage and rejection of foods is in many ways a subjective assessment on to which more objective and scientific assessments have been grafted.

Spoilage and degradation of foodstuffs is caused mainly by the metabolism and growth of micro-organisms. Knowledge of the principles and agents involved in food degradation is essential in order to prevent or retard spoilage. This chapter serves to introduce the topic in terms of the ecology of micro-organisms and the factors which can influence their growth in and on foods; more comprehensive discussions can be found elsewhere [2, 4, 7, 12].

How important is spoilage in the present world? It has been estimated that 25% of all harvested foods is spoiled by the activities of microbes. Indeed, for the more perishable seasonal foods, up to 40% of the harvest is lost, partly because the areas of the world in which foods are produced tend to be distant from the areas where the consumers live. Many of the losses can be avoided by the application of known technologies but in many parts of the world this is uneconomic or impractical: it is rare for gluts to be coped with adequately. It is common for fish to be reduced to fish-meal for animal food rather than be used directly for human food, and for agricultural crops to be ploughed in instead of being harvested if a market collapses.

It must be emphasized that micro-organisms are not the sole means of spoilage. Mechanical damage causing bruising, squashing or crushing, and biological damage by the activities of insects and vermin on cereals can also render food unfit for human consumption.

Food spoilage is an example of an interdisciplinary subject in which microbiology plays an important part. Control or modification of the processes involved has been regarded as 'negative biotechnology' [8], the objective being to slow down or eliminate the activities of micro-organisms, or, in the case of fermentations, to direct them in order to preserve valuable materials.

4.3.1 PRE-SPOILAGE BIOCHEMICAL CHANGES

When an animal is killed for food or a plant or a fruit is harvested, its metabolism does not cease immediately but becomes unbalanced as the aerobic energy-yielding biochemical pathways slow down due to the lack of their electron acceptor, oxygen. In both large organisms and micro-organisms, catabolic and anabolic processes are integrated (see Fig. 4.3.1). The organisms take in nutrients which supply them with carbon, nitrogen and energy. These are used to build up and turn over tissue. On death or harvesting, the supply of nutrients and of energy ceases but the degradative part of the cycle continues. Catabolic enzymes such as proteases, lipases, phosphatases, amylases, pectinases, peroxidases and polyphenol oxidases remain active and cause self-digestion, or autolysis. The energy reserves of the system run down and the energy-dependent reactions maintaining the integrity of the organism fail. For example, cell membranes lose their ability to be selective such that small molecules can diffuse freely; cell compartmentalization ceases so that enzymes, e.g. from lysosomes, are released.

Such processes are of interest here for two reasons: first, because they are often necessary to make the food edible, and, second, they prepare the way for microbial attack. An example of the former is rigor mortis. After death, the fibres of muscle tissue contract by means of the energy of hydrolysis of ATP. When all the ATP has been used up the fibres relax. The sequence of reactions is

$$\text{ATP} \rightarrow \text{ADP} \rightarrow \text{IMP} \rightarrow \text{ribose} + \text{hypoxanthine} + \text{phosphate}$$

The texture of meat cooked before rigor is tough; if it has been through rigor, it is softer and much more acceptable. The other compounds listed have strong flavour notes. IMP (inosine monophosphate) has a strong meaty flavour while hypoxanthine tastes bitter. Thus, as the

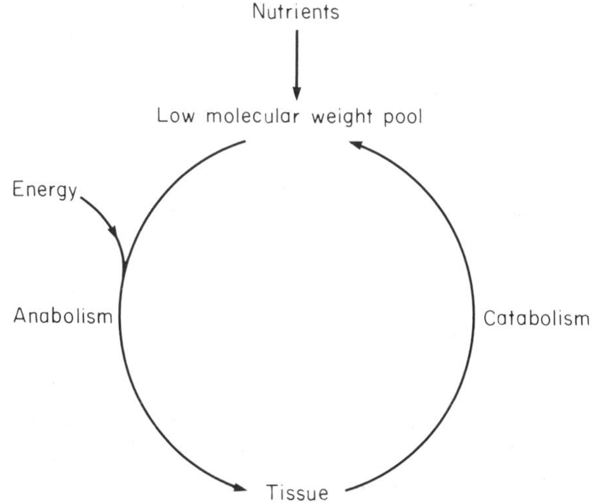

Fig. 4.3.1. Metabolism in higher organisms.

reactions proceed, desirable flavours appear and are then superseded by undesirable ones. Tissue enzymes have an important role in the development of food flavour and texture but do not cause sufficient changes to render the foods inedible or unacceptable.

Fruits and vegetables are usually still alive after harvesting and their metabolism produces carbon dioxide, water and heat, which have all been involved in the ripening process. Their metabolism then goes out of control so that macromolecules are degraded, causing softness and greater susceptibility to damage. In contrast, the seeds of cereals are relatively stable after harvest, since the parent plant has shut down their metabolism and they have developed a cuticle to resist adverse conditions.

4.3.2 FACTORS AFFECTING SPOILAGE

The interior tissues of healthy plants and animals are usually sterile, i.e. devoid of micro-organisms. Animals maintain the sterility by means of various biological, chemical and physical defence systems, e.g. the cornified layers of skin with their secretions of antimicrobial compounds and the action of antibodies and phagocytes on micro-organisms which have penetrated the organism (see Section 4.3.3). Plants may also produce antimicrobial compounds but depend on the effectiveness of their thick recalcitrant skins. Many defence systems depend on a continuous supply of energy for their effectiveness and so, when this ceases, micro-organisms can invade the interior tissues. Accordingly, the number of micro-organisms in fresh foods is generally low, usually of the order of 10^4 g^{-1}, and they have little effect. By the time the food is spoiled, numbers may have reached 10^8–10^{10} g^{-1}. The time scale for this to occur can be as little as a few hours for fish stored at 30°C or several years for dried vegetables. Thus, because of the long periods that foodstuffs may have to be stored, micro-organisms whose mean generation times under the prevailing conditions are measured in days or weeks can play a significant part in the eventual spoilage of the food. It should be noted that it is usually the activities of the micro-organisms rather than their numbers alone which determine their effects.

Even when food is spoiled, micro-organisms constitute a virtually insignificant volume of the food. The actual volumes occupied by various masses of bacteria of average size are shown in Fig. 4.3.2. Assuming that the food is rejected by the consumer when the total microbial load is 10^{10} g^{-1}, then on a weight basis only 0.1% of the food is

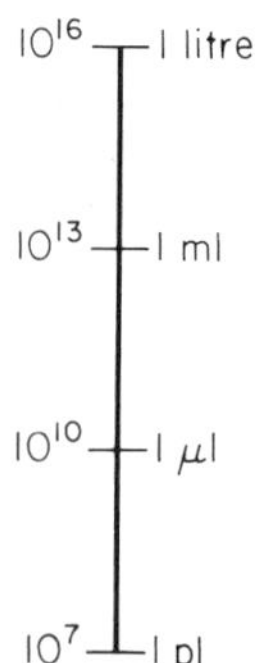

Fig. 4.3.2. The volume occupied by various numbers of bacteria.

microbial biomass. However, although micro-organisms are only a small proportion of the food, their metabolic activity has caused the chemical changes which render the foodstuff obnoxious to the consumer. In proteinaceous foods, micro-organisms cause the accumulation of metabolic end-products such as ammonia, hydrogen sulphide, indole and amines which can affect the odour and flavour of the food. Spoiled fish and meat can contain 1 mg ammonia g^{-1}. The time taken for this to accumulate depends on the number of competent organisms present but high numbers are necessary for rapid spoilage (see Table 4.3.1). As a general principle, microbial activity becomes a significant spoilage factor when numbers exceed 10^6 g^{-1}. Similarly, for the production of fermented products by the deliberate addition of microbes (e.g. starter cultures for cheese and yoghurt production), the inoculum must be sufficient to give a culture in excess of this value for the fermentation to proceed in reasonable time.

Even when spoiled food is eaten, cooked or uncooked, the ingestion of large numbers of spoilage organisms rarely causes any ill health to the consumer. The concept of ptomaine poisoning is probably only applicable in foods which are so putrid as to repel most consumers. Many bacterial pathogens are, however, transmitted via foods; these are discussed in Section 4.3.5.

Table 4.3.1. Time taken for the production of 1 mg NH_3 per g food assuming that each bacterium can produce 10^{-12} g day^{-1}

Cells g^{-1}	Time (days)
10^3	1 000 000
10^6	1 000
10^9	1

Contaminating micro-organisms may come from many sources. All food material has a natural surface flora and this, as well as the intestinal flora of animals, may be spread to the food by knives, hands, clothing, table surfaces, aerosols, air movement, etc. In addition, even under strict hygienic conditions, the storage and processing rooms always contain a microflora well adapted to the food products or materials and to the prevailing temperature and humidity.

The microflora of spoiling foods is heterotrophic; other micro-organisms may be found but do not thrive because the conditions favour the heterotrophic life style. It is not possible in this context to give any detail concerning individual types but such information is readily available in the references.

The environmental factors controlling microbial colonization and growth are usually classified into intrinsic and extrinsic [9].

4.3.3 INTRINSIC FACTORS

Chemical composition

The chemical composition of the food determines both the survival and the growth of the flora. The major constituent of many foods (see Table 4.3.2) is water, which has a decisive role in microbial activity. Flesh foods contain very little carbohydrate and so the micro-organisms which gain their energy principally from carbohydrate metabolism are at a disadvantage if those which can use amino acids and proteins are present. On the other hand, there are considerable quantities of carbohydrates in fruits and vegetables, but much is in the form of polysaccharides such as pectins which many micro-organisms

Table 4.3.2. Some selected values for the overall composition of some raw foods (g $(100\ g)^{-1}$)

	Protein	Carbohydrate	Fat	Water
Chicken	21	0	7	71
Beef	18	0	17	64
White fish	18	0	1	80
Eggs	12	0	11	75
Flour	10	79	1	10
Apples	0	15	0	85
Oranges	1	12	1	86
Potatoes	2	19	0	78
Lettuce	1	3	0	95
Butter	2	1	81	16

Table 4.3.3. Some nitrogenous extractives of protein foods

	Cod	Herring	Dogfish	Lobster	Chicken	Beef
Total nitrogenous extractive (g-kg)	12	12	30	55	12	35
Free amino acids (mM)	7	30	10	300	45	35
Creatine (mM)	3	3	2	0		4
Betaine (mM)	0	0	2	1		
Trimethylamine oxide (mM)	5	3	10	2	0	0
Anserine (mM)	1	0	0	0	1	1
Carnosine (mM)	0	0	0	0	1	1
Urea (mM)	0	0	33	0		1

cannot use. The fat or lipid components are often liable to chemical oxidation, leading to rancidity which may alone be sufficient to cause rejection by the consumer.

While the gross composition as shown in Table 4.3.2 is of interest in indicating which micro-organisms may be favoured, of more immediate interest in the initial stages of spoilage are the low molecular weight nutrients immediately available to the organisms. These include the free sugars and amino acids, fatty acids, etc. which are easily extracted with simple ionic solutions. Some nitrogenous extractives of proteinaceous foods are shown in Table 4.3.3. Micro-organisms tend to use these compounds in preference to proteins, as they do not have to synthesize and secrete degradative enzymes.

It is of interest to note that dogfish is extremely susceptible to spoilage by any bacterium possessing the enzyme to convert urea to ammonia (urease). Also lobsters, and many other shellfish, are extremely vulnerable to bacterial spoilage due to their very high level of free amino acids; indeed they constitute a very good nutrient medium for bacteria.

Water activity

Water content (i.e. the amount of water, g water (100 g food)$^{-1}$) is not usually a good indicator of the effect of the water on microbial growth. A much better measure is the water activity, which relates to the free water available to the microbes. Water activity (a_w) is defined as the water vapour pressure of the food (p) divided by that of pure water (p_0):

$$a_w = p/p_0$$

The difference between water content and water activity can be illustrated as follows. Suppose that it is possible to take all the ingredients of a food except the water and to mix them. Then some of the water is added. This will be taken up as the water of hydration around each ionic species and be tightly bound. If some more water is added, second and subsequent hydration layers are formed until there is some free water available as solvent. For meat or fish this occurs at a water content of 20%. This is shown in Fig. 4.3.3. Additional water is then available for solution purposes, to carry nutrients to the cells and to dilute toxic metabolic products.

Foods such as butter have a low water content but a high water activity because virtually all the water is free. Thus, in spoilage terms, it behaves more like a material with high rather than low water content.

Numerically, water activity is one hundredth of the equilibrium relative humidity (ERH). A food placed in an atmosphere at a humidity above the ERH will take up moisture and so gain weight, and vice versa (this is one way to measure a_w). Water activity is one of the colligative properties of solutions and therefore is dependent not on the weight of solute but on the number of particles present. Thus molar solutions of simple salts and sugars all yield solutions with the same a_w, despite the large differences in the weights of solutes. As a result, it is easy to be misled by the data in composition tables.

Water activity has an overriding influence in determining which micro-organisms will grow in a food (see Table 4.3.4). Bacteria can grow only at high a_w values. At lower a_w values, first the growth of bacteria is inhibited, the Gram-negative strains usually before the Gram-positives. Then the growth of yeasts, moulds, the highly specialized halophiles and eventually the xerophilic moulds are in turn inhibited. The micro-organisms dominant in a food are those which can grow fastest at

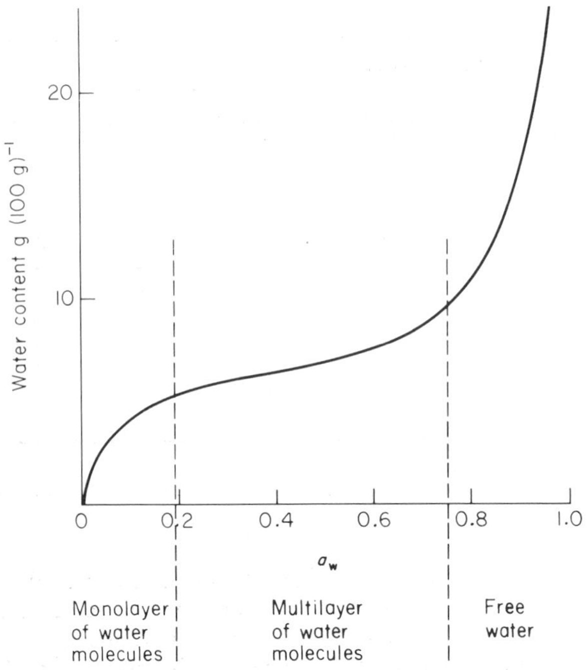

Fig. 4.3.3. The relationship between a_w and water content — a typical water sorption isotherm.

the prevailing a_w. Accordingly, in foods with high a_w, such as meat and fish, the Gram-negative bacteria usually outgrow the yeasts, assuming of course that no other selective factor is present. Even among the bacteria, there is some selectivity. If there is a layer of water on the surface of the food, caused for example by washing the product or by a change in the humidity of the storage environment, motile strains predominate over non-motile forms, possibly because they can move towards nutrients through chemotactic responses rather than be dependent on the rate of diffusion of the nutrients.

The most common end-product of microbial metabolism is water. Therefore in any food permitting microbial activity the a_w will rise, perhaps very locally, to create a micro-environment of higher a_w, which may permit some faster-growing organisms to grow and so produce even more water.

Many of the processes used in food technology effectively lower the a_w of the material from its native state so that the endogenous flora is inhibited but the nature of the food is not changed too much. It must be stressed, however, that under conditions of low a_w the growth and metabolism of micro-organisms is only inhibited; the organisms are still alive and many can survive extremely long periods under very dry conditions. Also, foods can still be contaminated from the environment during processing and storage operations and when the moisture conditions become favourable such organisms, as well as the original flora, can become active again.

The concept of water activity is similar to water potential, which is used in terrestrial systems (see Section 3.1.2).

pH and buffer capacity

The range of pH permitting the growth of micro-organisms present must be compatible with the pH of the food (see Table 4.3.5). Acidic foods therefore only harbour organisms which tolerate acidic conditions. Putrefactive bacteria which produce basic end-products of metabolism do not proliferate or have much activity below pH 6.0. One mechanism for this effect of pH is the transport of nutrients across the microbial cell membrane; this occurs only within a relatively limited pH range, which dictates the ionic form of the substrate. One result of shifts in ionic form is that the odours and flavours of many compounds

Table 4.3.4. The water content (g (100 g)$^{-1}$) and activity (a_w) and favoured micro-organisms of some foods

Foods	Micro-organisms	a_w range
Meat, fish, poultry, fresh fruits and vegetables containing > 50% water, < 25% sucrose or < 3.5% salt	Most micro-organisms have optimum growth rates in this range; Gram-negative types most active	0.98–1.00
Bread, continental sausages, hams, cheese; 5–10% salt, 25–50% sucrose	Gram-negative types inhibited; Gram-positive lactic acid bacteria, bacilli, micrococci predominate	0.93–0.98
Hard cheese, bakery goods, rice flour; 10–20% salt, 65% sucrose, 15% water	Gram-positive cocci, yeasts and moulds can grow	0.85–0.93
Jams, rolled oats, salt fish (25% NaCl) and meat, 10% water	Only xerophilic fungi and osmophilic yeasts can grow	0.60–0.85
Dried spices, milk powder, cereal products, biscuits, <10% water	No microbial growth	<0.60

Table 4.3.5. Spoilage associations of some foods

	pH	a_w	Microbial association					
			Gram-negative rods*	Cocci	Lactobacilli	Bacilli	Moulds	Yeasts
Fresh meat, fish, poultry and eggs	>4.6	>0.98	+++	+	0	0	+	0
Cooked sausages	>4.5	0.95	0	+	+	++	+	0
Fresh vegetables	>4.5	>0.98	+++	±	+	+	+	0
Dried vegetables	>4.5	<0.90	0	0	0	0	+++	0
Cereals, pulses	>4.5	<0.90	+	+	+	+	+++	+
Fruits and juices	<4.5	>0.95	±	+	++	0	++	+
Dried fruits	<4.5	<0.90	0	0	±	0	++	++
Bread, rolls, cakes	>4.5	0.95	0	0	0	+	++	±
Milk	>4.5	>0.95	+	+	±	++	0	0
Butter, margarine	>4.5	0.96	0	±	0	0	+	+

Occurrence: +++, predominant; ++, dominant; +, significant; ±, occasional; 0, rare.
* 'Gram-negative rods' include non-fermenters (belonging to the genera *Pseudomonas*, *Alteromonas*, *Acinetobacter*, *Flavobacterium*) and fermenters.

are noticeable only over a limited pH range. At values below pH 7.0 ammonia and amines are present as salts and have no odour, but hydrogen sulphide is volatile and odorous; the converse occurs at higher pH.

Usually during metabolism waste products are produced and these are rarely neutral compounds; some are acidic, such as acetic and lactic acids, while others are basic, such as ammonia and amines. The ability of a food to maintain constant pH in the presence of such compounds is its buffer capacity. Protein foods such as meat have a high buffering capacity in the important pH range, pH 5.5–7.5. About one-third of the buffering is due to the proteins, a third to phosphates, and the rest to small organic molecules such as amino acids and the peptides anserine and carnosine. Vegetables have very low buffer capacity.

Redox potential

The terminal electron transport systems of micro-organisms operate over a restricted redox range only. The redox potential prevailing in the food will determine whether the active flora will be using aerobic, micro-aerophilic, facultatively anaerobic or totally anaerobic terminal electron acceptor pathways. During the life of animals and plants aerobic conditions exist in most of the tissues but after death or harvesting the redox potential falls to a value dependent on the types and quantities of redox systems present. Of particular importance are the thiol (−SH) groups of proteins in flesh foods and the ascorbic acid and other reducing sugars in plant foods. It is important to note that the redox potential of unprocessed food is rarely uniform throughout its bulk. For example, the exposed surfaces of flesh foods are aerobic but the high demand for oxygen by the aerobic flora means that, even 2–3 mm below the surface, conditions are anoxic and so either a different population can flourish or the population has to adapt if possible to an alternative means of regenerating reduced cofactors. This is a situation analogous to the redox discontinuity layer in sediments.

Marine fish contain large amounts of trimethylamine-N-oxide (TMAO, see Table 4.3.3.), an odourless compound thought to be involved in osmoregulation. Some heterotrophic bacteria, including the types found on spoiling fish, can reduce it to trimethylamine (TMA), which has a strong distinctive stale fish odour at pH values above pH 7.0. The redox value for the TMAO/TMA couple is +19 mV (see the values in Table 2.5.1, Section 2.5.2). The bacteria can obtain energy for growth from this reaction. It enables them to maintain an aerobic style of metabolism, i.e. a tricarboxylic acid cycle centred metabolism, under anoxic conditions and so obtain relatively large amounts of energy compared with that from substrate level phosphorylation. Besides accounting for the redox value found in spoiling fish, the presence of TMAO and its reduction to TMA may well be the reason for the relatively fast rate of spoilage of fish compared with that of other similar foods.

Anti-microbial constituents

Many foods contain biologically active compounds left over from the organism's defence systems. For example, the enzyme lysozyme is present in animal secretions and egg white, some plants contain alkaloids and allylic compounds, and citrus fruits have anti-microbial lipids. Milk has many active components including: immunoglobulins and phagocytes akin to those in the animal; an efficient peroxidase–thiocyanate–hydrogen peroxide system; the bacteriocide H_2O_2; and various iron-chelating or -trapping proteins, such as lactoferrin and transferrin, which deprive the microbes of the element essential for the biosynthesis of many enzymes and cytochromes. Such proteins can be present in significant quantities; for example there are 3–9 mg ml^{-1} lactoferrin in milk, and in egg white there are 20 mg ml^{-1} conalbumin, another iron scavenger.

Biological structure

Foods cannot in general be regarded simply as a homogeneous culture medium for the growth of micro-organisms because, unlike laboratory media, they have structure, which impedes both the ingress and the activities of the organisms.

In animals, the main defence structure is the skin, whose surface consists of dead cells which are continuously being sloughed off. They provide nutrients for commensal bacteria during life but the microbes are also controlled by certain secretions of the host, e.g. sebum. When the animal is slaughtered, secretions cease and even the underlying cells die so there are ample nutrients for the micro-organisms to multiply. There are still physical barriers to the invasion of the flesh but it seems that sufficient substrates leak or diffuse from the tissue to the surface so that microbes can grow to a spoilage level without actually having to enter the tissues.

One of the most fascinating structures for resisting microbial attack is the hen's egg, illustrated in Fig. 4.3.4. The egg provides the food for the developing embryo, and so there is ample nutrient for any invading micro-organism, but it must also protect the embryo from damage and infection. To do this, the egg relies on multiple physical barriers composed of biological membranes, each with particular properties to augment the anti-microbial constituents mentioned earlier. A full description is given by Tranter and Board [13].

Many of the losses incurred in marketing fruits and vegetables are due primarily to mechanical damage resulting in breaks in the structure of their skin. Indeed, varieties are often chosen for commerce because of their thick skins. However, once the skin is broken, microbial growth can be rapid and invasion of the interior is much easier than with animal tissue. It is estimated that 30% of all fruit decay is caused by growth of *Penicillium* spp. on damaged specimens, and 36% of vegetable losses are caused similarly by soft rot bacteria.

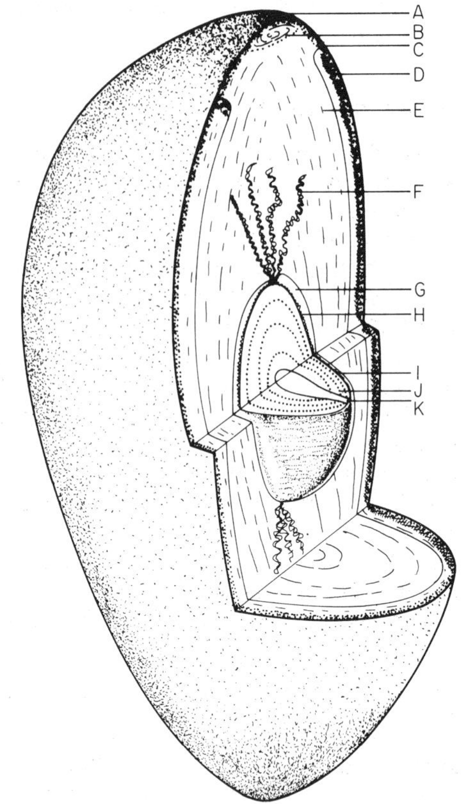

Fig. 4.3.4. The structure of the hen's egg. Reproduced with permission from Board [1]. A, Shell; B, air cell; C, inner shell membranes; D, outer thin white; E, albuminous sac; F, chalazae; G, inner thin white; H, chalaziferous membrane; I, vitelline membrane; J, yolk; K, latelra.

4.3.4 EXTRINSIC FACTORS

How can the effectiveness of the intrinsic factors be enhanced to reduce the spoilage rate of foods without employing procedures which would drastically reduce the desirability of the material as food? It follows from the points already made that one way to do this would be by lowering the a_w to below 0.6. This would eliminate all microbial spoilage but, as protein denaturation and the irreversible precipitation of many molecules would also occur, the consumer would not want to eat such tough unappetizing foods. As a compromise between ultimate safety, storage life and acceptability as food, gentler methods are used.

Storage temperature and time

The microbial flora found on animals is well adapted to growth at body temperature; if the temperature of the flesh were not lowered quickly, the contaminating micro-organisms would reduce the carcass to a putrid mass within a day. Chilling to 0°C is probably the most effective simple means used as it decreases both the rate of autolysis and of microbial growth.

There is even one major change in the predominant flora, especially of flesh foods, as they change from a living animal to food. The surface flora has been composed of mainly Gram-positive types favoured by the dry warm surfaces; this is superseded by Gram-negative strains, which are better adapted for the cooler wetter conditions.

However, the environment in which the food is stored always contains micro-organisms that are better adapted to the conditions, and because of the inevitable contamination the benefits of chill storage are less than expected. The chilling of fish is less effective than the chilling of meat because fish caught in temperate waters have grown at 5–10°C and so harbour bacteria able to grow at 0°C. Yet chilling is still effective, as shown by the data in Table 4.3.6. Traditionally, it has been assumed that the effect of temperature on microbial growth and food spoilage follows the kinetics of chemistry but Ratkowsky *et al.* [10] have shown that they are better described by the following equation:

$$\sqrt{r} = b\,(T - T_0)$$

where r is the spoilage rate, T the absolute temperature, as shown in Fig. 4.3.5, rather than the Arrhenius equation which incorporates an exponential function.

To stop microbial growth, low temperatures are needed. The temperature has to be reduced to at least −10°C and preferably −20°C for this to occur. Besides the effect of such low temperatures on the rates and the activation energies of chemical and biochemical reactions, the freezing of water out of the food has a major effect on the resulting water activity. The water activities for flesh foods under freezing conditions are shown in Table 4.3.7.

The true limiting factor for stopping microbial growth under freezing conditions is still a matter for debate. The minimum growth temperatures reported for bacteria are around −5°C, but yeasts and moulds have been reported to grow down to −15°C. However, even at lower temperatures, tissue and microbial enzymes can still be active. To combat this, before freezing the vegetables are given a short high temperature treatment called blanching to destroy polyphenol oxidase which would cause subsequent discoloration during frozen storage.

Freezing and cold storage do not kill micro-organisms, although there may be a slight reduction in the total numbers. However, the formation of ice crystals in the food may rupture membranes and other internal structures so that upon thawing nutrients become more available than previously. If this thawed food is stored, micro-organisms may multiply rapidly.

Table 4.3.6. Time for cod to become inedible at different temperatures

Storage temperature (°C)	Days
0	16
5	8
10	4
20	1

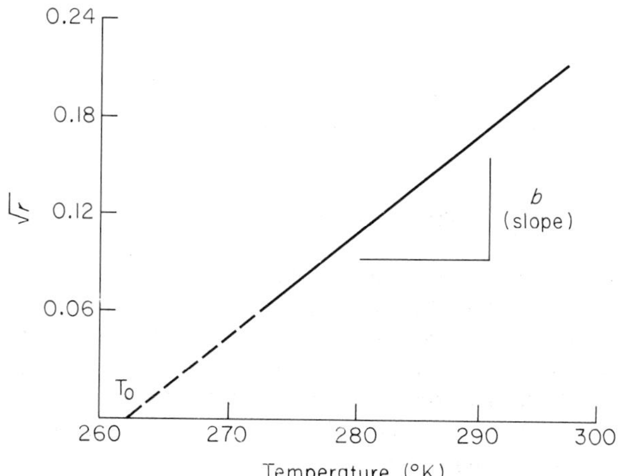

Fig. 4.3.5. Relationship between the spoilage rate (r) and the absolute temperature. After Ratkowsky *et al.* [10].

Table 4.3.7. Some a_w values of frozen food

Temperature (°C)	a_w
0	1.00
−5	0.95
−10	0.91
−20	0.82
−30	0.75
−50	0.62

Heat

All micro-organisms show an optimum growth temperature or temperature range, but the kinetics and mechanisms of the decline in growth rate at higher temperatures are complex. The decrease in numbers in a microbial population due to heat is usually expressed in terms of D, the time (in minutes) required for the population numbers to decrease by a factor of 10 (or 1 log cycle) at a constant temperature. Twelve times D ($12D$) is the treatment time necessary for sterilization at the same temperature. For pasteurization, a lower multiple of the D value, often 3, is used.

Given sufficient heat, all micro-organisms are killed as well as their spores, but drastic treatments adversely affect the product. In the canning of foods, the amount of heat applied depends on which microbes are expected to be present and the subsequent effects if they, their spores or their toxins survive. Thus the normal flora of meat and fish is heat-sensitive but these foods may harbour *Clostridium botulinum* (see Section 4.3.5) and so the heat treatment must be sufficient to ensure destruction of its heat-resistant spores, whereas the treatment required for acidic fruits juices, in which any surviving pathogens or their spores would fail to grow or germinate, is much less. The only serious problems with canned foods in recent years have arisen because micro-organisms have entered the cans during the cooling process through mechanical defects in the cans. There is much research by industry to find ways of making micro-organisms more heat-sensitive so that milder heat treatments would be effective and the flavour and texture of the food would be more like those of the fresh material.

Direct lowering of a_w

Many traditional methods of food processing involve lowering the a_w directly by drying the produce in the open air or indirectly by the addition of solutes such as salt or sugar to lower the vapour pressure and hence the a_w. Drying can only occur when the relative humidity of the air is less than the ERH of the food. The rate of drying is temperature-dependent and so the effectiveness of the process depends on the relative rates of microbial growth and of drying at the chosen temperature. The overall limiting factor for speed of drying is the rate of diffusion of water from the interior to the exterior.

Drying of many foods is preceded by salting. Salting is carried out by covering the food with approximately equal quantities of NaCl or soaking it in a saturated solution of NaCl (brining). It has two main effects on microbial growth: first, the membrane processes in the cells are inhibited by the high ionic strength and, second, the a_w is lowered, especially on the surface where, in fresh foods, most of the organisms are found. The rate of salt penetration is temperature-dependent and so high temperatures, which favour microbial growth, also assist salt uptake.

Although the normal flora of the food is inhibited by the salt, a new flora compatible with the new environment becomes established. Salted meat and fish may contain 25% NaCl (w/w) (see Table 4.3.4) and be a suitable medium for halophiles. Halophilic organisms are usually present in the salt, especially if the salt is produced in solar lagoons. Such salted products, even if subsequently dried, are extremely liable to moisture uptake unless they are well packaged and kept in atmospheres of low humidity. If halophiles are able to grow, they rapidly destroy the food because they secrete powerful lyases such as collagenase.

The manufacture of fruit preserves, jams and jellies involves similar principles. The a_w is lowered by the large amounts of added sugar and the indigenous flora killed by the heat treatment. If spoilage occurs, it is caused by osmophilic moulds whose spores have probably survived the processing.

The techniques of smoking foods involve elements of the salting and drying processes. While the traditional methods severely modified the appearance and flavour of the raw material and gave a product with long shelf-life, most modern smoking methods are used more for enhancing flavour rather than for extended preservation. A procedure for fish is as follows. First, pieces of fish are soaked in a salt solution so that the salt content is raised to about 2.5% (w/w). This is sufficient to inhibit much of the Gram-negative flora. The salt also draws protein to the surface and denatures it so that, when the fish are then placed in a smoke kiln or chamber, constituents of the smoke are deposited on the surface and of course diffuse into the tissue. One component of the smoke, phenolics, is antimicrobial. The smoking procedure is carried out either at low temperature, 30°C, to yield products which require cooking before consumption such as smoked cod fillets and kippers, or at high temperatures, 80°C, to give products ready to eat such as smoked mackerel or salmon. In both cases there is some drying of the surface, which, together with the other anti-microbial effects, is sufficient to inhibit the normal flora of the raw fish. Any spoilage is generally by moulds which come from the air and from the wood shavings used in the smoke generator. In advanced spoilage, there has usually been enough water produced to raise the a_w; this permits the growth of bacteria, as the remaining phenolic compounds have been chemically bonded.

Changing the pH

Foodstuffs which are low in buffering capacity can be easily preserved by lowering the pH; indeed, this can be done by the indigenous flora in some cases. Vegetables such as cabbages and gherkins can be converted to sauerkraut and pickles, respectively, simply from the acid produced by the action of the lactobacilli on the carbohydrates. This acid lowers the pH and inhibits the rest of the flora. To achieve a similar preservative effect with meat and fish, which have a high buffer capacity, it is necessary to add 3% acetic acid.

A traditional method of preserving milk is to convert it to yoghurt or other acidified products. Yoghurt is a complicated product with many variations, but essentially it is produced by inoculating milk with a mixed culture of *Streptococcus thermophilus* and *Lactobacillus bulgaris* and incubating. These bacteria produce sufficient acid to inhibit the normal flora and eventually even their own growth. Enzymes cause the accompanying textural changes. If the two strains are grown separately in milk, they produce less acid than when grown together, an example of symbiosis.

High pH treatments of raw materials are used in some local products, for example, in some alkaline treatments of fish, but they are not generally acceptable due to the soapy flavours and the destruction of protein.

Irradiation

Irradiation with electromagnetic waves of various wave lengths has also been used to kill or inhibit micro-organisms. Ultraviolet light is bacteriocidal but it has very poor penetration into foods and so has little value for food preservation. Ultraviolet lamps are used in food processing establishments to reduce the number of microbes on surfaces and in air. Infra-red waves cause heat and cooking. Irradiation of food with gamma rays from a Co^{60} source has been thoroughly investigated as the rays are known to kill micro-organisms. At dose levels capable of sterilizing the products, i.e. when all microbes and spores are killed, unacceptable flavour changes occur. However, the products have been exhaustively tested and shown to be safe for human consumption. At lower doses, pasteurized products can be made in which most of the commensal organisms that cause spoilage are killed. It is fortunate that the bacteria most active in the spoilage of the flesh foods are the most irradiation-sensitive and the survivors have a much poorer spoilage potential as they do not produce such copious quantities of obnoxious metabolic end-products. Foods pasteurized by irradiation can have three times the shelf-life of the raw materials stored under the same conditions. The dose given is sufficient to reduce numbers by a factor of 1000, i.e. equivalent to a 3D process. For the process to be commercially effective, only high-quality material, that with a low microbial load, should be treated and subsequent storage conditions must be optimal to retain the advantages in shelf-life. The process has been illegal in most countries but, since the World Health Organization has now given its approval to the process, its use is spreading.

Miscellaneous

Controlled gas atmospheres have been in use for many years in the bulk distribution of fruit in order to regulate the rate of ripening and indirectly to limit the activities of the indigenous flora.

The techniques have been further refined and are being used in consumer packs. The headspace of packs of meat, for example, is filled with gas mixtures such as 30% CO_2, 30% O_2 and 40% N_2. This has various effects on the flora of chill-stored products. First, as the CO_2 dissolves in the surface layers, the pH is lowered and the indigenous Gram-negative bacteria are inhibited. Second, the CO_2 has other inhibitory effects but the site(s) of action is not known. Third, the high O_2 levels maintain a high redox potential which affects the microbial flora and enhances the colour of meat.

Microbial growth is also inhibited by other chemicals. Sulphur dioxide is widely used to inhibit unwanted fermentation in carbohydrate-rich materials such as soft fruits used for the manufacture of jams, jellies and alcoholic beverages and to prevent the growth of enteric micro-organisms in the fresh British sausage during summer months. It also inhibits the enzymic browning of cut plant tissue by polyphenol oxidases. Sorbic acid is used to prevent the growth of yeasts, moulds and some bacteria in dairy products, bakery products, mayonnaise, margarines (especially the low-fat types), smoked foods, etc. It is most effective at pH 6.0–6.5, and its mode of action is thought to be a physicochemical interaction with microbial membrane lipids.

In the 1950s, antibiotics were used to control microbial

spoilage, especially of flesh foods, but their use is now generally illegal. However, there is increasing scope for the use of other antimicrobial compounds produced by the natural microbial flora of foods. For example, nisin is produced by some cultures of *Streptococcus lactis* used in cheese-making. Additional amounts are used to help prevent spoilage by *Clostridium* spp.

4.3.5 FOOD-POISONING ORGANISMS

Some bacterial diseases are transmitted by food. These are of two types, those in which the food serves merely as a carrier for the inoculum which then grows in the consumer, for example *Salmonella* spp., and others in which the organisms grow in the food and produce toxin which then affects the consumer soon after consumption, for example *Bacillus cereus*. In the former case, only low numbers (100 g^{-1}) are present but in the latter large numbers (10^6 g^{-1}) are involved — they have been killed during processing but the toxin remains. The limiting values for some of the intrinsic and extrinsic factors affecting growth and survival for some common pathogens is shown in Table 4.3.8. In recent years some other organisms have become important in food poisoning.They include *Campylobacter* spp., *Yersinia* spp. and *Bacillus cereus*.

The complete removal of pathogenic micro-organisms from foods is impossible since many of the pathogens are ubiquitous in the food-growing and -processing environments. However, in order to survive and multiply, the pathogens must compete with the normal commensal saprophytic flora in and on the food for nutrient and, since the environmental conditions usually favour the growth of the latter, proliferation of the pathogens is relatively rare. Often it occurs simply because of some contamination of the food after cooking or some other bacteriocidal treatment. Many saprophytes can inhibit the growth of some pathogens. For example, *Staphylococcus aureus* is inhibited by *Streptococcus* spp., *Leuconostoc* spp., *Lactobacillus* spp. and *Pseudomonas* spp., while *Clostridium botulinum* is inhibited by the lactic acid bacteria. Such interactions deserve more research effort.

Table 4.3.8. Effect of temperature and water activity (a_w) on the growth of food-poisoning bacteria

	Vegetative cell		Germination and outgrowth of	Toxin production			Minimum a_w for growth or spore germination	
Organism	Range (°C)	Optimum (°C)	spore — range (°C)	Range (°C)	Optimum (°C)	Storage at −20°C	a_w	Equivalent NaCl/sugar (g $(100g)^{-1}$)
Salmonella spp.	5.3–45	37	—	—	—	Survive for months	0.95	6% NaCl
Staphylococcus aureus	6.7–45	37	—	15–40	37	Survive for months	0.85 (aerobic) 0.90 (anaerobic)	
Clostridium perfringens	20–50	37	—	—	—	Spores survive indefinitely; cells slowly die		
Clostridium botulinum								
type A and B	12.5–50	37	15–42	12.5–?	29	Spores survive for months	0.92	50% sugar 10% NaCl
type E	3.3–45	35	5–37	3.3–?	26	Spores survive for months	0.96	4% NaCl
Campylobacter spp.	20–42		—	—	—	Survive 3 months	0.98	1.5% NaCl
Yersinia enterocolitica	4–42	28	—	—	—		0.96	5% NaCl
Vibrio parahaemolyticus	5–44	35	—	—	—	Survive	0.94	8% NaCl
Bacillus cereus	10–48	30	−1–59	15–?			0.95	7% NaCl

4.3.6 TYPES OF SPOILAGE OF FOODS

Some examples of the types of micro-organisms found on various foods and active in their spoilage are shown in Table 4.3.5. Why are the pseudomonads and similar bacteria so widespread in the spoilage of foods? Simply because they are so versatile in their nutrient requirements and can grow at significant rates under relatively adverse conditions. They grow at low temperatures, even below 0°C, they can use a wide variety of simple molecules such as amino acids and other non-protein nitrogenous compounds, and they can produce extracellular lipases and proteolytic enzymes. Their fast aerobic growth rates on wet surfaces give them numerical dominance and, as they can use alternative electron acceptors under anoxic conditions, they can grow under a wide range of redox conditions. They are outgrown by other microbes if the pH is outside the range of 6.0–8.0 and if the NaCl level exceeds 3% (w/w).

When significant numbers of bacteria are present, yeasts are found only in low numbers. But, if the a_w is reduced or the pH is low, they will predominate, assuming that the other conditions do allow microbial growth.

The spoilage of some foods will be considered briefly to illustrate the effects of some of the factors discussed earlier.

Fish

Marine fish caught in temperate waters have a flora consisting mainly of non-fermentative Gram-negative bacteria belonging to the *Pseudomonas, Alteromonas, Flavobacteria* and related genera with a few fermenters such as *Vibrio* spp. [6] and some Gram-positive coryneforms. When the fish are stored at chill temperatures, i.e. 0–5°C, the pseudomonads and alteromonads predominate as their growth rate is highest, but if the temperature is over 10°C then the Gram-positive strains take over. If the fish are stored under increased CO_2, the pH is lowered to near pH 6.0, the normal flora is partly inhibited and some lactobacilli may be detected.

Under chill conditions, the bacteria remain on the surface. They only invade the tissue when spoilage is advanced and noticeable. Presumably the skin acts as a mechanical barrier to invasion until it autolyses but it allows the free diffusion of substrates for the bacteria from the deep tissue and the diffusion of metabolic end-products back into the flesh. These give rise to the spoilage odours and flavours. Because of the high rate of activity of the spoilage flora, fish preservation by freezing is widely practised but enzymic changes can still occur even in the frozen state. Smoked fish are treated with salt, which inhibits the growth of the pseudomonads, and with heat and smoke, which kills and inhibits the growth of micro-organisms in general.

Meat

The interior tissues of healthy cattle are generally sterile but they can become contaminated during the slaughter procedure by means of shock reaction, which may allow microbes to penetrate the intestinal walls and be distributed throughout the carcass. The bacterial count on commercial carcasses at this stage is about 10^4 cm^{-2}, much of it arising from the environment including the workers and their clothing. After the resolution of rigor mortis, the pH of the muscle can be as low as 5.5 due to anaerobic glucose metabolism.

If the meat is still warm, i.e. 30°C, then spoilage by *Clostridium perfringens* is rapid. These bacteria produce copious amounts of gas and secrete powerful proteolytic enzymes. Thus the tissue integrity is destroyed and obnoxious odours typical of the metabolism of clostridia are produced. At lower temperatures, 15–20°C, some clostridia are still active along with bacilli; during spoilage the pH remains low due to acid formation. At chill temperatures, the spoilage process is similar to that of fish; it is caused by Gram-negative heterotrophs which particularly attack the nitrogenous compounds and accumulate basic end-products of metabolism such as amines and ammonia. The pH is in due course alkaline. Only when high numbers of such bacteria are present (10^9 g^{-1}) does significant proteolysis occur leading to the breakdown of the structure. The meat would already have been rejected by the consumer due to the accumulation of bad odours. If portions of meat are vacuum-packed to eliminate air, the spoilage is predominantly by Gram-positive strains which cause souring due to acid production. The Gram-negative bacteria of chilled meat are inhibited by 15% CO_2; thus, in modified atmosphere packs, the gas mixtures used contain at least this amount. The microbiology of meat has been described by Brown [3].

Grain

Much of the losses of cereal crops occur before harvesting due to plant pathogens and various infections such as rusts. If grain is harvested after full ripening and is dry, or is properly dried soon after, there is little possibility of microbial spoilage as the a_w is too low to support growth.

If the grain subsequently becomes wet, then moulds grow and not only is the material lost as food but there is the chance of mycotoxin production. Storage moulds can tolerate lower a_w values than the field strains; in view of the long storage periods involved, the ERH has to be less than 70% to prevent growth. Grain stores can also become infested with insects and vermin which are known to introduce bacterial pathogens, especially enteric strains, in their excreta. Thus grain stores must be well constructed and regularly fumigated to kill moulds, fungi and insects practised.

Eggs

The defence systems of an egg include the structure of the shell, membranes and protein gel and the antimicrobial constituents such as lysozyme, iron-binding proteins, avidin, which has anti-biotin activity, and a high pH. Indeed, if freshly laid eggs are collected, washed and cooled, they have a very long shelf-life so that rotten eggs are rarely encountered.

Fruit

Fruit tissues are generally high in carbohydrate and low in protein and may contain some lipids which have antimicrobial properties, the so-called essential oils. Most are of low pH so that only moulds and fungi are potential spoilers of the fresh product. They possess thick skins, which not only act as good barriers to mechanical damage but are also effective in reducing spoilage. Food-poisoning problems are virtually unknown so fruits may be eaten raw without any microbiocidal treatment. They can be contaminated from their environment during their growth, harvesting and distribution but many of the contaminants are inactive. The shelf-life of fruits is extended by proper storage temperature, relative humidity, control of the surrounding gas atmosphere and treatment with fungicidal chemicals. Long-term preservation of fruits can be achieved by canning where the low pH allows for only a mild heat treatment, by simple drying and by freezing. However, freezing is not suitable for fruits with extremely high water content where the structure virtually collapses.

4.3.7 MODELLING FOOD SPOILAGE

It is relatively easy to obtain sterile preparations of various foods in the laboratory for experimental purposes and to use them as the substrate for model spoilage studies. When typical isolates from spoiled proteinaceous foods are inoculated on to such preparations and then incubated at the temperature of the spoilage, it is found that only a low proportion, about 10–20%, of the cultures show typical production of spoilage odours. In fact what is called spoilage is related to various compounds often present in very low amounts because the threshold for their detection by man is low, perhaps only one part per million or less. The most important compounds are the amines, sulphydryls (mercaptans and sulphides) and lower volatile fatty acids (acetic, lactic, caprylic).

The role of the rest of the microbial population is poorly understood; whether it has a synergistic or antagonistic influence or simply scavenges waste products from the spoilers is not known.

4.3.8 FERMENTATION

The presence of high numbers of microbes in foods does not always equate with spoilage nor does the presence of some of the spoilage-related compounds mentioned earlier. Fermentation of foods is in effect the deliberate enhancement of a part of the flora under conditions which favour its growth. The resulting products may or may not still contain the microbes and are often characterized by strong flavours which if present in other foods would be grounds for its rejection. For example, the amines found in a ripe cheese and termed desirable would not be tolerated in fresh fish or meat.

The main classes of food prepared by fermentation are those derived from dairy products (cheeses, yoghurts, acidophilous milk, cultured creams), bakery products derived from cereals, products from plants fermented by lactic acid bacteria, alcoholic beverages, and various oriental foods involving all types of raw food materials. They are described in detail by Reed [11].

The term cheese covers an enormous variety of types but in essence all cheeses are made from milk. The casein is coagulated with rennin, an enzyme from calves' stomach, and the whey is separated. This eliminates the antimicrobial components of milk so that the addition of the starter culture is effective. It has been said that it is fortunate that milk contains lactose, which is fermented to lactic acid, rather than sucrose or other carbohydrates, which are fermented to ethanol, and indeed microbes can carry out only one or the other fermentation. The conditions produced by the starter cultures make it difficult for other microbes to gain access to the material but surface mould growth, essential for the ripening of many cheeses, does occur. Microbial food poisoning due to cheese ingestion is rare but the product does allow for the growth of and toxin formation by *Staphylococcus aureus* and if the starting material is contaminated with enteric organisms it is possible for them to survive the process and on consumption cause disease.

Many problems with cheese-making on an industrial scale have been related to failure of the starter or inoculum and it has been found that not only are many of the key enzymes of cheese-making carried on phages rather than being chromosomal but the phages can be lytic. To overcome such difficulties and potential economic loss, starter cultures are prepared separately in bulk, tested and, if suitable, frozen so that they are not carried from one fermentation to the next.

The bacteria involved in the production of yoghurt have been mentioned earlier. Although the basic product is relatively stable if kept chilled, spoilage problems arise from other ingredients added to make it more desirable to consumers. Fruit yoghurts are made by the addition of fruit or fruit pulp and may bring in contaminants able to grow and to survive the acidic environment.

Historically, alcoholic beverages can be regarded as a method of preserving some nutritional benefits from perishable carbohydrate agricultural products by a fermentation often carried out by the indigenous microbial flora, a controlled spoilage resulting in an acceptable product. Also, before the nature of disease and its transmission were known, the consumption of milk and water were unhealthy practices as milk spread milk fever (tuberculosis) in the young and water-borne epidemics were common. Thus, beers and wines of sufficiently low pH to inhibit enteric pathogens and with antimicrobial constituents or additives such as hops were important in disease prevention. Some of the fermentations practised in the Orient produced sauces which increased the protein and vitamins in an otherwise inadequate diet and added flavour to bland cereal products. In effect, the ecological factors affecting microbial growth were being exploited before the discovery of the microbes themselves.

The principles and practices of food fermentation have been adapted for the production of other materials such as enzymes, pharmaceuticals, flavouring materials and energy sources (methane and ethanol). These are often produced from materials which are undesirable themselves as food or are the waste materials from food processing. Thus, many of the techniques and end products of modern biotechnology owe their origins and philosophy to the accumulated experiences of the food industries [5].

Foods may be regarded as biological systems in which biochemical changes caused by tissue and microbial enzymes give rise to both desirable and undesirable end-products. By the application of biochemical knowledge and techniques, the favoured reactions can often be encouraged and the others inhibited or eliminated. The next stage of food biotechnology ought to be the use of microbes and biologically derived energy to reduce the high energy requirements of food processing and to control the rate of the reactions deleterious to the product quality.

4.3.9 REFERENCES

[1] Board R.G. (1969) The microbiology of the hen's egg. *Adv. Appl. Microbiol.* **11**, 245–81.

[2] Board R.G. (1983) *A Modern Introduction to Food Microbiology*. Blackwell Scientific Publications, Oxford.

[3] Brown M.H. (ed.) (1982) *Meat Microbiology*. Elsevier Applied Science Publishers, Barking.

[4] Hayes P.R. (1985) *Food Microbiology and Hygiene*. Elsevier Applied Science Publishers, Barking.

[5] Higgins I.J., Best D.J. & Jones J. (1985) *Biotechnology Principles and Applications*. Blackwell Scientific Publications, Oxford.

[6] Hobbs G. & Hodgkiss W. (1982) The bacteriology of fish handling and processing. In *Developments in Food Microbiology — 1*, p.71 (ed. R. Davies). Elsevier Applied Science Publishers, Barking.

[7] ICMSF (International Commission on Microbiological Specifications for Foods) (1986) *Microbial Ecology of Foods*, Vols. 1 & 2. Academic Press, London.

[8] Jarvis B. & Paulus K. (1982) Food preservation: an example of the application of negative biotechnology. *J. Chem. Tech. Biotechnol.* **32**, 233–50.

[9] Mossel D.A.A. (1982) *Microbiology of Foods. The Ecological Essentials of Assurance and Assessment of Safety and Quality*, 3rd edn. University Press, Utrecht.

[10] Ratkowsky D.A., Olley J., McMeekin T.A. & Ball A. (1982) Relationship between temperature and growth rate of bacterial cultures. *J. Bact.* **149**, 1–5.

[11] Reed G. (ed.) (1982) *Prescott and Dunn's Industrial Microbiology*, 4th edn. Macmillan, London.

[12] Roberts T.A. & Skinner F.A. (eds.) (1983) *Food Microbiology: Advances and Prospects*. Academic Press, London.

[13] Tranter H.S. & Board R.G. (1982) The antimicrobial defense of avian eggs: biological perspective and chemical basis. *J. Appl. Biochem.* **4**, 295–338.

4.4 Water pollution and its prevention

Human communities require water for many different and sometimes conflicting purposes. A river, for example, is often a source of potable water and fish, yet that same river may also be required to carry away sewage. Pathogens from this sewage render the water unsuitable for drinking and the sewage can ultimately make the river unable to support a healthy fish population. To avoid such problems requires management both of our aquatic systems and the pollutants entering them; a knowledge of the role of micro-organisms is necessary if this management is to be successful.

4.4.1 THE EXTENT OF THE PROBLEM

Nowadays it is axiomatic that waste from human, industrial or farm sources is a threat to health and the maintenance of an unpolluted environment, but this has not always been the case.

The problems of waste disposal began with the formation of small agricultural communities. Wastes were disposed of on land or in a convenient watercourse. Disposal on land can help in maintaining soil fertility, but it can also have deleterious effects by causing, for example, the development of anoxic conditions and an increase in heavy metals. With the growth of these settlements into large villages and towns, waste disposal became a major problem. The amount of land suitable for the disposal of waste was limited, and so increasing use was made of rivers and estuaries. The streets of an average 16th century town were usually mud tracks littered with garbage, dung and offal, while the rivers were open sewers [37].

The work of Snow and Budd was crucial in developing the public awareness of the importance of controlling water pollution. In 1855 Snow [59] proved that the cholera outbreaks in London were spread by drinking water contaminated with sewage. Budd showed a similar cause for some typhoid outbreaks. Cholera and typhoid, which are caused by the bacteria *Vibrio cholerae* and *Salmonella typhi*, were pandemic in Europe during the mid-19th century. This period of British history is marked by the start of great improvements in both the disposal of sewage and the provision of potable water. Some authorities have suggested that the impetus for these improvements was not the work of Budd and Snow but the 'great stink' of 1858. Parliament seriously considered moving out of London because of the smell which arose from the seriously polluted River Thames.

The advances in the control of water pollution which have been made during the last century must rank as one of the greatest contributions ever made to the health and quality of life of the human population. There is, however, still cause for concern. The last century has also witnessed a great increase in the human population and the rates of urbanization and industrialization. These events have increased the stress which man's activities impose on the aquatic environment. There has also been an increase in the farm animal population, which has been coupled with the adoption of intensive farming practices, with farms carrying tens or hundreds of thousands of birds or animals on small ground areas. Increase in the area of land covered by roads, car parks and other paved areas has increased runoff to the sewers and so the volume of municipal sewage to be treated.

The cost of installing systems which treat domestic, farm or industrial wastes before discharge into a river, or which bring water up to a potable quality, is high and, for many communities, prohibitive. Even in the 'advanced' countries many rivers and estuaries are still badly polluted. In 1970 over 85% of the populations of the rural areas of developing countries did not have access to reasonably uncontaminated water [63] and the percentage has probably changed little since that time. Water-borne disease is still bringing incapacity and death to the inhabitants of many countries.

4.4.2 MICRO-ORGANISMS AND WATER POLLUTION

For the purpose of this book, water is considered as polluted when it is unsuitable for an intended purpose because of contamination by certain micro-organisms or chemical compounds. This definition implies that water has many different uses which do not necessarily require the same standards and that man's activities can make water unfit for certain uses. Water can also be made unfit through a variety of natural processes, but these are not usually regarded as pollution. For example, a period of drought can increase the salt content of fresh water so that it becomes undrinkable.

Pollution can be caused by a wide variety of organic and inorganic compounds [21] and micro-organisms often play a major role in determining the extent of this pollution and in making pollution obvious. Contamination by sewage, which contains pathogenic micro-organisms and non-toxic organic matter, illustrates three important microbiological aspects of water pollution:

1 Water can act as a vector for the transmission of pathogenic micro-organisms.

2 The addition of organic matter to an aqueous system changes the patterns of microbial activity. The symptoms of pollution noted by the public, e.g. dead fish and unpleasant smells, are often the result of changed patterns of microbial activity.

3 Micro-organisms are the basis of most of the treatment methods which attempt to remove from sewage the potential for causing pollution.

4.4.3 THE TRANSMISSION OF DISEASE

Water-borne pathogenic micro-organisms

A number of human pathogenic micro-organisms are, or can be, transmitted by water. Some of these organisms are pathogenic only to humans, e.g. *Salmonella typhi*, while others can infect animals as well, e.g. *Shigella* spp. A major group, including many of the *Salmonella* strains, consists principally of animal pathogens which can occasionally infect man.

The mode of transmission

The major source of pathogenic micro-organisms in water is the faeces of people suffering from enteric diseases. A distinction must be made, however, between people who are temporary excreters of pathogens and those who are chronic excreters. The former excrete the organism only during an attack of the disease and during the subsequent convalescence; the latter, however, excrete erratically for very long periods and possibly for their entire lives.

For a number of reasons it is not possible to prevent the transmission of water-borne disease solely by separating the source of potable water from the waste of humans or domestic animals. Some human pathogenic micro-organisms can occur in the faeces of wild animals. In addition, fish can harbour some of these pathogens and transport them to uncontaminated water [29]. Not all water-borne diseases are transmitted by the drinking of contaminated water; a well-documented route is via the preparation of food in utensils which have been washed with contaminated water. Another route is via the filter-feeding shellfish which use micro-organisms for food. Eating shellfish from contaminated water can cause diseases such as infectious hepatitis or typhoid. Still another way is illustrated by strains of *Leptospira* which gain access to their host via skin abrasions and mucous membranes.

Water is not the natural environment of most pathogenic micro-organisms, but rather their means of passage to a new host. In fact these organisms are unable to multiply, or survive for very long, in the aquatic environment. They decline in numbers with time. That the decline is not due solely to dilution can be shown by measuring over time the numbers of bacteria and the concentrations of a recalcitrant tracer molecule. As the numbers of bacteria are reduced at a faster rate than the concentration of the tracer, factors other than dilution must be involved in the disappearance of the bacteria. Sedimentation, especially if the bacteria are attached to particles, starvation, sunlight, pH, temperature, plus competition with, and predation by, other organisms, are possible factors. The death rate of pathogenic bacteria in the aquatic environment is considered to be a function of time modified by a marked temperature coefficient [43].

Bacteriological standards

The wide range of pathogenic, water-borne micro-organisms, their low numbers when compared with non-pathogenic bacteria in most sewage, and the need to assay large numbers of samples combine to make the routine isolation of pathogens difficult, if not impossible. In water-quality control, non-pathogenic bacteria present in faeces are used to indicate the occurrence of faecal contamination and hence of the possibility that pathogenic micro-organisms are present. The indicator organisms must: (a) be present whenever pathogens are present, and in much greater numbers; (b) be as resistant or more resistant to disinfectants and the aqueous environment than the pathogens; (c) be easy to grow on a selective artificial medium; and (d) have easily identified characteristics. The most frequently used indicator organisms are the faecal coliforms, followed by the faecal streptococci and *Clostridium perfringens*. The presence of indicator organisms indicates faecal contamination and the possibility that pathogens are present. Their presence has no other relationship to water quality.

The coliform bacteria are part of the Enterobacteriaceae but various authorities have used different and sometimes conflicting characteristics to define these bacteria. In the examination of water, coliforms are arbitrarily defined in the USA [1] as aerobic and facultatively anaerobic, Gram-negative, non-spore-forming, rod-shaped bacteria which ferment lactose with gas formation within 48 hours at 35°C. In the UK [12], the term coliform organism refers to Gram-negative, oxidase-negative, non-spore-forming rods capable of growing aerobically on an agar medium containing bile salts and able to ferment lactose within 48 hours at 37°C with the production of both acid and gas. The counts obtained by incubation at 35°C or 37°C are usually said to be of total coliforms.

The basis of using the coliforms as indicators of faecal contamination is that the primary natural habitat of these bacteria is the intestines of warm-blooded animals. This assumption is probably true for *Escherichia coli* and large numbers of these bacteria are found in faeces; the numbers of *E. coli* in wet human faeces may approach 10^9 g^{-1}. However, other coliforms such as *Klebsiella pneumoniae* and *Enterobacter aerogenes* can grow in non-animal environments such as the soil, plant surfaces and even industrial effluents [13].

Coliform counts used as standards of water quality

In many countries the total and faecal coliform counts are used to assess the health risks posed by pathogenic micro-organisms in water. Water is often brought up to a potable standard by chlorination. Chlorinated water should contain no total or faecal coliforms per 100 ml, while unchlorinated potable water should also contain no faecal coliforms, although the presence of up to three coliforms may be tolerated in occasional samples [12]. The chlorination required to remove all coliform bacteria may in some cases be insufficient for the removal of viruses. For example, in Delhi, India, during 1955–56 there were over 20 000 cases of water-borne infectious hepatitis, yet there was no increase in enteric bacterial diseases and the water supply apparently conformed to the coliform standards. The absence of coliforms does not, of course, imply the absence of all bacteria; many bacteria, for example the pseudomonads, can grow in potable water.

For non-potable water different coliform standards are applied. For instance, to guard against transmission of enteric disease to humans eating shellfish grown in sewage-polluted water, shellfishing is permitted in the USA only if the coliform counts are below 70 (100 ml of water)$^{-1}$; if the coliforms are not of faecal origin, shellfishing can be permitted up to counts of 2300 coliforms (100 ml of water)$^{-1}$.

Faecal streptococci as standards of water quality

The streptococci normally occurring in human and animal faeces and therefore most likely to be found in polluted waters are *S. faecalis*, *S. equinus*, *S. durans* and *S. bovis*. These bacteria are able to grow at 45°C in the presence of 40% bile and in concentrations of sodium azide which are inhibitory to the coliforms and most other Gram-negative bacteria. These growth characteristics are used to isolate and enumerate the faecal streptococci in water, using either the multiple tube or membrane filtration methods.

S. bovis and *S. equinus* are not commonly found in man and therefore are a specific indicator of non-human, warm-blooded animal pollution which might arise from farms or from factories processing animal products. The ratio of faecal coliforms to faecal streptococci in fresh human faeces is above 4, while for all other warm-blooded animals it is below 0.6 (see Table 4.4.1). This ratio can be used to identify the source of the pollution [15], although considerable caution has to be applied because the ratio will change because of the different death-rates of the two populations. *S. bovis* and *S. equinus* die off rapidly, but *S. faecalis* will generally persist for a longer period than the faecal coliforms.

Clostridium perfringens as a standard of water quality

Clostridium perfringens is a Gram-positive, spore-forming rod found in human and animal faeces, which sporulates in nature but rarely on laboratory media. In water recently contaminated with sewage there will be more *E. coli* than *C. perfringens*, but the spores of the latter organism will survive in water for considerable periods and will be present when all the other faecal bacteria have died. *C. perfringens* is therefore a useful indicator of remote or intermittent pollution. To isolate and enumerate *C. perfringens* in water, use is made of the ability of spores, but not the majority of vegetative bacteria, to survive heating at 75°C for 10 minutes; however, membrane filtration followed by growth on solid selective media is now gaining favour. After heating, the water sample is used in inoculate a differential medium containing sulphite, which is reduced by *C. perfringens* to form an easily recognizable precipitate of black ferrous sulphide. The presence of *C. perfringens* is then confirmed by other tests.

4.4.4 WATER POLLUTION DUE TO MICROBIAL DEGRADATION OF ORGANIC MATTER

The addition of domestic, farm or industrial waste to a river, lake or estuary can often cause marked changes in the flora and fauna of the system (see Fig. 4.4.1). These

Table 4.4.1. Faecal coliform and faecal *Streptococcus* density relationships in faecal discharges from warm-blooded animals (data abstracted from Geldreich [15])

	Densities g^{-1} of faeces (median values)		
Faecal source	Faecal coliform	Faecal *Streptococcus*	Ratio FC : FS
Man (USA)	13 000 000	3 000 000	4.33
Cow	230 000	1 300 000	0.177
Pig	3 300 000	84 000 000	0.039
Sheep	16 000 000	38 000 000	0.421
Horse	12 600	6 300 000	0.002
Duck	33 000 000	54 000 000	0.611
Chicken	1 300 000	3 400 000	0.382
Turkey	290 000	2 800 000	0.104
Cat	7 900 000	27 000 000	0.293
Dog	23 000 000	980 000 000	0.024
Field mouse	330 000	7 700 000	0.043
Rabbit	20	47 000	0.0004
Rat	180 000	78 900 000	0.0023
Chipmunk	148 000	6 000 000	0.002
Elk	5100	760 000	0.007
Robin	25 000	11 700 000	0.002
English sparrow	25 000	1 000 000	0.025
Starling	10 000	11 800 000	0.0009
Red-winged blackbird	9000	11 250 000	0.0008
Pigeon	10 000	11 500 00	0.0009

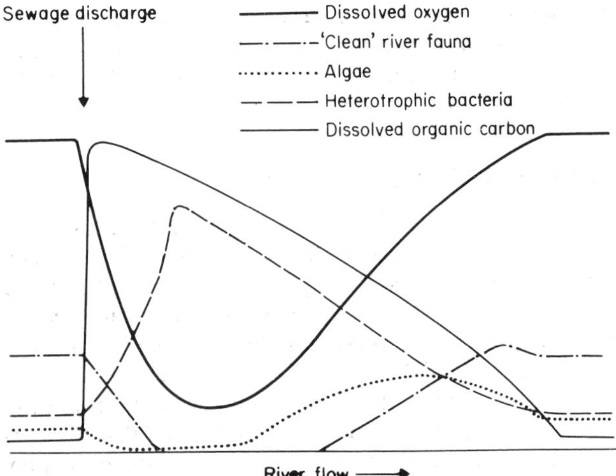

Fig. 4.4.1. The chemical and biological changes in a river caused by the discharge of sewage.

may be permanent in enclosed water such as a lake or temporary in a river in that the waste will be gradually diluted out or broken down and some way below the discharge the river will return to normal. In many cases these changes are not the result of discharges of toxic compounds with the waste, but are the by-products of the increased microbial activity caused by the addition of nutrients to the aquatic system. The aquatic environment is so changed by the increased microbial activity, particularly with respect to dissolved oxygen concentration, that the original flora and fauna cannot survive.

The importance of oxygen

Oxygen is essential for the survival of most macro-organisms. The oxygen concentration of a trout stream, for example, should not fall below 4 mg l^{-1} if the fish are to remain healthy. Susceptibility of fish to toxic chemicals is also increased at low oxygen concentrations. Oxygen has a relatively low solubility in water, and solubility is decreased by increasing temperature and salinity (see Table 4.4.2).

Assuming that no toxic compounds, such as heavy metals, are present, the dissolved oxygen concentration is a very useful indication of the quality of a river. As a rule of thumb the classification as to quality shown in Table 4.4.3 can be made. Ensuring that there is adequate dissolved oxygen in a river or other aquatic system is an important factor in the successful management of water resources.

Table 4.4.2. The effect of temperature and salinity on the solubility of oxygen in water

	Oxygen concentration (mg l^{-1}) at the following chloride concentrations (g l^{-1}) *				
Temperature (°C)	0	5	10	15	20
0	14.6	13.8	13.0	12.1	11.3
10	11.3	10.7	10.1	9.6	9.0
15	10.2	9.7	9.1	8.6	8.1
20	9.3	8.7	8.3	7.9	7.4
25	8.5	8.0	7.6	7.2	6.7

* The chloride concentration of sea water is 19.4 g l^{-1}.

Table 4.4.3. The quality of river water as determined by the oxygen concentration

Quality	% of the saturation concentration of oxygen at the ambient temperature and salinity
Good	90
Fair	75–90
Doubtful	50–75
Badly polluted	50

Calculating the oxygen balance

To calculate the oxygen balance for a river it is necessary to devise an equation expressing the various sources of, and demands for, oxygen. This equation has also to account for the changes in temperature and salinity, the inflow of unpolluted water and the turbulence, as these factors can have a profound effect on the result [43].

Oxygen can enter a body of water by diffusion from the atmosphere, a process which is aided by turbulence. Oxygen is also produced within the water by algal and macrophyte photosynthesis (see Sections 2.5.5 and 3.2.3). Photosynthesis requires light energy, but a high concentration of suspended solids in the water, or the presence of compounds such as the dark brown lignosulphonates which are found in the effluents from certain types of pulp mill, can prevent the transmission of light to the organisms and hence the production of oxygen [44].

The demand for oxygen is due to the aerobic respiration of micro- and macro-organisms found in the water column and sediment and to the abiotic chemical oxidation of reduced compounds. The reduced compounds can be present in the waste or can arise as a result of the metabolic activity of micro-organisms in the system —

for example, the sulphide produced by the sulphate-reducing bacteria (see Sections 2.5.3 and 3.2.6). The terms 'chemical oxygen demand COD' and 'biochemical oxygen demand BOD' are applied to the results of specific analyses [1] and do not refer to the oxygen demand due solely to the chemical and biological components of the system.

Five-day biochemical oxygen demand (BOD_5)

This method was originally designed to determine the amount of oxygen required by the aerobic microbes to utilize the organic matter in a sample of water. In the method, the oxygen concentration of the water sample, or a dilution of it, is determined after incubation in the dark (to prevent photosynthesis) at 20°C for 5 days. Mineral nutrients and a bacterial seed are often added. It is assumed that the microbial community within the incubation vessel has the same ability to degrade the organic matter as the natural community. The presence of compounds which could inhibit microbial activity should be determined before the BOD_5 analysis is carried out.

The growth of the micro-organisms during the BOD_5 test occurs in a closed system with carbon as the limiting nutrient. During the incubation period a succession therefore develops both in the nature and quantity of the organic matter present, as well as in the composition of the microbial community. In the first stage of the succession the micro-organisms use the readily available organic matter, and the concentration of organic matter therefore decreases while microbial biomass increases. The BOD_5 method is usually a measure of the oxygen demand of the process. Thus the method does not estimate the oxygen demands of compounds which are difficult to degrade, e.g. cellulose, or the microbial activity which occurs after 5 days.

The increase in microbial biomass allows increased activity by predacious micro-organisms, particularly the protozoa, and by microbial populations which can use the metabolic products of the so-called first-stage micro-organisms. There is also an increase in the activity of the aerobic nitrifying organisms which can use ammonia released during the mineralization of the organic matter (see Sections 2.5.4 and 3.2.6). This is often called second-stage uptake or the nitrogenous oxygen demand. In the natural aquatic system the mineralization process can also provide the nutrients for algal biosynthesis: this process increases the concentration of organic matter which can be utilized by the heterotrophic bacteria and hence exerts a further oxygen demand on the system. As a result of all these complications of resistant compounds and successions, the BOD_5 can often be less than 50% of the ultimate biological oxygen demand.

The use of BOD_5 in pollution control

For all its faults, the BOD_5 test of wastewaters remains one of the major tools in the control of pollution. The reason is that it is technically easier and arguably more relevant to demand that a sample of waste water, rather than the receiving waters, comply with regulatory standards (see Table 4.4.4).

The Royal Commission decided on a 5-day incubation period for the BOD test as it was considered that this was the maximum period required for British rivers to flow to the sea. The incubation temperature was set at 65°F (18.3°C) as this was considered to be the average summer temperature of British rivers. This incubation temperature was later set to 20°C. Although these BOD_5 standards have been widely applied throughout the world, they were developed for the British Isles and must be applied with caution to other countries with different climates [38].

Table 4.4.4. Royal Commission standards for sewage discharged into British rivers

Available dilution by clean river water	Standards (mg l^{-1})	
	BOD_5	Suspended solids
500	NS	NS
300–500	NS	150
150–300	NS	60
8–150	20	30
8	<20†	<30†

NS, no standard recommended.
Clean, usually means a BOD_5 of 2 or less.
† Exact standard depends on local circumstances.

4.4.5 MICROBIAL CHANGES IN A RIVER CAUSED BY THE DISCHARGE OF WASTES

Sewage fungus

In polluted rivers every surface, from rocks to tree branches, is often covered by the white threads of the aerobic bacterium *Sphaerotilus natans*. Its filaments are attached to the river bed and rocks and form the matrix for other bacteria and for diatoms and protozoa; this community is often called 'sewage fungus'. *S. natans* can grow at low oxygen tensions but not in anoxic water, though the organism can survive periods of anoxia. In acidic environments *S. natans* is replaced by aquatic fungi, e.g. *Fusarium aquaeductum*. The strands of sewage fungus efficiently remove nutrients and oxygen from the overlying water. It has been estimated that a square metre of active sewage fungus exerts an oxygen demand of 2.1 g O_2 h^{-1} [11]. A bloom of sewage fungus in a river is an aesthetically unpleasant example of the increase in heterotrophic microbes which occurs when excess organic matter is discharged into a river.

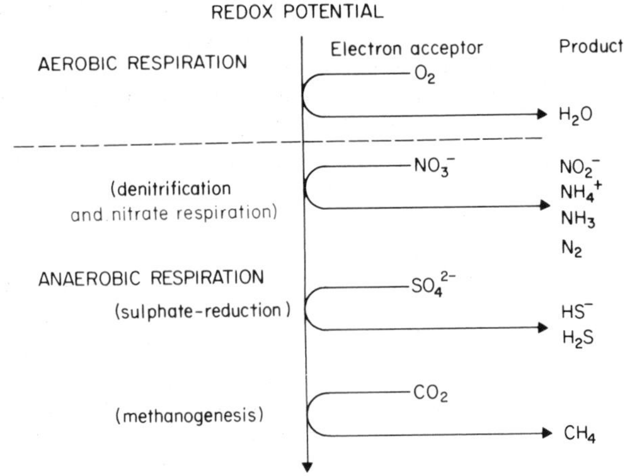

Fig. 4.4.2. The redox potential of a sediment can indicate which inorganic electron acceptor is being used by the microbial community. At a neutral pH sulphate reduction generally occurs at −150 to −200 mV, whereas methanogenesis occurs at −250 to −300 mV.

Anaerobic micro-organisms

If the demand for oxygen exceeds its supply the river or its sediments become anoxic and anaerobic micro-organisms, both facultative and obligate, become dominant. Even when oxygen is apparently abundant there will be anoxic micro-environments (see Section 3.1.2) where anaerobic bacteria such as the sulphate reducers are active [31]. Sediments often become anoxic. The extent of depletion of oxygen depends on the balance between the rate of microbial metabolism and the rate of diffusion of oxygen from the water column. The diffusion rate of oxygen depends, in turn, on the physical structure of the sediment and the activities of burrowing animals. The physical structure of the sediment, and so the diffusion of oxygen, can be altered by the addition of such particulate matter as cellulose fibres. The burrowing activities of the sediment-dwelling animals increase the surface area between the water and the sediment and so increase the diffusion of oxygen. As oxygen levels in the sediment decrease the animals become smaller and tend to dwell on the surface. Only a few animals survive when all the oxygen disappears [45].

The redox potential of sediments (see Fig. 4.4.2), like that of soils (see Section 3.1.6), can be a useful indicator of the inorganic electron acceptor being used by the anaerobic bacteria (see Sections 2.5.3 and 3.2.6).

The escape from the system of the gases N_2, H_2S and CH_4, which are the end-products of anaerobic metabolism, can increase the rate at which organic matter is degraded since these reduced products might otherwise be toxic or inhibitory to enzymic reactions. Hydrogen is also important in the anaerobic degradation of organic matter; however, these reactions are only thermodynamically possible at very low concentrations of H_2. Hydrogen-utilizing anaerobes such as sulphate-reducing and methanogenic bacteria therefore play a very important environmental role by keeping the H_2 concentration low and thus preventing the inhibition of further organic matter degradation. Other fermentation products, e.g. organic acids, may have a similar effect to that of H_2 but at higher concentrations.

As sulphate is a major anion in sea water, bacterial sulphate reduction is an important reaction in anoxic estuarine and marine environments. Both the sulphate-reducing and methanogenic bacteria can utilize only a limited number of organic compounds and hydrogen. In sediments of low E_h, sulphate-reducing and methanogenic bacteria compete for products (e.g. acetate, propionate, butyrate and hydrogen) of the degradation and fermentation of waste organic matter by other anaerobic bacteria. The anaerobic degradation of organic matter thus requires the interdependent metabolism of a complex community of interacting micro-organisms. In the presence of high concentrations of sulphate, methane production is inhibited and sulphate-reducing bacteria predominate [54, 61]. Sulphide formed by the sulphate-reducing bacteria reacts with any available heavy metal to form insoluble salts (the black colour of anoxic sediments is due mainly to ferrous sulphide). If there are no heavy metals available the sulphide will diffuse from the sediment and escape into the overlying water column and atmosphere as H_2S; the release of H_2S depends on the pressure, temperature and pH (at a pH of 7, sulphide will exist as $H_2S \rightleftharpoons HS^-$). The sulphide can be used as a source of energy by the sulphur-oxidizing bacteria (see Section 3.2.6), or as an electron donor by some of the photosynthetic bacteria (see Section 2.5.5), or it can be chemically oxidized. The chemical oxidation of sulphide can be an important factor in maintaining an anoxic environment by utilizing oxygen that enters the system [48]. Hydrogen sulphide is partly responsible for the unpleasant smell of an anoxic estuarine or marine environment. Sulphide is very toxic to both eukaryotic and prokaryotic organisms, even to the sulphate-reducing bacteria [3]. Hence it is likely that its concentration can influence microbial activity and so the degradation rate of organic matter in the field, as well as the activity of the macro-fauna. Sulphide produced in sewer pipes can cause considerable corrosion; moreover, the H_2S produced can reach concentrations in the enclosed system which are fatal to man.

4.4.6 TREATMENT OF ORGANIC WASTES

A variety of physical, chemical and biological treatments are used to reduce the amount of organic matter discharged into the environment. The effluent from industry and intensive animal farms can pollute just as much as that originating from domestic sewers (see Table 4.4.5).

Some industrial wastes have been economically recycled, e.g. the mash from a distillery has traditionally been used as an animal feed. However, any processing of distillery and other wastes which uses energy has become less economic in recent years as the cost of fossil fuels has increased. As a result, microbiological processes, which use little energy, are becoming increasingly attractive for the treatment of a variety of wastes. In some processes a useful end-product is obtained. The starch wastes from potato-processing plants, for example, can be converted by the symbiotic action of two yeasts (the Symba process) into single-cell protein [58]. In the process the amylase-producing *Endomycopsis fibuliger* degrades the starch to sugars which the non-amylase-producing *Candida utilis* then utilizes for growth. Cellulosic or starchy wastes can be utilized by some anaerobic bacteria to produce ethanol, which can be used as a fuel; other bacteria can produce acids or higher alcohols for industrial use. While waste-treatment plants are designed to economically remove or lessen the pollution potential of a particular effluent, the production of a saleable by-product (e.g. animal feed or biogas) influences the overall economics of the process.

Table 4.4.5. Polluting power of various wastes (in terms of COD and BOD, mg l^{-1})*

Domestic sewage	BOD 200–600
Fruit and vegetable processing	BOD 50–15 000
Meat and poultry processing	BOD 150–2500
Piggery effluents	BOD 25 000; COD 100 000
Cattle shed effluents	BOD 20 000; COD 100 000

* The figures in this table give an approximate idea of the values. For an extensive compilation of values for agricultural and food-processing wastes, see [34].

The biological methods for the treatment of organic wastes can be divided into those utilizing aerobic respiration and those utilizing anaerobic respiration or fermentation.

Aerobic treatments

A variety of batch or continuous cultures are used.

Composting is the aerobic breakdown of solid waste in a batch culture. A succession of mesophilic and thermophilic organisms degrade the material to carbon dioxide and water and produce a relatively innocuous residue of microbial cells plus the undegradable portions of the feedstock. The 'trickling-filter' method (see Fig. 4.4.3C) is aerobic continuous culture with an immobilized mass of organisms on a support medium over which wastewaters flow. The other aerobic processes use 'stirred-tank' continuous cultures in which wastewaters are passed through a container which provides a certain retention time.

Aerobic lagoons, which are large shallow ponds, are dependent on natural diffusion of air and on thermal and wind mixing; consequently they have a slow rate of microbial activity and a long retention time. They are therefore infrequently used in modern waste-treatment plants.

In the mechanically aerated tanks and ditches aeration increases the diffusion of oxygen from the atmosphere into the water and so ensures vigorous growth of aerobic bacteria. The container may be an oval or circular ditch around which the liquid is propelled and aerated (the 'Pasveer ditch', see Fig. 4.4.3 A and B), or a rectangular or circular tank. These systems may or may not have a feedback of bacteria. The 'activated sludge' system is a continuous culture with feedback and is used in many present-day domestic sewage works with either impeller or sparged-air aeration (see Fig. 4.4.3D).

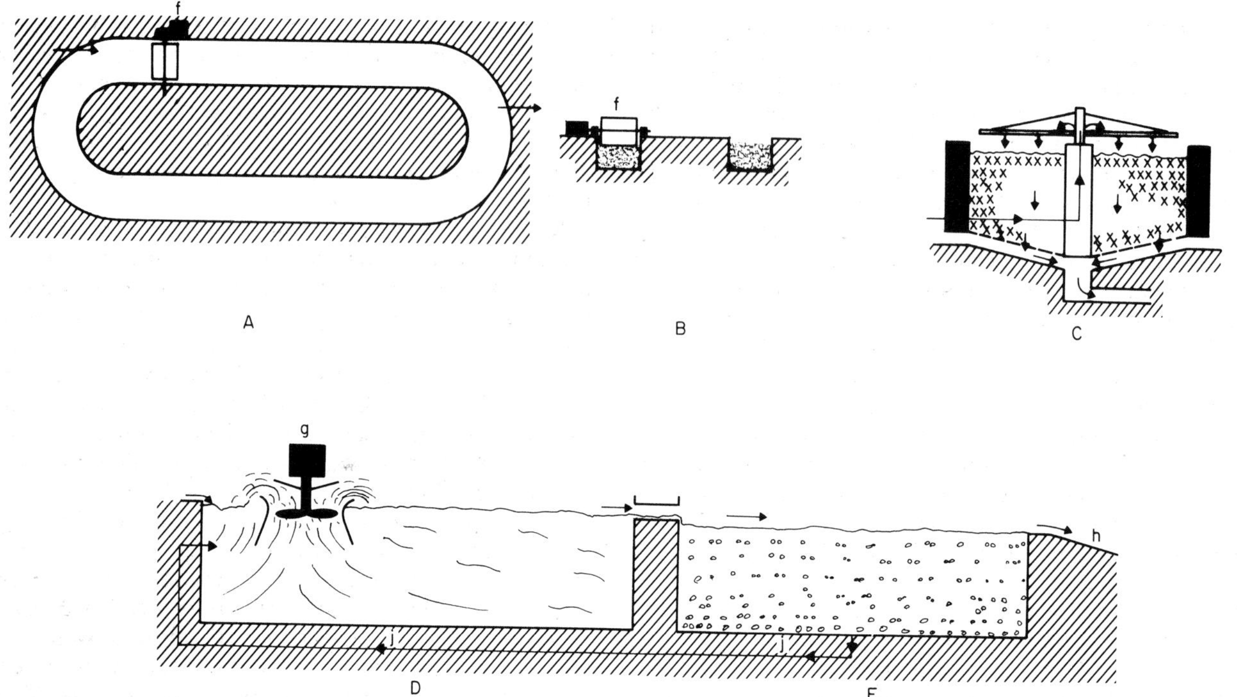

Fig. 4.4.3. Diagrammatic representation of three aerobic sewage treatment plants. Oxidation ditch, A, in section, B; trickling filter, C; activated sludge tank, D, with sludge-settling tank, E.

Arrows indicate direction of sewage flow in pipes into and out of systems. Long-line shading shows typical ground levels. The rotor (f) circulates sewage round the ditch and aerates the liquid. In C the sewage flows down over the filter medium and is collected in the drain at the bottom; air can also enter the filter at the bottom. The tank in D is aerated by the motor-driven, draught-tube aerator (g); a number of these would be used in a big tank or air can be sparged from the bottom. The liquid passes to the settling tank E. Settling tanks are rectangular or circular and have various types of conical or sloping bottoms, weirs, scrapers to remove settled sludge, and other equipment not indicated. The clarified water finally runs over the weir to a river or other outlet at h. Settled sludge, is returned to the aerobic treatment via a pipe system and pumps indicated at j. The relative volumes of activated sludge and settling tanks are adjusted to provide different retention times in the aeration and settling tanks.

MICRO-ORGANISMS AND COMPOSTING

Composting is the term given to the aerobic degradation of heterogeneous organic matter in the solid state by a microbial community which includes thermophiles. Composting, though used by gardeners throughout the world, has not yet been widely accepted for the large-scale treatment of wastes, in part because of the problem of controlling the flow of air. Composting of sewage sludges or animal excreta, which usually have a water content of 80–90%, is impossible. Composting of poultry excreta, which after collection has a water content near 50%, is possible. However, this material tends to ball up into masses impermeable to air, and mixing with some drier porous material, such as domestic garbage, is generally required. Domestic garbage itself can be composted to make a material suitable for landfill or for use as fertilizer.

For any waste treatment involving the growth of micro-organisms, sources of carbon, energy (e.g. carbohydrates) and nitrogen (e.g. proteins and ammonia) are obviously necessary and ideally these should be in proportions which provide just enough carbohydrate for the conversion of the waste nitrogen into microbial nitrogen. This ideal situation is rarely, if ever, achieved in practice since the sources of carbon, energy and nitrogen are complex molecules such as carbohydrates, lipids, proteins and lignin, which are used by the microbial community at varying rates and which may not be completely available to the micro-organisms. Hence, the carbon : nitrogen ratio must be used with caution to predict microbial activity in a waste unless the chemical nature of the carbon and nitrogen is also known. In many cases the waste must be treated as it comes, but it may be possible to mix certain wastes in order to obtain a more balanced substrate for the micro-organisms. An example is domestic garbage, which after removal of glass, metal and plastics, has a high concentration of carbohydrates (in the form of paper and vegetable wastes) but a low level of nitrogen. This poor carbon : nitrogen ratio can be improved by the addition of dewatered sewage sludge, which is high in nitrogen. Some 10% by weight of sludge is used and the resultant material is 'damp' (a water content of 50–60% is near optimum) but of sufficiently porous structure to allow air penetration (some 30% air space is needed). The aeration not only provides oxygen for continued microbial respiration but also removes carbon dioxide and water, the end-products of microbial metabolism. If too much water accumulates or the material is initially too wet, undesirable anaerobes become dominant.

The material can be composted in piles or 'windrows' built in such a fashion that air can circulate through the pile. The piles are often turned mechanically to improve aeration. Composting is also carried out in mechanically rotated drums, with or without forced aeration.

During the composting process there are predictable changes in the composition of the microbial community, in the availability of nutrients, and in the temperature and pH of the compost. In the first, or mesophilic, stage temperature increases as a result of the generation of heat by microbial activity, the low airflow, and the insulating effect of the material. Later, organisms adapted to the higher temperature build up during the thermophilic second stage.

Several studies on heat production by micro-organisms are particularly relevant to composting. The heat output of *Escherichia coli*, growing at 37°C on asbestos fibre soaked in different nutrient broths and well aerated, varied from 0.07 to 0.95×10^{-12} cal cell^{-1} s^{-1} (0.3 to 3.9×10^{-12} J cell^{-1} s^{-1}) depending on the medium [52]. Good growth of the bacteria was obtained and the heat output from the cultures was up to 39×10^{-4} cal s^{-1} ml^{-1} (164×10^{-4} J s^{-1} ml^{-1}) with 3.7×10^{10} bacteria ml^{-1}. In other experiments the maximum heat output from micro-organisms growing on wet straw was found to be about 26×10^{-4} cal s^{-1} (g dry straw)$^{-1}$ (109×10^{-4} J s^{-1} (g dry straw)$^{-1}$) with bacterial populations of 1.6×10^9 g^{-1} at 40°C [4]. Thermophilic bacteria caused a second maximum heat output of about 15×10^{-4} cal s^{-1} (63×10^{-4} J s^{-1} g^{-1}) at 60°C. The heat output of the natural thermophilic bacteria in wet wool varied with temperature, being 6×10^{-4} cal s^{-1} (g wool)$^{-1}$ (25×10^{-4} J s^{-1} (g wool)$^{-1}$) at 60°C, 3.4×10^{-4} cal s^{-1} g^{-1} (14.3×10^{-4} J s^{-1} g^{-1}) at 70°C and zero at 78°C [51]. Thus thermophilic bacteria, not chemical heating, could easily account for the heating of wet wool.

Results such as these demonstrate the ability of the metabolic heat of micro-organisms to raise the temperature of moist organic material. Mesophilic organisms will heat a mass of material up to about 44°C, but then, as the temperature continues to rise, they become inhibited. The thermophilic organisms begin to grow in the transition range from about 44 to 52°C and continue to heat the mass. Thermophilic growth then continues to above 70°C. The heating of damp haystacks is primarily caused by microbial metabolism (see Section 3.5.6).

During the initial mesophilic and thermophilic stages of composting the readily available carbon sources (e.g. simple sugars and proteins) are utilized by the micro-organisms. In these stages the rate of heating depends, therefore, on the composition of the organic wastes and the availability of nutrients, as well as on the water content, the particle size and the degree of aeration. These factors also determine the composition of the microbial community.

In a compost of domestic garbage plus sewage sludge many microbial species originating from soil, water and human excrement will be present. All will not grow and the relative dampness of the compost will favour the proliferation of bacteria, rather than fungi. These bacteria, particularly the mesophilic bacteria in the interior of the mass, will cause the initial temperature increase. Fungi may proliferate on the outside of the mass where it is cooler and often drier than the inside. Yet viable fungi have been found in composts at temperatures up to about 65–67°C [14]. Some of the heating-up of composts may thus be attributable to fungal growth in the mesophilic and lower thermophilic temperature ranges; the fungi do not appear to contribute to the final heating to temperatures of around 75°C. However, fungi were virtually absent during the composting of dewatered sewage sludge, perhaps because of the moisture content [42].

Thermophilic actinomycetes, in particular *Thermoactinomyces* and *Thermomonospora* spp., appear to be common in composts [9]. Their growth seems to come after that of the thermophilic bacteria, is dependent on good aeration of the compost and is possible up to about 70°C. The actinomycetes and fungi are regarded as the major degraders of cellulose.

The temperature in the compost will eventually become so high that all microbial activity is inhibited. As the compost cools a further cycle of activity will begin when a more equable temperature is reached and the compost is reinoculated from the environment. Thus, the cooling that occurs when compost is turned increases microbial growth. Eventually, when all readily available nutrients are used up, microbial activity will decline and the compost will cool. As the temperature drops, micro-organisms, either present in the interior or exterior of the mass or brought in by the air, will start to grow and utilize the remaining substrates such as lignified cellulose. These compounds are only slowly utilized by micro-organisms and the rate of heat generation cannot balance the heat loss; the compost therefore continues to cool down. Although temperatures during the cool-down stage are similar to those of the heat-up stage, the active organisms are different. This is a reflection of the different types of nutrients available during the two stages. Finally, during the maturing phase, macro-organisms such as worms and insects appear and there is considerable predation. The composted material is in practice considered mature when it fails to heat up on turning and does not go anoxic. Compost maturity can also be judged by the amount of humic substances present [50].

MICRO-ORGANISMS AND THE AEROBIC FILTER

In the aerobic (often called trickling) filter process, the surface for the development of the microbial film, which oxidizes the organic material, is provided by a matrix of coke, stones or plastic pieces. For domestic sewage the matrix is usually contained in a column about 2 m high and 7 m wide (see Fig. 4.4.3C). During operation, sewage water (the solids having already been physically removed) is sprayed evenly over the top of the filter and percolates slowly through the filter. Inlets at the sides and bottom of the filter allow upward diffusion of air. Trickling filters can be used for many wastewaters of low suspended-solid content such as those from certain factories and dairies.

Lists of microbial species occurring in waste-treatment plants must be treated with considerable caution. The isolation methods are inherently selective and there are natural variations of the microbial populations with season, climate and the wastes entering the treatment plant. Over 200 species of bacteria, algae, protozoa, worms and insects have been reported in trickling filters [7]. The most frequently isolated bacteria are *Beggiatoa alba*, *Sphaerotilus natans* and *Achromobacter*, *Alcaligenes*, *Flavobacterium*, *Pseudomonas* and *Zoogloea* spp. Under these conditions many of the bacteria produce capsules and zoogloeal growth, which could contribute to the polysaccharide nature of the microbial slime. This slime efficiently entraps the small particles present in the sewage.

The development of the microbial film on a stone support is shown in the scanning electron micrographs of Fig. 4.4.4. After 5 days the particles of sewage debris, which sorbed on to the surface during the first few hours of exposure, were embedded in a polysaccharide slime containing bacteria of various morphological forms. Amoebae and other protozoa, especially ciliates, were engulfing the bacteria. In the developed film the protozoa were confined to the lower layers where they could wander through the polysaccharide matrix feeding on bacteria. As the film developed, a layer of algae began to grow over the bacterial slime. The algal layer was variable in thickness but not continuous, and the bacteria-containing slime had finger-like projections extending up through the algae. In the film near the open top of the filter, mixed cocci and rods grew in colonies forming hollow cylinders with their open ends at the film surface. Each colony had a slit-like opening where water could drain out. The microbial film at the surface had a mean thickness of 2.1 mm, but at a depth of 30 cm the thickness was reduced to 0.01–0.02 mm and the tube-like bacterial colonies had been replaced by spherical colonies of rod-shaped bacteria. This decrease in microbial biomass was probably due to the lack of oxygen within the filter.

In other trickling filters fungi and algae were found to form the basal layer, with a secondary development of bacteria and protozoa [8].

One model of microbial activity in a column containing an inert support for microbial growth has been proposed by Pirt [47]. The column contains n theoretical compartments, each of which contains a liquid overlying a theoretical film of micro-organisms. In contact with the liquid is a volume of air. Nutrients in the liquid will diffuse into the microbial film. The liquid is assumed to be equally distributed between the theoretical compartments and to percolate down the column so that it spends an equal time in each compartment. Within each compartment the liquid is assumed to be completely mixed, but there is no mixing of liquid between different compartments.

Micro-organisms utilize the nutrients in the liquid and develop a film on the support. The thickness of the film will increase with time. The thickness of the active layer of organisms will, however, be limited by diffusion of nutrients. Eventually the biomass will consist of an active

A

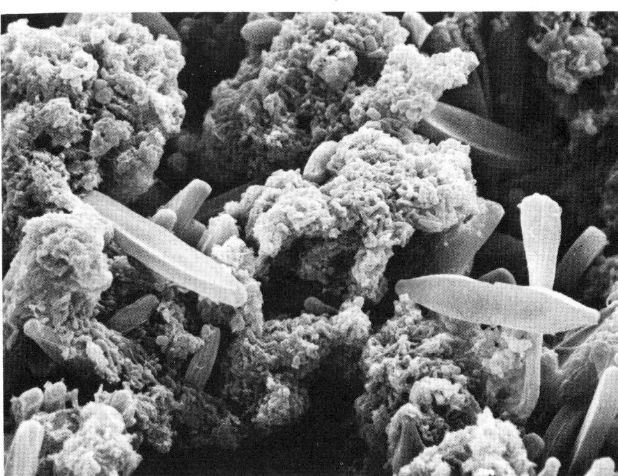

B

Fig. 4.4.4. The microbial film on a trickling filter. (A) The mass of algae, diatoms and bacterial colonies on the upper surface of the film. Magnification × 120. (B) A scanning micrograph looking down into one of the cylindrical bacterial colonies found at the surface. Magnification × 1300. Reproduced with permission from Mack *et al.* [36].

layer in contact with the liquid, with a layer below which is virtually inactive as all the nutrients are consumed by the surface layer. The active layer for the utilization of any nutrient will be only a fraction of a millimetre thick and if oxygen is limiting it will be even thinner. Equations were developed [47] to express the stepwise removal of nutrients from the liquid as it percolated through the compartments and the consequent production of biomass and oxygen uptake.

In practice the microbial film develops for a certain time and then sloughs off. The process then starts again, but detachment of the film increases the amount of organic material in the treated water (humus sludge). Sloughing off usually occurs in spring after a winter build-up of the film, but minor cycles of build-up and sloughing every few weeks can occur. Various factors can contribute to the detachment of the film, e.g. physical action of the flowing water coupled with gas formation due to anaerobic microbial metabolism in the inner, oxygen-deficient layers of the film [17]. Cell lysis in the layer of film next to the supporting matrix can also weaken attachment [20]. Some mathematical models have attempted to account for sloughing of the film [28].

MICRO-ORGANISMS AND ACTIVATED SLUDGE

In this process the sewage or factory wastewater, minus the large particulate matter, is run through tanks where it is aerated either by mechanical impellers or by compressed air. The aeration increases aerobic microbial activity and hence the mineralization of the sludge. Conventional activated-sludge plants have a maximum oxygen transfer rate from air of 0.25 mmol O_2 l^{-1} h^{-1}, though some 'high-rate' processes manage a transfer rate of 0.6 mmol O_2 l^{-1} h^{-1}. By comparison, industrial fermentation plants usually have transfer rates of 30–80 mmol l^{-1} h^{-1}, while some single-cell-protein plants transfer up to several hundred mmol O_2 l^{-1} h^{-1}. Pure oxygen is sometimes used now in activated-sludge plants, and other systems such as the 'deep-shaft' fermentor have been used to increase oxygen transfer rates.

The microbial biomass and any remaining small particulate matter must be removed in settling tanks before the effluent can be discharged into the river. Much of the settled sludge is fed back to the aeration tanks (see Fig. 4.4.3D). The process is thus a continuous culture with feedback of organisms; this allows a long residence time for the organisms but a short residence for the liquid. An important part of the process is the rapid settlement of micro-organisms caused by floc-forming bacteria. Encouraging the formation of, and then maintaining, compact flocs is a main occupation of the sewage engineer.

The matrix of the bacterial floc is mainly polysaccharides. Surface charge is believed to be a major factor causing the flocculation of polysaccharide-encapsulated organisms. Within the previously mentioned limits of analysis for this type of habitat the dominant aerobic bacteria do not appear to be markedly different from those found in the trickling filter, i.e. a community dominated by the Gram-negative rods belonging to the genera *Pseudomonas, Achromobacter, Alcaligenes* and *Flavobacterium*. In the floc, some bacterial colonies form a branching, tubular structure similar to that found on the trickling filter. Single cells isolated from inside the tubular structure can form a similar structure when grown in a pure culture. Other bacterial members of the floc just grow in a diffuse mass. The branching structures are more common in trickling filters than in aerated sludge flocs.

At times the activated sludge fails to flocculate and settle in the tanks, a process called 'bulking'. One major cause of poor flocculation is the growth of loose structures of filamentous bacteria, mainly *Sphaerotilus* spp. but other genera are often involved such as *Thiothrix, Nocardia, Beggiatoa, Bacillus, Leucothrix* and the fungus *Geotrichum*. The filamentous growth can also trap air bubbles which results in foaming. Bulking poses a major problem to the successful operation of an activated-sludge plant and is a frequent cause of inefficient sewage treatment. Many factors can contribute towards bulking, including a low dissolved oxygen concentration and the combined effects of high ratios of carbon to nitrogen and phosphorus. The presence of hydrogen sulphide in septic sewage encourages the growth of *Thiothrix* spp.; this aerobic autotrophic bacterium uses H_2S as its energy source.

Protozoa are common in the liquid and on the flocs and walls of the tanks. Around 50 000 cells ml^{-1} are present [46]. Most of these are ciliates, but amoebae and flagellates can occasionally become important. The protozoa feed on the bacteria and one of their major functions is to scavenge free-living bacteria. Protozoa are therefore important contributors to the clarification of the final effluent water.

Although the trickling-filter and the activated-sludge plants are not designed to remove pathogens, the feeding habits of protozoa are an important factor contributing to the removal of pathogenic bacteria [10]. In addition, viruses are removed by adsorption on the surface of flocs.

Anaerobic treatments

DIGESTION OF SLUDGES

The activated-sludge and trickling-filter plants are designed to treat very large volumes of dilute waste with negligible concentrations of solids, but in the process they generate large volumes of residual sludges; these, added to the sludges from the primary settling tanks of the sewage works, have to be disposed of. If left in tanks, these very wet sludges will become anaerobic and putrefy, and even after sewage sludges have been dewatered and coagulated to a filter-cake consistency there are still objections to their being spread on land as fertilizer or being dumped at sea. However, the anaerobic bacterial reactions which, uncontrolled, can lead to putrefaction and pollution can, when controlled, be used to stabilize and reduce the polluting properties of the sludges.

In sewage treatment the anaerobic digester is the direct descendant of the septic tank and cesspit once used to treat sewage. The septic tank is now used mainly to treat the waste from one or a few houses by a combination of sludge settlement and slow anaerobic digestion. The commercial anaerobic digester is now found in various forms from those in large municipal sewage works to those, developed from the farm digester, suitable for the volume of sludge from small sewage works. These can be used for factory waste sludges and different designs of digester can be used for factory-produced wastewaters of high volume and low suspended solids content.

The advantages of anaerobic digestion over aerobic processes are that residual microbial sludges are of low volume and waste residues are stabilized, and that biogas is produced that can be used as a fuel to offset process costs or produce a profit in powering the factory, just as the digester gas is used for heat or power in the sewage

Table 4.4.6. Some results of the anaerobic digestion of farm slurries *

	Input				Output			
Slurry from	TS	BOD	COD	VFA	TS	BOD	COD	VFA
Pigs	3.3	14 672	46 920	3697	2.1 (36)	2494 (83)	22 052 (53)	259 (93)
Dairy cattle	7.3	18 472	80 276	4778	5.9 (19)	3842 (79)	66 494 (17)	1219 (75)
Poultry	6.2	33 598	227 519	10 519	4.8 (23)	5340 (81)	126 752 (44)	1760 (83)

* Units mg l^{-1} except for TS, % w/v. TS, total solids; BOD, COD, as in text; VFA, volatile fatty acids. Digester retention times (days): pig, 10; cattle, 21; poultry, 20. Digester temperatures 35°C. Gas production (m^3 (kg TS)$^{-1}$ in input): pig, 0.30, 56% CH_4; cattle, 0.26, 56% CH_4; poultry, 0.38, 70% CH_4. In brackets % removal.

works. It is also an advantage that plant and animal pathogens are killed by anaerobic digestion. On the other hand, the treated waters may not be of Royal Commission standards for BOD (see Table 4.4.6).

All feedstock basically contain carbohydrates, proteins, fats, non-protein nitrogenous compounds and salts, with more or less of undegradable 'rubbish'. This latter may be very low in some factory wastewaters, but municipal sludges, farm wastes and other feedstocks can contain considerable amounts of grit, bits of wood, plastics and other debris. This material passes through the digester to form an inert component of the final residues. The degradable materials are broken down by a heterogeneous bacterial population which comprises groups, of different species and varieties, which carry out essentially the same reactions whatever the feedstock. The main reactions are shown in Fig. 4.4.5, but there are other reactions such as conversions of hydrogen plus carbon dioxide to acetate.

The principal difference between feedstocks, and so in types of digester, lies in the reactions labelled A in Fig. 4.4.5. In sewage sludges, farm animal wastes and some factory wastes (e.g. from slaughter houses or the peels and solid residues of fruit and vegetable processing) the components of A are contained in particles. In wastewaters the components are dissolved, or sometimes in very fine particulate or colloidal forms.

When solids are degraded, bacteria attach to the surfaces of the solids and the substrate is slowly broken down by hydrolytic enzymes which are either extracellular or on the bacterial surface (see Fig. 4.4.6). In these solids the principal energy source for bacteria is cellulosic fibres. While the chemical structure of all carbohydrate fibres is essentially the same, cellulose and hemicellulose, the physical structure of these polyhexoses or polypentoses and the amount and structure of attached non-degradable substrates, principally lignin, differ. Thus, while the enzymic hydrolysis of the glucose chains of a pure cellulose fibre is slow but complete, lignification of a natural fibre

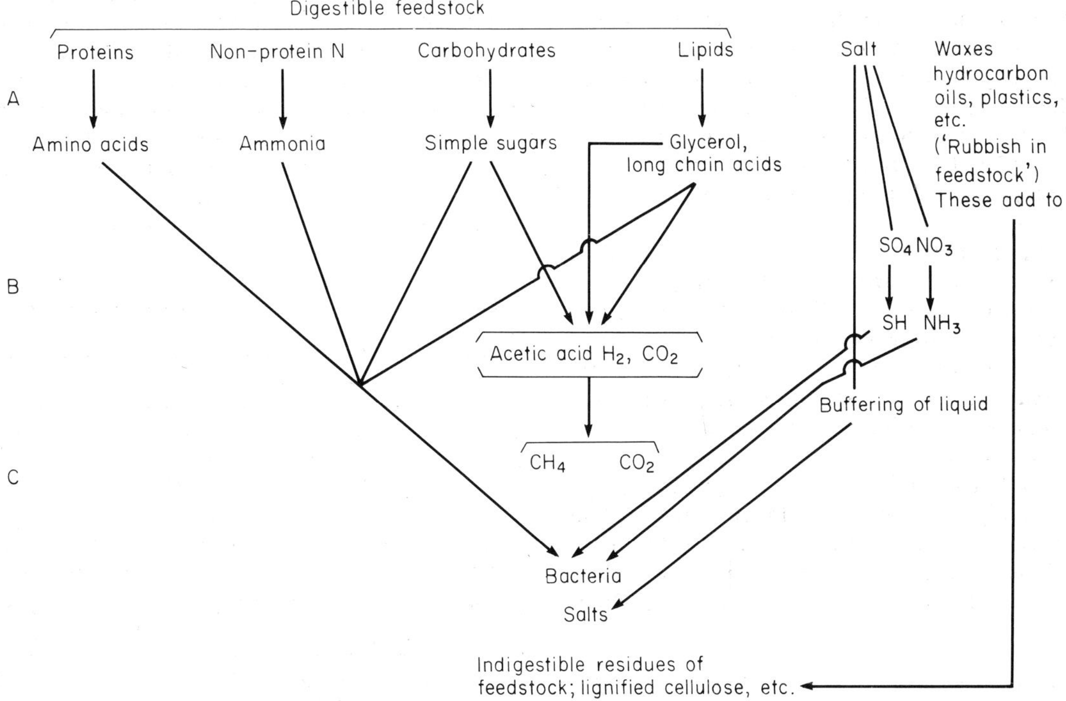

Fig. 4.4.5. Outline of reactions in anaerobic digestion. (A) The stage of hydrolysis of polymers in digestible feedstock. (B) Conversion of hydrolysed material to (principally) acetic acid, hydrogen and carbon dioxide. (C) Methanogenesis from products of B. These reactions contribute monomers and energy for formation of bacterial cells, which with indigestible residues of the feedstock form the 'digested sludge'.

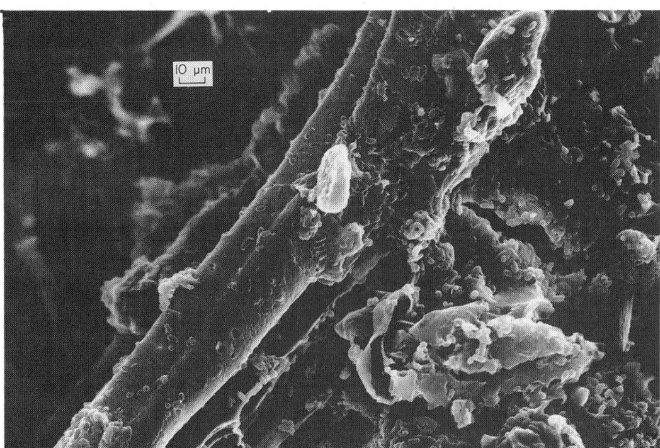

Fig. 4.4.6. Bacteria attacking the surfaces of cellulosic fibre particles in a pig-waste, mesophilic, anaerobic digester. Scale bar = 10 μm.

adds another factor governing both rate and extent of the cellulose hydrolysis, and the hydrolysis of fibres becomes the rate-limiting process in digestion of particulate feedstocks.

Digesters are usually heated to run at about 30–39°C (mesophilic populations) or are unheated and run at ambient temperatures. Thermophilic digester populations are possible, but there are energetic and other problems with thermophilic digesters and they are seldom used. The digester for particulate feedstocks is essentially a single-stage, stirred-tank, chemostat continuous culture (for more details of construction, etc. see [25]). Solids and liquids pass together through the digester (unlike the rumen, see Section 3.3.3) and the retention time of the feed is governed by the rate of breakdown of the solids. In a sewage digester, fibrous solids are largely particles of papers which have been manufactured and treated to reduce the lignin content and make them 'soft' and 'soluble' (toilet and handkerchief tissues, etc.). Vegetable matter in sewage is the remains of human food. Kitchen waste is vegetable remains which have been chopped or macerated and cooked, or even if raw are from fruits and vegetables which are young and relatively low in lignin. Faeces contain vegetable residues, essentially cellulosic fibres, which have resisted degradation in the human gut. The more soluble polysaccharides of the plant cell structure, pectins and fructosans, are digested.

Overall, these materials are relatively easily attacked by the digester bacteria and a minimum working retention time for a sewage digester is about 12 days. 'Working' is used here in the sense that the extent of degradation of the solids increases with increasing retention time. However, the retention time for a given flow of feedstock is related to digester size; for this reason, if 95% of the degradation at 'infinite' retention time can be attained in 12 days, it may not be economic to double the size of the digester to give a retention time of 24 days to obtain a further 3% of the possible degradation.

Waste from farms which is to be treated is that from animals kept in confinement for meat or milk production, and modern, intensive farming units with from a few hundreds to tens of thousands of pigs or cattle, or even millions of poultry, can each produce waste which in pollution terms is the equivalent of a small town. (One pig or cow is the equivalent of four or five humans in the pollution properties of its excreta.) The carbohydrates of animal feeds are virtually unprocessed. The constituents are also generally higher in lignified vegetable matter than those used in human foods. With pigs fed on grains (e.g. whole barley) the starch is metabolized in the digestive tract leaving a residue of cellulosic husks in the faeces. With cattle, starch and the more easily degraded plant fibres are digested by microbial action in the rumen and caecum (see Section 3.3.3) leaving in the faeces only residues very resistant to microbial action. The faecal residues vary with the feed of the animals, but a working digester retention time of 12–15 days is usually needed for pig waste and 20–25 days for whole cattle wastes. Because of the higher degree of lignification in the feed, perhaps only 25% of the organic material in cattle wastes is digestible, compared with 60% in sewage sludge.

The polysaccharide polymers are hydrolysed to simple sugars (glucose, xylose, etc.) which are then rapidly fermented, not only by hydrolytic but also by non-hydrolytic bacteria, to acids and gases (see Fig. 4.4.5.). Although consortia of bacteria in the digester populations will degrade to methane aromatic ring compounds of the type making up the monomers of lignin and low molecular weight fragments of lignin [18, 5], and there is evidence for a very slow degradation of native lignins [2], relatively rapid degradation of lignin takes place only under aerobic conditions. So in anaerobic digesters lignin is virtually undegraded and acts as a barrier to hydrolysis of the cellulose and hemicellulose carbohydrates with which it is associated.

Proteins, mainly in the faecal bacteria, are degraded to amino acids and peptides and these are deaminated to give ammonia, which is the main nitrogen source for the digester bacteria. Non-protein nitrogenous compounds are ammonia and compounds such as urea from urine, which can readily be degraded to ammonia in the digester. Overall, there is little, if any, change in total nitrogen and ammonia concentrations in farm waste and sewage digesters. Hydrolysis of proteins and production of ammonia is balanced by utilization of ammonia for bacterial synthesis [39, 60].

Lipids in digester feedstocks are hydrolysed to glycerol or galactose and the liberated long-chain fatty acids are hydrogenated; but, unlike the process in the rumen, the acids are degraded to acetic acid, hydrogen and eventually methane. However, microbial degradation of long-chain fatty acids by a process akin to β-oxidation is thermodynamically impossible unless hydrogen is removed by another bacterium [35]. Thus the process involves two bacteria. One is a fatty-acid degrading bacterium and the other a methanogenic or a sulphate-reducing bacterium which uses hydrogen. In most digesters the methanogenic bacteria predominate. As detailed earlier, the methanogenic bacteria are now classified as members of a distinct kingdom, the Archaebacteria [62].

In digesters, methanogens use two substrates, either hydrogen plus carbon dioxide (or the equivalent formate) or acetate. While utilization of hydrogen is a property of virtually all methanogens, only a few can utilize acetate. Utilization of hydrogen pulls the fermentation products of the individual carbohydrate-fermenting bacteria from higher volatile fatty acids, lactate, succinate and ethanol towards acetate and hydrogen. It also allows degradation of long-chain fatty acids and volatile fatty acids higher than acetic and removes hydrogen which can be inhibitory to cellulolytic bacteria. Bacteria converting acetate to methane are slow-growing and their minimum doubling time puts a rate limit to stirred-tank digestion of about 3–4 days retention time. The degradation of long-chain fatty acids is also slow and limits the minimum retention time of sewage-sludge and other digesters containing large amounts of feedstock fats to 6–10 days.

As a first approximation, a sewage-sludge or animal-waste digester can be considered as a number of separate chemostat continuous cultures and modelled by Monod kinetics [22, 23, 40] (see Section 2.4.2). In effect, the solids are treated as if they were in solution and values for Y_{max}, K_m and residual solids determined. Since solids from different wastes vary, the kinetics of digestion vary. Values for constants have been obtained by generalization (lumping all results from wastes of the same class together) [22]. Another model considered pig and cattle wastes to be each made up of two solid components of different degradabilities [23]. However, even this latter is a simplification in both kinetics of degradation and number of components and a later model considers growth of the bacteria on the surfaces of solid particles (see Fig. 4.4.6.) [24]. Despite all this, the 'Monod' models give reasonable values for residual solids in steady-state running of digesters at different retention times, and model washout retention times.

In contrast to the treatment of solids, hydrogen and acetate are in solution and chemostat kinetics can be more accurately applied to modelling the conversion of these to methane [23, 40, 41]. The models show that at long retention times steady-state concentrations of hydrogen and acetate are low but they rapidly increase as the washout retention times of the hydrogen- and acetate-utilizing bacteria are approached. In fact, one of the first signs of malfunction of a digester is an increase in gaseous hydrogen concentration (in equilibrium with hydrogen in solution) above the normal steady-state value; this leads to propionate and butyrate accumulation and the cessation of other digester functions (see above).

Although the digester bacterial reactions can be modelled by considering the bacteria as functional groups each behaving as a pure culture, in actual fact these groups are made up of many species. It appears that there are smaller numbers of active bacteria in digesters than in the superficially similar rumen system, presumably because of the lower substrate concentrations. On the other hand, the types of bacteria in each group in digesters are more varied than in the rumen and rumen bacteria do not seem to occur even in cattle-waste digesters. Some half of the population of sewage of animal-waste digesters is composed of bacteria which are influenced in type by the predominant bacteria of the intestines from which the waste faecal matter comes (see previous sections on intestinal bacteria) [26]. These bacteria can help to reduce the digester contents but they are inactive in the main digester reactions. The active cellulolytic, hemicellulolytic, proteolytic and deaminative bacteria are, unlike those in the rumen, preponderantly spore-forming organisms, such as *Clostridium, Sporolactobacillus*, and sporing species of *Sarcina* and other cocci and rods [55, 56, 57]. The ability to form spores is probably of survival value in a slow-growing culture with a long retention time and recalcitrant substrates. There are inhibitory and stimulatory interactions amongst these bacteria which lead to the population being in a dynamic rather than a true steady state [56]. The presence of the long-chain acid-degrading bacteria and acetate-utilizing methanogenic bacteria in digesters but not in the rumen is probably also determined by the long digester retention times which allow establishment of these slow-growing bacteria.

DIGESTERS FOR WASTEWATERS

Wastewaters are usually dilute and produced at high flow rates. While there is no need for the long retention times necessary for the digestion of fibres in sludges, a minimum retention of about four days is required for multiplication of the acetate-using methanogenic bacteria in a stirred-tank digester. With a feedstock of high flow rate this implies a very large digester.

A solution to this problem has been found in the retained-biomass digester, the first type of which was the 'contact' digester. This is a stirred tank in which the bacteria in the effluent are separated from the liquid in a separating tank and returned to the main digester tank. Thus, like the trickling filter, the residence time of the bacteria in the digester can be increased to a matter of days while that of the liquid can be only a few hours [53].

Problems associated with the separation of bacteria from the digester effluent have led to the development of the upflow-sludge-blanket digester. Here the bacteria grow in clumps which are retained in the middle of the tank by a combination of gravity settling and the lifting effect of the upwards flow of wastewater. Baffles also prevent undue loss of bacterial flocs, although a small proportion does continuously escape. The net result is again a long bacterial residence time with a short hydraulic residence time. This type of digester is being used to treat wastewaters from fruit- and vegetable-processing plants [33].

In a similar form of digester the bacteria do not grow as clumps but on the surface of small glass beads or other particles (millimetre or less). The beads are suspended by the upward flow of the feedstock (hence the name 'fluidized-bed' digester). Since the particles are denser than purely bacterial clumps, extra speed has to be imparted to the upflowing feedstock by recirculating part of the outflow to the bottom of the tank. This type of digester is designed mainly for the treatment of food-processing wastewaters, but treatment of sewage waters of low suspended solids is possible using this or the sludge-blanket type of digester. More details can be found elsewhere [30].

An early form of the retained-biomass digester was the upflow anaerobic filter [64]. This, as its name implies, is a tank containing a filter medium such as stones or pieces of plastic with inflow of wastewater at the bottom and outflow at the top. The bacteria grow attached to the filter medium or trapped in the spaces between the stones. As with the previous two digesters, trapping of the bacteria in the system is not complete and thus the cells have a very long, though finite, retention time in the digester compared with the hydraulic retention time. Escape of cells means that some settling system has usually to be put into the effluent stream to give a reasonably clear water. A more recent version of the anaerobic filter uses vertical tubes of pottery or plastic to pack the tank and the wastewater flows downwards through and around the tubes [32].

The retained-biomass digesters have been used at temperatures suitable for mesophilic organisms and can produce useful results even at temperate ambient temperatures (factory wastewaters are often hot in any case). It should be noted that the concentration of degradable substrate in the effluent does not follow chemostat theory as in the stirred-tank digester and as a result the BOD of the effluent liquids can be much lower than that of the effluent from a stirred-tank digester. It will, however, be obvious that this type of digester cannot be used when the feedstock contains a large amount of suspended solids, but it is suitable as the second stage of a two-phase digester or of a two-stage digester as described below.

TWO-PHASE DIGESTERS

The only way to separate the reactions of digestion is to have two linked fermentors in which the conditions in one prevent growth of particular bacteria which can multiply under the conditions imposed in another.

The only digester conditions which can be altered are retention time and, to some extent, pH. However, retention times needed for hydrolysis of cellulosic fibres in farm wastes or sewage sludges (10–15 days) are sufficient to allow methanogenic bacteria in the feedstocks to grow, and the optimum pH for fibre hydrolysis (6.5 or rather above) is within the pH range of growth of the methanogens. It thus follows that methanogenesis cannot be excluded from a digester in which fibres are being hydrolysed.

Farm-animal slurries will hydrolyse and ferment to some extent if stored at ambient temperature and the VFA concentration will build up to perhaps 9000 mg l^{-1}. The reactions are eventually stopped by accumulation of acids and fall in pH. This reaction has been used in a system where the acid-containing supernatant from stored and settled piggery slurry is run off through an anaerobic filter where the acids are converted to methane and carbon dioxide [6]. However, the overall retention time in this two-phase system is greater than that in a mesophilic stirred-tank digester as the hydrolytic reactions take place at a lower temperature. In addition, the overall yield of acids from the hydrolysis is usually less (because the reactions do not go to completion) than that from the stirred tank. Gas yield is also theoretically lower as some methane and hydrogen is formed in the hydrolysis tank and escapes from the system.

It is also possible to run a two-phase digester with particular sewage sludges that contain little recalcitrant fibre material and where hydrolysis and fermentation can take place in a stirred tank of 12–24 hours' retention time. The acid liquids are then run into a stirred tank with about 6 days' retention time where methanogenic bacteria can convert the acids to methane [16].

TWO-STAGE DIGESTERS

A different system, but one with which there is some confusion, is the 'two-stage' digester. Here, there are two linked digesters but each carries out the same reactions. These are really two stirred-tank chemostats connected in series and the advantage is that the residual substrate leaving the first tank can be greatly reduced by passage through the second tank. In addition, the second tank is continuously inoculated by bacteria from the first tank, so it can be of smaller volume and shorter retention time than the first tank. The Monod kinetics model of the chemostat can be extended to two (or more) chemostats in series, as was done by Herbert and Powell [19, 49]. However, the mathematics becomes complicated for more than two tanks, although consideration of a large number of tanks in series leads to a mathematical description of a tubular fermentor.

This description of the two-stage digester actually fits the usual sewage or farm-waste digester. The effluent from sewage digesters passes to open tanks where the residual solids settle and so thicken the sludge over a period of some weeks. But at the same time the digester bacteria continue to work at ambient temperature and slowly gasify some of the remaining solids and acids. The same thing happens to farm digester effluents, which are usually stored until weather and crops allow spreading of the digested sludge as fertilizer.

Experiments have also been carried out with two-stage, heated and stirred farm-waste digesters. As expected, the residual acids from the first digester are lowered at long first-stage retention times [27]. However, when shorter first-stage retention times are used, some residual solids from the first stage remain to be digested and so add to the acids from the first stage, and fibres flowing from the first to second stage are already colonized by bacteria and so may not be mathematically akin to a dissolved residual substrate from the first stage.

The second stage could be a tubular digester, or it could be a retained-biomass digester used to polish first-stage liquids after removal of solids. Two retained-biomass digesters in series might give a better effluent, at lower retention time, than one large one.

All these configurations offer methods of cleaning anaerobically digested wastewaters without recourse to often energy-intensive and expensive aerobic treatments, and they can improve biogas yields. However, one must remember that in sludge digestion a multi-stage digester cannot produce fibre degradation greater than that obtained in a single-stage digester running at long retention times, i.e. mathematically approaching infinity. Whether a system is worth putting into practice depends on many mechanical and economic considerations.

4.4.7 CONCLUSION

In this chapter only a brief summary of the theoretical and practical aspects of cleaning up man's wastes has been given. There are prospects for improving the purification processes by mechanical methods and by the newer methods of genetic engineering to produce bacteria able to degrade new substances or with improved degradative activity for the present substrates. But all these processes are carried out by mixed cultures liable to contamination and under very uncontrolled conditions compared with laboratory cultures. Until the microbial ecologist knows much more about the constituent bacteria and their place in reaction schemes, about the methods of internal control of these systems, and about their natural variation with time, there will be great difficulty in applying the sophisticated methods of the industrial, pure-culture microbiologist to improving water-pollution control.

4.4.8 REFERENCES

[1] American Public Health Association (1980) *Standard Methods for the Examination of Water and Waste Water*, 15th edn. APHA, Washington.

[2] Benner R., Maccubin A.E. & Hodson R.E. (1984) Anaerobic biodegradation of the lignin and polysaccharide components of lignocellulose and synthetic lignin

by sediment microflora. *Appl. Environ. Microbiol.* **47**, 998–1004.

[3] Brown D.E., Groves G.R. & Miller J.D.A. (1973) pH and E_h control of cultures of sulphate-reducing bacteria. *J. Appl. Chem. Biotechnol.* **23**, 141–9.

[4] Carlyle R.E. & Norman A.G. (1941) Microbial thermogenesis in the decomposition of plant materials. III. Factors involved. *J. Bacteriol.* **41**, 699–724.

[5] Colberg P.J. & Young L.Y. (1985) Anaerobic degradation of soluble fractions of [^{14}C-lignin] lignocellulose. *Appl. Environ. Microbiol.* **49**, 345–9.

[6] Colleran E., Barry M., Wilkie A. & Newell P.J. (1982) Anaerobic digestion of agricultural wastes using the upflow anaerobic filter design. *Process Biochem.* **17** (2), 12–17.

[7] Cooke W.B. (1959) Trickling filter ecology. *Ecology* **40**, 273–91.

[8] Cooke W.B. & Hirsch A. (1958) Continuous sampling of trickling filter populations. II. Populations. *Sewage and Ind. Wastes,* **30**, 138–56.

[9] Cross T. & Goodfellow M. (1973) Taxonomy and classification of the Actinomycetes. In *Actinomycetales: Characteristics and Practical Importance* (ed. G. Sykes & F.A. Skinner). Academic Press, London.

[10] Curds C.R. & Fey G.J. (1969) The effect of ciliated protozoa in the activated sludge process. *Water Res.* **3**, 853–67.

[11] Curtis E.J.C. (1972) Sewage fungus in rivers in the United Kingdom. *Wat. Pollut. Control* **6**, 673–83.

[12] Department of Health and Social Security (1983) *The Bacteriological Examination of Water Supplies 1982*. Reports on Public Health and Medical Subjects No.71. HMSO, London.

[13] Duncan D.W. & Razzel W.E. (1972) *Klebsiella* biotypes among coliforms isolated from forest environments and farm products. *Appl. Microbiol.* **24**, 933–8.

[14] Firstein M.S. & Morris M.L. (1975) Microbiology of municipal solid waste composting. *Adv. Appl. Microbiol.* **19**, 113–51.

[15] Geldreich E.E. (1976) Fecal coliform and fecal streptococcus density ratios in waste discharge and receiving waters. *CRC Crit. Rev. Environ. Cont.* **6**, 349–69.

[16] Ghosh S. & Klass D.L. (1978) Two phase anaerobic digestion. *Process Biochem.* **13** (4), 15–24.

[17] Hawkes H. A. (1957) Film accumulation and grazing activity in the sewage filters at Birmingham. *J. Proc. Inst. Sewage Purif.* 88–110.

[18] Healy J.B. & Young L.Y. (1979) Anaerobic biodegradation of eleven aromatic compounds to methane. *Appl. Environ. Microbiol.* **38**, 84–9.

[19] Herbert D. (1964) Multi-stage continuous culture. In *Continuous Culture of Microorganisms*, p.23 (ed. I. Malek, K. Beran, Z. Fencl, V. Munk, J. Řičica & H. Smrčková). Academia, Prague.

[20] Heukelekian H. & Crosby E.J. (1956) Slime formation in polluted water. *Sewage Ind. Wastes* **28**, 78–92.

[21] Higgins I.J. & Burns R.G. (1975) *The Chemistry and Microbiology of Pollution*. Academic Press, London.

[22] Hill D.T. (1983) Simplified Monod kinetics of methane fermentation of animal wastes. *Agric. Wastes* **5**, 1–16.

[23] Hobson P.N. (1983) The kinetics of anaerobic digestion of farm wastes. *J. Chem. Tech. Biotechnol.* **33B**, 1–20.

[24] Hobson P.N. (1985) A model of anaerobic bacterial degradations of solid substrates in a batch digester. *Agric. Wastes* **14**, 255–74.

[25] Hobson P.N., Bousfield S. & Summers R. (1981) *Methane Production from Agricultural and Domestic Wastes*. Applied Science Publishers, London.

[26] Hobson P.N. & Shaw B.G. (1974) The bacterial population of piggery waste anaerobic digesters. *Water Res.* **8**, 507–16.

[27] Hobson P.N., Summers R. & Harries C. (1984) Single- and multi-stage fermenters for treatment of agricultural wastes. In *Microbial Methods for Environmental Biotechnology*, p.119 (ed. J.M. Grainger & J.M. Lynch). Academic Press, London.

[28] Howell J.A. & Atkinson B. (1976) Sloughing of microbial film in trickling filters. *Water Res.* **10**, 307–15.

[29] Jansen W.A. & Myers C.D. (1968) Fish: serological evidence of infection with human pathogens. *Science* **159**, 547–8.

[30] Jewell W.J., Switzenbaum M.S. & Morris J.W. (1982) Municipal waste water treatment with the anaerobic attached microbial film expanded bed process. *J. Water Poll.* **53**, 482–90.

[31] Jørgensen B.B. (1977) Bacterial sulphate reduction within microniches of oxidised marine sediments. *Marine Biol.* **41**, 7–17.

[32] Kennedy K.J. & van den Berg L. (1982) Anaerobic digestion of piggery waste using a stationary fixed film reactor. *Agric. Wastes* **4**, 151–8.

[33] Lettinga G., van Velson S.W., Hobma W., de Zeeuw W. & Klapwyk A. (1980) Use of the upflow sludge blanket (USB) reactor concept for biological waste water treatment, especially for anaerobic treatment, *Biotech. Bioeng.* **22**, 699–734.

[34] Loehr R.C. (1984) *Pollution Control for Agriculture*, 2nd edn. Academic Press, London.

[35] McInerney M.J., Bryant M.P., Hespell R.B. & Costerton J.W. (1981) *Syntrophomonas wolfei* gen. nov. sp. nov., an anaerobic, syntrophic, fatty acid-oxidizing bacterium. *Appl. Environ. Microbiol.* **41**, 1029–39.

[36] Mack W.N., Mack J.P. & Anderson A.O. (1975) Microbial film development in a trickling filter. *Microbial*

Ecol. **2**, 215–26.

[37] McLaughlin T. (1971) *Coprophilia or a Peck of Dirt.* Cassell, London.

[38] Mara D. (1976) *Sewage Treatment in Hot Climates.* Wiley, London.

[39] Melchior J.L., Binot R., Perez I.A., Naveau H. & Nyns E-J. J. (1982) Biomethanation: its future development and the influence of the physiology of methanogenesis. *J. Chem. Tech. Biotechnol.* **32**, 189–97.

[40] Mosey F.E. (1983) Mathematical modelling of the anaerobic digestion process. Regulatory mechanisms for the formation of short-chain fatty acids from glucose. *Water Science Technol.* **15**, 209–32.

[41] Mosey F.E. & Fernandes X.A. (1984) Mathematical modelling of methanogenesis in sewage sludge digestion. In *Microbiological Methods for Environmental Biotechnology*, p.159 (ed. J.M. Grainger & J.M. Lynch). Academic Press, London.

[42] Nakasaki K., Sasaki M., Shoda M. & Kubota H. (1985) Change in microbial numbers during thermophilic composting of sewage sludge with reference to CO_2 evolution rate. *Appl. Environ. Microbiol.* **49**, 37–41.

[43] Nemerow N.L. (1974) *Scientific Stream Pollution Analysis.* McGraw-Hill, New York.

[44] Parker R.R. & Sibert J. (1976) Responses of phytoplankton to renewed solar radiation in a stratified inlet. *Water Res.* **10**, 123–8.

[45] Pearson T.H. & Rosenberg R. (1976) A comparative study of the effects on the marine environments of wastes from cellulose industries in Scotland and Sweden. *Ambio.* **5**, 77–9.

[46] Pike E.B. & Curds C.R. (1971) The microbial ecology of the activated sludge process. In *Microbial Aspects of Pollution*, p.123 (ed. C.R. Curds & H.A. Hawkes). Academic Press, London.

[47] Pirt S.J. (1973) A quantitative theory of the action of microbes attached to a packed column: relevant to trickling filter effluent purification and to microbial action in soil. *J. Appl. Chem. Biotechnol.* **23**, 389–400.

[48] Poole N.J., Wildish D.J. & Kristmanson D.D. (1978) The effects of the pulp and paper industry on the aquatic environment. *CRC Crit. Rev. Environ. Cont.* **8**(2), 153–95.

[49] Powell E.O. & Lowe J.R. (1964) Theory of multi-stage continuous cultures. In *Continuous Culture of Microorganisms*, p.45 (ed. I. Malek, K. Beran, Z. Fencl, V. Munk, J. Řičica & H. Smrčková). Academia, Prague.

[50] Roletto E., Chiono R. & Barberis E. (1985) Investigation on humic matter from decomposing popular bark. *Agric. Wastes* **12**, 261–72.

[51] Rothbaum, H.P. (1961) Heat output of thermophiles occurring in wool. *J. Bacteriol.* **81**, 165–71.

[52] Rothbaum H.P. & Stone H.M. (1961) Heat output of *Escherichia coli. J. Bacteriol.* **81**, 172–7.

[53] Schroepfer G.J. & Ziemke N.R. (1959) Development of the anaerobic contact process. I. Pilot plant investigations and economics. *Sewage Ind. Wastes* **31**, 164–90.

[54] Senior E., Lindström E.B., Banat I.M. & Nedwell D.B. (1982) Sulphate reduction and methanogenesis in the sediment of a saltmarsh on the east coast of the United Kingdom. *Appl. Environ. Microbiol.* **43**, 987–96.

[55] Sharma V.K. & Hobson P.N. (1985) Isolation and cellulolytic activities of bacteria from a cattle waste anaerobic digester and the properties of some *Clostridium* species. *Agric. Wastes* **14**, 173–96.

[56] Sharma, V.K. & Hobson, P.N. (1986). Interactions among cellulolytic bacteria from an anaerobic digester. *Microb. Ecol.* **12**, 232–5.

[57] Siebert M.L. & Torien D.F. (1969) The proteolytic bacteria present in the anaerobic digestion of raw sewage sludge. *Water Res.* **3**, 241–50.

[58] Skogman H. (1976) Production of Symba yeast from potato wastes. In *Food from Wastes*, p.167 (ed. G.G. Birch, K.J. Parker & J.T. Worgan). Applied Science Publishers, London.

[59] Snow J. (1855) *On the Mode of Communication of Cholera.* Churchill, London.

[60] Summers R. & Bousfield S. (1980) A detailed study of piggery waste anaerobic digestion. *Agric. Wastes* **2**, 61–78.

[61] Winfrey M.R. & Ward D.M. (1983) Substrates for sulfate reduction and methane production in intertidal sediments. *Appl. Environ. Microbiol.* **45**, 193–9.

[62] Woese C.R., Magrum L.J. & Fox G.E. (1978) Archaebacteria. *J. Mol. Evol.* **11**, 245–52.

[63] World Health Organization (1976) *Surveillance of Drinking Water Quality.* WHO, Geneva.

[64] Young J.C. & McCarty P.L. (1967) The anaerobic filter for waste treatment. *Proc. 22nd Purdue Indust. Waste Conf.* 559–74.

4.5 The problem of xenobiotics and recalcitrance

4.5.1 INTRODUCTION

The terms xenobiotic and pollutant are often used loosely to describe compounds which are produced by large-scale industrial processes and which, at some stage, are deposited in the environment. By strict definition xenobiotics are compounds considered to be unnatural. A wider definition in current usage, and one which will be used here, extends to man-made compounds which, although they occur naturally, are detected in parts of the environment at unnaturally high concentrations [49, 65]. Metals and certain products of the oil industry, such as phenols and polycyclic aromatics, are examples of naturally occurring compounds which should, by this definition, be considered as xenobiotics. It follows that pollutants, compounds which contaminate the environment and can disrupt the normal functioning of the biosphere, constitute a particularly important group of xenobiotics.

'Pollution potential' is a popular phrase which is associated with measurement of the relative dangers of compounds contaminating the environment. Algorithms, based on a variety of parameters (aquatic toxicity, mammalian toxicity, carcinogenicity, levels of production, reactivity, bioaccumulation and persistence or recalcitrance), are used to measure the 'pollution potential' of xenobiotics and to distinguish priority pollutants — compounds which present an immediate threat to the biosphere [54]. Table 4.5.1 summarizes the major categories of xenobiotics considered by the US Environmental Protection Agency to be priority pollutants. Undoubtedly, listings of this sort are useful and can be important in defining existing pollution problems. However, bioaccumulation and persistence or recalcitrance, two important factors used to identify priority pollutants, are not easily subjected to analysis and cannot be predicted solely on the basis of the physicochemical properties of the potential pollutant.

In this chapter we shall examine the relationship between biodegradation and recalcitrance. The former term is generally used to describe biological activities which result in the breakdown of organic compounds: biotransformations which are usually catalysed by catabolic enzymes. The central role of micro-organisms in the continuous operation of the carbon cycle is generally accepted [1, 22], and it is the metabolic versatility of the bacteria and some fungi which determines the potential for de-

Table 4.5.1. Categories of chemical pollutants (the most abundant chemicals from each group are shown in parentheses; see [54])

Category	Number of classified chemicals
Chloroaliphatics (chloroform)	31
Pesticides (polychlorinated biphenyls)	26
Polycyclic aromatic hydrocarbons (anthracene, phenanthrene)	17
Chloroaromatics (2,4,6-trichlorophenol)	15
Simple aromatics (*bis* (2-ethylhexyl) phthalate)	13
Nitrogen-containing compounds (cyanides, 2-nitrophenol)	13
Metals (copper)	13

gradation of xenobiotic compounds in the environment. Consequently, we shall emphasize aspects of biochemistry and genetics which relate to the metabolism of xenobiotics by micro-organisms and the fate of xenobiotics in the environment. Biodegradation, like other microbial activities in the biosphere, involves interactive mixed populations, and different types of metabolic interaction which account for the efficient catabolism of xenobiotic compounds by microbial communities will be illustrated. The extent to which micro-organisms can adapt to degrade xenobiotics has direct implications for recalcitrance. Therefore, we shall consider some of the mechanisms of microbial adaptation and the evolution of metabolic pathways. Finally, recent developments in applied research in biodegradation will be outlined and some of the strategies currently being developed for the biotreatment of industrial effluents and dispersed wastes containing recalcitrant xenobiotics will be examined.

4.5.2 EXPERIMENTAL APPROACHES TO THE STUDY OF BIODEGRADATION

Two main lines of investigation are open to microbiologists studying biodegradation: the environmental or ecological approach and the molecular approach. Environmental studies are frequently concerned with biodegradation by an undefined microflora in a defined environment [1, 2]. Typically, biodegradation is assessed by measurement of biological oxygen demand (BOD), chemical oxygen demand (COD) and, sometimes, the disappearance of the test substrate and detection of putative metabolic intermediates [103]. More sophisticated studies may require a 'microcosm' of the environment which can be constructed in the laboratory and used to determine the effects of abiotic variables on the rate of biodegradation. Continuous-flow culture techniques using apparatus such as the chemostat have broadened the scope of the microbial ecologist in defining both the environment and the microflora simultaneously [97]. Growth conditions can be manipulated and steady-state cultures can (theoretically) be maintained and studied for weeks on end. Micro-organisms for molecular studies of biodegradation are usually isolated from batch (closed) or continuous (open) enrichment cultures, and need to be purified in order that biodegradative pathways, enzymes and genes can be studied in isolation.

The extent to which experiments with pure cultures of micro-organisms growing under carefully controlled, often optimized, conditions can contribute to an understanding of 'real world' pollution problems has been disputed [1, 24]. Isolation and purification are essential prerequisites for biochemical and genetic studies, and without these steps little useful detailed information can be obtained because of interference from other organisms. Dagley noted that biochemists attempting to present a coherent picture of human metabolism faced criticism from sceptical clinicians who believed that the complexities of human physiology could not be defined in molecular terms [24]. In the light of the experience of those molecular biologists, it seems reasonable to suppose that the molecular approach will make a significant contribution to the study of recalcitrance and biodegradation in the environment.

The limitations of this method of investigation have been clearly stated elsewhere [1] and will not be discussed in detail here. Suffice it to say that experimental evidence suggests that different organisms are isolated in enrichments under different conditions; *Bacillus* species predominate in high temperature enrichments on xenobiotics, and a variety of micro-aerophilic and anaerobic species can be isolated under the appropriate culture conditions excluding oxygen.

Two points should be considered in assessing the relevance of conventional enrichments and characterization of isolated microbial cultures to biodegradation in the biosphere. First, it seems likely that pseudomonads and related bacteria, the most commonly studied species, constitute significant populations in the environments and they are not artefacts selected in laboratory enrichments. Second, major qualitative differences in biodegradation have been seen only in comparisons of oxygenated and anoxic environments. The concentration of research effort on biodegradation by aerobic mesophilic bacteria has been important in developing a base from which studies of biodegradation under non-optimized conditions can be undertaken.

Most of the better characterized metabolic pathways in micro-organisms which participate in the transformation of xenobiotic compounds cause the complete oxidation of the carbon skeleton to carbon dioxide and the mineralization of other substituents. In these circumstances the micro-organisms can usually utilize the substrate as a source of carbon and energy. However, substrate mineralization is not a prerequisite for biodegradation, and micro-organisms may carry out limited metabolism of xenobiotics whilst their growth is supported by an alternative source of carbon and energy (co-metabolism). Thus, the biodegradation of a xenobiotic requires only interation with enzymes which catalyse its chemical transformation, however limited. An important feature of the transformations of xenobiotic organic compounds by micro-organisms is that they are almost wholly associated with catabolic and not anabolic pathways. In addition, transformations are usually catalysed by intracellular enzymes, so that transmembrane permeation of the substrate and induction or derepression of genes encoding enzymes associated with substrate transformation are normally required for biodegradation.

4.5.3 THE BIOCHEMISTRY OF XENOBIOTIC BIODEGRADATION

C_1 compounds

Most of the catabolic pathways which will be described in this chapter result in the mineralization of organic compounds comprising two or more carbon atoms. However, there are a number of C_1 xenobiotics which deserve consideration, notably: carbon monoxide, the cyanides and the halogenated methanes. Despite the metabolic constraints that are associated with utilization of C_1 compounds as carbon sources, many different species which can carry out this function have been characterized. The methanotrophs or methane-utilizers oxidize methane to CO_2 via formaldehyde, an intermediate which can be channelled into the biosynthetic serine pathway and used to produce cell materials [48]. The methane mono-oxygenase system oxidizes methane to methanol and can transform a wide range of substrate analogues including halomethanes. The enzyme, which comprises three components, catalyses the oxidation of methane by the incorporation of one atom of molecular oxygen to form methanol. Water is formed by combination of the second oxygen atom with reducing equivalents from NADH. The broad specificity of methane mono-oxygenases has prompted speculation that methanotrophs may be involved in biotransformations of a range of xenobiotics, including hydrophobic and chlorinated hydrocarbons [47]. The physiological role of the methane mono-oxygenase system and the ecological role of methanotrophs in the transformation of xenobiotics remains uncertain. In this context, it is interesting to note that dichloromethane was reported to be utilized as carbon and energy source by several microbial genera but that catabolism did not involve oxygenases. Instead, dechlorination, resulting in the formation of formaldehyde, was catalysed by glutathione-dependent hydrolases [60], active in the absence of molecular oxygen.

A group of bacteria called the carboxydotrophs utilize carbon monoxide as sole carbon and energy source under anaerobic conditions [69]. The key enzymes in this conversion are carbon monoxide oxidases which catalyse the transfer of oxygen to CO from H_2O, forming CO_2. Reducing equivalents from H_2O are transferred to an oxidizing substrate such as NAD^+; the transformation does not require molecular oxygen. The carboxydotrophs can assimilate the CO_2 formed by the CO oxidase, and can utilize energy from the oxidation. Several genera of anaerobic lithotrophic bacteria are known which can utilize CO as an energy source, producing CO_2 with either CH_4 or H_2. An alternative transformation carried out by the acetogenic autolithotrophs is the oxidation of CO to acetate.

Cyanide is produced biologically as well as by industry, and consumption of this compound in the biosphere results from several distinct biological transformations mostly carried out by fungi and bacteria. A well-studied biotransformation is that catalysed by fungal cyanide hydratases which convert HCN to formamide. Bacteria such as *Pseudomonas fluorescens* transform cyanide to NH_3, a utilizable nitrogen source, and CO_2 in a reaction catalysed by a novel enzyme, cyanide mono-oxygenase. The mineralization of nitriles (by nitrilases) and limited breakdown of metal–cyanide complexes has also been described and seems to involve similar enzymes [58].

Aliphatic hydrocarbons

A major source of saturated and unsaturated hydrocarbons is crude oil, and chronic oil spillages have introduced these compounds as xenobiotics into a variety of habitats. In general, *n*-alkanes are considered to be the most easily degraded components of petroleum mixtures [4, 11] and catabolism of alkane chain lengths up to n-C_{44} has been reported. Some useful generalizations for assessing biodegradability of aliphatic hydrocarbons have been proposed [11] and are outlined below.

1 Long-chain *n*-alkanes are transformed more rapidly than are short-chain homologues. This may reflect the higher solubility and toxicity to micro-organisms of short-chain alkanes.

2 Saturated aliphatic hydrocarbons are transformed more readily than unsaturated analogues.

3 Branching of the aliphatic chains is associated with lower rates of biotransformation.

Aerobic microbial catabolism of alkanes, whether by fungi or bacteria, is usually initiated by mono-oxygenase-catalysed terminal oxidation of the substrate, and the formation of a carboxylic acid from the oxidized methyl group. The fatty acid product can be channelled into β-oxidation. Subterminal oxidation of *n*-alkanes and hydrolysis of the ester product also yield substrates for β-oxidation. 1-Alkene oxidation can be carried out in a variety of ways and different pathways have been identified in pure cultures; oxides, diols and primary, secondary and unsaturated alcohols are products of the initial stages of 1-alkene biotransformation. Once again, various mono-oxygenases have been implicated as catalysts for some of these reactions and products are channelled into β-oxidation pathways.

Recalcitrance is commonly observed with highly methylated alkanes. For example, the quaternary carbon-containing hydrocarbon, pivalic acid (2,2-dimethylpropionic acid), which is produced during the biodegradation of methylated alkanes (e.g. 2,2-dimethylheptane) and aromatics (e.g. *tert*-butylbenzene), does not appear to undergo biotransformation readily [11, 17].

Alicyclic hydrocarbons

Numerous natural compounds contain the alicyclic ring including products as diverse as waxes from plants, crude oil and microbial lipids. Xenobiotics such as petroleum products and pesticides are also members of this group. Recent reviews by Trudgill provide a comprehensive summary of the biodegradation of the alicyclic ring by aerobic bacteria [107, 108]. Important general observations are that the unsubstituted cyclohexane ring can be

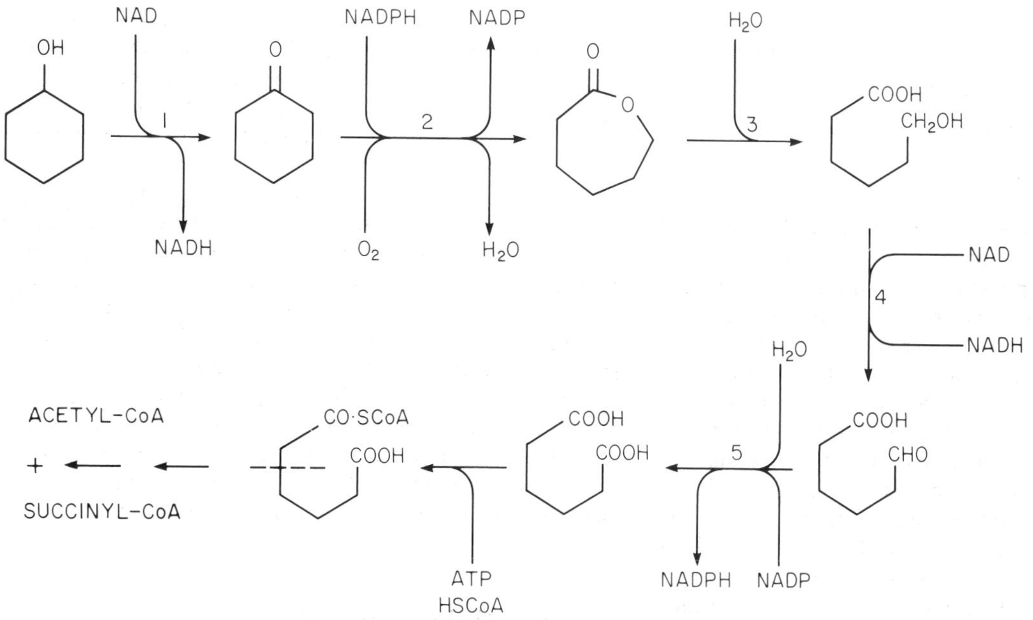

Fig. 4.5.1. The pathway of cyclohexanol degradation in *Acinetobacter* sp. Reaction (2) is a Baeyer–Villiger type mono-oxygenation [107].

activated by mono-oxygenase-catalysed hydroxylation, and that ring cleavage can be potentiated by mono-oxygenases which produce lactones from alicyclic ketones, in a reaction analogous to chemical Baeyer–Villiger oxygenation. Hydrolysis of the lactone results in ring cleavage, a reaction that may be spontaneous or enzyme-catalysed depending on the effect of other ring substituents present (see Fig. 4.5.1). Ring fission of cycloalkyl carboxylic acids is catalysed by hydrolases rather than mono-oxygenases, and frequently these products are degraded by β-oxidation.

Aromatic hydrocarbons

There are several excellent reviews on the utilization of benzenoid compounds by bacteria and fungi [17, 22, 25, 36, 44, 88], and the reader is referred to these for detailed discussion of the biochemistry of aromatic hydrocarbon catabolism. The benzene nucleus is one of the commonest chemical structures found in natural organic products. Despite its stability which is derived from the large resonance energy of delocalized electrons around the benzene unit, it is a relatively simple matter to isolate micro-organisms which cleave the aromatic ring and mineralize its carbon skeleton; indeed, bacterial growth on benzene was first reported over 70 years ago. Aerobic catabolism of the benzene ring requires that ring-cleavage be potentiated by energy-consuming hydroxylations; oxygenases catalyse these crucial reactions. The di- and sometimes tri-hydroxylated benzenoid intermediates thus formed are substrates for aromatic ring cleavage, a reaction which is usually catalysed by dioxygenases without consumption of energy (see Fig. 4.5.3). A useful, though quite arbitrary distinction may be made between 'upper' pathways, which convert aromatic hydrocarbons to ring-cleavage substrates, and 'lower' pathways, responsible for aromatic ring cleavage and channelling of products into central metabolism. Thus, the upper pathway for benzene conversion to catechol comprises only one other metabolite, *cis*-1,2-dihydroxycyclohexa-3,5-diene, which is formed by dioxygenase-catalysed insertion of molecular oxygen into the benzene ring [44]. Hydroxylated ring-cleavage substrates (Fig. 4.5.2) represent points of convergence for different 'upper' pathways. For example, catechol is an intermediate in the aerobic catabolism of naphthalene, salicylate, toluene, benzene, phenol, nitrophenol and aniline.

Three modes of aromatic ring cleavage have been described, each associated with distinct catabolic pathways. The pathways shown in Fig. 4.5.3 serve to indicate

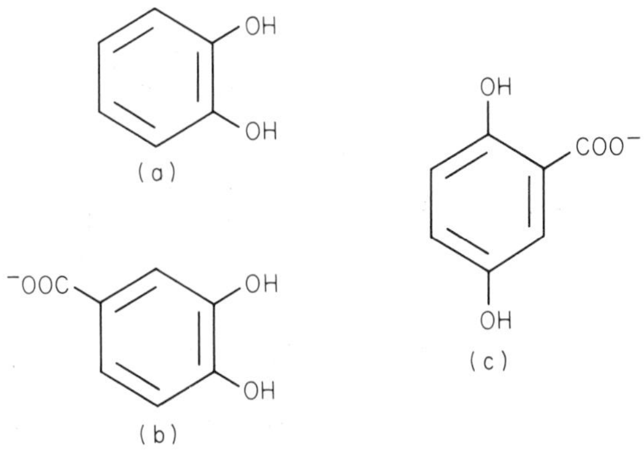

Fig. 4.5.2. Common aromatic ring-cleavage substrates (a) Catechol. (b) Protocatechuate. (c) Gentisate.

the main variations and the reader is referred to reviews by Dagley [25] and Fewson [36] for detailed information. The ring-cleavage dioxygenases catalyse the insertion of molecular oxygen (activated as an electrophile) into aromatic substrates which carry two or three hydroxyl groups positioned either *ortho* or *para* to each other. *Ortho* (intradiolic) ring-cleavage dioxygenases (pyrocatechases) cleave the bond between adjacent carbon atoms both carrying hydroxyl substituents (Fig. 4.5.3a–c) [25, 99] whereas *meta* (extradiolic) ring-cleavage dioxygenases (metapyrocatechases) cleave the bond between a carbon atom carrying one of the hydroxyls and the adjacent non-hydroxylated carbon atom (Fig. 4.5.3d–g) [6, 25]. A different type of ring-cleavage dioxygenase (e.g. gentisate 1,2-dioxygenase) cleaves the bond between the hydroxylated carbon atom and the carbon carrying the carboxylate group of gentisate or its homologue homogentisate (Fig. 4.5.3h and j) [6, 25].

Although ring-cleavage substrates possessing similar chemical structures are frequently mineralized by the same type of catabolic pathway, isofunctional reactions in bacterial pathways are usually catalysed by different enzymes. Thus, in *Pseudomonas* spp. different sets of genes encoding distinct catabolic pathways may be induced for the dissimilation of homologous substrates (e.g. gentisate and homogentisate; gallate and 3-0-methylgallate) [25]. Given the variety of options for the metabolism of aromatic substrates, bacteria seem to be reasonably conservative in their 'choices'. To some extent generalizations allow predictions on the likely route of catabolism when considering specific organisms and pure substrates; however, it should be emphasized that no

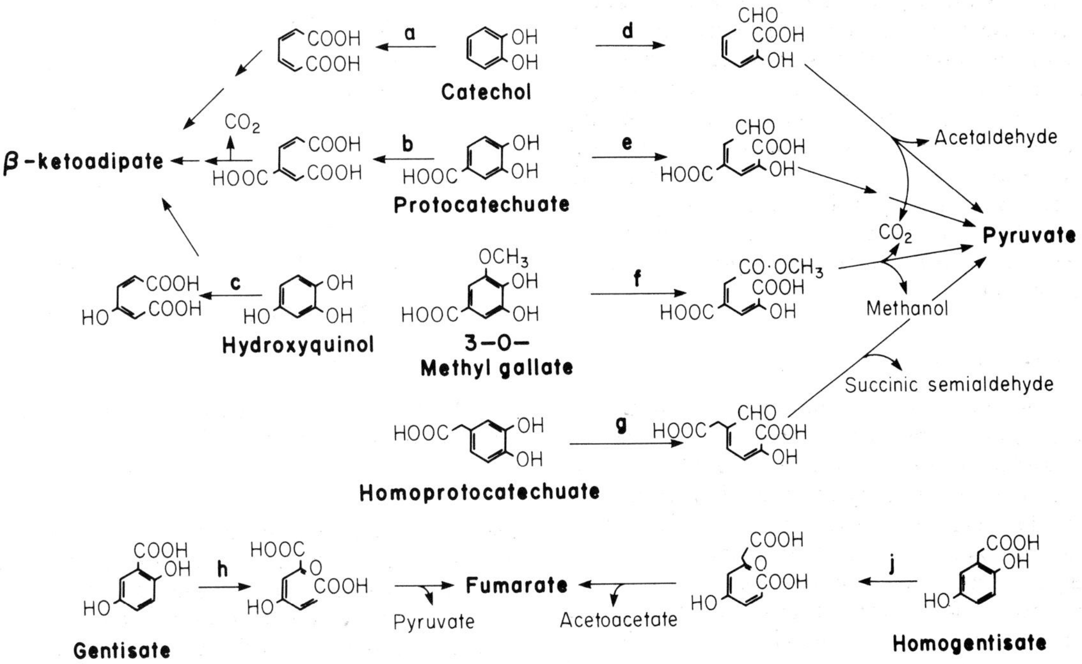

Fig. 4.5.3. Summary of the different aerobic aromatic ring-cleavage ('lower') pathways described in micro-organisms: (a)–(c) are *ortho*-cleavage pathways; (d)–(g) are *meta*-cleavage pathways; (h) and (j) are gentisate pathways [25].

rules apply. Phenol is usually oxidized to catechol and mineralized by a *meta* pathway in *Ps. putida* (but mutant strains have been isolated which use an *ortho* pathway for catabolism of phenol), benzoate by an *ortho* pathway, and naphthalene and methylaromatics, which produce methyl catechols, by *meta* pathways [25, 36].

The specificities of oxygenative ring cleavage and lower pathway enzymes are restricted to the extent that only rarely is a ring-cleavage dioxygenase capable of catalysing the oxidation of more than one of the substrates illustrated in Fig. 4.5.3. Limited cross-reactivity of this sort is a general feature of catabolic enzymes which discriminate between closely related, sometimes isomeric compounds, and it leads to the production of distinct pathways for the catabolism of similar compounds. However, as pointed out by Dagley [23, 24], catabolic enzymes of narrow specificity encoded by genes which are tightly controlled, so that expression occurs only in the presence of a narrow range of inducers, might limit metabolic versatility. In a laboratory microbial culture provided with only one substrate, broad-specificity enzymes and pathways may metabolize it less efficiently than highly evolved, specific enzymes and might, therefore, be considered energetically wasteful. However, the biodegradation of unnatural compounds is dependent on evolution of appropriate catabolic enzymes, and the inherent flexibility of broad-specificity enzymes is likely to provide a selective advantage in evolving catabolic pathways (see Section 4.5.6).

In addition, micro-organisms which contain broad-specificity catabolic enzymes and pathways may be able to utilize substrate mixtures more efficiently than those containing distinct pathways for each substrate. The TOL plasmid-encoded *meta* ring-cleavage pathway (see Section 4.5.5) can transform a variety of alkyl-substituted catechols, whereas the chromosomally-encoded *ortho* ring-cleavage pathway in the plasmid-containing host, *Ps. putida* strain PaW1, can transform only unsubstituted catechol. In energetic terms the chromosomal *ortho* pathway dissimilates catechol and, therefore, benzoate more efficiently than the *meta* pathway. When strain PaW1 was grown on benzoate, cured strains were selected which had lost the TOL plasmid; consequently, catechol 2,3-dioxygenase and the *meta* ring-cleavage pathway did not interfere with catabolism of catechol [25]. Thus, an alternative catabolic pathway for the mineralization of a variety of related aromatic substrates was sacrificed in favour of efficient utilization of a single substrate.

Studies with *Trichosporon cutaneum* indicate that this yeast has sacrificed catabolic efficiency whilst evolving broad-specificity enzymes and pathways. For example, the fungus produces relatively few different ring-cleavage dioxygenases (and no *meta*-cleaving enzymes), but those which are produced have a broader specificity than isofunctional bacterial enzymes. As a result, the aromatic 'upper' pathways in *T. cutaneum* are markedly convergent, limiting the number of available catabolic routes and constraining the efficiency with which aromatic compounds are converted to ring-cleavage substrates [25]. Despite these limitations the catabolic versatility of *T. cutaneum* is comparable even to the pseudomonads, and it has been suggested that the broad-specificity fungal enzymes which are encoded by loosely regulated genes could give *T. cutaneum* a selective advantage in environments containing mixtures of structurally related aromatic substrates [23].

Despite a lack of direct evidence it is generally accepted that the extensive catabolism of aromatic substrates by eukaryotic micro-organisms, as described in *T. cutaneum* [82], is atypical, and that these micro-organisms have a less significant role in the aerobic mineralization of benzenoid compounds. Fungi and algae are, however, able to transform various aromatic hydrocarbons and catalyse hydroxylation reactions involving epoxide intermediates, with broad-specificity mono-oxygenases [44].

Polycyclic aromatic hydrocarbons (PAHs) and alkylated derivatives are widely distributed in soil and aquatic habitats, as natural and xenobiotic products from the combustion of organic matter. Aerobic catabolism of PAHs follows similar pathways to those described for individual benzenoid units. Biodegradation of naphthalene, anthracene and phenanthrene proceeds by oxygenase-catalysed hydroxylation of one aromatic ring and the sequential breakdown and cleavage of the rings. These di- and tricyclic aromatic hydrocarbons can be mineralized by bacterial species [15, 44], whereas the more complex fused ring structures such as the fluorenes, coronenes and pyrenes may be subject only to limited degradation. Cerniglia and his colleagues described the role of different fungal species in the oxidation of PAHs [15]. In general, fungi do not seem to be able to utilize these compounds as growth substrates, but fungal oxygenases, linked to cytochrome P-450 systems, can transform various PAHs, frequently producing *trans*-dihydrodiol derivatives. It has been suggested that fungal metabolism of PAHs serves to detoxify these compounds in an analogous manner to the detoxification of xenobiotics by mammalian P-448 systems. Certain fungal biotransformations of PAHs have been reported which convert the substrates to more serious pollutant products; for example, benzo[*a*]pyrene is converted to a carcinogen by *Cunninghamella elegans* and the polymerization of 1-naphthol is potentiated by the laccase of *Rhizoctonia pracicola* [36].

Halogenated compounds

The halogen substituent is frequently encountered in organic pollutants; the degree of halogenation of organic molecules and their persistence in the environment are correlated. As a group, halogenated (mainly chlorinated) organic compounds constitute the majority of the listed priority pollutants. Detoxification and extensive biodegradation of halogenated compounds usually requires mineralization of the halogen substituent at some stage during catabolism, and a variety of enzymes catalysing dehalogenation reactions have been described. The isolation from unpolluted habitats of micro-organisms which can utilize haloaromatic and haloaliphatic compounds as sole sources of carbon and energy, indicates that a significant number of naturally occurring halogenated compounds are distributed throughout the environment [102], and that dehalogenating enzymes have some natural role. In addition, it is likely that anthropogenic increases in the environmental concentrations of the more widely used halogenated hydrocarbons (pesticides, solvents, etc.) during this century have selected organisms containing novel dehalogenating enzymes.

Biodegradation of haloaliphatic compounds by bacteria is well documented [72, 84]. Specific hydrolytic dehalogenases catalyse the dehalogenation of haloalkanoic acids [53, 113] and haloalkanes [52, 60], although preliminary evidence suggests involvement of oxygenases in the dechlorination of α,ω-chloroalkanes. Dechlorination is the best studied dehalogenase-catalysed reaction; however, many dehalogenases can bind and eliminate fluoro-, bromo-, and iodo-substituents. The dehalogenated products are usually fed into central metabolism, either directly or after further catabolism.

The effects of halogen substitutions on the biodegradation of haloaromatic compounds have been discussed by Reineke [84]. However, the chemical properties of arene carbon–halogen bonds (bond energy, polar and steric effects, etc.) have been found to be of limited use in predicting biodegradability of haloaromatic compounds. Aerobic catabolism of haloaromatic compounds closely follows the pathways described for unsubstituted analogues, and dehalogenation may occur either before or after aromatic ring cleavage.

One of the best studied haloaromatic degrading bacteria is *Pseudomonas* sp. strain B13, which can utilize fluoro- and chlorobenzoates as sole sources of carbon and energy [84]. Defluorination of 2-fluorobenzoate by strain B13 before ring cleavage occurred fortuitously as a result of labilization of the halogen substituent by benzoate 1,2-dioxygenase, an enzyme that normally catalyses a different reaction. The formation of an unstable dihydrodiol by the enzyme resulted in spontaneous elimination of fluoride and CO_2 and formation of catechol [84]. A two-component dioxygenase produced by *Pseudomonas* sp. strain CBS3 catalysed a similar reaction to dechlorinate 4-chlorophenylacetate and produced protocatechuate. In this case, hydroxylation of the C_4 carbon probably caused spontaneous chloride release and the dechlorinated product was mineralized by a normal *meta* pathway [68]. The 4-chlorophenylacetate 3,4-dioxygenase did not catalyse oxidation of non-halogenated substrates and, therefore, dehalogenation cannot be described as a fortuitous reaction of this enzyme. Enzyme-catalysed hydrolytic dehalogenation of 4-chlorobenzoate to form 4-hydroxybenzoate was also observed in *Pseudomonas* sp. strain CBS3 [75] and other species [67, 74]. Preliminary evidence from the purification of the unusual aromatic dehalogenase from strain CBS3 suggested that it was soluble and required Mn^{2+}.

Knackmuss and his colleagues (see [84]) demonstrated that *Pseudomonas* sp. strain B13 converted most halobenzoates to halocatechols, and that a modified *ortho* pathway was required for mineralization of the ring-cleavage substrates (Fig. 4.5.4). The chloro-substituents of 3-chloro- and 4-chlorocatechol were spontaneously eliminated during enzyme-catalysed cycloisomerization, producing a 'dienelactone' (BV and CV in Fig. 4.5.4) which was converted to β-ketoadipate via maleylactate. 3,5-Dichlorocatechol was also mineralized via this pathway, but the chlorinated 'dienelactone' produced after cycloisomerization required further dechlorination, which probably occurred during the reduction of chloromaleylacetate to β-ketoadipate. In strain B13 the chlorocatechol pathway was found to operate independently of a conventional catechol *ortho* (β-ketoadipate) pathway; the former was plasmid encoded and separately induced. Although two of the enzymes in the catechol and chlorocatechol *ortho* pathways were isofunctional, and unsubstituted catechol was mineralized by the chlorocatechol pathway enzymes when induced, the two pathways were not identical. Divergence of the pathways occurred after cycloisomerization; dienelactone hydrolase and maleylacetate reductase were present in the chlorocatechol pathway, whereas lactone isomerase and a different hydrolase were present in the catechol pathway. Catabolism and dehalogenation of chlorocatechols derived from other substrates, such as mono- and dichlorobenzenes [26, 86], chloroanilines [120, 122] and chlorophenoxyacetate herbicides [35], have been reported to require a similarly modified *ortho* pathway.

Crawford *et al.* [21] showed that a strain of *Bacillus brevis* used a modified gentisate ring-cleavage pathway for the utilization of 5-halosalicylates as carbon and energy sources. Two exceptional features of this catabolic pathway were reported: first, the transformation of monohydroxylated halobenzoates, but not gentisate or salicylate, by the ring-cleavage dioxygenase, and, second, enzyme-catalysed dechlorination of the chlorolactone, formed after ring cleavage, to produce maleylacetate.

Meta ring cleavage of 3-substituted chlorocatechols produces acylhalides, reactive intermediates which can acylate and irreversibly inhibit the producing enzymes. However, some chloroaromatic compounds have been reported to be mineralized by *meta* pathways; for example, 5-chloroprotocatechuate, the ring-cleavage substrate produced during catabolism of 5-chlorovanillate by soil bacteria, was converted to an acylhalide by *meta* cleaving protocatechuate 4,5-dioxygenase, but the product spontaneously cyclized to form a lactone and eliminated the halogen [55].

Some interesting general features have emerged from studies of the biodegradation of halogenated hydrocarbons.

1 Catabolism of isomeric substrates may require distinct pathways, even though one pathway may be capable of mineralizing the different isomers.

2 Dehalogenation need not necessarily occur at the first step of catabolism; consequently broad-specificity enzymes are required to catalyse transformations of halogenated intermediates. Incomplete biotransformation of halogenated compounds is frequently observed in the environment, probably reflecting the limited extent to which these compounds are metabolized by broad-specificity enzymes in the absence of enzymes catalysing critical reactions such as dehalogenation and aromatic ring cleavage,

3 Bacterial strains may produce more than one dehalogenating enzyme with similar or different specificities [113]. *Pseudomonas* sp. strain CBS3, for example, carries separately regulated genes encoding two aliphatic de-

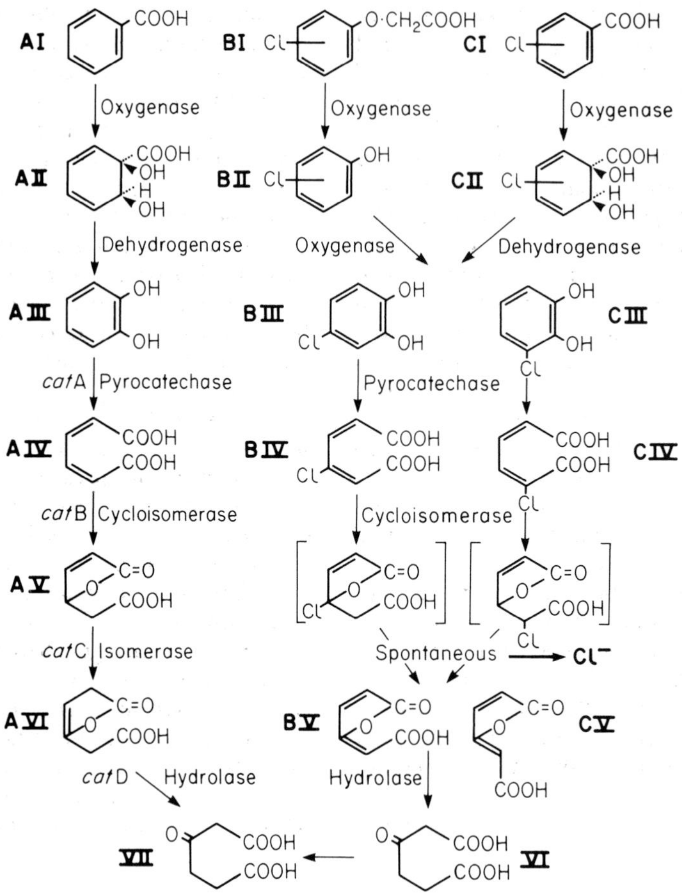

Fig. 4.5.4. Converging *ortho*-ring-cleavage pathways for the catabolism of: benzoate (AI), chlorophenoxyacetates (BI) and chlorobenzoates (CI) and channelling of these substrates into central metabolism via β-ketoadipic acid. The following intermediates are marked: 3,5-cyclohexadiene-1,2-diol-1-carboxylic acid or DHB (AII), chlorophenol (BII), chloroDHB (CII), catechol (AIII), 4-chlorocatechol (BIII), 3-chlorocatechol (CIII), *cis,cis*-muconic acid (AIV), 3-chloro-*cis,cis*-muconic acid (BIV), 2-chloro-*cis,cis*-muconic acid (CIV), 4-carboxymethylbut-2-en-4-olide (AV), 4-carboxymethylbut-3-en-4-olide (AVI), *cis*-4-carboxymethylenebut-2-en-4-olide (BV), *trans*-4-carboxymethylenebut-2-en-4-olide (CV), maleylacetic acid (VI), and β-ketoadipic acid (VII). Enzymes catalysing the conversions are indicated. Intermediates in parentheses are unstable and spontaneously dechlorinate.

halogenases as well as the two aromatic dehalogenases mentioned above [74].

Highly chlorinated aromatic compounds such as priority pollutants polychlorinated biphenyls (PCBs), once a component of transformer electrolytes, the herbicide 2,4,5-trichlorophenoxyacetate (245-T) and pentachlorophenol (PCP), a broad-spectrum biocide, can be degraded by micro-organisms. Surprisingly, it is not difficult to isolate bacteria which can utilize these compounds as sole sources of carbon and energy. The mineralization of PCP by *Flavobacterium, Arthrobacter* and *Pseudomonas* has been reported [100], as has extensive aerobic catabolism of 245-T and PCB mixtures by several Gram-negative species [41, 43, 112]. To date, the mechanisms for enzymatic dechlorination of the polychlorinated substrates have not been elucidated, although it seems likely that some interesting aromatic dehalogenases could be isolated from these bacteria. As indicated in Fig. 4.5.5, growth of *Alcaligenes* sp. and *Acinetobacter* sp. on 4-chlorobiphenyl results in the excretion of 4-chlorobenzoate, a product which cannot be catabolized by these bacteria. The catabolism of other PCB isomers and the commercial PCB mixtures (Aroclors, Kaneclors) seems to proceed by a similar pathway, with dioxygenase-catalysed hydroxylation of one ring (unsubstituted if available) and *meta* ring cleavage [42]. Data from environmental studies and laboratory studies with pure cultures of bacteria consistently indicate that the rate and extent of biodegradation of PCB isomers is correlated with the number and relative positions of chlorine substituents on the two aromatic rings. Highly substituted PCB isomers are degraded slowly, if at all, and catabolic enzymes preferentially attack and cleave the less substituted ring [41].

Biodegradation of nitrogen- and sulphur-containing compounds by micro-organisms can be associated with their utilization as sole nutrient sources, with or without utilization of the mineralized carbon. Despite reports indicating that many nitrogen- and sulphur-containing compounds are biodegradable this *ad hoc* group contains several economically important and persistent pollutants. In many cases biodegradation of these compounds depends upon catabolic pathways for the dissimilation of natural products such as amino acids, purines, pyrimidines and vitamins [14].

The nitrogen in aliphatic amines has been reported to be mineralized by microbial hydrolytic amidases and oxidative deaminases, with production of NH_4^+ [19]. Nitroalkanes may be reduced to form amines, but oxidase or oxygenase-catalysed release of nitrate is likely to be a more significant biodegradative route [17]. Aromatic amines are commonly observed products and intermediates in the biodegradation of azo-dyes and aniline-based herbicides (e.g. Propanil), and oxidative deamination of variously substituted anilines, before aromatic ring cleavage, has been observed by several authors (see [120, 122]). Similarly, substituted nitrophenols were shown to be converted to catechols by a broad-specificity NADPH-requiring mono-oxygenase which released nitrite from the substrates [121].

Further biodegradation of nitrogen-containing aromatics after mineralization of the nitrogen substituent is dependent on the presence of appropriate ring-cleavage catabolic pathways (see earlier sections). The biodegradation of heterocyclic sulphur- and nitrogen-containing compounds has been reviewed by Callely [14] and Ensley [33]. Extensive catabolism and ring cleavage of heterocyclic compounds is normally required for mineralization of sulphur or nitrogen heteroatoms. Again, oxygenases catalyse important ring-activating oxygenations and in some cases are required for ring cleavage. For example, the pyridine ring of nicotinic acid and *iso*-nicotinic acid (a photolytic decomposition product of the herbicide Paraquat) is activated and cleaved by various hydroxylations catalysed by oxygenases. By contrast, the pyridine

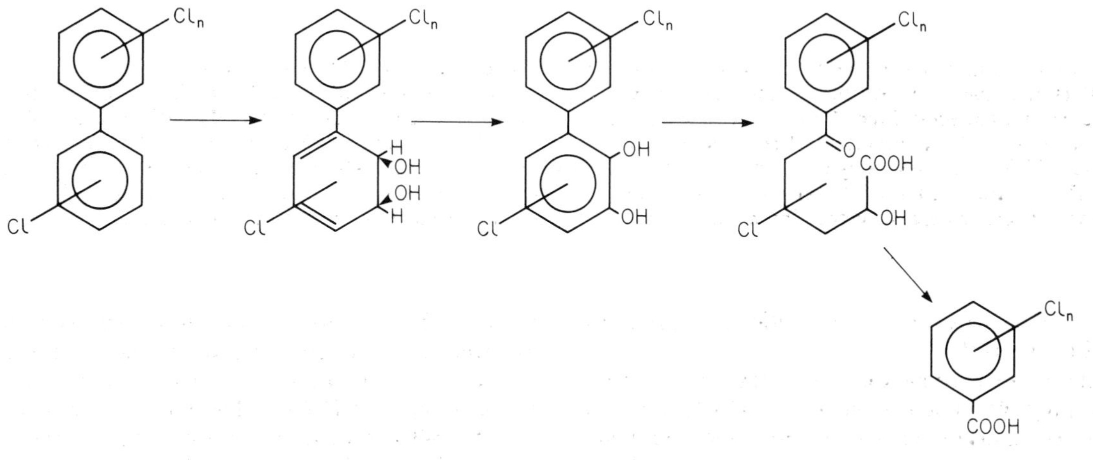

Fig. 4.5.5. Catabolism and *meta* cleavage of polychlorinated biphenyls by *Alcaligenes* sp. showing production of chlorobenzoates [42].

and s-triazine heterocyclic rings, containing two and three nitrogen atoms, respectively, are usually cleaved by hydrolytic enzymes [14, 20].

Apart from the sulphur-containing heterocyclics, such as the thiophenes found in sulphurous crude oils, the most important sulphur-containing xenobiotics are alkylsulphates and arylsulphonates produced commercially as surfactants. It is interesting to note that linear alkylbenzene sulphonates were introduced by the chemical industry to replace the recalcitrant branched chain cationic detergents (see above). Oxygenase-catalysed mineralization of SO_3^{2-} from alkylsulphates by a range of microbial primary and secondary sulphatases has been studied by several authors [13, 27].

Polymers

Polymeric organic compounds are among the most resistant to microbial attack; yet laboratory studies indicate that micro-organisms are able to degrade a variety of polymers. Okada's group isolated a *Flavobacterium* species capable of utilizing a nylon oligomer as the sole source of carbon and energy [78, 79]. The cyclic dimer was depolymerized by two novel, plasmid-encoded hydrolases [78], the evolutionary origins of which have been investigated. *Alcaligenes* spp. and *Pseudomonas* spp. strains have been reported to utilize styrene dimers and, to a limited extent, polystyrene as carbon sources [110]. Vulcanized rubber was also shown to be depolymerized by a strain of *Nocardia* sp. which, intriguingly, utilized latex glove rubber as the sole source of carbon and energy [109]. It will be interesting to discover the biochemical and genetic characteristics of these biodegradative reactions.

The biodegradation of a natural polymer, lignin, has been the subject of intense investigation [57]. A broad-specificity exocellular 'ligninase' was recently purified from the white-rot basidiomycete, *Phanerochaete chrysosporium*, and was shown to be a peroxide-requiring oxygenase [105]. The enzyme catalysed several different types of reactions including: (a) carbon–carbon bond cleavage resulting in depolymerization of lignin to form various methoxylated phenolics and other substituted benzenoid compounds; (b) hydroxylation and oxidation; and (c) aromatic ring cleavage. Involvement of 'ligninases' in the mineralization of unrelated recalcitrant environmental pollutants, such as Lindane (hexachlorocyclohexane), DDT (1,1-*bis*(4-chlorophenyl)-2,2,2-trichlorohexane), polychlorinated biphenyls, benzo[*a*]pyrene and 2,3,7,8-tetrachlorodibenzo-*p*-dioxin, has been suggested although the physiological and environmental significance of these reactions has not yet been reported [12, 29].

Anaerobic biodegradation

The significance of anaerobic biotransformations has, perhaps, been underestimated in view of the important contribution of anaerobic micro-organisms to the carbon cycle. It is to be expected that the upsurge in research will improve our understanding of biodegradation in anoxic environments. Not surprisingly, rates of biodegradation are generally lowered by the exclusion of oxygen, and the persistence of some compounds in sediments and submerged soils, notably lignin and polycyclic aromatic hydrocarbons, suggests that they are not metabolized by anaerobes.

Few anaerobic catabolic pathways for the dissimilation of xenobiotics have been elucidated, but those which have clearly shown that anaerobic and aerobic biodegradation are qualitatively different. The range of aromatic substrates now reported to be metabolized by enriched cultures of anaerobic bacteria attests to their versatility.

However, details of catabolic pathways are not usually available and the considerable difficulties of working with obligate anaerobes, their slow growth rates and complex syntrophic associations, have generally limited the rate of progress in this research. Shelton and Tiedje [92] have described a simple method for estimating anaerobic biodegradation using digested sewage sludge and measuring gas production over a period of weeks. Amongst the better studied anaerobic biodegradative pathways is the mineralization of benzoate by facultative denitrifying bacteria (*Pseudomonas* spp.), obligately anaerobic phototrophs (*Rhodospirillaceae*) and methanogenic mixed cultures. The group of sulphate-reducing bacteria contains some catabolically versatile species [114] also capable of mineralizing the aromatic carbon skeleton.

Research in the laboratories of Evans [34] and Young [119] has shown that mechanistically related pathways are used for the anaerobic catabolism of simple aromatic substrates, whether associated with anaerobic denitrification, photosynthesis or methanogenesis. For example, in the absence of oxygen and the presence of NO_3^- as electron-acceptor, the mineralization of benzoate by a facultative anaerobe, *Pseudomonas stutzeri* (Fig. 4.5.6), was shown to be mediated by enzyme-catalysed reductions of the aromatic nucleus and reductive or β-oxidation pathways for cleavage of the alicyclic ring and catabolism of the alkanoate ring-cleavage product [34].

The study of mixed cultures or communities in anaerobic biodegradation is important, particularly when catabolism is linked with fermentation or methanogenesis, and reducing equivalents (hydrogen) need to be recycled. The associations between the species which constitute

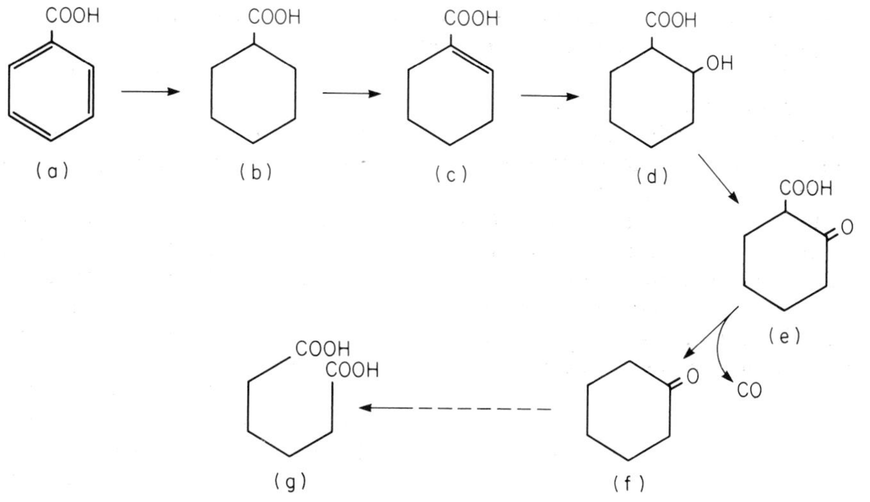

Fig. 4.5.6. Anaerobic biodegradation of benzoate by a denitrifying pseudomonad. (a) Benzoic acid. (b) Cyclohexane carboxylic acid. (c) Cyclohexene-1-carboxylic acid. (d) 2-Hydroxycyclohexane carboxylic acid. (e) 2-Oxocyclohexane carboxylic acid. (f) Cyclohexanone. (g) Adipic acid [34].

such mixed cultures are frequently complex, and it is sometimes difficult to distinguish organisms involved in mainstream catabolism from those syntrophically involved in stabilizing the community [119]. The isolation of methanogenic mixed cultures capable of mineralizing chlorobenzoates and chlorophenols (including PCP) with the suggested involvement of reductive dechlorination [8, 70, 93], and the anaerobic catabolism of anthranilate by denitrifying pseudomonads [10], indicate that anoxic environments contain micro-organisms carrying out important biotransformations of xenobiotics.

4.5.4 MICROBIAL COMMUNITIES AND BIODEGRADATION

The limitations imposed by the use of pure cultures in studies of biodegradation were indicated in Section 4.5.2. Environments in which single species grow in isolation from other populations are uncommon and tend to be associated with extreme physical conditions, such as extremes of temperature or pH. Normally, micro-organisms occupy habitats with other species and the interactions, both physical and chemical, which occur can lead to the establishment of stable mixed microbial communities representing a variety of catabolic capabilities [96, 97]. The enrichment of xenobiotic-degrading micro-organisms in liquid culture using conventional methods almost always results in the isolation of mixed populations. The characterization of some of these mixed cultures has added an extra dimension to the study of biodegradation and has demonstrated the importance of secondary utilizers in microbial populations. We may define secondary utilizers as those micro-organisms which are unable to initiate catabolism of a substrate but which may utilize intermediates derived from incomplete catabolism of the compound by primary utilizers. Alternatively, they may grow on products from primary utilizers which are not derived directly from the compound (e.g. excretory products). Secondary utilizers may not be mere 'hangers-on'; they can support the growth of primary utilizers in a number of ways; indeed the primary utilizers may be dependent upon them for growth. A general observation indicating the importance of interactions between component species in microbial communities is that growth rates and rates of substrate utilization are frequently higher in enriched mixed cultures than in pure cultures isolated from the mixture. Table 4.5.2 lists some of the interactions which have been identified in different xenobiotic-utilizing microbial communities [96], and more detailed examples of some are described below.

Provision of co-factors and other growth-stimulating nutrients

This type of interaction is exemplified by a mixed culture containing two pseudomonads which together utilize

Table 4.5.2. Types of interactions observed in microbial communities degrading xenobiotics [96]

Interaction	Example	
	Substrate*	Observations
Provision of specific co-factor/nutrient	Trichloroacetate (2)	1°-utilizer required vitamin B_{12}
	Polyvinylalcohol (2)	See text
Removal of toxic product	Methane (4)	Removal of inhibitory methanol by 1°-utilizer
Modification of growth parameters	Orcinol (3)	2°-utilizers affected K_s and K_i of 1°-utilizer for substrate
Concerted metabolism	Linear alkyl benzene sulphonates (4)	Several 1°-utilizers. Rate of substrate degradation higher in mixed culture than combined rates of isolated primary population
	2,2-Dichloropropionate (8)	As above (see text)
	4-Chlorobiphenyl (2)	2°-utilizer mineralized, 4-chlorobenzoate excreted by 1°-utilizer as a product of substrate catabolism
	Dodecylcyclohexane (2)	1°-utilizer converted substrate to cyclohexanacetate; 2°-utilizer grew on this product
Co-metabolism	Parathion (>4)	See text
	Cyclohexane (2)	1°-utilizer converted substrate to cyclohexanone when growing on propane; 2°-utilizer grew on cyclohexanone
Gene transfer	2-Chlorobutanoate	Transposition, see text
	Chlorobenzoates	Plasmid transfer, see text

* Numbers of different strains identified in each mixed culture are shown in parentheses.

polyvinyl alcohol (PVA) as the sole source of carbon and energy [95]. Neither of the *Pseudomonas* spp., designated VM15A and VM15C (see Fig. 4.5.7), was able to transform PVA in pure culture. It was found that strain VM15C catabolized PVA and that strain VM15A utilized excretory products from strain VM15C, but the primary utilizer, strain VM15C, required an unusual coenzyme called pyrroloquinoline (PQQ), which was supplied by strain VM15A (Fig. 4.5.7). The growth rate of the mixed culture, dependent upon the rate of utilization of PVA by strain VM15C, was shown to be limited by the rate of production of PQQ by strain VM15A [95].

Co-metabolism and microbial communities

Co-metabolism is an imprecise term which is frequently used to indicate transformation of a compound by metabolic reactions which do not contribute to the growth of the organism carrying out the transformation [1]. A wide range of factors can account for co-metabolism and, as a result, semantic arguments have clouded the more important issue of determining the role of co-metabolism in biodegradation. Because co-metabolic transformations of xenobiotic compounds are not directly linked to their utilization as nutrient or energy sources, co-substrates are required to support the growth of the co-metabolizing micro-organism. In addition, the co-substrate may participate directly in catabolism; for example, the co-substrate may be responsible for the expression of the co-metabolic pathway or may supply co-factors for the enzymes of the pathway or energy for transport of the xenobiotic.

The utilization of the organophosphate insecticide Parathion (*o,p*-nitrophenylphosphorothioate) by a microbial community isolated in Hseih's laboratory and maintained in continuous-flow culture for several months [76, 96] elegantly illustrates the role of co-metabolism in the

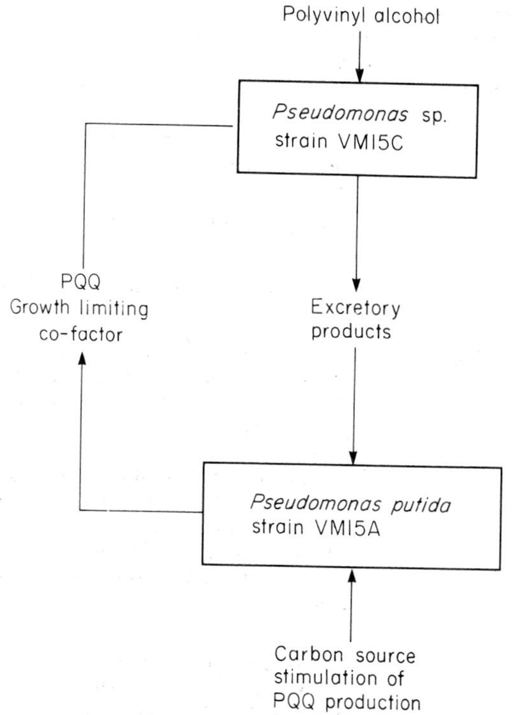

Fig. 4.5.7. Structure of a simple microbial community degrading polyvinylalcohol [95].

biodegradation of a xenobiotic (Fig. 4.5.8). This community comprised at least four bacterial species but none was capable, in isolation, of utilizing Parathion as the sole source of carbon and energy; however, together the four bacteria mineralized the insecticide. Two biotransformations were found to be central to mineralization of Parathion: first, the hydrolysis of Parathion by *Pseudomonas stutzeri* forming diethylthiophosphate and *p*-nitrophenol, and, second, the utilization of one of these products, *p*-nitrophenol, by *Ps. aeruginosa*. The *Ps. aeruginosa* strain, in addition to utilizing *p*-nitrophenol as sole carbon and energy source, could support the growth of *Ps. stutzeri* and other species in the community by providing utilizable excretory metabolites and products from cell lysis. Thus, the initial hydrolysis of Parathion was a co-metabolic reaction upon which the community was dependent.

Two other features of the biodegradation of Parathion by this community were noted. First, the diethylthiophosphate produced by the Parathion hydrolase of *Ps. stutzeri* was not metabolized. This may have resulted from simplification of the community structure during enrichment in the chemostat; organisms capable of utilizing diethylthiophosphate might have been present initially and were later excluded by continuous-flow culture. Second, the Parathion hydrolase was later shown to be encoded by genes carried on a transmissible plasmid ([90], see Section 4.5.5). Presumably the transfer of this plasmid to the *Ps. aeruginosa* population would have resulted in transconjugants capable of utilizing Parathion as the sole source of carbon and energy. However, the community retained the same structure over 2 years of continuous culture — an indication either that transfer was not possible (see Chapter 2.3) or that transconjugants containing the complete pathway for Parathion mineralization had no selective advantage under these conditions.

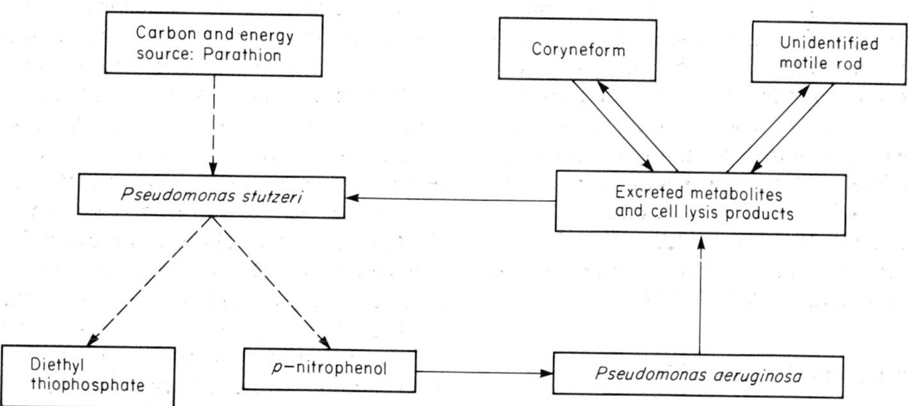

Fig. 4.5.8. Parathion-utilizing microbial community isolated in chemostat culture. Co-metabolic production of diethylthiophosphate and *p*-nitrophenol from Parathion is indicated by broken arrows (see [96]).

Evolution and genetic exchange in microbial communities

Aspects of gene transfer in microbial populations have been discussed in Chapter 2.3, and will only briefly be elaborated upon here. Secondary utilizers in chemostat cultures of microbial communities enriched with xenobiotic compounds are continuously being exposed to a potential substrate. These conditions select for competitive primary utilizers which can arise out of the population of secondary utilizers by mutation or gene transfer. Mutation of a secondary utilizer within a seven-membered community isolated by continuous-flow enrichment on the herbicide Dalapon (2,2-dichloropropionate) resulted in the expression of genes encoding dehalogenases and acquisition of the ability to catabolize Dalapon. The mutant strain was found to compete with other primary utilizers and also displaced the population of secondary utilizers from which it was derived [96].

Transfer of plasmids carrying dehalogenase genes was observed in an artificially constructed two-membered mixed culture. The primary utilizer, *Ps. putida* strain PP3R, was grown on 2-chloropropionate and converted a co-substrate, 2-chlorobutanoate, to 2-hydroxybutanoate, a product which it could not utilize. The secondary utilizer, *Pseudomonas* sp. strain HB2001, did not contain a dehalogenase but was able to grow on the 2-hydroxybutanoate produced co-metabolically by strain PP3. Transfer of the plasmid-encoded dehalogenase from strain PP3 into strain HB2001 gave rise to a population of 2-chlorobutanoate primary utilizers derived from the latter strain.

4.5.5 CATABOLIC PLASMIDS AND THE GENETICS OF BIODEGRADATION

The isolation and preliminary characterization of the first plasmids carrying genes encoding catabolic enzymes was reported in the early 1970s by research groups in the UK and USA (also see Chapter 2.3) [39]. Since that time the number of catabolic plasmids shown to be associated with transformations of xenobiotics has increased steadily. Table 4.5.3 lists most of the plasmids which have been reported to encode specific catabolic enzymes or pathways. Undoubtedly, this list represents only a small sample of the range of catabolic plasmids present in the soil and aquatic microflora. However, detailed studies of

Table 4.5.3. Plasmid involvement in the biodegradation of xenobiotics and related compounds

Compounds	Plasmid*	Size (kbp)	Original host	Reference
Alkanes (e.g. octane, decane)	OCT	>500	*Pseudomonas putida*	81, 91
Aniline	pCIT1	~100	*Pseudomonas* sp.	3
Camphor	CAM (T)	>500	*Pseudomonas putida*	59, 87
Chlorobenzoates	pAC25 (T)	117	*Pseudomonas* sp.	18
	pWR1 (T)	111	*Pseudomonas* sp.	18, 84
Chlorobiphenyls	pKF1	81	*Acinetobacter* sp.	40
	pSS50 (T)	56	*Alcaligenes* sp.	94
Chlorophenols, chlorophenoxyacetates	pJP4 (T)	80	*Alcaligenes eutrophus*	28
Haloalkanoates	pUO1 (T)	69	*Moraxella* sp.	53
Dibenzothiophene	Not designated (T)	83	*Pseudomonas* sp.	71
2,6-Dichlorotoluene	Not designated (T)	96	*Pseudomonas cepacia*	111
Isopropylbenzene	pRE4	105	*Pseudomonas putida*	30
Naphthalene	NAH (T)	83	*Pseudomonas putida*	117
Nicotine	NIC (T)	nd	*Pseudomonas* sp.	104
	pAO1	160	*Arthrobacter oxidans*	9
Nylon oligomer	pOAD2	44	*Flavobacterium* sp.	79
Parathion	pCS1	65	*Pseudomonas diminuta*	90
	Not designated (T)	66	*Flavobacterium* sp.	73
Salicylate	SAL (T)	85	*Pseudomonas putida*	39
Styrene	pEG (T)	37	*Pseudomonas* sp.	7
Toluene, xylene, toluate, ethylbenzene, trimethyltoluene	TOL (T)	117	*Pseudomonas putida*	37, 77, 115
Toluidine	pTDN1 (T)	75	*Pseudomonas putida*	66

* (T) indicates conjugal plasmids.

some of these plasmids (only a handful of plasmids have been characterized to the extent of mapping catabolic genes) have made a significant contribution to our understanding of the genetics of biodegradation.

The interactions between plasmid and chromosomally-encoded catabolic genes are crucial to the expression of the catabolic phenotype. Many catabolic plasmids carry genes encoding incomplete catabolic pathways, and the utilization of a xenobiotic as sole carbon and energy source may require complementation of plasmid genes by host chromosomal genes which encode enzymes linking the plasmid pathway with energy-yielding central metabolic pathways. Clustering and operon control are features of the structure and regulation of plasmid catabolic genes. In some cases, two or more functionally linked operons (regulons) have been identified. Most catabolic plasmids are large enough to carry the genes required for conjugal transfer (see Chapter 2.3), and many of the plasmids from Gram-negative species have been shown to be self-transmissible and to have a fairly broad host range.

The predominance of Gram-negative aerobic bacteria, especially *Pseudomonas* spp., in the list of plasmid-containing hosts is probably as much a reflection of the selective nature of laboratory enrichments from which xenobiotic-utilizing micro-organisms are isolated, as the distribution of catabolic plasmids between the groups of bacteria. Only one catabolic plasmid, pA01, which carries genes encoding enzymes catalysing the first steps of nicotine catabolism [9], has been identified in a Gram-positive species, but there is every reason to suppose that catabolic plasmids play the same role in Gram-positive genera such as *Bacillus* and *Nocardia* as in the aerobic Gram-negative bacteria.

The TOL plasmid

One of the first catabolic plasmids to be isolated was TOL, and the more recent work with this plasmid has uncovered some intriguing characteristics relating to the organization, regulation and control of catabolic genes [37, 38, 45, 46, 77]. A brief description of the TOL plasmid, outlining its role in the catabolism of methyl-substituted benzenes, is given below.

TOL plasmids were isolated in *Pseudomonas* species by several groups around the world in the early 1970s, and subsequently one variant, pWWO, has been studied in great detail. As indicated in Table 4.5.3, the archetypal TOL plasmid, pWWO, is a large (117 kbp) self-transmissible plasmid carrying regulated genes encoding enzymes catalysing the conversion of 1,2,4-trimethylbenzene, *m*- and *p*-xylenes (dimethylbenzenes), toluene, 3-ethylbenzene (and corresponding acid, aldehyde and alcohol analogues) into simple aliphatic carboxylic acids and aldehydes (Fig. 4.5.9). To date, 15 catabolic genes have been mapped on plasmid pWWO [45, 46, 77] including two, *xyl*S and *xyl*R, which encode regulators for two positively controlled promoters. The structural genes are organized into two separate operons: *xyl*CAB and *xyl*XYZLEGFJIH. The *xyl*CAB operon is associated with the oxidation of the aryl or methyl substituent to a carboxylic acid. The conversion of this metabolite to intermediates of central metabolism is mediated by a *meta* ring-cleavage pathway encoded by the *xyl*X-H operon. The *xyl*S-encoded regulator interacts with inducers such as 3-methylbenzoate to switch on the *xyl*X-H operon by binding at operator-promoter region OP_2, thereby stimulating transcription. Conversely, the *xyl*R-encoded regulator binds to another operator-promoter, OP_1, to switch on the *xyl*CAB operon in the presence of xylenes, toluene and methylbenzylalcohol. In addition, the *xyl*R-encoded regulator can switch on the *xyl*X-H operon when it binds inducers such as xylene, but only in the presence of the *xyl*S gene product [77]. In summary, the TOL-encoded catabolic pathway comprises two operons controlled by two regulatory genes which probably interact in some way for full induction of both operons.

There are at least four other catabolic plasmids encoding pathways which dissimilate xenobiotics and which are organized into separate operons: the NAH7 [116, 118], SAL1 [116], OCT [81] and pJP4 [28] plasmids associated with catabolism of naphthalene, salicylate, alkanes and chlorinated phenoxyacetate herbicides, respectively. In these plasmids and in the TOL plasmid each separate operon encodes enzymes which constitute one part of their respective catabolic pathway. At the moment, there is little direct evidence for the attractive idea that the catabolic gene organization observed in these plasmids reflects their having evolved by modular assembly of complementary operons. There is, however, compelling evidence to suggest that catabolic genes can be carried on mobile genetic elements, such as transposons [16, 31, 43, 50, 98], which can move from one genetic locus to another by site-specific recombination.

4.5.6 MICROBIAL ADAPTATION AND THE EVOLUTION OF XENOBIOTIC CATABOLIC PATHWAYS

Evidently, micro-organisms are capable of transforming a remarkable variety of chemical structures and their

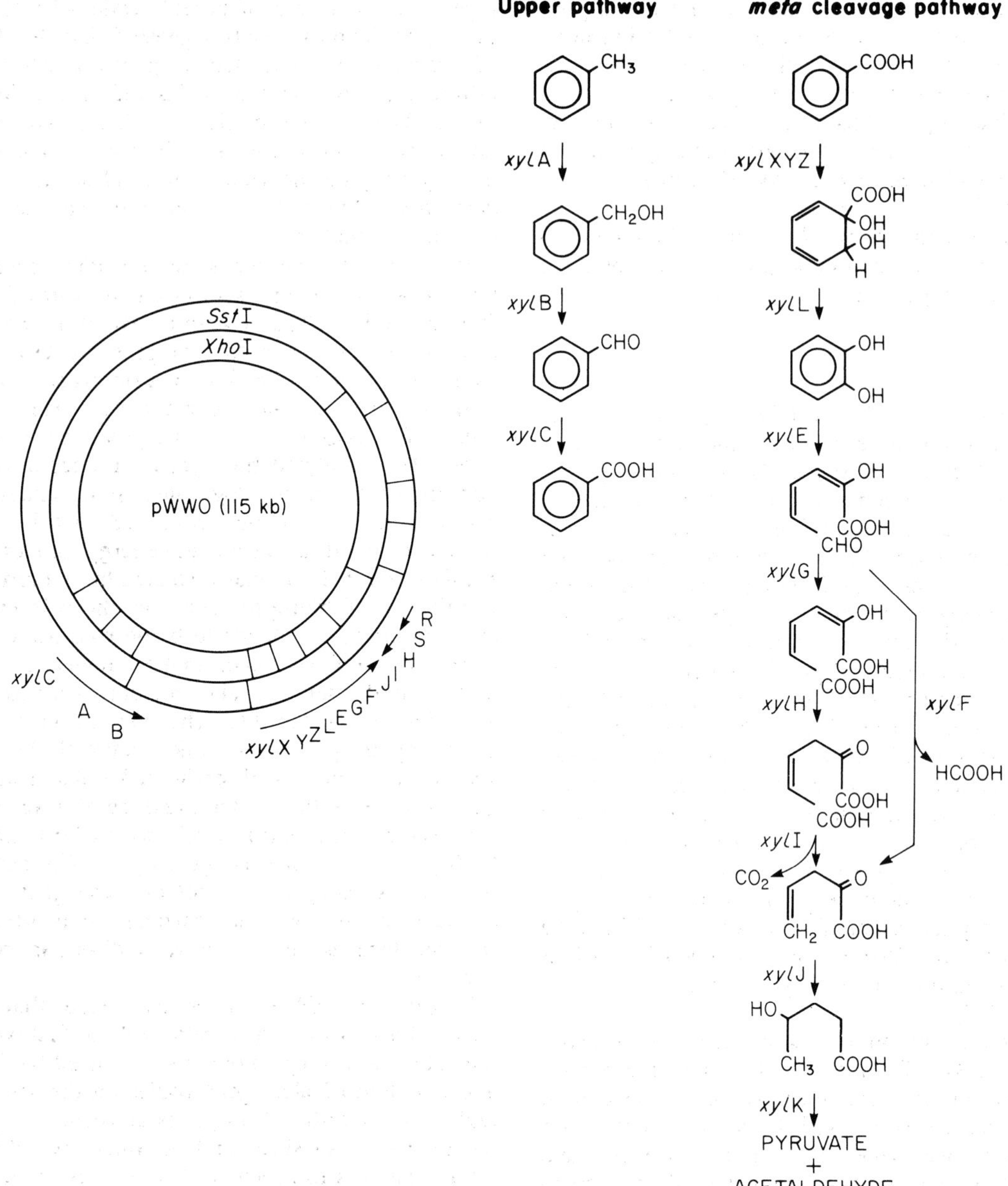

Fig. 4.5.9. Structure of the TOL plasmid, pWWO. See text for details [45, 46].

metabolic versatility, for the main part, results from genetic plasticity and fast generation times. An important mechanism of adaptation in microbial populations, which was mentioned in Section 4.5.5, is the transfer of catabolic plasmids encoding biodegradative enzymes or complete pathways. Broad host-range catabolic plasmids and transposons may facilitate rapid evolution in mixed microbial populations by horizontal gene transfer from a primary host to recipients in the host's vicinity. The spread of plasmid genes encoding enzymes catalysing the biodegradation of a xenobiotic through a microbial population is likely to occur in response to a selection pressure. For example, all xenobiotic organic compounds represent a potential carbon and energy source to heterotrophic microbial populations. In some cases direct selection pressures may not be so obvious, and acquisition of a mobile genetic element *per se* may result in a competitive advantage for a new host.

Acclimation of natural microbial populations by continuous exposure to a xenobiotic probably involves the type of genetic changes described below.

Enzyme evolution

The questions arise as to how catabolic genes, especially genes encoding novel biodegradative functions, evolve *de novo* and how they are subsequently assembled to form regulated catabolic operons. The pioneering studies on bacterial aliphatic amidases by Clarke and her colleagues was aimed at answering the first part of this question: where do new enzymes come from [19]? They found that *Pseudomonas aeruginosa* strain PAC1 produced an acetamidase which hydrolysed a limited number of aliphatic amides. Thus, acetamide and propionamide were hydrolysed by the enzyme and were utilized as carbon, nitrogen and energy sources by strain PAC1. Clarke demonstrated that an understanding of the basic biochemistry and genetics of the acetamidase system could be used to apply defined selection pressures upon the producing organism in closed and open cultures. In response to these pressures, mutations developed which resulted in the evolution of new enzymatic activities. For example, a family of broad-specificity phenylacetamidases, all derived from the acetamidase of strain PAC1, were obtained by various methods involving point mutations in structural and regulatory genes [19].

The directed evolution of azoreductases in *Pseudomonas* sp. strain KF1 illustrates an alternative mechanism by which novel enzyme activities can be obtained [61]. Chemostat cultures were used to select spontaneous mutants of strain KF1 which could utilize Orange I and Orange II, sulphonated dye analogues of dicarboxyazobenzene (DCAB). Strain KF1 was maintained in continuous-flow culture on DCAB and exposed to progressively higher concentrations of Orange II and/or Orange I. The mutants isolated in this way were found to contain one or two novel azoreductases, the enzymes which catalysed the cleavage of the diazo bond. The new enzymes, azoreductases types B and C, were unrelated to each other and showed no obvious relationship with the type A azoreductase of parent strain KF1 [61]. In addition, preliminary results suggested that the three types of enzymes were encoded by genes which mapped to different genetic loci in plasmids and on the chromosome of the *Pseudomonas* species [19]. Thus evolution of the new enzymes did not seem to be based upon an existing activity or gene, as seen with amidase mutants of *Ps. aeruginosa*, but resulted from the expression of new genetic information.

It is possible that the genes encoding azoreductases types B and C were present in a non-expressible, cryptic form in strain KF1, and were activated during the selection of mutant strains on Orange I and Orange II. An example of 'decryptification' was observed in the activation of genes encoding aliphatic dehalogenases in *Ps. putida*. Dehalogenases I and II, produced by *Ps. putida* strain PP3, catalysed the hydrolytic dechlorination and debromination of α-halogenated short chain (C_2-C_4) alkanoic acids, enabling strain PP3 to utilize some of these compounds as carbon and energy sources [113]. *Ps. putida* strain PP3, originally isolated from a mixed culture utilizing 2,2-dichloropropionate as the sole carbon and energy source, was shown to be derived from strain PP1 (formerly S3); the chromosomal dehalogenase genes were cryptic in the latter strain and only very low enzyme activities were detectable. The mechanism of activation of the cryptic genes was studied indirectly by examining spontaneous mutants of strain PP3 isolated at high frequencies ($10^{-3}-10^{-5}$ in the presence of a toxic substrate such as dichloroacetate), which switched off either one or both of the dehalogenase genes [98]. The data showed that the switching on and off of these genes was associated with gross genetic rearrangements which moved the dehalogenase genes between different loci on the chromosome.

Negoro and Okada have suggested that plasmid-encoded hydrolases which catalysed the first two steps in the catabolism of nylon oligomer, evolved by duplication and mutation of silent gene copies on the same plasmid [79]. Studies with hybrid genes constructed by *in vitro* recombination of silent and activated hydrolase genes, led the authors to conclude that conversion of the silent *nyl*B′ gene to *nyl*B (encoding the active oligomer hydrolase) only required expression of the gene by linkage with a promoter and alterations to less than five codons [78].

Pathway evolution

The mechanisms which operate in the evolution of catabolic pathways are poorly understood. Until recently, the only data available have been those obtained from comparative studies of existing pathways in different micro-organisms. For example, Ornston and his colleagues have examined the β-ketoadipate pathway (the *ortho* ring-cleavage pathway for dissimilation of catechol and protocatechuate) in a variety of soil bacteria [99]. Evolutionary relationships between isofunctional enzymes of pathways in different species, and between enzymes catalysing sequential catabolic steps in the same pathway, were indicated by *N*-terminal amino-acid sequence homologies, comparisons of enzyme structure and serological cross-reactions [117]. It was proposed that one common ancestral gene might have evolved by a complex series of gene amplifications and mutations within duplicated genes, giving rise to the different pathway enzymes. The hypothesis that enzymes in the same pathway catalysing totally unrelated reactions, such as oxygenolytic ring cleavage and cycloisomerization, co-evolved from the same ancestral gene is challenging and controversial. It has been noted by Clarke and Slater [19] that recombinant DNA techniques should ease the task of data gathering and testing the validity of this hypothesis with respect to the evolution of other established catabolic pathways.

The isolation and characterization of catabolic plasmids in parallel with developments in recombinant gene manipulation have provided some general observations which may have implications for the evolution of xenobiotic-transforming catabolic pathways. It is difficult to escape the conclusion that the preponderance of catabolic plasmids associated with xenobiotic biodegradation is directly linked to the evolution of novel catabolic genes. Plasmids are effectively minichromosomes on which genes may evolve without directly affecting the functioning of essential chromosomal genes. Teleologically, a plasmid may be regarded as a melting pot, recombining genes and facilitating rapid evolution of ancillary functions. Two aspects of plasmid biology are probably important in this respect: first, plasmid transfer between hosts of different species which broadens the 'pool' of genes for selection, and, second, interactions between plasmid and host chromosome, whether mediated by general or site-specific recombination [39]. The clustering of catabolic genes, observed in most of the well-characterized plasmids (e.g. TOL, SAL, NAH, pJP4) corresponds with the notion of functionally related genes condensing into coordinately expressible units. The evolution of regulatory genes, later on the evolutionary time-scale, to form distinct, coordinately controlled operons could account for the operon structure observed in the TOL and NAH plasmids, where regulatory genes *xyl*R and *nah*R are distantly separated from the transcriptional units which they control [77, 118]. Regulation of chromosomal genes encoding enzymes of the β-ketoadipate pathway also seems to have evolved independently in different micro-organisms and after the formation of organized structural gene units. For example, *Ps. putida* and *Acinetobacter calcoaceticus* produce closely related isofunctional enzymes, but control of gene expression is very different in the two species [80].

In addition to catalysing a reaction, a highly evolved enzyme imposes restrictions on the reaction pathway so that the enzyme's active site reacts with only one or a limited number of substrates and usually only one of a number of possible products is formed in significant quantities. Recent experiments with cloned and plasmid catabolic genes have suggested that xenobiotic catabolic pathways may evolve in a piecemeal fashion by combining complementary broad-specificity enzymes from different pathways. Thus, broad-specificity enzymes which are active in one pathway may be recruited to perform a new function and either constitute a new pathway or augment an existing one. In an elegant series of experiments Reineke and Knackmuss [85, 86] demonstrated that *Pseudomonas* sp. strain B13 was limited to utilizing only one chlorobenzoate isomer, 3-chlorobenzoate (3CB), because of the relatively high specificity of the benzoate 1,2-dioxygenase catalysing the first step in the mineralization of 3CB (see Section 4.5.3). By transferring plasmid genes encoding a broader-specificity dioxygenase (the toluate 1,2-dioxygenase encoded by the TOL plasmid's *xyl*XYZ genes) into strain B13, transconjugants were isolated which could mineralize a wider range of chlorobenzoates. For example, 4-chlorobenzoate, which was not a substrate for strain B13's benzoate 1,2-dioxygenase, was converted by the toluate 1,2-dioxygenase to 4-chloro-1,2-dihydroxycyclohexa-3,5-diene-1-carboxylate (4-chloroDHB). This product was mineralized by the 3-chlorobenzoate pathway enzymes in an analogous way to 3-chloroDHB, via the chlorocatechol *ortho* ring-cleavage pathway (see Fig. 4.5.4). Thus, the block at the initial dioxygenase in strain B13 was relieved by recruitment of the broad-specificity dioxygenase encoded by the TOL plasmid, and expansion of strain B13's broad-specificity chlorocatechol pathway allowed it to utilize 4-chlorobenzoate as a carbon and energy source [85].

In addition to the transfer of the TOL plasmid into *Pseudomonas* sp. strain B13, mutations were involved in the acquisition of the new growth phenotypes. Growth on 4-chlorobenzoate was only observed in strain B13 transconjugants where the *meta* ring-cleavage enzyme encoded by the TOL plasmid (*xyl*E) had been inactivated by gross alterations in the structure of the TOL plasmid. The inactivation of *xyl*E removed the interference caused by the channelling of 4-chlorocatechol (from 4CB) down the *meta* instead of the *ortho* ring-cleavage pathway [5]. This was confirmed by the introduction of a recombinant plasmid containing only the *xyl*XYZ genes (no *xyl*E) into strain B13, which resulted in recruitment of the *xyl*XYZ gene product (toluate 1,2-dioxygenase) and utilization of 3- and 4-chlorobenzoate by the recombinant strain [64]. In addition, transfer of the *xyl*XYZ genes from the TOL plasmid into strain B13 facilitated the isolation of mutants capable of utilizing 3,5-dichlorobenzoate (35DCB). Growth on 35DCB also required removal of the interfering *meta* ring-cleavage activity encoded by *xyl*E. In addition, a mutation in *xyl*S, the positive regulator of *xyl*XYZ, was necessary in order that 35DCB could induce the genes and trigger expression of the toluate 1,2-dioxygenase. In recombinant derivatives of *Pseudomonas* sp. strain B13 containing only the *xyl*XYZ and the mutant *xyl*S, toluate 1,2-dioxygenase was produced and recruited for the conversion of 35DCB to 3,5-dichloroDHB, which was mineralized via the host's chlorocatechol pathway [83].

The same strategy — recruitment of broad-specificity enzymes to extend chlorocatechol pathways in other bacteria — has been used to construct laboratory strains which can grow on a range of chlorinated aromatic compounds [62, 89, 112]. It is very likely that recruitment of this sort, mediated by plasmid transfer, probably occurs in natural microbial populations; however, there are no data available to estimate its effect in terms of adaptation and natural selection. The identification of homologous DNA sequences in genes encoding the *meta* ring-cleavage pathways of independently isolated TOL, NAH and SAL plasmids [63] suggests that recruitment of genes is an important mechanism for the evolution of xenobiotic degrading bacteria.

4.5.7 BIODEGRADATION IN THE ENVIRONMENT

A central problem to be addressed in ecological studies of biodegradation is to explain the observed differences between laboratory and field data. In general, rates of degradation of xenobiotic compounds in the biosphere are often very slow and do not reflect the versatility of micro-organisms isolated in the laboratory by selective culture methods. The general question of microbial fallibility — whether or not all known xenobiotics can be transformed by micro-organisms isolated in the laboratory — is not a primary issue in dealing with current pollution problems. Many priority pollutants are biodegradable, and a substantial number can be mineralized by either mixed or pure microbial cultures. The problem is that in polluted environments correspondingly high rates of transformation are usually not apparent nor do they evolve rapidly. Practical approaches to the treatment of wastes containing pollutant xenobiotics have, therefore, to proceed on the basis of case-by-case experimental studies, where the relative effects of abiotic and biological components of a particular habitat upon the persistence of a xenobiotic can be examined [1].

Solubility, chemical reactivity and adsorptivity are the major physicochemical properties of a xenobiotic determining its availability to the microflora. In addition, the microflora and, therefore, the biodegradative potential of a given habitat will be selected, in part, by abiotic variables such as temperature, pH, oxygen tension and salinity. For example, quantitative and qualitative differences in biodegradation are observed when aerobic and anaerobic habitats are compared. The rapid biodegradation of xenobiotics in aerobic environments probably reflects the efficiency of aerobic metabolism and the importance of oxygenase enzymes. However, it is now clear that many anaerobic micro-organisms, once presumed to catabolize only simple carbon compounds, are metabolically versatile, and that certain biotransformations are more likely to occur in anoxic rather than aerated habitats. The insecticide Lindane (γHCH) was reported to be almost completely dechlorinated by *Clostridium* spp. [51], an observation which is consistent with data showing that γHCH was more rapidly degraded in anaerobic soils than in aerobic upland soils.

The published data suggest that only quantitative and not qualitative differences are observed in biodegradation of xenobiotics at different temperatures; however, only a few studies of xenobiotic catabolism by thermophiles or psychrophiles have been reported.

Attempts to correlate chemical structure with biodegradability must be carried out with caution. The temptation to attribute recalcitrance of a particular compound to one or two features or substituents of the molecule is often irresistible — a number of such generalizations have been made in this chapter. Chapman [17] suggested that generalizations about relationships between structure and recalcitrance can result in untenable conclusions, and noted that, although extensive branching of alkane substituents is frequently associated with resistance of the structure to biodegradation, several easily degradable compounds contain one or more quaternary carbon atoms (e.g. (+) camphor and α-pinene). On the other hand Cain [13] reported that small changes in the structure of a biodegradable molecule, CMOS (carboxymethyloxysuccinate, a 'builder' component of detergents), which would not be expected to affect biodegradability, e.g. replacement of $-H$ with $-OH$ or $-O-CH_2COOH$, gave synthetic products which were highly recalcitrant.

The concentration of biodegradable but toxic chemicals in the environment can have a direct effect on their persistence. At high concentrations toxic compounds are liable to inhibit the metabolic activities of a significant proportion of the exposed microflora and thereby disrupt communities which might normally participate in detoxification. Conversely, the affinity of microbial enzymes and transport systems required for biodegradation may not be sufficiently high to scavenge those compounds which are deposited in the environment at relatively low concentrations [1, 2]. Under these conditions a compound can persist at low levels which may still be toxic and could be magnified by bioaccumulation.

4.5.8 PRACTICAL APPROACHES TO THE TREATMENT OF WASTES CONTAINING XENOBIOTICS

The treatment of chemical pollutants in the environment is a problem requiring an understanding of the ecological, physiological, genetic and biochemical principles of biodegradation. The application of laboratory-based studies is likely to be most successful where industrial effluents can be treated at source, under controlled conditions. However, the effective optimization of waste treatment processes with an active microbial population being used to degrade a well-defined waste liquor is likely to be feasible in only a few specialized applications. It should be noted that mixtures of xenobiotics are common in industrial wastes but that the biodegradation of mixed substrates by both pure and mixed cultures is poorly understood. In some cases one component of a substrate mixture may strongly inhibit the degradation of another. For example, the catabolism of chloroaromatic compounds via the *ortho* ring-cleavage pathway is arrested in the presence of an active *meta* ring-cleavage pathway (which may be induced if phenol or methylaromatic substrates are present) and toxic products derived from chlorocatechols accumulate [84]. In the biodegradation of mixtures from crude oil, a phenomenon analogous to catabolite repression, called 'sparing', has been described [4]. Here the accumulation of metabolic intermediates reduces the rate of utilization of certain components of the oil.

Microbial inoculants

Recently, industrial interest in the use of microbial inoculants for the treatment of dispersed chemical wastes has become apparent, and there are a number of products on the market for which more or less extravagant claims are made. The introduction of a new species into a polluted habitat containing an indigenous microflora presents several problems, especially if the inoculant is to be effective in degrading a mixture of xenobiotics. For certain applications, however, factors such as long-term survival of the inoculated species, its resistance to starvation and abiotic stress, physical barriers to colonization and to transformation of the xenobiotic(s), and predation may not need to be considered. Laboratory experiments indicate that microbial inoculants can be effective in removing compounds such as chlorinated phenols (including pentachlorophenol) and 245T from soils and water [32, 56, 101]. Prospects for the use of microbial inoculants containing genetically engineered micro-organisms (GEMS) has been widely discussed [2, 83, 112]. Recent reports clearly emphasize the need for more detailed information on the biochemistry and genetics of relevant catabolic pathways if the construction of recombinant pathways for the transformation of xenobiotics is to become a realistic option. Currently, the most successful strategies involve the manipulation of genes encoding broad-specificity enzymes to expand existing catabolic pathways [64, 83]. The use of cloned genes facilitates a defined and rational approach to the manipulation of biodegradative functions. Specific regulatory and structural genes can be spliced and the effects of modulating gene expression on the flux of metabolites through the recombinant pathway can be studied under controlled conditions. The basic aims of most studies are to understand well-defined microbiological or biochemical problems. The suggestion that our understanding of biodegradation has now reached the stage where genetically engineered 'superbugs' can be produced which are capable of degrading a wide range of xenobiotics at high rates and in stable association with the indigenous microflora seems to be unduly optimistic.

Finally, a more pragmatic approach to the treatment of recalcitrant dispersed wastes, which are biodegradable by microbial cultures isolated in the laboratory, is to stimulate biodegradation by the indigenous microflora. This may be achieved by supplying appropriate nutrients, aerating the site and, if necessary, making the pollutant(s) physically available to the micro-organisms. Consequently, the manipulation of polluted environments requires the combined attention and efforts of engineers and biologists.

4.5.9 CONCLUSION

It is reasonable to assume that industrial production of xenobiotic compounds will increase in the foreseeable future and that the amounts of recalcitrant chemicals in the environment will rise proportionately. Concern about these trends is not misplaced and it is appropriate that a high priority should continue to be given to the type of basic research described here. In this way an understanding may be reached of the ecological determinants of recalcitrance. At the same time, an unavoidable conclusion of the data which are currently available is that industrial society needs urgently to consider balancing the benefits of chemical production against the long-term, deleterious effects of pollution.

4.5.10 REFERENCES

[1] Alexander M. (1981) Biodegradation of chemicals of environmental concern. *Science* **211**, 132–8.

[2] Alexander M. (1984) Ecological constraints on genetic engineering. In *Genetic Control of Environmental Pollutants*, p.151 (ed. G.S. Omenn & A. Hollaender). Plenum Press, New York.

[3] Anson J.G. & Mackinnon G. (1984) Novel *Pseudomonas* plasmid involved in aniline degradation. *Appl. Environ. Microbiol.* **48**, 868–9.

[4] Atlas R.M. (1981) Microbial degradation of petroleum hydrocarbons: an environmental perspective. *Microbiol. Rev.* **45**, 180–209.

[5] Bartels I., Knackmuss H.-J. & Reineke W. (1984) Suicide inactivation of catechol 2,3-dioxygenase from *Pseudomonas putida* mt-2 by 3-halocatechols. *Appl. Environ. Microbiol.* **47**, 500–5.

[6] Bayly R.C. & Barbour M.G. (1984) The degradation of aromatic compounds by the meta and gentisate pathways: biochemistry and regulation. In *Microbial Degradation of Organic Compounds*, p.253 (ed. D.T. Gibson). Marcel Dekker, New York.

[7] Bestetti G., Galli E., Ruzzi, M., Baldacci G., Zennaro E. & Frontali L. (1984) Molecular characterization of a plasmid from *Pseudomonas fluorescens* involved in styrene degradation. *Plasmid* **12**, 181–8.

[8] Boyd S.A. & Shelton D.R. (1984) Anaerobic biodegradation of chlorophenols in fresh and acclimated sludge. *Appl. Environ. Microbiol.* **47**, 272–7.

[9] Brandsch R., Faller W. & Schneider K. (1986) Plasmid pAO1 from *Arthrobacter oxidans* encodes 6-hydroxy-D-nicotine oxidase: cloning and expression of the gene in *Escherichia coli*. *Mol. Gen. Genet.* **202**, 96–101.

[10] Braun K. & Gibson D.T. (1984) Anaerobic degradation of 2-aminobenzoate (anthranilic acid) by denitrifying bacteria. *Appl. Environ. Microbiol.* **48**, 102–7.

[11] Britton L.N. (1984) Microbial degradation of aliphatic hydrocarbons. In *Microbial Degradation of Organic Compounds*. p.89 (ed. D.T. Gibson). Marcel Dekker, New York.

[12] Bumpus J.A., Tien M., Wright D. & Aust S.D. (1985) Oxidation of persistent environmental pollutants by a white rot fungus. *Science* **228**, 1434–6.

[13] Cain R.B. (1981) Microbial degradation of surfactants and 'builder' components. In *Microbial Degradation of Xenobiotics and Recalcitrant Compounds*, p.326 (ed. T. Leisinger, A.M. Cook, T. Hütter, & J. Nüesch). Academic Press, London.

[14] Callely A.G. (1978) The microbial degradation of heterocyclic compounds. *Prog. Ind. Microbiol.* **14**, 205–81.

[15] Cerniglia C.E. (1984) Microbial metabolism of polycyclic aromatic hydrocarbons. *Adv. Appl. Microbiol.* **30**, 31–65.

[16] Chakrabarty A.M., Friello D.A. & Bopp L.H. (1978) Transposition of plasmid DNA segments specifying hydrocarbon degradation and their expression in various microorganisms. *Proc. Natl. Acad. Sci. USA* **75**, 3109–12.

[17] Chapman P.J. (1979) Degradation mechanisms. In *Workshop: Microbial Degradation of Pollutants in Marine Environments*, p.28 (ed. A.W. Bouquin & P.H. Pritchard). US Environmental Protection Agency, Florida.

[18] Chatterjee D.K. & Chakabarty A.M. (1983) Genetic homology between independently isolated chlorobenzoate-degradative plasmids. *J. Bacteriol.* **146**, 639–46.

[19] Clarke P.H. & Slater J.H. (1986) Evolution of enzyme structure and function in *Pseudomonas*. In *The Bacteria. Vol. 10: The Biology of Pseudomonas*, p.71 (ed. J.R. Sokatch). Academic Press, New York.

[20] Cook A.M., Beilstein P., Grossenbacher H. & Hütter R. (1985) Ring cleavage and degradative pathway of cyanuric acid in bacteria. *Biochem. J.* **231**, 25–30.

[21] Crawford R.L., Olson P.E. & Frick T.D. (1979). Catabolism of 5-chlorosalicylate by a *Bacillus* isolated from the Mississippi River. *Appl. Environ. Microbiol.* **38**, 379–84.

[22] Dagley S. (1975) A biochemical approach to some problems of environmental pollution. *Essays Biochem.* **11**, 81–138.

[23] Dagley S. (1981) New perspectives in aromatic catabolism. In *Microbial Degradation of Xenobiotics and Recalcitrant Compounds*, p.181 (ed. T. Leisinger, A.M. Cook, R. Hütter & J. Nüesch). Academic Press, London.

[24] Dagley S. (1984) Introduction. In *Microbial Degradation of Organic Compounds*, p.1 (ed. D.T. Gibson). Marcel Dekker, New York.

[25] Dagley S. (1986) Biochemistry of aromatic hydrocarbon degradation in pseudomonads. In *The Bacteria. Vol. 10: The Biology of Pseudomonas*, p.527 (ed. J.R. Sokatch). Academic Press, New York.

[26] deBone J.A.M., Vorage M.J.A.Q., Hartmans S. & van den Tweel W. (1986) Microbial degradation of 1,3-dichlorobenzene. *Appl. Environ. Microbiol.* **52**, 677–80.

[27] Dodgson K.S. & White G.E. (1982) Some microbial enzymes involved in biodegradation of sulphated surfactants. In *Topics in Enzyme and Fermentation Technology*, Vol. 7, p.90 (ed. A. Wiseman). Horwood, Chichester.

[28] Don R.H., Weightman A.J., Knackmuss H.-J. & Timmis K.N. (1985) Transposon mutagenesis and cloning analysis of the pathways for degradation of 2,4-dichlorophenoxyacetic acid and 3-chlorobenzoate in *Alcaligenes eutrophus* JMP134(pJP4). *J. Bacteriol.* **161**, 85–90.

[29] Eaton D.C. (1985) Mineralization of polychlorinated biphenyls by *Phanerochaete chrysosporium:* a ligninolytic fungus. *Enzyme Microb. Technol.* **7**, 194–6.

[30] Eaton R.W. & Timmis K.N. (1986) Characterization of a plasmid-specific pathway for catabolism of isopropylbenzene in *Pseudomonas putida* RE204. *J. Bacteriol.* **168**, 123–31.

[31] Eaton R.W. & Timmis K.N. (1986) Spontaneous deletion of a 20-kilobase DNA segment carrying genes specifying isopropylbenzene metabolism in *Pseudomonas putida* RE204. *J. Bacteriol.* **168**, 428–30.

[32] Edgehill R.U. & Finn R.K. (1983) Microbial treatment of soil to remove pentachlorophenol. *Appl. Environ. Microbiol.* **45**, 1122–5.

[33] Ensley B.D. (1984) Microbial metabolism of condensed thiophenes. In *Microbial Degradation of Organic Compounds*, p.309 (ed. D.T. Gibson). Marcel Dekker, New York.

[34] Evans W.C. (1977) Biochemistry of the bacterial catabolism of aromatic compounds in anaerobic environments. *Nature (Lond.)* **270**, 17–22.

[35] Evans W.C., Smith B.S.W., Fernley H.N. & Davies J.I. (1971) Bacterial metabolism of 2, 4-dichlorophenoxyacetate. *Biochem. J.* **122**, 543–51.

[36] Fewson C.A. (1981) Biodegradation of aromatics with industrial relevance. In *Microbial Degradation of Xenobiotics and Recalcitrant Compounds*, p.141 (ed. T. Leisinger, A.M. Cook, R. Hütter, & J. Nüetch). Academic Press, London.

[37] Franklin F.C.H., Bagdasarian M., Bagdasarian M.M. & Timmis K.N. (1981) Molecular and functional analysis of the TOL plasmid pWWO from *Pseudomonas putida* and cloning of the genes for the entire regulated aromatic ring *meta*-cleavage pathway. *Proc. Natl. Acad. Sci. USA* **78**, 7458–62.

[38] Franklin F.C.H., Lehrbach P.R., Lurz R., Rueckert B., Bagdasarian M. & Timmis K.N. (1983) Localization and functional analysis of transposon mutations in regulatory genes of the TOL catabolic plasmid. *J. Bacteriol.* **154**, 676–85.

[39] Frantz B. & Chakrabarty A.M. (1986) Degradative plasmids in *Pseudomonas*. In *The Bacteria. Vol. 10: The Biology of Pseudomonas*, p.295 (ed. J.R. Sokatch). Academic Press, New York.

[40] Furkawa K. & Chakrabarty A.M. (1982) Involvement of plasmid in the total degradation of chlorinated biphenyls. *Appl. Environ. Microbiol.* **44**, 619–25.

[41] Furkawa K. & Matsumura F. (1976) Microbial metabolism of polychlorinated biphenyls. Studies on the relative degradability of polychlorinated biphenyl components by *Alcaligenes* sp. *J. Agric. Food Chem.* **24**, 251–6.

[42] Furkawa K., Tomizuka N. & Kamibayashi A. (1983) Metabolic breakdown of Kaneclors (polychlorinated biphenyls) and their products by *Acinetobacter* sp. *Appl. Environ. Microbiol.* **46**, 140–5.

[43] Ghosal D., You I.-S., Chatterjee D.K. & Chakrabarty A.M. (1985) Microbial degradation of halogenated compounds. *Science* **228**, 135–42.

[44] Gibson D.T. & Subramanian V. (1984) Microbial degradation of aromatic hydrocarbons. In *Microbial Degradation of Organic Compounds*, p.181 (ed. D.T. Gibson). Marcel Dekker, New York.

[45] Harayama S., Lehrbach P.R. & Timmis K.N. (1984) Transposon mutagenesis analysis of *meta*-cleavage pathway operon genes of the TOL plasmid of *Pseudomonas putida* mt-2. *J. Bacteriol.* **160**, 251–5.

[46] Harayama S., Rekik M. & Timmis K.N. (1986) Genetic analysis of a relaxed substrate specificity aromatic ring dioxygenase, toluate 1,2-dioxygenase, encoded by TOL plasmid pWWO of *Pseudomonas putida*. *Mol. Gen. Genet.* **202**, 226–34.

[47] Higgins I.J., Best D.J. & Hammond R.C. (1980) New findings in methane-utilizing bacteria highlighting their importance in the biosphere and their commercial potential. *Nature (Lond.)* **286**, 561–4.

[48] Higgins I.J., Hammond R.C. & Scott D. (1984) Transformation of C_1 compounds by microorganisms. In *Microbial Degradation of Organic Compounds*, p.43 (ed. D.T. Gibson). Marcel Dekker, New York.

[49] Hutzinger O. & Veerkamp W. (1981) Xenobiotic chemicals with pollution potential. In *Microbial Degradation of Xenobiotics and Recalcitrant Compounds*, p.3 (ed. T. Leisinger, A.M. Cook, T. Hütter & J. Nüesch). Academic Press, London.

[50] Jacoby G.A., Rogers J.E., Jacob A.E. & Hedges R.W. (1978) Transposition of *Pseudomonas* toluene-degrading genes and expression in *Escherichia coli*. *Nature (Lond.)* **274**, 179–80.

[51] Jagnow G., Haider K. & Ellwardt P.C. (1977) Anaerobic dechlorination and degradation of hexachlorocyclohexane isomers by anaerobic and facultative anaerobic bacteria. *Arch. Microbiol.* **115**, 285–92.

[52] Janssen D.B., Scheper A., Dijkhuizen L. & Witholt B. (1985) Degradation of halogenated aliphatic compounds by *Xanthobacter autotrophicus* GJ10 *Appl. Environ. Microbiol.* **49**, 673–7.

[53] Kawasaki H., Yahara H. & Tonomura K. (1984) Cloning and expression in *Escherichia coli* of the haloacetate dehalogenase genes from *Moraxella* plasmid pUO1. *Agric. Biol. Chem.* **48**, 2627–32.

[54] Keith I.H. & Telliard W.A. (1979) Priority pollutants — a perspective. *Environ. Sci. Technol.* **13**, 416–23.

[55] Kersten P.J., Chapman P.J. & Dagley S. (1985) Enzymatic release of halogens or methanol from some substituted protocatechuic acids. *J. Bacteriol.* **162**, 693–7.

[56] Kilbane J.J., Chatterjee D.K. & Chakrabarty A.M. (1983) Detoxification of 2,4,5-trichlorophenoxyacetic acid from contaminated soil by *Pseudomonas cepacia*. *Appl. Environ. Microbiol.* **45**, 1697–700.

[57] Kirk T.K. (1984) Degradation of lignin. In *Microbial Degradation of Organic Compounds*, p.399 (ed. D.T. Gibson). Marcel Dekker, New York.

[58] Knowles C.J. & Bunch A.W. (1986) Microbial cyanide metabolism. *Adv. Microb. Physiol.* **27**, 73–111.

[59] Koga H., Aramaki H., Yamaguchi E., Takeuchi K., Horiuchi T. & Gunsalus I.C. (1986) *cam*R, a negative regulator locus of the cytochrome P-450_{cam}. *J. Bacteriol.* **166**, 1089–95.

[60] Kohler-Staub D. & Leisinger T. (1985). Dichloromethane dehalogenase of *Hyphomicrobium* sp. strain DM2. *J. Bacterial.* **162**, 676–81.

[61] Kulla H.G., Krieg R., Zimmerman T. & Leisinger T. (1984) Biodegradation of xenobiotics. In *Current Perspectives in Microbial Ecology*, p.663 (eds. M.J. Klug & C.A. Reddy). Am. Soc. Microbiol., Washington DC.

[62] Latorre J., Reineke W. & Knackmuss H.-J. (1984) Microbial metabolism of chloroanilines: enhanced evolution by natural genetic exchange. *Arch. Microbiol.* **140**, 159–65.

[63] Lehrbach P.R., McGregor I., Ward J.M. & Broda P. (1983) Molecular relationships between *Pseudomonas* INC P-9 degradative plasmids TOL, NAH, and SAL. *Plasmid* **10**, 164–74.

[64] Lehrbach P.R., Zeyer J., Reineke W., Knackmuss H.-J. & Timmis K.N. (1984) Enzyme recruitment *in vitro*: use of cloned genes to extend the range of haloaromatics degraded by *Pseudomonas* sp. strain B13. *J. Bacteriol.* **158**, 1025–32.

[65] Leisinger T. (1983) Microorganisms and xenobiotic compounds. *Experientia* **39**, 1183–91.

[66] McClure N.C. & Venables W.A. (1986) Adaptation of *Pseudomonas putida* mt-2 to growth on aromatic amines. *J. Gen. Microbiol.* **132**, 2209–18.

[67] Marks T.S., Wait R., Smith A.R.W. & Quirk A.V. (1984) The origin of the oxygen incorporated during the dehalogenation/hydroxylation of 4-chlorobenzoate by an *Arthrobacter* sp. *Biochem. Biophys. Res. Commun.* **124**, 669–74.

[68] Markus A., Klages U., Krauss S. & Lingens F. (1984) Oxidation and dehalogenation of 4-chlorophenylacetate by a two-component enzyme system from *Pseudomonas* sp. strain CBS3. *J. Bacteriol.* **160**, 618–21.

[69] Meyer O. & Rohde M. (1984) Enzymology and bioenergetics of carbon monoxide-oxidizing bacteria. In *Microbial Growth on C_1 Compounds. Proceedings of the 4th International Symposium*, p.26. Am. Soc. Microbiol., Washington, DC.

[70] Mikesell M.D. & Boyd S.A. (1986) Complete reductive dechlorination and mineralization of pentachlorophenol by anaerobic microorganisms. *Appl. Environ. Microbiol.* **52**, 861–5.

[71] Monticello D.J., Bakker D. & Finnerty W.R. (1985) Plasmid-mediated degradation of dibenzothiophene by *Pseudomonas* species. *Appl. Environ. Microbiol.* **49**, 756–60.

[72] Motosugi K. & Soda K. (1983) Microbial degradation of synthetic organochlorine compounds. *Experientia* **39**, 1214–20.

[73] Mulbry W.W., Karns J.S., Kearney P.C., Nelson J.O., McDaniel C.S. & Wild J.R. (1986) Identification of a plasmid-borne parathion hydrolase gene from *Flavobacterium* sp. by southern hybridization with *opd* from *Pseudomonas diminuta. Appl. Environ. Microbiol.* **51**, 926–30.
[74] Müller R. & Lingens F. (1986) Microbial degradation of halogenated hydrocarbons: a biological solution to pollution problems? *Angew. Chem. Int. Ed. Eng.* **25**, 779–89.
[75] Müller R., Thiele J., Klages U. & Lingens F. (1984) Incorporation of [^{18}0] water into 4-hydroxybenzoic acid in the reaction of 4-chlorobenzoate dehalogenase from *Pseudomonas* spec. CBS3. *Biochem. Biophys. Res. Commun.* **124**, 178–82.
[76] Munnecke D.M. & Hseih D.P.H. (1977) Parathion utilization by bacterial symbionts in a chemostat. *Appl. Environ. Microbiol.* **34**, 175–84.
[77] Nakazawa T. & Inouye S. (1986) Cloning of *Pseudomonas* genes in *Escherichia coli*. In *The Bacteria. Vol. 10; The Biology of Pseudomonas*, p.357 (ed. J.R. Sokatch). Academic Press, New York.
[78] Negoro S., Nakamura S., Kimura H., Fujiama K., Zhang Y.-Z, Kanzaki N. & Okada H. (1984) Construction of hybrid genes of 6-aminohexanoic acid-oligomer hydrolase and its analogous enzyme. *J. Biol. Chem.* **22**, 13648–51.
[79] Negoro S., Nakamura S. & Okada H. (1984) DNA–DNA hybridization analysis of nylon oligomer-degradative plasmid pOAD2: identification of the DNA region analogous to the nylon oligomer degradation gene. *J. Bacteriol.* **158**, 419–24.
[80] Ornston L.N. & Parke D. (1976) The evolution of induction mechanisms in bacteria: insights derived from the study of the β-ketoadipate pathway. *Curr. Top. Cell. Regul.* **12**, 209–62.
[81] Owen D.J. (1986) Molecular cloning and characterization of sequences from the regulatory cluster of the *Pseudomonas* plasmid *alk* system. *Mol. Gen. Genet.* **203**, 64–72.
[82] Powlowski J.B. & Dagley S. (1985) β-Ketoadipate pathway in *Trichosporon cutaneum* modified for methyl-substituted metabolites. *J. Bacteriol.* **163**, 1126–35.
[83] Ramos J.L., Stolz A., Reineke W. & Timmis K.N. (1986) Altered effects on specificities in regulators of gene expression: TOL plasmid *xyl*S mutants and their use to engineer expansion of the range of aromatics degraded by bacteria. *Proc. Natl. Acad. Sci. USA* **83**, 8467–71.
[84] Reineke W. (1984) Microbial degradation of halogenated aromatic compounds. In *Microbial Degradation of Organic Compounds*, p.319 (ed. D.T. Gibson). Marcel Dekker, New York.
[85] Reineke W. and Knackmuss H.-J. (1979) Construction of haloaromatic utilizing bacteria. *Nature (Lond.)* **277**, 385–6.
[86] Reineke W. & Knackmuss H.-J. (1984) Microbial metabolism of haloaromatics: isolation and properties of a chlorobenzene-degrading bacterium. *Appl. Environ. Microbiol.* **47**, 395–402.
[87] Rheinwald J.G., Chakrabarty A.M. & Gunsalus I.C. (1973) A transmissible plasmid controlling camphor oxidation in *Pseudomonas putida. Proc. Natl. Acad. Sci. USA* **70**, 885–9.
[88] Ribbons D.W. & Eaton R.W. (1982) Chemical transformations of aromatic hydrocarbons that support the growth of microorganisms. In *Biodegradation and Detoxification of Environmental Pollutants*, p.59 (ed. A.M. Chakrabarty). CRC Press, Florida.
[89] Schwien U. & Schmidt E. (1982) Improved degradation of monochlorophenols by a constructed strain. *Appl. Environ. Microbiol.* **44**, 33–9.
[90] Serder C.M. & Gibson D.T. (1985) Enzymatic hydrolysis of organophosphates. Cloning and expression of parathion hydrolase gene from *Pseudomonas diminuta. Bio/Technology* **3**, 567–71.
[91] Shapiro J.A., Owen D.J., Kok M. & Eggink G. (1984) *Pseudomonas* hydrocarbon oxidation. In *Genetic Control of Environmental Pollutants*, p.229 (ed. G.S. Omenn & A. Hollaender). Plenum, New York.
[92] Shelton D.R. & Tiedje J.M. (1984) General method for determining anaerobic biodegradation potential. *Appl. Environ. Microbiol.* **47**, 850–7.
[93] Shelton D.R. & Tiedje J.M. (1984) Isolation and partial characterization of bacteria in an anaerobic consortium that mineralizes 3-chlorobenzoic acid. *Appl. Environ. Microbiol.* **48**, 840–8.
[94] Shields M.S., Hooper S.W. & Sayler G.S. (1985) Plasmid-mediated mineralization of 4-chlorobiphenyl. *J. Bacteriol.* **163**, 882–9.
[95] Shimao M., Fujita I., Kato N. & Sakazawa C. (1985) Enhancement of pyrroloquinoline quinone production and polyvinyl alcohol degradation in mixed continuous cultures of *Pseudomonas putida* VM15A and *Pseudomonas* sp. strain VM15C with mixed carbon sources. *Appl. Environ. Microbiol.* **49**, 1389–91.
[96] Slater J.H. & Lovatt D. (1984) Biodegradation and the significance of microbial communities. In *Microbial Degradation of Organic Compounds*, p.439 (ed. D.T. Gibson). Marcel Dekker, New York.
[97] Slater J.H. & Somerville H.J. (1979) Microbial aspects of waste treatment with particular attention to the degradation of organic compounds. In *Microbial*

Technology: Current Status and Future Prospects, Soc. Gen. Microbiol. Symp. 29, p. 221 (ed. A.T. Bull, D.C. Ellwood & C.R. Ratledge). Cambridge University Press, Cambridge.

[98] Slater J.H., Weightman A.J. & Hall B.G. (1985) Dehalogenase genes of *Pseudomonas putida* PP3 on chromosomally located transposable elements. *Mol. Biol. Evol.* **2**, 557–67.

[99] Stanier R.Y. & Ornston L.N. (1973) The β-ketoadipate pathway. *Adv. Microb. Physiol.* **9**, 89–151.

[100] Stanlake G.J. & Finn R.K. (1982) Isolation and characterization of a pentachlorophenol-degrading bacterium. *Appl. Environ. Microbiol.* **44**, 1421–7.

[101] Steiert J.G. & Crawford R.L. (1985) Microbial degradation of chlorinated phenols. *Trends Biotechnol.* **3**, 300–5.

[102] Suida J.F. & DeBernarids J.F. (1973) Naturally-occurring halogenated organic compounds. *Lloydia* **36**, 107–43.

[103] Tabak H.H., Quave S.A., Mashni C.I. & Barth E.F. (1981) Biodegradability studies with organic priority pollutant compounds. *J. Water Pollut. Fed.* **53**, 1503–18.

[104] Thacker R., Rorvig O., Kahlon P. & Gunsalus I.C. (1978) NIC, a conjugative nicotine-nicotinate degradative plasmid in *Pseudomonas convexa. J. Bacteriol.* **135**, 289–90.

[105] Tien M. & Kirk T.K. (1983) Lignin-degrading enzyme from the hymenomycete *Phanerochaete chrysosporium* Burds. *Science* **221**, 661–3.

[106] Tien M. & Kirk T.K. (1984) Lignin-degrading enzyme from *Phanerochaete chrysosporium*: purification, characterization, and catalytic properties of a unique H_2O_2-requiring oxygenase. *Proc. Natl. Acad. Sci. USA* **81**, 2280–4.

[107] Trudgill P.W. (1984) Microbial degradation of the alicyclic ring: structural relationships and metabolic pathways. In *Microbial Degradation of Organic Compounds*, p.131 (ed. D.T. Gibson). Marcel Dekker, New York.

[108] Trudgill P.W. (1986) Terpenoid metabolism by *Pseudomonas*. In *The Bacteria. Vol. 10: The Biology of Pseudomonas*, p.483 (ed. J.R. Sokatch). Academic Press, New York.

[109] Tsuchii A., Suzuki T. & Takeda K. (1985) Microbial degradation of natural rubber vulcanizates. *Appl. Environ. Microbiol.* **50**, 965–70.

[110] Tsuchii A., Suzuki T. & Takhara Y. (1977) Microbial degradation of styrene oligomer. *Agric. Biol. Chem.* **41**, 2417–21.

[111] Vandenbergh P.A., Olsen R.H. & Colaruotolo J.F. (1981) Isolation and genetic characterization of bacteria that degrade chloroaromatic compounds. *Appl. Environ. Microbiol.* **42**, 737–9.

[112] Weightman A.J., Don R.H., Lehrbach P.R. & Timmis K.N. (1984) The identification and cloning of genes encoding haloaromatic catabolic enzymes and the construction of hybrid pathways for substrate mineralization. In *Genetic Control of Environmental Pollutants*, p.47 (ed. G.S. Omenn & A. Hollaender). Plenum, New York.

[113] Weightman A.J., Weightman A.L. & Slater J.H. (1982) Stereospecificity of 2-monochloropropionate dehalogenation by the two dehalogenases of *Pseudomonas putida* PP3: evidence for two different dehalogenation mechanisms. *J. Gen. Microbiol.* **128**, 1755–62.

[114] Widdel F., Kohring G.-W. & Mayer F. (1983) Studies on dissimilatory sulfate-reducing bacteria that decompose fatty acids. *Arch. Microbiol.* **134**, 286–94.

[115] Williams P.A. & Murray K. (1974) Metabolism of benzoate and methylbenzoates by *Pseudomonas putida* (*arvilla*) mt-2. Evidence for the existence of a TOL plasmid. *J. Bacteriol.* **120**, 416–23.

[116] Yeh W.K. & Ornston L.N. (1980) Origins of metabolic diversity: substitution of homologous sequences into genes for enzymes with different catalytic activities. *Proc. Natl. Acad. Sci. USA* **77**, 5365–9.

[117] Yen K.-M, & Gunsalus I.C. (1985) Regulation of naphthalene catabolic genes of plasmid NAH7. *J. Bacteriol.* **162**, 1008–13.

[118] Yen K.-M., Sullivan M. & Gunsalus I.C. (1983) Electron microscope heteroduplex mapping of naphthalene oxidation genes on the NAH7 and SAL1 plasmids. *Plasmid* **9**, 105–11.

[119] Young L.Y. (1984) Anaerobic degradation of aromatic compounds. In *Microbial Degradation of Organic Compounds*, p.487 (ed. D.T. Gibson). Marcel Dekker, New York.

[120] Zeyer J. & Kearney P.C. (1982) Microbial metabolism of propanil and 3,4-dichloroaniline. *Pest. Biochem. Physiol.* **17**, 224–31.

[121] Zeyer J. & Kearney P.C. (1984) Degradation of *o*-nitrophenol by *Pseudomonas putida. J. Agric. Food Chem.* **32**, 238–42.

[122] Zeyer J., Wasserfallen A. & Timmis K.N. (1985) Microbial mineralization of ring-substituted anilines through an *ortho*-cleavage pathway. *Appl. Environ. Microbiol.* **50**, 447–53.

Part 5
Epilogue. Microbial Ecology in Biotechnology

My own feeling is that the most spectacular rewards will accrue to those who are now doing some lateral thinking. And one of the most prolific strands of thought could mean switching attention from desirable end-products (interferon, insulin . . .) and processes (efficient biomass conversions . . .) to the astronomical range of putative producers among the microbial talent already available to use. In other words, as well as exploiting recombinant DNA manipulations to improve the yield of predictable materials, we should be reviewing the extraordinary creativity of submicroscopic life and contemplating possible uses for skills which lie well beyond those of human chemists. Most of the winners, I predict, will be those biotechnology teams which have had a microbial ecologist on board.

Bernard Dixon
The Need for Microbial Ecologists (1983)
In the first issue of *Bio/Technology* **1**, pp. 45, 131

BERNARD DIXON clearly describes the need for a microbial ecology approach within biotechnology teams and proceeds to give illustrations where this has been applied successfully in finding new antibiotics and in the generation of a 'new' sugar, isomaltulose, use of which could prevent dental caries. He then lists unexploited microbial talent such as the methylation of tin by estuarine micro-organisms, anaerobic degradation of haloaromatics in lake sediment, bacteria which thrive at 105°C in the Mediterranean seabed or in the frozen Antarctic, a virus pathogenic to rice but which promotes the growth of other plants, and bacteria which act as the propulsive system for protozoa in the hindgut of the wood termite. There are, of course, many other illustrations in our present volume.

Dr Dixon has frequently commented on the value of microbial ecology before, but in reviewing the volume to which this book is a sequel (*New Scientist*, 5 July 1979) he pointed to the dangers of forgetting the great pioneers (Serge Winogradsky and Selman Waksman) of the study of mixed populations of micro-organisms as they occur naturally. Waksman, having spent his early career as a soil microbiologist and mushroom scientist, probably made a greater contribution to what we would now define as biotechnology than most modern groups of 'biotechnologists', which was of course recognized with the award of the Nobel Prize in Physiology and Medicine in 1952 for the discovery of streptomycin. But definitions of biotechnology can be difficult and some of these have been catalogued [1] with a preferred definition stated as: *The application of organisms, biological systems or biological processes to manufacturing industries and agriculture and their service industries.*

This definition, however, may not cover adequately the consequences of biotechology in nature. True, there are many opportunities for harnessing organisms from nature and if necessary increasing their activities by genetic manipulation. But what is the consequence of re-introducing these organisms into nature at elevated levels? Clearly toxicological testing must be carried out, but the ecologist should also come back into his or her own for assessing environmental impact. For example, what is the frequency of plasmid transfer from introduced biocontrol bacteria, is it any greater than for organisms naturally present and does it have a consequence anyway? There will probably be many negative responses to these questions, but, extending Bernard Dixon's views, the smart biotechnology groups will certainly have microbial ecologists on board to establish the facts.

REFERENCE

[1] Grainger J.M. & Lynch J.M. (1984) Introduction: towards an understanding of environmental biotechnology. In *Microbiological Methods for Environmental Biotechnology*, p.1 (ed. J.M. Grainger & J.M. Lynch). Academic Press, London.

Index